533.6 GOL

AF183135

DATE DE RETOUR

14 10 99		

NUMERICAL SIMULATION OF VISCOUS SHOCK LAYER FLOWS

FLUID MECHANICS AND ITS APPLICATIONS
Volume 33

Series Editor: R. MOREAU
MADYLAM
Ecole Nationale Supérieure d'Hydraulique de Grenoble
Boîte Postale 95
38402 Saint Martin d'Hères Cedex, France

Aims and Scope of the Series

The purpose of this series is to focus on subjects in which fluid mechanics plays a fundamental role.

As well as the more traditional applications of aeronautics, hydraulics, heat and mass transfer etc., books will be published dealing with topics which are currently in a state of rapid development, such as turbulence, suspensions and multiphase fluids, super and hypersonic flows and numerical modelling techniques.

It is a widely held view that it is the interdisciplinary subjects that will receive intense scientific attention, bringing them to the forefront of technological advancement. Fluids have the ability to transport matter and its properties as well as transmit force, therefore fluid mechanics is a subject that is particulary open to cross fertilisation with other sciences and disciplines of engineering. The subject of fluid mechanics will be highly relevant in domains such as chemical, metallurgical, biological and ecological engineering. This series is particularly open to such new multidisciplinary domains.

The median level of presentation is the first year graduate student. Some texts are monographs defining the current state of a field; others are accessible to final year undergraduates; but essentially the emphasis is on readability and clarity.

For a list of related mechanics titles, see final pages.

Numerical Simulation of Viscous Shock Layer Flows

E. N. S. H. M. G.
BIBLIOTHÈQUE
RM 95-95

by

YURI P. GOLOVACHOV
A.F. Ioffe Physico-Technical Institute,
Russian Academy of Sciences,
St. Petersburg, Russia

KLUWER ACADEMIC PUBLISHERS
DORDRECHT / BOSTON / LONDON

10243034

A C.I.P. Catalogue record for this book is available from the Library of Congress.

ISBN 0-7923-3626-7

Published by Kluwer Academic Publishers,
P.O. Box 17, 3300 AA Dordrecht, The Netherlands.

Kluwer Academic Publishers incorporates
the publishing programmes of
D. Reidel, Martinus Nijhoff, Dr W. Junk and MTP Press.

Sold and distributed in the U.S.A. and Canada
by Kluwer Academic Publishers,
101 Philip Drive, Norwell, MA 02061, U.S.A.

In all other countries, sold and distributed
by Kluwer Academic Publishers Group,
P.O. Box 322, 3300 AH Dordrecht, The Netherlands.

Printed on acid-free paper

All Rights Reserved
© 1995 Kluwer Academic Publishers
No part of the material protected by this copyright notice may be reproduced or
utilized in any form or by any means, electronic or mechanical,
including photocopying, recording or by any information storage and
retrieval system, without written permission from the copyright owner.

Printed in the Netherlands

Contents

Preface

The book is concerned with mathematical modelling of supersonic and hypersonic flows about bodies. Permanent interest in this topic is stimulated, first of all, by aviation and aerospace engineering. The designing of aircraft and space vehicles requires a more precise prediction of the aerodynamic and heat transfer characteristics. Together with broadening of the flight condition range, this makes it necessary to take into account a number of gas dynamic and physical effects caused by rarefaction, viscous-inviscid interaction, separation, various physical and chemical processes induced by gas heating in the intensive bow shock wave.

The flow field around a body moving at supersonic speed can be divided into three parts, namely, shock layer, near wake including base flow, and far wake. The shock layer flow is bounded by the bow shock wave and the front and lateral parts of the body surface. A conventional approach to calculation of shock layer flows consists in a successive solution of the inviscid gas and boundary layer equations. When the afore-mentioned effects become important, implementation of these models meets difficulties or even becomes impossible. In this case, one has to use a more general approach based on the viscous shock layer concept. In this approach, which is the topic of the book, the whole of the flow ahead of the body is calculated using viscous gas dynamics equations, which allows one to avoid matching procedures, to take into account an interaction between the flow regions and to investigate complex stationary and non-stationary flows including inner shocks, separation and recirculation zones.

Excepting the simplest problems, the viscous shock layer models are implemented with numerical methods. It is worth noting here that the role of numerical simulation in solving gas dynamic problems is rapidly increasing due to impressive progress in computer capacity and efficiency of numerical algorithms. At relatively low expenses, numerical simulation provides comprehensive data on the flow under study and allows one to investigate a wide range of flight conditions, including those which can not be reproduced in laboratory experiments.

The first viscous shock layer calculations were carried out for hypersonic stagnation region flows ahead of blunt bodies in the early sixties. Later on, this approach was substantially developed. A number of viscous shock layer models have been suggested which differ from each other in details of the flow description, range of validity and mathematical properties. Appropriate numerical methods have been elaborated as well. In spite of a wide use of the viscous shock layer approach, it has not been expounded in a systematic way. Dealing with a wider range of problems, available books on supersonic and computational gas dynamics include only fragments of this approach. The author attempts to make up for this deficiency.

The book presents a description of the viscous shock layer models and appropriate numerical methods. The basic ideas of mathematical modelling are elucidated with a number of calculation examples which demonstrate also some interesting and important features of supersonic and hypersonic flows about bodies. Our attention is focused mainly on the flows about blunt bodies, since even if such nose tip is not made especially beforehand, it is usually generated in a flight due to aerodynamic heating. It should be stressed here that the author did not intend to give a comprehensive analysis of all relevant gas dynamic phenomena. The main purpose was to demonstrate application of the viscous shock layer approach.

The book consists of five chapters. The first of them presents a formulation of the problem with the use of full Navier–Stokes equations. Modelling of turbulent flows with Reynolds equations and applicability of the continuous medium approach to rarefied flow regimes are also discussed. The chapter contains a review of finite-difference methods for solving full Navier–Stokes equations, description of the implicit unidirectional method suggested by the author and discussion of some grid generation techniques. Chapter 2 presents viscous shock layer models using the reduced Navier–Stokes equations. The appropriate numerical methods are considered including the time-marching, global relaxation and space-marching procedures. The self-similar solutions are discussed as well. To clarify basic ideas and their numerical implementation, they are expounded through the first two chapters with reference to a two-dimensional problem. Chapter 3 presents calculation results for some axisymmetric and plane flows (blunt-body problem in non-uniform flow fields, flows with mass injection, unsteady flows). Chapter 4 extends the above viscous shock layer models and numerical methods to three-dimensional flows. Some kinds of degenerate three-dimensional flows and approximate methods for prediction of aerodynamic characteristics are also discussed. Chapter 5 deals with physical and chemical processes which are important at hypersonic flight speed.

The book is intended mainly for researchers working in the field of super- and hypersonic computational gas dynamics. It can be of use for post- and undergraduates of the proper specialities. The reader is supposed to be acquainted with the fundamentals of the finite-difference numerical methods, gas dynamics and physics of high-temperature gases.

The book is based, to a great extent, on the results obtained by the author and his colleagues at the Ioffe Physico–Technical Institute. I thank them for useful collaboration and I thank other researchers both in Russia and abroad for kind permission to reproduce some of their results. I am much obliged to Professor G A Tirskii for perusal the manuscript and helpful remarks. Many thanks are also due to A M Kuzmin and T V Serova for preparation of the camera ready copy of the manuscript.

Yu P Golovachov
1 May 1995

Basic notations and scales

1.1 LATIN LETTERS

a, b, c, d	parameters of the oncoming stream non-uniformity
C_f	skin friction coefficient
C_H	heat transfer coefficient (Stanton number)
C_p	wall pressure coefficient
C_x	drag coefficient
c	speed of sound; speed of light
c_i	mass fraction of the i-th species
c_j^*	mass fraction of the j-th element
c_p	specific heat capacity at constant pressure
c_v	specific heat capacity at constant volume
$\boldsymbol{D}$	bow shock velocity with regard to the body surface
D	diffusion coefficient
Da	Damkohler number
E	total specific energy; energy of the electronic state
e	specific internal energy
G	mass injection rate
g	determinant of the metric tensor; degeneracy of the electronic state
g_{ij}, g^{ij}	covariant and contravariant components of the metric tensor
H	total specific enthalpy; Lamé coefficient; altitude; average body surface curvature
h	specific enthalpy; Planck constant
$\mathbf{I}$	identity matrix
I	radiation intensity
$\boldsymbol{J}_i$	mass diffusion flux of the i-th species
$\boldsymbol{J}_j^*$	mass diffusion flux of the j-th element
J	Jacobian of the coordinate transformation
K_P	Planck mean absorption coefficient
K_R	Rosseland mean absorption coefficient
Kn	Knudsen number
K_{pi}	equilibrium constant of the i-th reaction
k	wall-to-stagnation temperature ratio; Boltzmann constant
k_f, k_r	forward and reverse reaction rate constants

k_w	effective wall catalycity coefficient
l	mean free path of the particles; distance between the bodies
M	Mach number
m	molar mass
N_A	Avogadro number
N_e	electron number density
n	distance from the body surface
$\mathbf{P}$	stress tensor
Pr	Prandtl number
p	pressure
$\boldsymbol{q}$	heat flux
q^i	general curvilinear coordinates
R	body nose radius; universal gas constant
R^*	specific gas constant
Re	Reynolds number
Sc	Schmidt number
Sh	Strouhal number
s	distance along the body contour
$\mathbf{T}$	viscous stress tensor
T	temperature
t	time
u, v, w	gas velocity components
$\boldsymbol{V}$	gas velocity vector
$\dot{w}_i$	mass production rate of the i-th species
x^α	Cartesian coordinates

1.2 GREEK LETTERS

α	angle of attack
β	angle of side-slip; sweep angle
γ	gas specific heat ratio; intermittency coefficient
δ	boundary layer thickness; hypersonic flow parameter
δ^i_j	Kronecker function
ε	bow shock distance from the body surface; body surface emissivity
ζ	normalized Dorodnitsin's variable
η	Dorodnitsin's variable
θ_c	cone half-angle
κ	body surface curvature; absorption coefficient
λ	heat conductivity; radiation wavelength
μ	viscosity
$\boldsymbol{\nu}$	unit normal on the bow shock surface
ν	stoichiometric coefficient; radiation frequency
ξ	distance from the body surface normalized by the shock layer thickness; Dorodnitsin's variable

ρ	density
σ	bow shock slope to the oncoming stream direction; effective particle collision diameter; Stephan-Boltzmann constant
τ	wall shear stress; optical coordinate
Ω	angular velocity; collision integral; solid angle
$\boldsymbol{\omega}$	velocity of the moving coordinate system

1.3 SUBSCRIPTS

s	just behind the bow shock
∞	free stream
w	body surface
o	non-uniform oncoming stream axis
ν	normal to the bow shock; radiation frequency

1.4 SUPERSCRIPTS

$*$	free stream stagnation parameters; excited electronic state
o	stagnation point
m	maximum
s	separation point; sonic point

1.5 SCALES

Unless otherwise specified, the non-dimensional quantities are defined as follows: distances are normalized by the body nose radius R, gas velocity components by V_∞, time by R/V_∞, density by ρ_∞, pressure by $\rho_\infty V_\infty^2$, temperature by $V_\infty^2/2R^*$, specific enthalpy and energy by $V_\infty^2/2$, heat flux by $\rho_\infty V_\infty^3$, mass diffusion flux by $\rho_\infty V_\infty$, mass production rate of the species by $\rho_\infty V_\infty/R$, skin friction coefficient is defined as the wall shear stress-to-$\rho_\infty V_\infty^2$ ratio.

1

Simulation of supersonic flow around a body using the Navier–Stokes equations

The book is concerned with flows which can be investigated within the continuous medium approach. The most general mathematical model for such flows is given by the set of Navier–Stokes equations. These equations can be obtained either in a phenomenologic way, assuming a linear dependence of the flux vectors on the gradients of gas dynamic functions, or by the methods of the kinetic theory of gases, taking into account two terms of the series expansion for the distribution function with respect to the parameter which is inverse to the particle collision frequency.

The Navier–Stokes equations apply to laminar and turbulent flows. However, because of an extremely wide range of time and length scales of turbulence, direct use of the Navier–Stokes equations for numerical simulation of turbulent aerodynamic flows requires unrealistic computer resources. Therefore, such flows are calculated mostly with the use of Reynolds equations which describe variations of the time-averaged values of gas dynamic functions. These equations are closed by semi-empirical theories of turbulent transport. With the conventional hypotheses for the turbulent stress tensor, one can write the Reynolds equations in the same form as the Navier–Stokes equations and use the same numerical methods for solving both sets.

The main field of application of the Navier–Stokes (Reynolds) equations in supersonic gas dynamics is a detailed investigation of flows including strong viscous-inviscid interaction, separation and recirculation regions. When calculating non-separated flows about smooth bodies, one has to use the full Navier–Stokes equations only at low Reynolds numbers.

1.1 FORMULATION OF THE PROBLEM

To elucidate the formulation of the problem, let us consider an axisymmetric blunt body flow of a perfect gas with constant specific heats. This flow is schematically shown in figure 1.1. When the free-stream Mach number $M \gg 1$ and the Reynolds number $Re \to \infty$, the extent of various parts of the flow field can be estimated

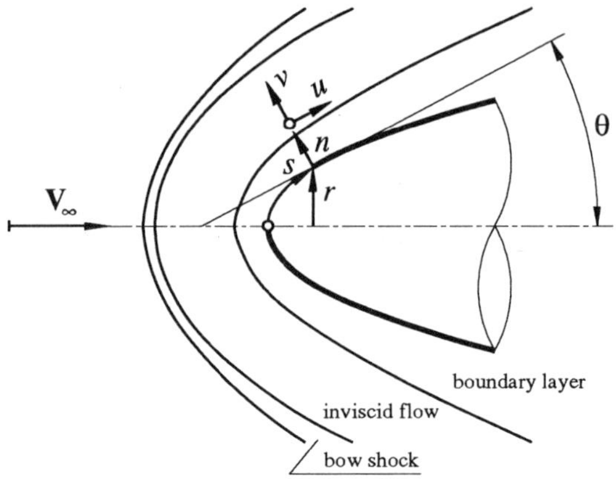

Figure 1.1 Sketch of the flow field and coordinate system.

as follows, see for example (Hayes and Probstein 1959). The thickness of the perturbed flow region near the stagnation streamline normalized by the body nose radius is approximately equal to the gas density ratio at the bow shock wave. Non-dimensional thickness of the bow shock region and the wall boundary layer are equal, in order of magnitude, to Re^{-1} and $Re^{-1/2}$, respectively. With decreasing Reynolds number, these two regions expand while the region of inviscid flow shrinks. At Reynolds numbers $Re \lesssim 10^2$, the molecular transport processes influence essentially the whole of perturbed flow field. Here we mean the Reynolds number defined with the gas parameters just behind the bow shock, which is equal, in the order of magnitude, to the ratio of the body nose radius to the mean free path of the molecules. Depending on the flight conditions, the flow under study can be calculated with various mathematical models, the most general one being given by the full Navier–Stokes equations.

1.1.1 Navier–Stokes equations

In the calculation of supersonic blunt body flows, it is often convenient to use the body-fitted curvilinear orthogonal coordinate system (s, n, φ), where n is the distance from the body surface, s is the distance from the stagnation point along the body surface in a meridional plane $\varphi = const$ shown in figure 1.1. The Lamé coefficients for this coordinate system are the following: $H_s = 1 + \kappa n$, $H_n = 1$, $H_\varphi = r + n \cos\theta$, where κ is the curvature of the meridional section of the body, $r(s)$ is the distance from the symmetry axis to the body surface, θ is the slope of the body generatrix to the direction of the oncoming stream.

For the above coordinate system, the Navier–Stokes equations of an axisymmetric flow without bulk forces are written as follows:
the continuity equation

$$\frac{\partial \rho}{\partial t} + \frac{1}{H_s}\frac{\partial}{\partial s}(\rho u) + \frac{\partial}{\partial n}(\rho v) + \rho(g_1 u + g_2 v) = 0 \tag{1.1}$$

the balance equation for the n-momentum component

$$\begin{aligned}
&\rho\left(\frac{\partial v}{\partial t}+v\frac{\partial v}{\partial n}+uV_1\right)+\frac{\partial p}{\partial n}\ \Bigg|\ -\left(\mu g_1+\frac{1}{H_s}\frac{\partial \mu}{\partial s}\right)\frac{\partial u}{\partial n}-\frac{\mu}{H_s}\\
&\times\frac{\partial^2 u}{\partial s\partial n}-\frac{2}{3}\frac{\partial}{\partial n}\left[\mu\left(2\frac{\partial v}{\partial n}-\frac{1}{H_s}\frac{\partial u}{\partial s}-g_1 u\right)\right]\ \Bigg\|\ -\mu g_1 V_1\\
&-2\mu\left(g_2\frac{\partial v}{\partial n}-\frac{\kappa}{H_s}V_2-\frac{\cos\theta}{H_\varphi}V_3\right)-\frac{1}{H_s}\frac{\partial}{\partial s}(\mu V_1)\\
&+\frac{2}{3}\frac{\partial}{\partial n}(\mu g_2 v)=0
\end{aligned}\tag{1.2}$$

the balance equation for the s-momentum component

$$\begin{aligned}
&\rho\left(\frac{\partial u}{\partial t}+v\frac{\partial u}{\partial n}+uV_2\right)+\frac{1}{H_s}\frac{\partial p}{\partial s}-\frac{\partial}{\partial n}\left(\mu\frac{\partial u}{\partial n}\right)\ \Bigg|\ -g_2\mu\frac{\partial u}{\partial n}\\
&-V_1\frac{\partial \mu}{\partial n}\ \Bigg\|\ -\mu\left[g_2V_1+2g_1\left(V_2-V_3\right)+\frac{1}{H_s}\frac{\partial^2 v}{\partial s\partial n}\right]\\
&-\frac{2}{3}\frac{1}{H_s}\frac{\partial}{\partial s}\left[\mu\left(2V_2-\frac{\partial v}{\partial n}-V_3\right)\right]=0
\end{aligned}\tag{1.3}$$

the energy balance equation

$$\begin{aligned}
&\rho\left(\frac{\partial h}{\partial t}+\frac{u}{H_s}\frac{\partial h}{\partial s}+v\frac{\partial h}{\partial n}\right)-\frac{\partial p}{\partial t}-\frac{u}{H_s}\frac{\partial p}{\partial s}-v\frac{\partial p}{\partial n}-\mu\left(\frac{\partial u}{\partial n}\right)^2\\
&-\frac{\partial}{\partial n}\left(\lambda\frac{\partial T}{\partial n}\right)\ \Bigg|\ -g_2\lambda\frac{\partial T}{\partial n}-2\mu V_1\frac{\partial u}{\partial n}\ \Bigg\|\\
&-\mu\left\{V_1^2+\frac{4}{3}\left[V_2^2+V_3^2-V_2V_3+\frac{\partial v}{\partial n}\left(\frac{\partial v}{\partial n}-V_2-V_3\right)\right]\right\}\\
&-\frac{1}{H_s}\left[g_1\lambda\frac{\partial T}{\partial s}+\frac{\partial}{\partial s}\left(\frac{\lambda}{H_s}\frac{\partial T}{\partial s}\right)\right]=0
\end{aligned}\tag{1.4}$$

Here ρ, T, p, h are the density, temperature, pressure and specific enthalpy; u, v are the gas velocity components in the s, n directions; μ, λ are the viscosity and heat conductivity; $g_i (i=1,2)$ and $V_i (i=1,2,3)$ are defined by the following formulae

$$g_1=\frac{1}{H_sH_\varphi}\frac{\partial H_\varphi}{\partial s}\qquad g_2=\frac{\kappa}{H_s}+\frac{\cos\theta}{H_\varphi},$$

$$V_1=\frac{1}{H_s}\left(\frac{\partial v}{\partial s}-\kappa u\right)\qquad V_2=\frac{1}{H_s}\left(\frac{\partial u}{\partial s}+\kappa v\right)\qquad V_3=g_1u+\frac{\cos\theta}{H_\varphi}v.$$

Vertical bars pertain to the reduced equations being discussed in the following chapter.

The set of equations (1.1) – (1.4) is closed by the thermodynamic relationships and the formulae for transport coefficients. Variation of pressure, density and temperature of a perfect gas is governed by the equation of state

$$p = \rho R^* T, \tag{1.5}$$

where R^* is the specific gas constant. Specific enthalpy is related to gas temperature by the equation

$$h = c_p T, \tag{1.6}$$

where c_p is specific heat at constant pressure.

Viscosity and heat conductivity can be obtained from the kinetic theory of gases provided that the data on the particle interaction potentials are available. In practice, the viscosity of a perfect gas with constant specific heats is often approximated by the power formula $\mu \propto T^\omega$, $0.5 \leqslant \omega \leqslant 1.0$. The extreme values of the power index ω correspond to the most rigid (rigid spheres) and most soft (Maxwellian molecules) potentials of the particle interaction. A more accurate prediction is provided by Sutherland's formula

$$\frac{\mu}{\mu_0} = \left(\frac{T}{T_0}\right)^{3/2} \frac{T_0 + T_s}{T + T_s}, \tag{1.7}$$

where T_s is Sutherland's constant; μ_0, T_0 are viscosity and temperature in the reference state. For air, Sutherland's formula reads

$$\mu = 14.65 \times 10^{-7} \frac{T^{3/2}}{T + 114} \quad \frac{\mathrm{kg}}{\mathrm{m \cdot s}} \tag{1.8}$$

where temperature is substituted in Kelvins. Heat conductivity of a perfect gas with constant specific heat is proportional to its viscosity. Instead of heat conductivity, the Prandtl number $Pr = \mu c_p / \lambda$ is often used which remains constant throughout the flow field.

1.1.2 Initial and boundary conditions

The non-stationary Navier–Stokes equations represent the set with hyperbolic and parabolic properties, for which the initial-boundary value problem is well-posed (Kovenya and Yanenko 1981). For stationary flows, the set of Navier–Stokes equations is elliptic and one has to solve the boundary value problem. Until nowadays, the initial and boundary conditions which would ensure existence and uniqueness of the solution of the above problems have not been formulated strictly.

The initial conditions for the non-stationary equations are usually given by the fixed function values at the initial moment. The boundary conditions are formulated as follows.

The computational domain is bounded either by the bow shock or by the surface located in the free stream. In the latter case the outer boundary conditions are given by the free stream parameters. Such formulation of the problem applies at low Reynolds numbers or when the processes occurring within the shock layer influence essentially the flow field ahead of the bow shock.

For most viscous shock layer problems, it seems to be expedient to consider the bow shock as the outer boundary of the computational domain. Since the Navier–Stokes equations are invalid within the intensive shock wave, this formulation does not deteriorate or shrink the information as compared with the use of bow shock capturing methods. At the same time, the bow shock fitting allows one to save the computer resources due to elimination of the flow region with maximum gradients from the computational domain.

When the computational domain is bounded by the bow shock wave, the gas parameters behind the shock front can be determined with the use of modified Rankine–Hugoniot relationships which were discussed in detail by Tolstykh (1964) and Magomedov (1970). They represent a set of algebraic and first-order ordinary differential equations resulting from integration of the local one-dimensional Navier–Stokes equations across the bow shock from the free stream. The molecular transport processes are taken into account only in the direction normal to the shock and the shock curvature is ignored. The error caused by these assumptions is estimated to be $O(Re^{-1})$. For an axisymmetric flow, the modified Rankine–Hugoniot relationships are written as follows:

$$\begin{aligned}
&\rho(V_\nu - D_\nu) = \rho_\infty(V_{\infty\nu} - D_\nu)\\
&\rho(V_\nu - D_\nu)(V_\tau - V_{\infty\tau}) = \mu\frac{\mathrm{d}}{\mathrm{d}\nu}(V_\tau - D_\tau)\\
&\rho(V_\nu - D_\nu)^2 + p = \frac{4}{3}\mu\frac{\mathrm{d}}{\mathrm{d}\nu}(V_\nu - D_\nu) + \rho_\infty(V_{\infty\nu} - D_\nu)^2 + p_\infty\\
&\rho(V_\nu - D_\nu)\left[h + \frac{(V_\nu - D_\nu)^2 + (V_\tau - D_\tau)^2}{2}\right] = \lambda\frac{\mathrm{d}T}{\mathrm{d}\nu}\\
&\quad + \mu(V_\tau - D_\tau)\frac{\mathrm{d}}{\mathrm{d}\nu}(V_\tau - D_\tau) + \frac{4}{3}\mu(V_\nu - D_\nu)\frac{\mathrm{d}}{\mathrm{d}\nu}(V_\nu - D_\nu)\\
&\quad + \rho_\infty(V_{\infty\nu} - D_\nu)\left[h_\infty + \frac{(V_{\infty\nu} - D_\nu)^2 + (V_{\infty\tau} - D_\tau)^2}{2}\right]
\end{aligned} \tag{1.9}$$

Here $\boldsymbol{D}$ is the bow shock velocity; subscripts ν, τ denote the vector components normal and tangent to the shock front, respectively; subscript ∞ denotes the free stream parameters. Without the terms responsible for the molecular momentum and energy transport, equations (1.9) turn into the conventional shock relationships.

The velocity vector components V_ν, V_τ and the derivative $\frac{\mathrm{d}}{\mathrm{d}\nu}$ are given by the formulae

$$V_\nu = \boldsymbol{V}\cdot\boldsymbol{\nu} \qquad V_\tau = \boldsymbol{V}\cdot\boldsymbol{\tau} \qquad \frac{\mathrm{d}}{\mathrm{d}\nu} = \boldsymbol{\nu}\cdot\boldsymbol{\nabla}$$

The s, n-projections of the unit vectors $\boldsymbol{\nu}$ and $\boldsymbol{\tau}$ which are normal and tangent to the bow shock are written as follows

$$-\nu_s = \tau_n = \frac{\varepsilon'_s}{k(1+\kappa\varepsilon)} \qquad \nu_n = \tau_s = \frac{1}{k}$$

where $\varepsilon(s,t)$ is the bow shock wave detachment from the body surface,

$$k = \sqrt{1 + [\varepsilon'_s/(1+\kappa\varepsilon)]^2}$$

The wall boundary conditions for high Reynolds number flows are given by fixed values of the velocity components, wall temperature or heat flux to the body surface. In calculation of flows with mass injection, the mass injection rate is often prescribed as one of the wall boundary conditions instead of the normal velocity component.

With decreasing Reynolds number, one has to take into account a deviation from the continuous flow model near the body surface. In this case, the Navier–Stokes equations are solved with the velocity slip and temperature jump wall boundary conditions. These boundary conditions are derived by matching the solution of the kinetic equation for the Knudsen layer adjacent to the wall with the solution of the Navier–Stokes equations for the outer flow. Using the model kinetic equation of a relaxation type and assuming diffusive reflection of gas molecules on the body surface with the Maxwellian distribution function, Kogan (1967) obtained the following velocity slip and temperature jump wall boundary conditions:

$$u = 1.431\frac{\mu}{\sqrt{\rho p}}\frac{\partial u}{\partial n} + 0.84\frac{\mu}{\rho T}\frac{\partial T}{\partial s} \tag{1.10}$$

$$T = T_w + \frac{0.627}{p}\sqrt{\frac{T}{R}}\frac{2-0.827\alpha}{\alpha}\lambda\frac{\partial T}{\partial n} \tag{1.11}$$

Here R is the universal gas constant, α is the energy accommodation coefficient of the body surface. For most body surface materials, the energy accommodation coefficient is close to unity. The second term of the right-hand side of (1.10) becomes significant only when the mean free path of the molecules in the shock layer reaches in an order of magnitude the characteristic length of the flow L, i.e. at Knudsen numbers $Kn = l/L$ approaching unity.

Formulating the problem for a flow bounded by the bow shock wave, one has to keep in mind that this boundary of the computational domain is not located beforehand but has to be found as a part of the solution together with the shock layer parameters. Since the total order of set (1.1) – (1.4) with respect to n is equal to seven, it is evident that four bow shock relationships together with three wall boundary conditions are not sufficient for solving the problem with an unknown location of one of the boundaries. To obtain an additional relationship which closes the problem, several ways were proposed (Tolstykh 1966, Breyev and Golovachov 1970, Pavlov 1971, Golovachov *et al* 1973). For instance, Breyev and Golovachov (1970) used the normal momentum balance equation written on the body surface subject to the boundary conditions for the velocity components. In the method by Golovachov *et al* (1973) which is discussed in detail in the next section, the closing relationship results from finite-difference approximation of the governing equations, which is performed in accordance with the order of magnitude analysis of the viscous terms.

The downstream boundary of the computational domain intersecting the body surface and the bow shock wave is disposed far enough from the stagnation region ensuring the velocity component normal to this boundary to exceed the speed of sound, except for a relatively thin layer adjacent to the body surface. With such locations of the downstream boundary, the effects of an upstream disturbance propagation are small, and therefore one can use the approximate boundary conditions

$$\frac{\partial^k f}{\partial s^k} = 0 \qquad k = 1, 2 \tag{1.12}$$

where f is any unknown function. Numerical implementation of (1.12) depends on the computational method. For instance, it can be taken into account in the finite-difference equations written on the downstream boundary using either one-sided difference formulae for the s-derivatives or central formulae which include the function values obtained by extrapolation from within the computational domain.

When calculating the flows with planes or axes of symmetry considered as the boundaries of the computational domain, the symmetry relationships are used as the proper boundary conditions.

1.1.3 Similarity criteria

Having written the above equations, initial and boundary conditions for the non-dimensional variables, one can conclude that the problem of supersonic viscous blunt body flow of a perfect gas with constant heat capacities without external bulk forces is characterized, besides the geometric parameters, by the following similarity criteria: Mach number, $M_\infty = V_\infty / c_\infty$; Reynolds number, $Re_\infty = \rho_\infty V_\infty R / \mu_\infty$; Prandtl number, $Pr = \mu c_p / \lambda$; gas specific heat ratio $\gamma = c_p / c_v$; power index, ω, of the viscosity dependance on gas temperature; wall temperature factor, $k = T_w / T^*$; Strouhal number, $Sh = R/(V_\infty \Delta t)$, where Δt is the characteristic time of the unsteady process.

The above criteria display a relative importance of various processes governing the shock layer flow, aerodynamic characteristics and heat transfer.

The Mach number M_∞ characterizes the gas compressibility. The square M_∞^2 is directly proportional to the ratio of the kinetic energy of translational motion of the molecules at speed V_∞ to the energy of their thermal chaotic motion at free-stream temperature T_∞.

The Reynolds number Re_∞ represents the ratio of convective to molecular momentum transfer and the Prandtl number Pr defines a correlation between molecular momentum and energy transfer.

The specific heat ratio γ depends on excitation of the internal degrees of freedom of the molecules. Its deviation from unity displays a correlation between the expansion work at constant pressure and the internal energy of a gas. In the case of a monatomic gas of molecules which have only translational degrees of freedom, $\gamma = 5/3$. For a diatomic gas with equilibrium rotational and frozen vibrational degrees of freedom, $\gamma = 7/5$. Further excitation of the internal degrees of freedom of the molecules diminishes the contribution of the expansion work to gas heat capacity at constant pressure and, therefore, the value of γ approaches unity.

The power index ω, the temperature factor k and the Strouhal number Sh characterize the particle interaction potential, wall-to-free stream stagnation temperature ratio and non-stationarity, respectively.

At high supersonic speed of the oncoming stream, the number of basic similarity criteria diminishes. Under these flight conditions, one can usually assume $k << 1$, $Sh << 1$. The Reynolds number Re_∞ is replaced by the Reynolds number Re_s which is evaluated with the body characteristic length and gas parameters just behind the bow shock. The latter excludes the Mach number M_∞ from the list of similarity criteria. The value of Re_s is inversely proportional to the Knudsen number, i.e. to the ratio of the mean free path of molecules in the shock layer to the body nose radius. If the gas viscosity $\mu \propto T^\omega$, then the Reynolds number Re_s is related to Re_∞ by the formula

$$Re_s = \frac{\rho_s v_s R}{\mu_s} = Re_\infty \left[1 + \frac{2(\gamma - 1)}{(\gamma + 1)^2 M_\infty^2} \left(M_\infty^2 - 1\right)\left(1 + \gamma M_\infty^2\right)\right]^{-\omega} \tag{1.13}$$

At a hypersonic speed, $Re_s \sim Re^*$ where Reynolds number Re^* is based on the body nose radius, the free stream velocity and density and the gas viscosity at the free stream stagnation temperature T^*. The above set of similarity criteria for hypersonic flow regimes has been substantiated by a number of experimental and calculation data on aerodynamic characteristics and heat transfer for a wide range of aerodynamic shapes, see for example the papers by Gusev *et al* (1970), Gusev (1993).

1.1.4 Modelling of turbulent flows

At Reynolds numbers $Re_\infty \gtrsim 10^5$, the flow near the body surface becomes turbulent. The transition to the turbulent regime is accompanied by a significant change of the flow characteristics and, in particular, by a drastic increase of the heat flux to the body surface.

Laminar-turbulent transition occurs within some region which is sometimes replaced by the transition point. Location of the transition region (transition point) depends on such factors as the Mach number, body shape, roughness of the surface and its temperature. To locate the transition region, Reynolds number Re_θ is often used which is calculated with the local value of momentum thickness. On the basis of experimental data, Batt and Legner (1983) proposed the following formulae for Reynolds numbers Re_θ^t at the beginning of the transition region

$$Re_\theta^t = \begin{cases} 500X^{-1.5}, & \text{if } X \geqslant 1; \\ 500, & \text{otherwise.} \end{cases} \tag{1.14}$$

Here

$$X = \frac{k}{\theta}\frac{T_e}{T_w}\frac{1}{1 + 350k/R}$$

R is the body nose radius, θ is the momentum thickness of the boundary layer, k is the roughness height value that is exceeded by no less than 30 per cent roughness elements, the subscripts e and w refer to the edge of the boundary layer and the body surface. Formulae (1.14) represent experimental data for $0.1 \leqslant X \leqslant 10$ accurate within $\pm 30\%$. The Reynolds number of laminar-turbulent transition varies for typical surface materials in the range $20 \lesssim Re_\theta^t \lesssim 500$. It is supposed that transition occurs in the subsonic region of the shock layer at Mach number $M_e < 1$. For a smooth surface, formulae (1.14) yield $Re_\theta^t = 500$. The transition region is followed by a region of developed turbulent flow.

Nowadays, investigation of turbulent flows for aerodynamic applications is based on the equations for average values of gas dynamic functions. Most research is carried out with the use of Reynolds equations which are derived from the Navier–Stokes equations by decomposing the dependent variables into mean and fluctuating components and application of the time-averaging procedure. For compressible flows it is convenient to use the mass-weighted time-averaging suggested by Favre (1965). The use of averaging procedures results in omission of such important characteristics of turbulence as frequency, phase and wave length of turbulent fluctuations. Fortunately, information on these characteristics is not urgent in most engineering applications.

The Reynolds equations contain extra terms as compared with the Navier–Stokes equations, which appear as a result of time-averaging due to non-linearity of the original mathematical model. Being time-averaged products of the fluctuating parameters, these terms represent the average momentum and energy transport in fluctuating motion. Depending on the problem under study, only the largest extra terms are usually taken into account which are written as the turbulent stress tensor $\boldsymbol{\tau}_t$ and the turbulent heat flux vector $\boldsymbol{q}_t$.

The set of Reynolds equations is closed by the relationships which express the above quantities through the mean flow characteristics. Closing expressions are given by the semi-empirical turbulence models. Most of them are based on Boussinesq's hypothesis postulating a linear dependence of the turbulent stress tensor on the tensor of the mean rates of strains. For theCartesian coordinate system, the above hypothesis gives

$$(\tau_t)_{ij} = -\overline{\rho u_i' u_j'} = \mu_t \left(\frac{\partial \overline{u}_i}{\partial x_j} + \frac{\partial \overline{u}_j}{\partial x_i} \right) - \frac{2}{3} \delta_{ij} \left(\mu_t \frac{\partial \overline{u}_k}{\partial x_k} + \rho \overline{k} \right) \tag{1.15}$$

where μ_t is the eddy viscosity, $\overline{k} = (\overline{u_i' u_i'})/2$ is the kinetic energy of turbulence, δ_{ij} is the Kronecker function, the prime and bar denote fluctuating and mean quantities, respectively.

Eddy viscosity is defined by models of turbulent momentum exchange. The simplest of them are the algebraic models based on the mixing length theory of Prandtl. According to this theory,

$$\mu_t = \rho l^2 \left| \frac{\partial \overline{u}}{\partial y} \right| \tag{1.16}$$

where l is the mixing length, $\overline{u}$ is the streamwise mean velocity component, y is the coordinate normal to the streamwise direction. The mixing length is determined by the relationships which depend on the type of flow under study and contain empirical parameters, see for instance (Lapin 1982, Loitsianskii 1987).

In more complicated models, the eddy viscosity is determined from the supplementary differential equations. Most of these models are founded on the Kolmogorov–Prandtl hypothesis relating eddy viscosity to kinetic energy and length scale of turbulence. The above models include at least one supplementary differential equation. A number of turbulent flow calculations have been carried out with $k-\varepsilon$ models in which the eddy viscosity is determined through the kinetic energy of turbulence k and its dissipation rate ε. To obtain these quantities, two supplementary differential equations are integrated along with mass, momentum and energy balance equations of the mean flow.

Within the framework of Boussinesq's hypothesis, the turbulent heat flux vector is presented as a product of the turbulent heat conductivity and the mean temperature gradient

$$\boldsymbol{q}_t = -\lambda_t \nabla \overline{T} \tag{1.17}$$

Instead of the turbulent heat conductivity, the turbulent Prandtl number $Pr_t = \mu_t c_p / \lambda_t$ is often used. This parameter characterizes the correlation between turbulent momentum and energy transport. In formulation of the turbulence models, it is assumed that both momentum and energy transports occur simultaneously due to fluctuating motion of the same gas volumes. Thus, deviation of the turbulent Prandtl number from unity is caused only by a different efficiency of the dynamic and thermal interactions between these volumes and the ambient gas in fluctuating motion. Near the wall this difference is small and, therefore, the turbulent Prandtl number is usually assumed to be equal to 0.9.

With Boussinesq's hypothesis, the governing equations become similar to the Navier–Stokes equations with effective transport coefficients which are the sums of proper molecular coefficients and turbulent coefficients corrected for laminar-turbulent interaction. In the region of laminar flow, the terms responsible for turbulent transport are supposed to be equal to zero and, therefore, the equations describe instantaneous variation of gas dynamic functions. In the region of turbulent flow the above terms are switched on and the equations describe variation of the mean flow characteristics.

If the finite length of the transition region is taken into account, the effective transport coefficients within this region are calculated using the intermittency coefficient γ

$$\mu_{\text{eff}} = \mu + \gamma \mu_t \qquad \left(\frac{\mu}{Pr}\right)_{\text{eff}} = \left(\frac{\mu}{Pr}\right) + \gamma \left(\frac{\mu}{Pr}\right)_t \tag{1.18}$$

To evaluate the intermittency coefficient, one can use the empirical formulae suggested, for instance, by Dhawan and Narasimha (1958), Chen and Thyson (1971), Feiereisen and Acharya (1986).

The principal difficulty which hampers investigation of turbulent flows with the Reynolds equations is the lack of a general closing model. The results provided by different turbulence models were compared by Viegas and Coakley (1977), Viegas and Horstman (1979), Cebeci *et al* (1986). For relatively simple flows most algebraic and differential models give close results which agree with the experimental data. However, the use of different turbulence models in simulation of separated flows can provide substantially different results and, moreover, complication of a turbulence model does not ensure, as a matter of fact, more generality and a more exact flow prediction. For this reason, the investigations of complex turbulent flows often use simpler algebraic models.

Transition to the turbulent regime happens at high Reynolds numbers when the main of shock layer flow is well described by the inviscid gas equations. So, the turbulent shock layer flows are usually calculated with the simplified mathematical models, instead of the full Reynolds equations. Such shock layer models are discussed in the next chapter. The bow shock boundary conditions are given by the conventional Rankine–Hugoniot relationships. The boundary conditions on other boundaries of the shock layer are the same as for the Navier–Stokes equations. Using the differential models of turbulence, one has to formulate the supplementary boundary conditions for turbulence characteristics. It is worth noting here that numerical implementation of the wall boundary conditions for turbulent flows is sometimes carried out with the use of wall functions (Launder and Spalding 1974). This approximate technique is founded on well-known data on the structure of the inner region of the turbulent boundary layer outside the vicinity of its incipiency and separation. With this technique, the boundary conditions are imposed at some distance from the wall, where the velocity profile for the 'law of the wall' coordinates is described by a log-linear function.

In conclusion, it should be stressed that numerical investigation of turbulent flows presents a more difficult problem as compared with calculation of laminar flows because of a more complicated behaviour of the gas dynamic functions. Turbulent profiles of gas dynamic functions in the near-wall region are substantially more filled in, which compels use of the finer computational grids to ensure accurate prediction of the heat flux and shear stress on the body surface. Beside that, numerical simulation of turbulent shock layer flows is hampered by abrupt variation of the flow parameters within the transition region.

1.2 FINITE-DIFFERENCE METHODS

In spite of a considerable progress in computer speed and numerical methods, integration of the Navier-Stokes and Reynolds equations remains to be an arduous problem, especially for three-dimensional flows. This augments the importance of efficiency of the numerical algorithms, which often determines not only the cost but even the possibility of numerical simulation. For the most part, the viscous shock layer flows are calculated with the finite-difference methods.

1.2.1 Brief review of the finite-difference methods

The first numerical solutions of the full Navier–Stokes equations for the problems of supersonic flows about bodies were obtained with the use of explicit finite-difference schemes for low Reynolds number flows (Pavlov 1968, Scala and Gordon 1968, Molodsov 1969). The main advantage of the explicit methods is the simplicity of calculation of the unknown functions at a new time step. However, the time increment is strongly restricted by the stability condition. With increasing Reynolds number, allowable time increment diminishes below the value consistent with the accuracy requirements, which results in a waste of computer resources. The above shortcoming is especially grave in solving stationary problems by the time-asymptotic technique.

Nowadays, most researchers prefer to solve the Navier–Stokes equations using more stable implicit methods. For multidimensional flows, alternative direction, unidirectional and splitted factored schemes are applied which are unconditionally stable within the framework of linear analysis. In these methods, the unknown functions at each time level are calculated through several stages each coming to the one-dimensional sets of difference equations. Below, some problems of numerical implementation of such methods are briefly discussed.

Briley and McDonald (1977), Beam and Warming (1978) suggested the implicit finite-difference schemes for solving the Navier–Stokes equations with the use of approximate factorization and coordinate splitting of the multidimensional operator. The Navier–Stokes equations were written in the conservation law delta form. Using Taylor series expansions for the flux vectors, the non-iterative implicit methods were formulated which resulted at each fractional step in the sets of linear difference equations with block tridiagonal matrices of coefficients. They were solved by block 'progonka' algorithm which is often referred to as block Thomas algorithm.

The implicit factored schemes elaborated by Kovenya and Yanenko (1981) include additional splitting in the physical processes. Additional splitting increases the number of fractional steps; however, at each step the difference equations are solved using only scalar procedures, which decreases the total amount of computations per time step.

It is worth noting here that in spite of some increase of the amount of computations per time step, it is often preferable to solve block sets of difference equations because such algorithms are more stable, more robust and require less time steps to achieve the stationary solution (McDonald and Briley 1981, Belotserkovskii *et al* 1982). It seems to be expedient to solve the difference equations for all unknown functions together when the number of these functions is relatively small. Otherwise, it results in an unacceptable increase of computations because one has to solve large sets of linear algebraic equations at each mesh point. A proper example is the calculation of non-equilibrium multicomponent flow. In such problems it may be expedient to solve jointly the groups of difference equations.

When applying the factored splitted schemes, one is faced with the problem of formulating the boundary conditions at the fractional steps. This problem, which is especially embarrassing for non-stationary flows, arises because the consistency of such schemes is the result of all fractional steps, while each of them is not consistent with the multidimensional equations. To obtain the boundary conditions required, one can apply a difference approximation of the governing equations at the boundaries with the use of the boundary conditions of the problem and function values at inner mesh points followed by splitting the operators thus obtained, see Kovenya and Yanenko (1981), Belotserkovskii *et al* (1982). Some other ways to formulate the boundary conditions for the implicit splitted finite-difference schemes are discussed in the monograph by Lapin and Strelets (1989).

Coordinate splitting is realized by approximate factorization of the multidimensional operators with the error which is proportional to the product of Courant numbers of the one-dimensional problems. The factorization error restricts substantially the allowable value of time increment. This shortcoming is removed in the implicit unidirectional methods suggested by Golovachov and Popov (1972a,b), Golovachov *et al* (1973), Karyakin and Popov (1977), Belousov *et al* (1991), in which the allowable value of time increment is limited only by the required accuracy in time and non-linearity of the difference equations. These methods can be considered as a generalization of the calculation procedure suggested by Babenko *et al* (1964), Lubimov and Rusanov (1970) for supersonic inviscid gas flows about bodies. Similar to the splitted factored schemes, the unknown functions at each time level are calculated through several stages (iterations). In each iteration, one-dimensional sets of linear difference equations are solved by 'progonka' algorithm, but in contrast to the factored schemes, the 'progonka' direction remains invariable, which is especially favourable for the use of parallel computational systems. Alternating 'progonka' directions, one has to redistribute the flow field data between the processors, which deteriorates the efficiency of parallel computations. Another advantage of the implicit unidirectional methods is a straightforward implementation of the boundary conditions. It is worth noting also that sometimes the solution at a new time level can be obtained without iterations. Then, the methods discussed turn to the [explicit-implicit methods]explicit-implicit line relaxation procedure which is conditionally stable even within the framework of linear analysis. Nevertheless, the time step allowed by the stability condition is often acceptable for the flows with smooth variation of gas dynamic functions in the direction of explicit approximation.

Other explicit-implicit numerical methods, combining the attractive features of both kinds of finite-difference approximations, have been also elaborated. For instance, Shang (1978) suggested a method in which the implicit calculations were carried out only in the direction of the most rapid variation of the gas dynamic functions and only within the region of steep gradients. The gas parameters in the rest of the flow were calculated with a more simple explicit algorithm. The disadvantage of this method consists in non-physical perturbations arising at the mesh points where the type of difference approximation is changed. The explicit-implicit method by Mac Cormack (1982) presents a modification of his explicit two-step method (Mac Cormack 1969). The modification, undertaken to relax the stability restriction for the time increment, includes the additional computations with the use of implicit formulae both at the predictor and corrector stages. These computations are organized similarly to the simplest implicit schemes for the wave and heat transfer equations with approximate accounting for convective and molecular transport processes. Approximate factorization of the implicit difference operator and coordinate splitting are applied to calculate multidimensional flows. The linear analysis proves an unconditional stability of the modified method. Its realization requires a relatively small amount of computations due to the use of explicit formulae in the flow regions where the CFL-stability criterion is satisfied on the one hand and the bidiagonal sets of difference equations in the rest of the flow field on the other hand. The drawbacks are the factorization error, decrease of allowable time increment in the regions with steep gradients of gas dynamic functions and approximate accounting for disturbance propagation by the implicit difference operators.

Most numerical methods for solving the Navier–Stokes equations use the central difference formulae for the viscous terms. In the first papers, see for example (Beam and Warming 1978), the central formulae were applied to the convective terms of the Navier–Stokes equations as well. It resulted in oscillations of the numerical solution, which were suppressed by the artificial dissipation. At present, the convective terms are often approximated with the formulae resulting from the flux-splitting procedures performed in accordance with local characteristic properties of the Euler equations with the use of local eigenvectors of the proper Jacobi matrices (Thomas and Walters 1985, Mac Cormack 1985, Shang 1985, Newsome *et al* 1987). To ensure the monotonicity, such methods generally use the idea of local reduction of the order of accuracy of the numerical solution in the regions of steep function gradients. This popular idea is quite consistent with the inviscid gas model but it is not appropriate to the Navier–Stokes equations which have to provide accurate results for flows including the shear layers. In this view, it seems to be more expedient to use uniformly accurate schemes (Belousov *et al* 1991, Golovachov and Zemlyakov 1994).

Comparison of the numerical methods presents a rather difficult problem for lack of clear efficiency criteria. For instance, Lomax (1982) suggests to evaluate the efficiency of numerical algorithms with the following criteria: 1) the rate of convergence to prescribed accuracy; 2) robustness i.e. applicability to a wide range of problems; 3) the simplicity of numerical implementation. It is evident that these requirements are contradictory to some extent and do not provide a simple quantitative characteristic of the numerical algorithm. It is worth noting also that in evaluating the efficiency of numerical algorithms, one has to take into account an extent of their adaptation to the computer architecture.

If the efficiency of a numerical algorithm is evaluated from computer resources required to achieve a prescribed accuracy, one can try to improve the efficiency using higher-order accurate finite-difference schemes. These schemes ensure the prescribed accuracy in the regions of continuous flow at fewer grid points, however, the number of operations per one grid point and storage requirements increase. If the order of accuracy is improved by expansion of the difference stencil, one has to solve a more complicated set of difference equations and is faced with difficulties in implementation of the boundary conditions. For these reasons, the use of compact higher-order finite-difference schemes seems to be more expedient. Such methods for solving the Navier–Stokes equations are considered in the monograph by Tolstykh (1990).

Another way to improve the accuracy of the numerical solution at a fixed number of grid points is to use spectral methods (Hussaini and Zang 1987). Available calculations of compressible flows show the advantage of the spectral methods over finite-difference schemes of second and even fourth order of accuracy within the regions of continuous flow. The use of the spectral approximation in the shock capturing methods leads to poorer results as compared with the finite-difference schemes (Shang 1985, Hussaini *et al* 1985, Hussaini and Zang 1987). Spectral methods are especially attractive in calculating stationary flows because they provide a high convergence rate to the stationary solution due to fast propagation of the information from the boundaries throughout the computational domain. Convergence to the stationary solution can be accelerated also by using the multigrid methods. The example of viscous flow calculation (Chima and Johnson 1985) demonstrates approximately a fivefold increase of the convergence rate. Simulating stationary flows, one can find the stationary iterative methods to be preferable as compared with the time-asymptotic technique. Such methods for calculating supersonic viscous gas flows about bodies were suggested, for instance, by Gromov and Kessel (1970), Pinchukov (1983).

1.2.2 Unidirectional method

A considerable part of numerical results discussed below has been obtained with the implicit unidirectional finite-difference schemes. Consider one of these schemes with reference to an axisymmetric supersonic flow about a blunted body of revolution assuming gas specific heats to be constant. The Navier–Stokes equations written for the (s, n)-coordinate system and the boundary conditions were presented in Section 1.1. The bow shock is supposed to be the outer boundary of the flow under study. Introducing a new independent variable $\xi = n/\varepsilon(s,t)$, where $\varepsilon(s,t)$ is the distance between the bow shock wave and the body surface, the shock layer region is mapped onto the rectangle $0 \leqslant s \leqslant S$, $0 \leqslant \xi \leqslant 1$ in the computational plane. For the coordinate system (t, ξ, s), differential equations (1.1) – (1.4) are written as follows

$$\mathsf{E}\frac{\partial \boldsymbol{X}}{\partial t} - \frac{1}{\rho}\frac{\partial}{\partial \xi}\left(\mathsf{A}\frac{\partial \boldsymbol{X}}{\partial \xi}\right) + \mathsf{B}\frac{\partial \boldsymbol{X}}{\partial \xi} + \mathsf{C}\boldsymbol{X} + \boldsymbol{D} = 0 \tag{1.19}$$

$$\mathbf{e}\frac{\partial \boldsymbol{X}}{\partial t} + \mathbf{b}\frac{\partial \boldsymbol{X}}{\partial \xi} + \mathbf{c}\boldsymbol{X} + \boldsymbol{d} = 0 \tag{1.20}$$

where $\boldsymbol{X} = \{u\ T\ v\ p\}^T$ is the column-vector of unknown functions. The first of the above matrix equations contains momentum balance equation (1.3) and energy balance equation (1.4), the second one contains momentum balance equation (1.2) and continuity equation (1.1). Notation of the Navier–Stokes equations as matrix

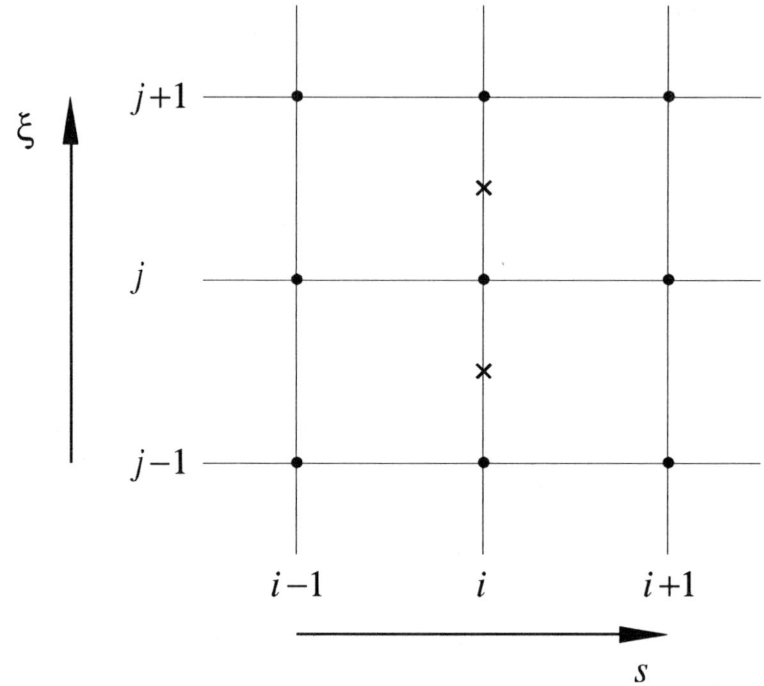

Figure 1.2 Computational grid for the unidirectional method.

equations (1.19) and (1.20) is consistent with an order of magnitude analysis of the viscous terms based on the conventional boundary layer theory with parameter $\delta = 1/\sqrt{Re}$ (Loitsianskii 1987). The viscous terms of the Navier–Stokes equations, which are in an order of magnitude less than unity throughout the shock layer, are included in vectors $\boldsymbol{D}$ and $\boldsymbol{d}$. These terms of equations (1.2)–(1.4) are written on the right of the single vertical bars. Supposing the above estimates to be valid, one can consider the n-momentum balance equation as a first-order equation with respect to n.

A computational grid is introduced in accordance with the formulae

$$\begin{aligned} &s_i = i\Delta s && \Delta s = S/N && i = 0, 1, \ldots, N \\ &\xi_j = j\Delta\xi && \Delta\xi = 1/M && j = 0, 1, \ldots, M \\ &t^n = n\Delta t && && n = 0, 1, \ldots \end{aligned} \tag{1.21}$$

The governing equations are approximated at the grid points located on the lines $s = s_i$ $(i = 0, 1, \ldots, N)$ between the $(n-1)$-th and the n-th time levels with time weight coefficient $0 \leqslant \alpha \leqslant 1$. Equations (1.19) and (1.20) are approximated at grid points $\xi_j = j\Delta\xi$ $(j = 1, \ldots, M-1)$ and $\xi_{j+1/2} = (j+1/2)\Delta\xi$ $(j = 0, 1, \ldots, M-1)$, respectively. These grid points are depicted by the dots and crosses in figure 1.2 which shows a fragment of the computational grid.

Let us introduce

$$X^n_{i,j} = X(i\Delta s,\ j\Delta\xi,\ n\Delta t) \qquad X^\alpha_{i,j} = (1-\alpha)X^{n-1}_{i,j} + \alpha X^n_{i,j}. \tag{1.22}$$

The difference formulae for derivatives with respect to t and ξ are the following: equations (1.19)

$$\begin{gathered}
\frac{\partial X}{\partial t} \sim \frac{\left(X^n - X^{n-1}\right)_{i,j}}{\Delta t} \\
\frac{\partial X}{\partial \xi} \sim \frac{\left(X_{j+1} - X_{j-1}\right)^\alpha_i}{2\Delta\xi} \\
\frac{\partial}{\partial\xi}\left(\mathsf{A}\frac{\partial X}{\partial\xi}\right) \sim \frac{\left[\mathsf{A}_{j+1/2}\left(X_{j+1} - X_j\right) - \mathsf{A}_{j-1/2}\left(X_j - X_{j-1}\right)\right]^\alpha_i}{(\Delta\xi)^2} \\
(i = 0, 1 \ldots, N; \quad j = 1, 2, \ldots, M-1)
\end{gathered} \tag{1.23}$$

equations (1.20)

$$\begin{gathered}
\frac{\partial X}{\partial t} \sim \frac{\left[\left(X_{j+1} + X_j\right)^n - \left(X_{j+1} + X_j\right)^{n-1}\right]_i}{2\Delta t} \\
\frac{\partial X}{\partial \xi} \sim \frac{\left(X_{j+1} - X_j\right)^\alpha_i}{\Delta\xi} \\
(i = 0, 1 \ldots, N; \quad j = 0, 1, \ldots, M-1)
\end{gathered} \tag{1.24}$$

Partial derivatives in the components of vector $\boldsymbol{D}$ and vector $\boldsymbol{d}$ are approximated by the following difference formulae:
equations (1.19)

$$\begin{gathered}
\frac{\partial X}{\partial s} \sim \frac{\left(X_{i+1} - X_{i-1}\right)^\alpha_j}{2\Delta s} \\
\frac{\partial}{\partial s}\left(\gamma\frac{\partial X}{\partial s}\right) \sim \frac{\left[\gamma_{i+1/2}\left(X_{i+1} - X_i\right) - \gamma_{i-1/2}\left(X_i - X_{i-1}\right)\right]^\alpha_j}{(\Delta s)^2} \\
\frac{\partial}{\partial s}\left(\gamma\frac{\partial X}{\partial \xi}\right) \sim \frac{1}{4\Delta s\Delta\xi}\Big[\gamma_{i+1,j}\left(X_{j+1} - X_{j-1}\right)_{i+1} \\
-\gamma_{i-1,j}\left(X_{j+1} - X_{j-1}\right)_{i-1}\Big]^\alpha \\
(i = 0, 1 \ldots, N; \quad j = 1, 2, \ldots, M-1)
\end{gathered} \tag{1.25}$$

equations (1.20)

$$\begin{aligned}
\frac{\partial \boldsymbol{X}}{\partial s} &\sim \frac{1}{4\Delta s}\left[(\boldsymbol{X}_{j+1}+\boldsymbol{X}_j)_{i+1}-(\boldsymbol{X}_{j+1}+\boldsymbol{X}_j)_{i-1}\right]^{\alpha} \\
\frac{\partial}{\partial \xi}\left(\gamma\frac{\partial \boldsymbol{X}}{\partial \xi}\right) &\sim \frac{1}{2(\Delta\xi)^2}\left[\gamma_{j+1}\left(\boldsymbol{X}_{j+2}-\boldsymbol{X}_j\right)\right. \\
&\qquad \left.-\gamma_j\left(\boldsymbol{X}_{j+1}-\boldsymbol{X}_{j-1}\right)\right]_i^{\alpha} \\
\frac{\partial}{\partial s}\left(\gamma\frac{\partial \boldsymbol{X}}{\partial s}\right) &\sim \frac{1}{2(\Delta s)^2}\left[\gamma_{i+1/2,j+1/2}\right. \\
&\qquad \times\left(\boldsymbol{X}_{i+1,j+1}+\boldsymbol{X}_{i+1,j}-\boldsymbol{X}_{i,j+1}-\boldsymbol{X}_{i,j}\right) \\
&\qquad -\gamma_{i-1/2,j+1/2} \\
&\qquad \left.\times\left(\boldsymbol{X}_{i,j+1}+\boldsymbol{X}_{i,j}-\boldsymbol{X}_{i-1,j+1}-\boldsymbol{X}_{i-1,j}\right)\right]^{\alpha} \\
\frac{\partial}{\partial s}\left(\gamma\frac{\partial \boldsymbol{X}}{\partial \xi}\right) &\sim \frac{1}{2\Delta s\Delta\xi}\left[\gamma_{i+1,j+1/2}\left(\boldsymbol{X}_{j+1}-\boldsymbol{X}_j\right)_{i+1}\right. \\
&\qquad \left.-\gamma_{i-1,j+1/2}\left(\boldsymbol{X}_{j+1}-\boldsymbol{X}_j\right)\right]^{\alpha} \\
&\quad (i=0,1\ldots,N;\quad j=0,1,\ldots,M-1)
\end{aligned} \tag{1.26}$$

To apply formulae (1.25) and (1.26) with $i = 0$ and $i = N$, the vectors $\boldsymbol{X}_{-1}$ and $\boldsymbol{X}_{N+1}$ are determined by extrapolation with the use of function values at the interior mesh points, symmetry relationships and approximate boundary conditions (1.12) with $k = 2$. At $j = 0$ and $j = M - 1$, the one-sided difference formulae

$$\begin{aligned}
\frac{\partial}{\partial \xi}\left(\gamma\frac{\partial \boldsymbol{X}}{\partial \xi}\right) &\sim \frac{1}{2(\Delta\xi)^2}\left[\gamma_1\left(\boldsymbol{X}_2-\boldsymbol{X}_0\right)\right. \\
&\qquad \left.-\gamma_0\left(-3\boldsymbol{X}_0+4\boldsymbol{X}_1-\boldsymbol{X}_2\right)\right]_i^{\alpha} \\
\frac{\partial}{\partial \xi}\left(\gamma\frac{\partial \boldsymbol{X}}{\partial \xi}\right) &\sim \frac{1}{2(\Delta\xi)^2}\left[\gamma_M\left(3\boldsymbol{X}_M-4\boldsymbol{X}_{M-1}+\boldsymbol{X}_{M-2}\right)\right. \\
&\qquad \left.-\gamma_{M-1}\left(\boldsymbol{X}_M-\boldsymbol{X}_{M-2}\right)\right]_i^{\alpha} \\
&\quad (i=0,1,\ldots,N)
\end{aligned} \tag{1.27}$$

are used.

The non-linear set of difference equations for the function values at a new time level n is solved by means of iterations. All coefficients of equations (1.19) and (1.20), including those which contain the derivatives of unknown functions, are calculated using the previous iteration data. As a result, in each iteration the total set of difference equations falls into the following uncoupled linear subsets on the lines $s = s_i$ $(i = 0, 1, \ldots, N)$

$$\begin{aligned}
&\mathbf{P}_j^{\alpha(q)}\boldsymbol{X}_{j+1}^{n(q+1)}+\mathbf{R}_j^{\alpha(q)}\boldsymbol{X}_j^{n(q+1)}+\mathbf{Q}_j^{\alpha(q)}\boldsymbol{X}_j^{n(q+1)}+\boldsymbol{\Phi}_j=0 \\
&(j=1,2,\ldots,M-1)
\end{aligned} \tag{1.28}$$

$$\begin{aligned}
&\mathbf{r}_{j+1/2}^{\alpha(q)}\boldsymbol{X}_{j+1}^{n(q+1)}+\mathbf{q}_{j+1/2}^{\alpha(q)}\boldsymbol{X}_j^{n(q+1)}+\boldsymbol{\phi}_{j+1/2}=0 \\
&(j=0,1,\ldots,M-1)
\end{aligned} \tag{1.29}$$

Here q denotes the iteration number, $\boldsymbol{X}^{n(q)}$ is vector $\boldsymbol{X}$ at time level n in the q-th iteration, $\boldsymbol{X}^{\alpha(q)} = (1-\alpha)\boldsymbol{X}^{n-1} + \alpha\boldsymbol{X}^{n(q)}$; subscript i is omitted.

The sets of difference equations (1.28) and (1.29) with block tridiagonal matrices of coefficients are solved along the lines $s = s_i$ by means of the block 'progonka' algorithm. Without the superscripts, the basic 'progonka' relation is written as follows

$$\mathbf{L}_{j+1}\boldsymbol{X}_{j+1} + \boldsymbol{X}_j + \boldsymbol{\Pi}_{j+1} = 0 \qquad (j = 0, 1, \ldots, M-1) \tag{1.30}$$

The first 'progonka' coefficients are obtained using the wall boundary conditions for the gas temperature and tangent velocity component along with difference equations (1.29) with $j = 0$. Applying no-slip wall boundary conditions, one obtains

$$\begin{aligned} \mathbf{L}_1 &= \left\{\begin{matrix} 1 & 0 & 0 & 0 \\ 0 & 1 & 0 & 0 \\ & \mathbf{q}_{1/2} & & \end{matrix}\right\}^{-1} \left\{\begin{matrix} 0 & 0 & 0 & 0 \\ 0 & -\varkappa & 0 & 0 \\ & \mathbf{r}_{1/2} & & \end{matrix}\right\} \\ \boldsymbol{\Pi}_1 &= \left\{\begin{matrix} 1 & 0 & 0 & 0 \\ 0 & 1 & 0 & 0 \\ & \mathbf{q}_{1/2} & & \end{matrix}\right\}^{-1} \left\{\begin{matrix} 0 \\ (\varkappa - 1)T_w \\ \phi_{1/2} \end{matrix}\right\} \end{aligned} \tag{1.31}$$

where $\varkappa = 0,\ 1$ correspond to prescribed wall temperature T_w and adiabatic wall, respectively. Implementation of the velocity slip and temperature jump wall boundary conditions meets no difficulties either. The next 'progonka' coefficients are determined by the recurrent formulae

$$\begin{aligned} \mathbf{L}_{j+1} &= \left\{\begin{matrix} \mathbf{R}_j - \mathbf{Q}_j\mathbf{L}_j \\ \mathbf{q}_{j+1/2} \end{matrix}\right\}^{-1} \left\{\begin{matrix} \mathbf{P}_j \\ \mathbf{r}_{j+1/2} \end{matrix}\right\} \\ \boldsymbol{\Pi}_{j+1} &= \left\{\begin{matrix} \mathbf{R}_j - \mathbf{Q}_j\mathbf{L}_j \\ \mathbf{q}_{j+1/2} \end{matrix}\right\}^{-1} \left\{\begin{matrix} \boldsymbol{\Phi}_j - \mathbf{Q}_j\boldsymbol{\Pi}_j \\ \phi_{j+1/2} \end{matrix}\right\} \\ & (j = 1, \ldots, M-1) \end{aligned} \tag{1.32}$$

In calculating the coefficients $\mathbf{L}_j$ and $\boldsymbol{\Pi}_j$, the wall boundary condition for the normal velocity component is transferred to the outer boundary of the shock layer. This boundary condition is written as

$$\boldsymbol{l}_0\boldsymbol{X}_0 + \pi_0 = 0 \tag{1.33}$$

where $\boldsymbol{l}_0 = \{0 \quad 0 \quad 1 \quad 0\}$, $\pi_0 = -v_w$. Using (1.30), it is easy to obtain the relations similar to (1.33) for all $j = 0, 1, \ldots, M$. The recurrent formulae for the coefficients are

$$\boldsymbol{l}_{j+1} = \beta^{-1}\boldsymbol{l}_j\mathbf{L}_{j+1} \qquad \pi_{j+1} = \beta^{-1}\left(\boldsymbol{l}_j\boldsymbol{\Pi}_{j+1} - \pi_j\right) \tag{1.34}$$

where β is the scaling factor equal to the length of vector $\boldsymbol{l}_j\mathbf{L}_{j+1}$.

Having calculated coefficients $\mathbf{L}_j$, $\boldsymbol{\Pi}_j$, $\boldsymbol{l}_j$ and π_j, one can determine the gas parameters behind the bow shock. The mass balance at the bow shock gives

$$D_\nu = \frac{\rho V_\nu - \rho_\infty V_{\nu\infty}}{\rho - \rho_\infty} \tag{1.35}$$

Using (1.35), let us eliminate D_ν from other Rankine–Hugoniot relations (1.9) and write them along with

$$\boldsymbol{l}_M \boldsymbol{X}_M + \pi_M = 0 \tag{1.36}$$

in the following form

$$\mathbf{F}_M^{n(q)} \left(\frac{\partial \boldsymbol{X}}{\partial \xi}\right)_M^{n(q+1)} + \boldsymbol{G}\left(\boldsymbol{X}_M^{n(q+1)}\right) + \boldsymbol{H}^{n(q)} = 0 \tag{1.37}$$

The derivative $\frac{\partial \boldsymbol{X}}{\partial \xi}$ in (1.37) is approximated by the one-sided difference formula including $\boldsymbol{X}_M$, $\boldsymbol{X}_{M-1}$ and $\boldsymbol{X}_{M-2}$. Eliminating $\boldsymbol{X}_{M-2}$, $\boldsymbol{X}_{M-1}$ with the use of (1.30) and applying Taylor series expansion to linearize vector $\boldsymbol{G}\left(\boldsymbol{X}_M^{n(q+1)}\right)$, one can derive the following formula for the unknown function vector at the bow shock wave in the $(q+1)$-th iteration

$$\begin{aligned} \boldsymbol{X}_M^{n(q+1)} &= -\left\{\mathbf{F}_M\left[3\mathbf{I} + \left(4\mathbf{I} + \mathbf{L}_{M-1}\right)\mathbf{L}_M\right] + 2\Delta\xi\left(\frac{\partial \boldsymbol{G}}{\partial \boldsymbol{X}_M}\right)\right\}^{-1} \\ &\times \left\{\mathbf{F}_M\left[\left(4\mathbf{I} + \mathbf{L}_{M-1}\right)\boldsymbol{\Pi}_M - \boldsymbol{\Pi}_{M-1}\right]\right. \\ &\left. + 2\Delta\xi\left(\boldsymbol{G} + \boldsymbol{H} - \frac{\partial \boldsymbol{G}}{\partial \boldsymbol{X}_M}\boldsymbol{X}_M\right)\right\} \end{aligned} \tag{1.38}$$

where $\mathbf{I}$ is the identity matrix. The right-hand side of (1.38) is calculated using the q-th iteration data.

Later on, the bow shock velocity D_ν is found from equation (1.35) and the bow shock distance from the body surface is specified

$$\varepsilon^{n(q+1)} = \varepsilon^{n-1} + \Delta t\left[\alpha\left(\varepsilon_t'\right)^{n(q+1)} + (1-\alpha)\left(\varepsilon_t'\right)^{n-1}\right] \tag{1.39}$$

where

$$\varepsilon_t' = \sqrt{1 + \left[\varepsilon_s'/(1+\kappa\varepsilon)\right]^2} D_\nu$$

Having the unknown function values at the bow shock, one can calculate them at the interior mesh points and on the body surface from equation (1.30).

The above calculations are carried out along all lines $s = s_i$ $(i = 0, 1, \ldots, N)$. On the symmetry axis $(i = 0)$, only three functions are calculated because the tangent velocity component u is equal to zero. Then, one can start the next iteration or the next time step calculations.

When slightly modified, the computer code applies to the calculation of both viscous and inviscid flows. Let us introduce the vector

$$\boldsymbol{X}_{i,j\beta}^\alpha = 1/4\left[\beta(1+\beta)\boldsymbol{X}_{j+1} + 2(2-\beta^2)\boldsymbol{X}_j + \beta(\beta-1)\boldsymbol{X}_{j-1}\right]_i^\alpha \tag{1.40}$$

where parameter β takes the values 0 or ± 1. The modification consists in employing the following difference formulae for the inviscid terms of equations (1.19)

$$\begin{aligned} \frac{\partial \boldsymbol{X}}{\partial t} &\sim \frac{1}{\Delta t}\left(\boldsymbol{X}^n - \boldsymbol{X}^{n-1}\right)_{i,j\beta} \\ \frac{\partial \boldsymbol{X}}{\partial s} &\sim \frac{1}{2\Delta s}\left(\boldsymbol{X}_{i+1} - \boldsymbol{X}_{i-1}\right)_{j\beta}^\alpha \\ \frac{\partial \boldsymbol{X}}{\partial \xi} &\sim \frac{1}{2\Delta\xi}\left[(1+\beta)\boldsymbol{X}_{j+1} - 2\beta\boldsymbol{X}_j + (\beta-1)\boldsymbol{X}_{j-1}\right]_i^\alpha \end{aligned} \tag{1.41}$$

With $\beta = 0$, these formulae give the same difference approximations which have been presented above for the Navier–Stokes equations. When $\beta = \pm 1$, the tangent momentum and energy balance equations are approximated at the mesh points $\xi = \xi_{j \pm 1/2}$ shown in figure 1.2 by crosses. With the use of (1.41) the following three alterations have to be made to obtain the algorithm for inviscid flow calculation:

1) assume the gas viscosity and heat conductivity to be equal to zero (bypass the calculation of relevant terms of the momentum and energy balance equations);

2) use difference equations (1.28) for all $j = 1, \ldots, M-1$ with $\beta = 1$;

3) use equations (1.28) with $j = 1$, $\beta = -1$ instead of the wall boundary conditions for the gas temperature and tangent velocity component in formulae (1.31) for the first 'progonka' coefficients.

The above algorithm for calculation of the inviscid gas flows does not differ from that described by Lubimov and Rusanov (1970) besides the artificial dissipation procedure.

The above finite-difference method is second-order accurate with respect to both spatial coordinates. Its accuracy with respect to time is first-order when $\alpha \neq 1/2$ and second-order when $\alpha = 1/2$. In calculation of both viscous and inviscid flows the bow shock shape may be distorted with the alternating sign error which is eliminated by the following smoothing procedure

$$\varepsilon_i^* = (1-2\beta)\varepsilon_i + \beta\,(\varepsilon_{i-1} + \varepsilon_{i+1}) \qquad i = 0, 1, \ldots, N-1 \qquad (1.42)$$

where the parameter β is usually assumed to be equal to 0.25.

The above numerical method reduces to solving the one-dimensional sets of equations. The analysis of the finite-difference scheme for one-dimensional equations with constant coefficients shows unconditional stability with $\alpha \geqslant 0.5$ and monotonicity in the regions of steep gradients provided that the local mesh Reynolds number $Re_h \lesssim 2$ (Golovachov and Koleshko 1988). The latter restriction is removed in the modified method (Belousov *et al* 1991) in which the tangent velocity component and gas temperature variation across the shock layer are approximated with the exponential difference formulae following from the local fundamental solution of the model equation. The results of numerical experiments agree, in general, with the above stability analysis but reveals the dependence of the allowable value of time increment on the number of iterations. A similar dependence was demonstrated by Lubimov and Rusanov (1970) in their method for solving the Euler equations. If integration in time is carried out without iterations, the allowable time increment has been found to be dependent on mesh spacing in the direction of explicit approximation of the spatial derivatives.

Figure 1.3 displays the rate of convergence to the stationary solution with a reference to the viscous shock layer flow ahead of a sphere. U_t is defined as

$$U_t = \max_{(i,j)} \left| \frac{1}{u} \frac{u^n - u^{n-1}}{\Delta t} \right|_{i,j} \qquad \begin{array}{l} i = 0, 1, \ldots, N \\ j = 1, 2, \ldots, M \end{array}$$

The linear function profiles on the lines $s = s_i$ were prescribed as the initial data. The number of grid points is equal to 30×10. The grid points were clustered to

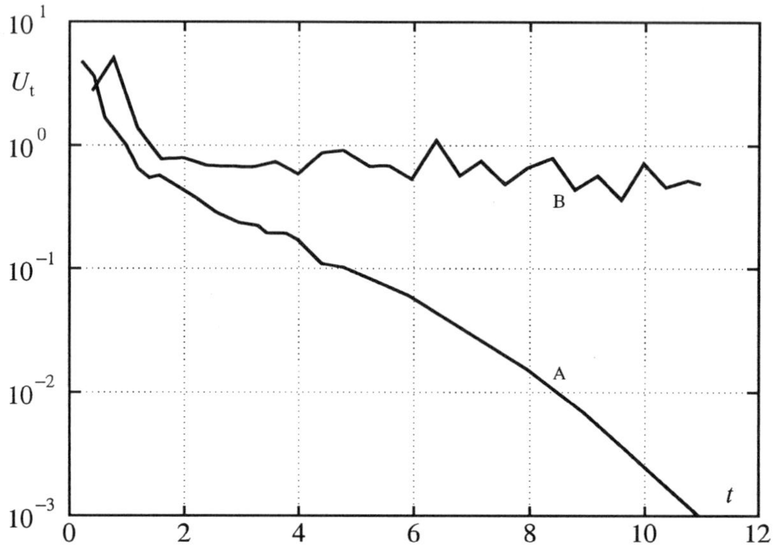

Figure 1.3 Convergence to the stationary solution for the shock layer flow ahead of a sphere. $M_\infty = 6$, $\gamma = 1.4$, $Re_\infty = 35500$, $T_w = 0.15T^*$. A, $\Delta t = 0.2$; B, $\Delta t = 0.4$.

the body surface with logarithmic transformation for the normal coordinate. The minimum mesh spacing $\Delta\xi$ near the body surface was approximately 75 times less than its maximum near the bow shock. The mesh spacing Δs was chosen to be equal to 10°. With time increment $\Delta t < 0.2$, the stationary solution is achieved approximately at the same moment as with $\Delta t = 0.2$. However, the number of time steps required to achieve the stationary solution increases. All these calculations were carried out with two iterations at each time step. It is evident that this number of iterations is insufficient when $\Delta t = 0.4$.

With decreasing Reynolds number, the viscous terms of the Navier–Stokes equations become more important and since some of them are calculated using previous iteration data, it deteriorates the convergence of the iterations. Nevertheless, numerical experiments have proven applicability of the above method to solving the Navier–Stokes equations in the whole range of validity of the continuous medium approach.

1.2.3 Transformation of governing equations and grid generation

When calculating supersonic flows about bodies with the bow shock fitting techniques, it proves to be convenient to normalize the transversal coordinate by the bow shock distance from the body surface. As a result, the shock layer region is mapped onto the unit segment, rectangle or parallelepiped, depending on the number of spatial coordinates. With this transformation which simplifies elaboration and numerical implementation of the finite-difference schemes, the calculations are carried out on the moving grid whose nodes are shifting in the physical domain in accordance with the bow shock motion.

Further transformations of the independent variables and governing equations can be necessitated by a non-homogeneity of the shock layer flow field which may contain the regions of steep function gradients. Such regions appear because of compressibility, viscosity, non-equilibrium effects, local energy release, abrupt variation of the body shape, etc. The characteristic length of these regions may differ by some orders of magnitude. Adequate numerical simulation of such flows with reasonable computer resources is possible only if the grid points are placed in the physical space in accordance with the flow structure.

In problems of a supersonic viscous gas flow past bodies, the gas dynamic functions often vary across the shock layer much more rapidly than in other directions. Let us suppose the distance from the body surface to be normalized by the shock layer thickness so that the transversal coordinate varies within the segment $0 \leqslant \xi \leqslant 1$, where $\xi = 0$ and $\xi = 1$ correspond to the body surface and the bow shock wave, respectively.

If the locations and length scales of the steep gradient regions are known a priori, one can generate a non-uniform grid beforehand using the coordinate transformations. Examples of such transformations are presented below.

Calculating viscous shock layer flows at high Reynolds numbers, one has to cluster the grid points to the body surface. With this end in view, Golovachov and Popov (1972a,b) used the transformation

$$x = \frac{\ln(1 + H\xi)}{\ln(1 + H)} \tag{1.43}$$

where x is the coordinate in the computational plane with uniform mesh spacing, the parameter H governs the grid point clustering in the physical plane.

In problems of non-equilibrium flows, it is often necessary to refine the computational grid not only near the body surface but near the bow shock as well. For this purpose, Golovachov (1979a), Golovachov and Karyakin (1980), Golovachov and Leontieva (1982) applied the transformation

$$x = \frac{1}{2}\left\{1 + \frac{\ln(1 + H_1\xi)}{\ln(1 + H_1)} - \frac{\ln\left[1 + H_2(1 - \xi)\right]}{\ln(1 + H_2)}\right\} \tag{1.44}$$

where the grid point distribution is governed by the parameters H_1 and H_2. The values of ξ for the equally spaced grid points in the x-coordinate are obtained from (1.44) by iterations.

In calculations of viscous shock layer flows with mass injection, it is expedient to ensure mesh refinement in the vicinity of some inner point of the unit segment.The appropriate transformation suggested by Vinokur (1983) is given by

$$\xi = \xi^* \left\{1 + \frac{\operatorname{sh}\left[(x - x^*)z\right]}{\operatorname{sh}(x^* z)}\right\} \tag{1.45}$$

where ξ^* locates the position of maximum grid point concentration in the physical plane. The value of z is obtained from the equation

$$\frac{1}{(H\xi^* z)^2} = \left[\frac{\operatorname{ch}(z) - 1 + 1/\xi^*}{\operatorname{sh}(z)}\right]^2 - 1 \tag{1.46}$$

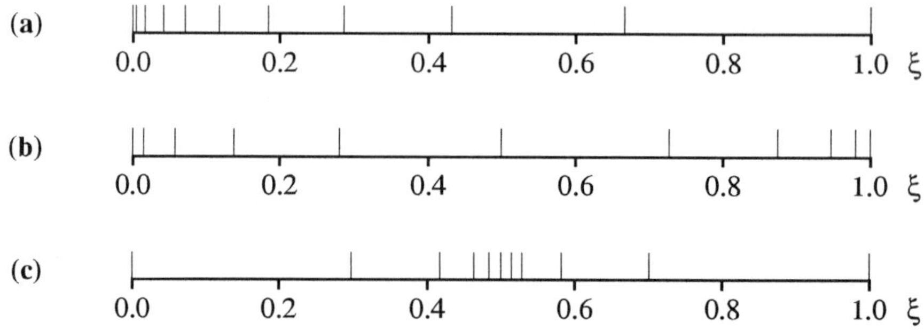

Figure 1.4 Computational grids obtained with transformations (1.43)–(1.45). (a), (1.43) $H = 50$; (b), (1.44) $H_1 = H_2 = 100$; (c), (1.45) $\xi^* = 0.5$, $H = 10$.

where H is the maximum value of the derivative x_ξ which characterizes the degree of mesh refinement. Equation (1.46) is solved by iterations. Later on, x^* is calculated

$$x^* = \frac{1}{2z} \ln \left[\frac{1 + \xi^* (e^z - 1)}{1 - \xi^* (1 - e^{-z})} \right] \tag{1.47}$$

The metric coefficients at the nodes of uniform grid in the computational plane are given by

$$x_\xi = \frac{1}{\xi^* z} \frac{\mathrm{sh}\,(x^* z)}{\mathrm{ch}\,[(x - x^*)z]} \tag{1.48}$$

Figure 1.4 shows the grid point distributions in the physical plane provided by transformations (1.43)–(1.45) with mesh spacing in the computational plane $\Delta x = 0.1$.

The effect of mesh refinement in the regions of steep gradients is clearly seen in figures 1.5 and 1.6. Here the temperature profiles in the near-wall region of the shock layer ahead of a sphere are presented. The calculations were carried out with $\mu \propto \sqrt{T}$, Prandtl number $Pr = 0.7$. The full lines depict the solution of the Navier–Stokes equations, the dashed lines depict the solution of the Euler equations. The dash-dotted curve in figure 1.6 shows the results of the boundary layer calculation for $Re_\infty = 71000$. The Navier–Stokes and Euler equations were solved using the implicit unidirectional finite-difference scheme presented in subsection 1.2.2. The computational grids for the Navier–Stokes equations contained 25 nodes between the body surface and the bow shock wave which clustered to the body surface with transformation (1.43). Inviscid flow calculations were carried out using the uniform grid with 11 nodes across the shock layer. The boundary layer flow was calculated with the numerical method suggested by Buirkin and Schennikov (1970).

Figure 1.5 presents solutions of the Navier–Stokes equations on the flow symmetry axis at $Re_\infty = 71000$ calculated with various mesh refinements near the wall. One can see a distortion of the temperature profile when mesh spacing is not sufficiently small ($H = 0.01$, $H = 1$). The profile calculated with $H = 100$ is monotonous. It is not affected by increasing the number of grid points and

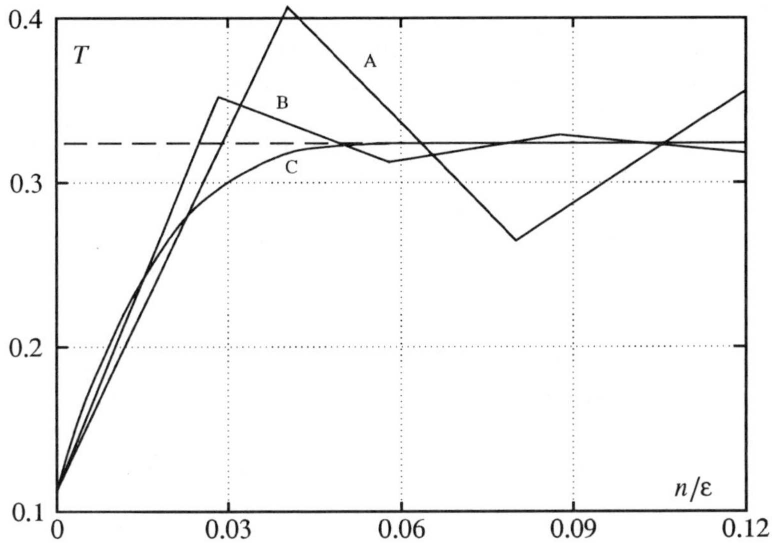

Figure 1.5 Effect of mesh refinement on the calculated temperature profile near the body surface on the stagnation streamline. $M_\infty = 6$, $\gamma = 1.4$, $Re_\infty = 71000$, $T_w = 0.35T^*$. A, $H = 0.01$; B, $H = 1$; C, $H = 100$.

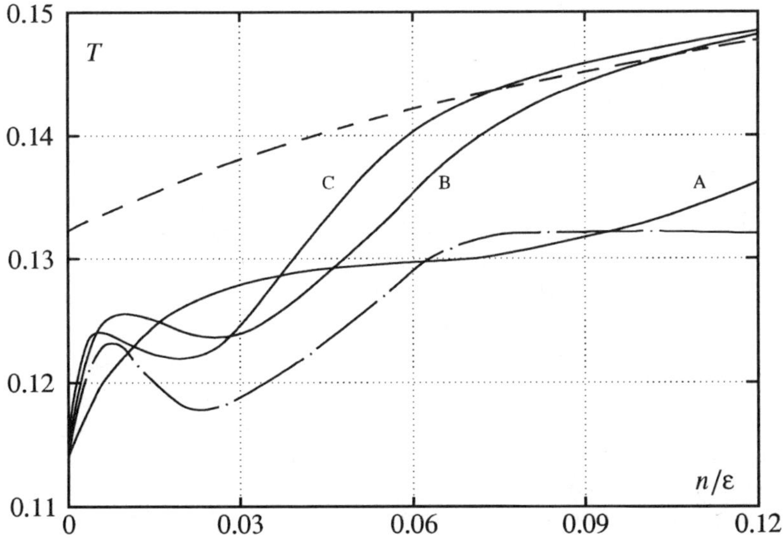

Figure 1.6 Temperature profiles near the body surface on the line $s = \pi/2$. $M_\infty = 6$, $\gamma = 1.4$, $T_w = 0.35T^*$. A, $Re_\infty = 3550$; B, $Re_\infty = 35500$; C, $Re_\infty = 71000$.

coincides with the boundary layer profile. The heat flux value at the stagnation point of a sphere at $H = 100$ agrees with the boundary layer calculations within

1 per cent.

Figure 1.6 shows the temperature profiles on the line $s = \pi/2$ calculated with $H = 100$ for three Reynolds numbers. It is seen that with increasing Reynolds number, the temperature profile near the wall becomes non-monotonous, which is predetermined by the simultaneous influence of the following three factors: body surface cooling, kinetic energy dissipation in the boundary layer, decrease of gas temperature in the outer supersonic expanding flow. A discrepancy with the boundary layer solution at $Re_\infty = 71000$ is caused by the use of the 'inviscid' wall temperature value (dashed curve) as the boundary condition at the outer edge of the boundary layer.

The above numerical results indicate that one can reproduce adequately even small details of the shock layer flow field near the body surface using the computational grids with a relatively small total number of mesh points.

In a general case, the shock layer flow may contain several regions of steep gradients with different and unknown beforehand length scales. These regions can arise, move and disappear. It seems to be expedient to solve such problems with the computational grids adapting to variation of the flow field structure. One of the ways for the adaptive grid generation is based on the use of a moving coordinate system in which the grid point motion in the physical plane is governed not only by the boundary motion but the shift of steep gradients of gas dynamic functions within the shock layer as well. In this case, one has to know the grid point speeds in the physical plane relative to the fixed coordinate system. Without any a priori information, this problem is solved using the available flow field data.

Depending on the way in which the grid point speed is determined, the methods for adaptive grid generation can be divided into two groups. The most general approach is founded on the use of complementary equations for the components of grid speed, which are integrated simultaneously with the mass, momentum and energy balance equations. The grid point coordinates at each time level are obtained by integrating their speeds. The equations governing the grid point motion can be derived by various means and, in particular, using the variational principle (Kovenya and Yanenko 1981, Yanenko *et al* 1977, Anderson *et al* 1984). This technique seems especially attractive in calculating non-stationary flows. In other methods, the problems of grid generation and solution of the balance equations at each time step are solved successively. Then one can start each time step calculation with determination of either the speed of the grid points or their coordinates.

Some ways to formulate the grid speed equations were discussed by Anderson *et al* (1984), Hawken (1987), Daryin and Mazhukin (1988). For instance, following the method by Rai and Anderson (1982) one determines the speed of the i-th mesh point evaluating the difference between the local error of numerical solution e_j at all other points ($j = 1, \ldots, i-1, i+1, \ldots, N$) and the average value e^*. It is also assumed that the larger the distance between any two points, the smaller the effect they have on each other. The above assumptions lead to the formula

$$(x_t')_i = k \left(\sum_{j=i+1}^{N} \frac{|e|_j - |e|^*}{r_{ij}^n} - \sum_{j=1}^{i-1} \frac{|e|_j - |e|^*}{r_{ij}^n} \right) \tag{1.49}$$

where r_{ij} is the distance between the i-th and j-th nodes, n is the power index, k is the damping factor. According to (1.49), the mesh points attract each other in regions where the local error of the numerical solution exceeds the average value, and they repel each other where it is less than the average value. Reciprocal influence of any two mesh points decreases proportional to the n-th power of the distance between them. The computational grid thus generated is supposed to ensure a uniform distribution of the error. To use the above procedure, one has to know the error of the numerical solution. The exact solution being unknown, evaluation of this quantity presents a fairly complicated problem that hampers essentially application of the above method.

If one starts the new time level calculations with determination of the grid point coordinates, their speed can be evaluated by means of numerical differentiation with respect to time of their coordinates at the neighbouring time levels. It is worth noting that in this case one can use interpolation procedures to obtain gas dynamic function values at the nodes of the latest grid instead of using the moving coordinates and calculating the grid point speed.

The computational grid at a new time level can be generated in a number of ways including the use of the above coordinate transformations. In this particular case, the new grid differs from the former one by the values of the parameters which locate the maximum grid point clustering in the shock layer and determine the extent of mesh refinement in these regions.

One of the most popular techniques for generating adaptive grids of the type discussed is based on the idea of equidistribution of the truncation error. According to this idea, the grid point coordinates are obtained from the condition

$$w(\xi)\xi_x = const \tag{1.50}$$

where the weight function $w(\xi) > 0$ controls the grid point distribution in the physical plane. It is evaluated using the previous iteration or previous time level data.

Grid point coordinates ξ in the physical plane are obtained by interpolation of the tabular function $\xi(x)$ whose argument is calculated with the formula

$$x = \int_0^\xi w(\xi)d\xi \Big/ \int_0^1 w(\xi)d\xi \tag{1.51}$$

Note that a simpler, at first sight, way to find the grid point coordinates in the physical plane from the formula

$$\xi = \int_0^x \frac{dx}{w(x)} \Big/ \int_0^1 \frac{dx}{w(x)} \tag{1.52}$$

leads to unsatisfactory results because retaining the interpolation, it restricts the grid point shift to the regions of steep gradients.

The above method for adaptive grid generation can be considered as a particular case of the variational approach. Indeed, it is easy to show that relationship (1.50) is the first integral of the Euler–Lagrange equation for the functional which characterizes grid non-uniformity in the physical domain; see Anderson *et al* (1984).

The key problem is the choice of weight function $w(\xi)$. Since grid adaptation is carried out with a view to minimize the truncation error and to distribute it uniformly over the computational domain, the most appropriate weight function would be just the truncation error. However, evaluation of this quantity meets the difficulties. Keeping in mind that the truncation error is usually a maximum in the vicinities of steep gradients, the combinations of gas dynamic functions and their gradients are often used as the weight functions, see for instance Tolstykh (1978), Dwyer *et al* (1980), Kovenya and Yanenko (1981).

Considering the adaptive grid application to the calculation of stationary flows, it is worth noting that, in this case, the exact values of the grid point speed are not required. Moreover, the shock layer flow can be calculated with the use of fixed grids or grids, whose nodes are moving merely due to the outer boundary motion. Grid adaptation reduces to the use of the sequence of grids which are generated through one or another number of iterations (time steps) in accordance with available flow field data. The gas dynamic functions at the points of each new grid can be obtained by interpolation.

Some results of numerical experiments on adaptive grid application to the problems of supersonic viscous flows about bodies were presented by Belousov and Golovachov (1992), Golovachov and Zemlyakov (1993a,b). The calculations were carried out using the implicit unidirectional methods. The grid point coordinates in the physical plane were determined from the equidistribution law (1.50) with weight functions $w(\xi)$ which were chosen to be combinations of the gas dynamic function gradients and various representations of the truncation error. The number of grid points between the body surface and the bow shock wave was equal to 50.

As an illustration of an adaptive grid for the stationary problem, figure 1.7 shows the grid generated in the calculation of a turbulent axisymmetric flow about a sphere set in a supersonic wake. The weight function was chosen to be the following

$$w = 1 + 2(|\nabla T| + |\nabla u|) \qquad (1.53)$$

where T, u are the non-dimensional temperature and tangent velocity component. Calculation results for this problem are discussed in detail in subsection 3.1.2.

Application of the adaptive grids to the calculation of non-stationary shock layer flows was considered with reference to the problem of supersonic laminar flow about an accelerating body. Abrupt increase of the flight speed gives rise to an appearance of the intensive compression and expansion waves propagating within the shock layer and interacting with the bow shock and the body surface. Grid adaptation was carried out by calculation of the grid point coordinates in the physical space at each time level with the weight function

$$w = |\nabla T/T_s|^{0.5} + 0.02 \qquad (1.54)$$

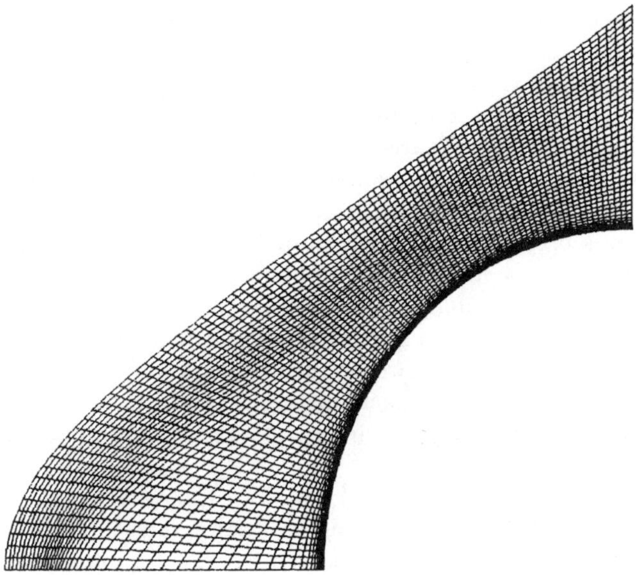

Figure 1.7 Converged adaptive grid for stationary turbulent flow about a sphere in supersonic wake.

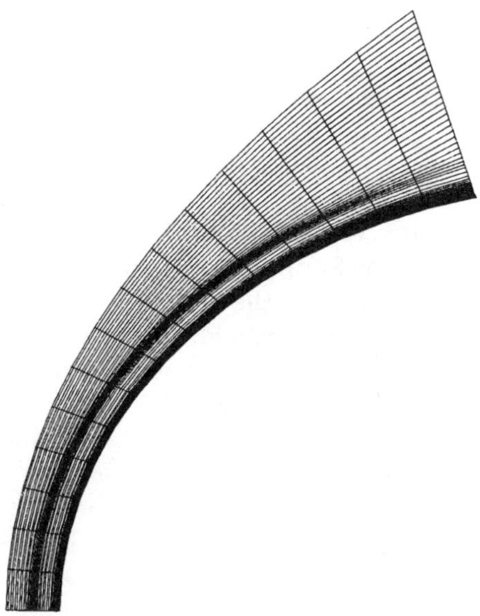

Figure 1.8 Adaptive grid for the problem of supersonic flow around an accelerating body.

Transition to each new computational grid was performed by means of interpolation. The calculations were also carried out with the use of a coordinate system moving at the speed of the steep gradient motion in the shock layer. At Strouhal numbers $Sh \gtrsim 1$, the latter grid adaptation procedure required an unacceptable decrease of time increment. Figure 1.8 displays the computational grid generated at one of the moments. One can see mesh refinement both near the body surface and in the vicinity of the inner shock. Detailed discussion of the calculation results for this problem is presented in subsection 3.5.1.

When solving viscous shock layer problems, one has, for the most part, to control the grid point distribution only across the shock layer. Nevertheless, in calculation of multi-dimensional flows, it may be expedient sometimes to use the grids which are non-uniform in two or three directions. The above methods for adaptive grid generation can be generalized, in principle, to this case by successive one-dimensional adaptations with respect to each coordinate; see for instance Dwyer and Onyejekwe (1985).

Adaptive grids for multidimensional flows can be generated on the basis of the variational principle. Following Brackbill and Saltzman (1982), the adaptive grid in the case of two spatial coordinates is generated by minimizing the functional which is a linear combination of the following integrals

$$
\begin{aligned}
I_1 &= \int_D \left[(\nabla x)^2 + (\nabla y)^2\right] ds \\
I_2 &= \int_D (\nabla x \cdot \nabla y)^2 J^3 ds \\
I_3 &= \int_D w J ds
\end{aligned}
\tag{1.55}
$$

where x, y are the coordinates in the computational plane, $J = \xi_x \eta_y - \xi_y \eta_x$ is the Jacobian of the coordinate transformation, w is the weight function. The first of the above integrals is responsible for the smoothness of the transformation, the second represents the measure of orthogonality of the computational grid and the third provides grid point clustering in the regions of steep gradients. The set of Euler–Lagrange equations for functions $\xi(x, y)$, $\eta(x, y)$ is written out by Kreis *et al* (1986), where one can find also some numerical results and recommendations on the choice of coefficients of the linear combination of the integrals $I_i (i = 1, 2, 3)$. Some other methods for generating adaptive grids for multidimensional problems are discussed by Godunov *et al* (1976), Anderson *et al* (1984).

1.3 SIMULATION OF RAREFIED FLOWS

The results of theoretical and experimental investigations of the rarefaction effects on supersonic flows about bodies are presented in available monographs, see for instance Belotserkovskii *et al* (1967, 1974), Kokoshinskaya *et al* (1980). So we restrict ourselves to a discussion of the issues concerning departure from the continuous medium approach and the limits of its validity.

1.3.1 Velocity slip and temperature jump effects

Departure from the continuous medium model manifests itself, first of all, near the body surface. For a weakly rarefied flow it can be taken into account with the wall boundary conditions modified to include the velocity slip and temperature jump. The effects of these boundary conditions have been investigated by many authors. However, not all of the viscous shock layer models exploited provide an adequate description of the flow when the above effects become important. In particular, the thin viscous shock layer model can result in large quantitative and even qualitative errors in the prediction of the flow characteristics at low Reynolds numbers (Jain and Adimurthy 1974).

Numerical results by Jain and Adimurthy (1974), Molodsov (1979), Pinchukov (1979), Tarnavskii (1983), Molodsov and Ryabov (1979) and others prove that the velocity slip and temperature jump effects depend to a great extent on the gas and body surface properties and wall temperature. Below, some numerical results by Golovachov and Kanailova (1987) are presented which demonstrate the dependence of the velocity slip and temperature jump effects on the above factors and the Reynolds number. The results demonstrate also the range of validity of the no-slip wall boundary conditions.

Consider supersonic flow of a perfect gas with constant heat capacity about a sphere. The computational domain is bounded by the bow shock wave where modified Rankine–Hugoniot relationships (1.9) are used. The boundary conditions on the body surface are given by relationships (1.10) and (1.11). Note that under the flight conditions considered ($Re_s \gtrsim 20$), the contribution of the second term of the right-hand side of (1.10) in the slip velocity does not exceed 2 per cent. The normal velocity component on the body surface was assumed to be equal to zero. For comparison, the calculations were carried out also with the wall boundary conditions

$$u\,|_{n=0} = 0 \qquad T\,|_{n=0} = T_w \tag{1.56}$$

Stationary solutions of the Navier–Stokes equations (1.1)–(1.4) were obtained by the time-asymptotic technique using the implicit unidirectional finite-difference scheme described in subsection 1.2.2. The computational grid contained 10 lines $s = const$ with 31 nodes between the body surface and the bow shock wave on each line. At low wall temperature, the grid points were clustered to the body surface by means of normal coordinate transformation (1.43) to resolve abrupt temperature variation in this region. The accuracy of the numerical solutions is estimated to be within a few per cent.

Calculations were carried out for air and argon flows using Sutherland's formula (1.7) for the gas viscosity. The velocity slip and temperature jump effects on the profiles of gas dynamic functions in the shock layer are illustrated in figure 1.9.

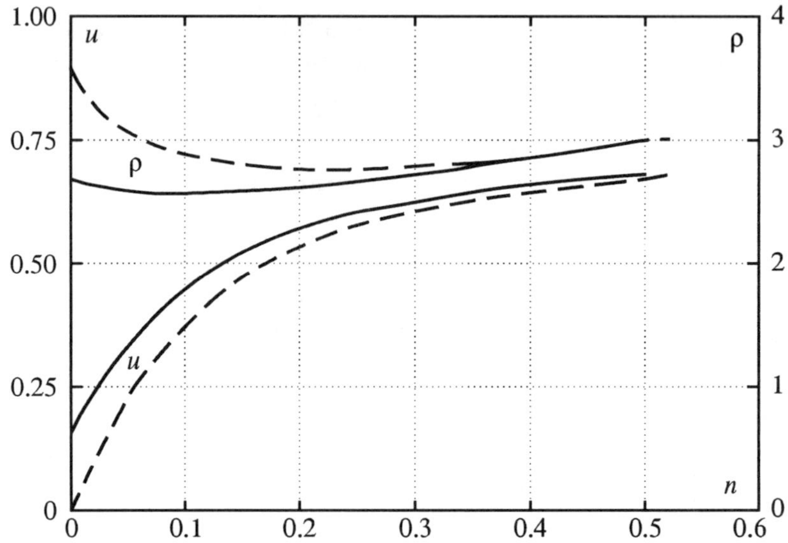

Figure 1.9 Profiles of gas density and tangent velocity component in the shock layer ahead of a sphere on the line $s = \pi/3$. Air, $M_\infty = 3$, $Re_\infty = 50$, $T_w = 0.48T^*$. Dashed curves, no-slip boundary conditions.

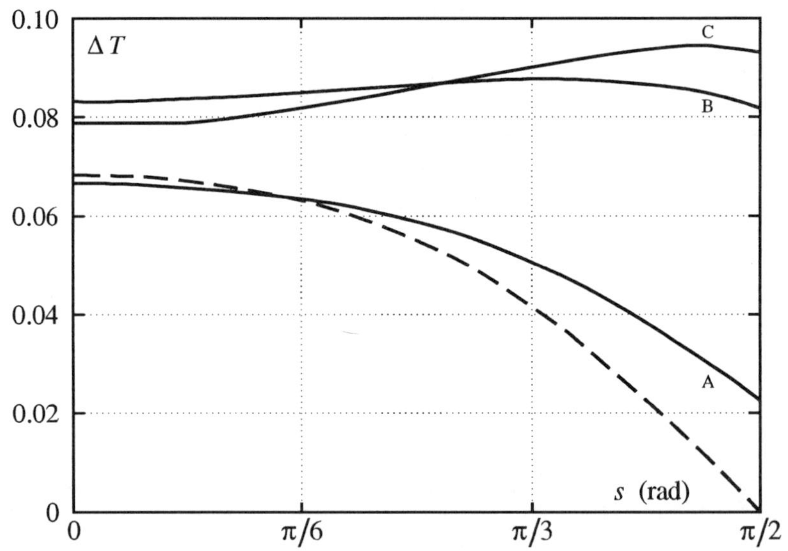

Figure 1.10 Variation of temperature jump over the sphere surface in air (full curves) and argon (dashed curve) flows. $M_\infty = 3$, $Re_s = 25$. A, $k = 0.47$; B, $k= 0.15$; C, $k=0.05$.

Figures 1.10 and 1.11 show variation of the temperature jump ΔT and slip velocity δu over the sphere surface in air and argon flows for three values of the

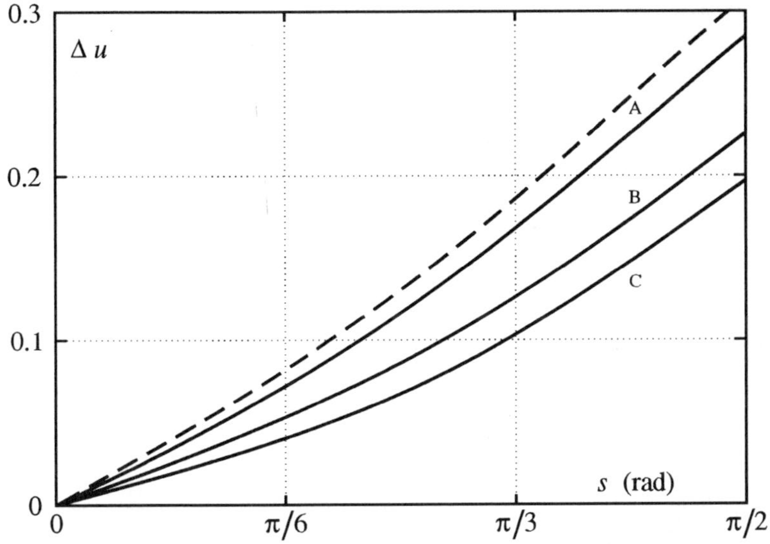

Figure 1.11 Variation of slip velocity over the sphere surface in air (full curves) and argon (dashed curve) flows. $M_\infty = 3$, $Re_s = 25$. A, $k = 0.47$; B, $k=0.15$; C, $k = 0.05$.

temperature factor $k = T_w/T^*$. One can see a non-monotonous ΔT dependence on the temperature factor in the stagnation region and its non-monotonous variation over the sphere surface at $k \lesssim 0.15$. The slip velocity in air is less than in argon flow and decreases monotonically with the body cooling.

The next figures demonstrate the velocity slip and temperature jump effects on the drag and heat transfer coefficients.

Figures 1.12 and 1.13 show the skin friction coefficient and heat flux variation over the sphere surface. The dashed lines depict the calculations with boundary conditions (1.56). At relatively high wall temperature, the velocity slip lowers noticeably not only the shear stress but the wall pressure as well. It results in reduction of the aerodynamic drag. Within the range of conditions considered ($Re_s \gtrsim 20$, $k \lesssim 0.5$), the maximum decrease of the drag coefficient of the sphere front ($0 \leqslant s \leqslant \pi/2$) was found to be equal to 12 per cent.

If the velocity slip is taken into account, the heat flux to the body surface is determined by the formula

$$q = \lambda \frac{\partial T}{\partial n} + \mu u \frac{\partial u}{\partial n} \tag{1.57}$$

Figure 1.13 shows that velocity slip and temperature jump reduce the heat flux to the body surface. The effect diminishes with distance from the stagnation point of a sphere because of augmentation of the second term of the right-hand side of (1.57). The dash-dotted curve in figure 1.13 shows the heat flux variation calculated ignoring this term.

All above results were obtained with energy accommodation coefficient $\alpha = 1$. This parameter influences substantially the heat flux to the body surface. For

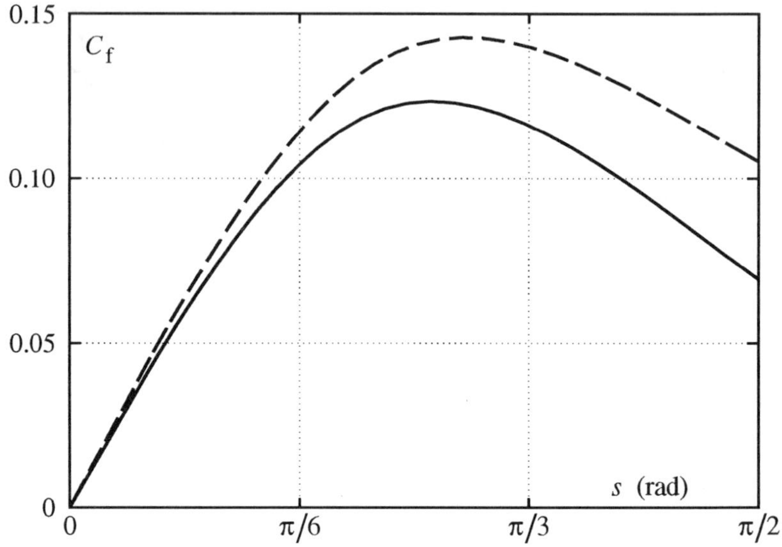

Figure 1.12 Variation of skin friction coefficient over the sphere surface. Air, $M_\infty = 3$, $Re_\infty = 50$, $T_w = 0.48T^*$. Dashed curve, no-slip boundary conditions.

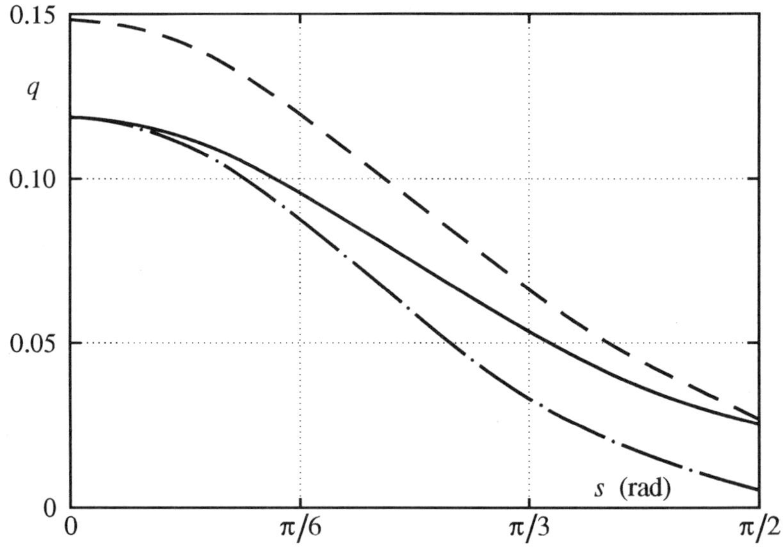

Figure 1.13 Heat flux variation over the sphere surface. Air, $M_\infty = 3$, $Re_\infty = 50$, $T_w = 0.48T^*$. Dashed curve, no-slip boundary conditions; dash-dotted curve, wall shear stress contribution ignored.

instance, in the air flow at $M_\infty = 3$, $Re_\infty = 50$, $k = 0.48$ the heat flux to the stagnation point of a sphere at $\alpha = 0.5$ is less by 35 per cent as compared to $\alpha = 1$.

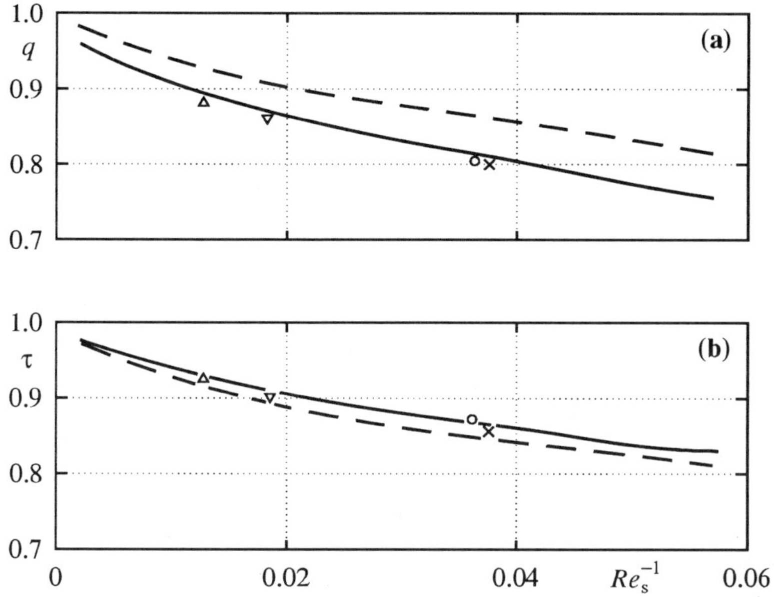

Figure 1.14 Dependence of the velocity slip and temperature jump effects on the Reynolds number for air (full lines) and argon (dashed lines) flows at $M_\infty = 3$, $T_w = 0.48T^*$. ○, ×, △, ▽: calculations for air flow at $M_\infty = 2, 4, 6, 10$, respectively.

The heat flux dependence on the energy accommodation coefficient lessens with a distance from the stagnation point due to an increase of the second term of (1.57) which is only slightly influenced by the value of α. The energy accommodation effect on the wall shear stress is negligible, especially at high wall temperature.

The next two figures display the dependence of the velocity slip and temperature jump effects on the Reynolds number and wall temperature. They present the variation of the heat flux (a) and shear stress (b) maximums on the sphere surface normalized by their maximums obtained with boundary conditions (1.56). The lines in figure 1.14 correspond to the air and argon flows at $M_\infty = 3$, $k = 0.48$. To demonstrate a similarity ensured by the use of the criterion of Re_s, the results for some other values of the Mach number are plotted as well. The results presented in figure 1.15 pertain to the air flow. It is seen that the wall shear stress slightly exceeds the value obtained with no-slip boundary conditions at $k \lesssim 0.075$, which may be explained by the viscosity increase due to the temperature jump on the body surface. With decreasing wall temperature, the temperature jump effect on the heat flux diminishes.

Figure 1.16 presents comparison of calculations of the heat transfer coefficient at the stagnation point of a sphere $C_H = q/\rho_\infty V_\infty(h^* - h_w)$ with experimental data by Gusev and Nikolskii (1971). One can see a good agreement between the calculation and experimental results.

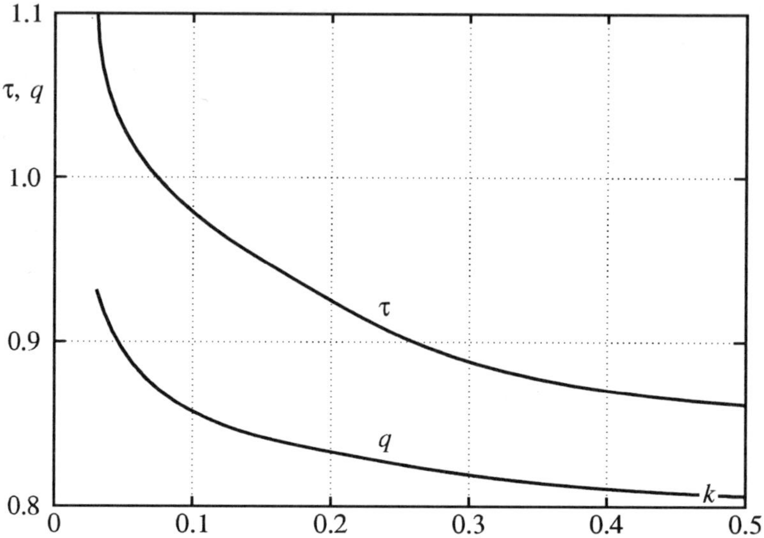

Figure 1.15 Dependence of the velocity slip and temperature jump effects on the wall temperature factor. Air, $M_\infty = 3$, $Re_\infty = 50$.

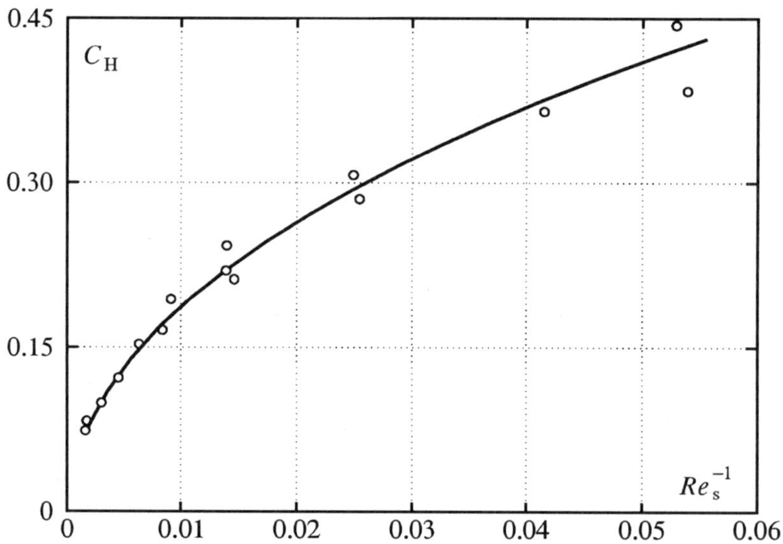

Figure 1.16 Comparison of the stagnation point heat transfer calculations with experimental data of Gusev and Nikolskii (1971). Air, $T_w = 0.19T^*$.

1.3.2 Range of validity of the Navier–Stokes equations

With increasing rarefaction, the mathematical model of the continuous medium given by the Navier–Stokes equations becomes invalid in the whole flow field about

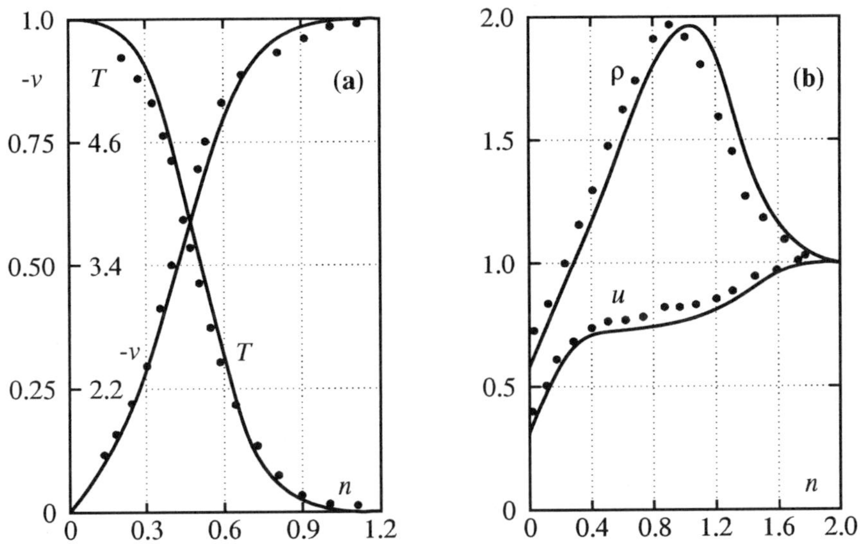

Figure 1.17 Profiles of gas dynamic functions ahead of a sphere on the stagnation line $s = 0$ (a) and the line $s = \pi/2$ (b). $M_\infty = 3.8$, $\gamma = 5/3$, $Re_\infty = 100$, $T_w = T^*$. Full curves, Navier–Stokes solution; dots, Monte–Carlo solution of the kinetic equation.

the body. To obtain quantitative data on the range of validity of these equations for the problems of supersonic flows about bodies, the solutions of the Navier–Stokes equations have been compared with the solutions of the Boltzmann and model kinetic equations and the experimental data. The Navier–Stokes equations were solved subject to the velocity slip and temperature jump wall boundary conditions. The outer boundary of the computational domain was placed in the undisturbed stream ahead of the body. Below, some of the above-mentioned comparisons are discussed.

Figure 1.17 reproduced from the paper by Tarnavskii (1983) depicts the profiles of gas dynamic functions ahead of a sphere calculated with the Navier–Stokes equations and the kinetic equation. Here the gas viscosity $\mu \propto \sqrt{T}$. All results are normalized by their free stream values and the distance from the body surface is normalized by the sphere radius.

Figure 1.18 presents the density variation along the stagnation streamline in front of a sphere (Jain and Adimurthy 1974). The self-similar solution of the Navier–Stokes equations is compared with the experimental data by Russell (1968). It is worth noting here that noticeable underestimation of the shock layer thickness in the calculations is caused by an error of the self-similarity assumption.

Artamonov and Arkhipov (1979) compared the solutions of the Navier–Stokes equations for hypersonic stagnation line flow with the results of the 'beam-continuum' model which is formulated assuming the distribution function of molecules near the body surface to be a superposition of the Navier–Stokes distribution function and the delta-function of the oncoming hypersonic rarefied stream.

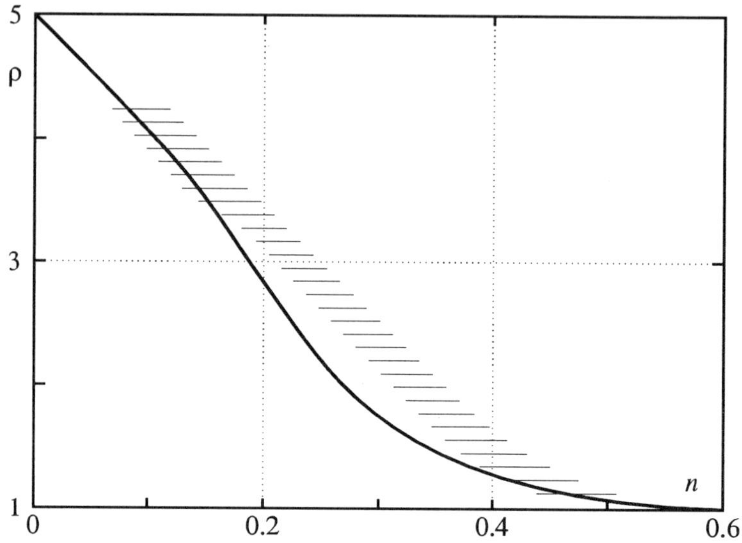

Figure 1.18 Density profile on the stagnation streamline. $M_\infty = 4.38$, $\gamma = 1.4$, $Re_s = 12.2$. Full line, self-similar solution of the Navier–Stokes equations (Jain and Adimurthy 1974); shading, experimental data of Russell (1968).

Since some of the oncoming molecules reach the body surface and rebound without collisions, the governing equations involve the distribution function of the rebounded molecules as well. The equations of the 'beam-continuum' model contain the terms responsible for mass, momentum and energy exchange between the above kinds of particles as well as the complementary equations describing variation of their concentrations. To close the problem, one has to prescribe the particle interaction potential and the law of particle rebounding on the body surface.

Figure 1.19 depicts the variation of the stagnation point heat flux with Knudsen number based on the free stream value of the mean free path of the molecules and the sphere radius

$$Kn_\infty = \frac{l_\infty}{R} \approx 2\sqrt{\gamma}\,\frac{M_\infty}{Re_\infty}$$

Heat flux is normalized by its free-molecular value subject to complete energy accommodation on the body surface. The particle interaction was described with the rigid sphere potential. The numerical results obtained with the Navier–Stokes equations, 'beam-continuum' model and direct statistical simulation (Bird 1976) are compared.

Figure 1.20, reproduced from the paper by Molodsov and Ryabov (1979), compares the calculation and experimental data on the sphere drag coefficient

$$C_x = \frac{F_x}{0.5\rho_\infty V_\infty^2 \pi R^2} = 4\int_0^\pi (p_w \cos s + C_f \sin s)\sin s\, ds \qquad (1.58)$$

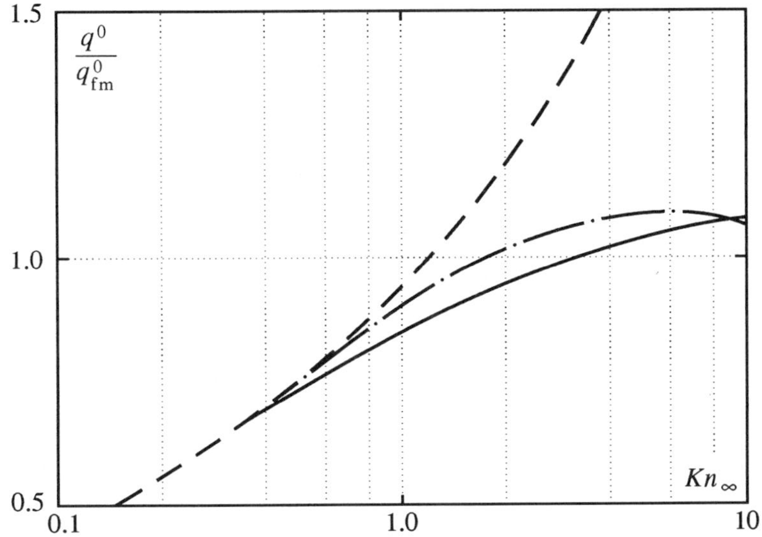

Figure 1.19 Comparison of the stagnation point heat flux calculations using the 'beam-continuum' model (full line), Navier–Stokes equations (dashed line) and direct statistical simulation (dash-dotted line). $M_\infty = 10$, $\gamma = 1.4$, $T_w = T_\infty$.

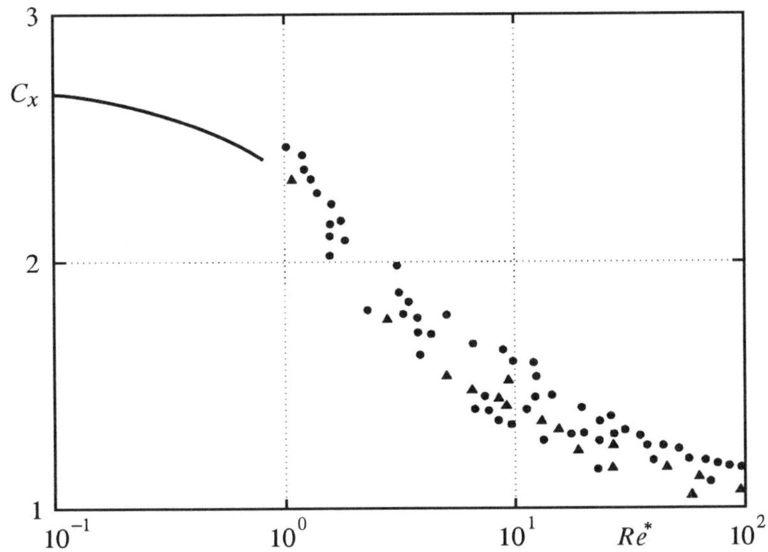

Figure 1.20 Comparison of calculated and experimental data on sphere drag coefficient. Filled circles, experimental data of Gusev *et al* (1970); triangles, solutions of Navier–Stokes equations at $3.8 \leqslant M_\infty \leqslant 12$, $6 \leqslant Re^* \leqslant 180$, $0.19 \leqslant k \leqslant 1.00$; full line, single collision model.

Here the circles denote the experimental data obtained by Gusev *et al* (1970), the triangles correspond to numerical solutions of the Navier–Stokes equations, the full line depicts Monte–Carlo calculations within the framework of the single collision model for the Boltzmann equation.

The above comparisons and other results of the afore-mentioned and other authors indicate that the Navier–Stokes equations are an adequate mathematical model for simulating supersonic flows about bodies at Reynolds numbers $Re_s \gtrsim 10$.

2

Viscous shock layer models and computational methods

The essence of the viscous shock layer approach is calculation of the whole perturbed flow field in front of the body with a single set of equations. Such calculations can be carried out not only with the full Navier–Stokes equations but also with more simple mathematical models whose implementation requires a substantially smaller amount of computations.

The governing equations of the simplified viscous shock layer models are obtained from the full Navier–Stokes equations by omitting some terms which are expected to be small. If the simplification does not change mathematical properties of the problem, the simplified equations are solved with the same numerical methods as the full Navier–Stokes equations. In this case, the amount of computations decreases only to the extent which would be required for calculating the omitted terms. Much more computer resources are saved when the boundary value problem is replaced by the initial-boundary value problem with the same number of independent variables. It is also expedient to take into account the similarity properties of the flow under study in order to decrease, if possible, the number of independent variables.

To clear up the basic ideas, the viscous shock layer models and their numerical implementation are discussed in this chapter with reference to a two-dimensional flow of a perfect gas with constant heat capacity. The accuracy and the range of validity of the viscous shock layer models are illustrated with calculation results.

2.1 REDUCED NAVIER-STOKES EQUATIONS

The equations under consideration differ from the full Navier–Stokes equations by lack of some terms responsible for the molecular momentum and energy transport. In the problems of supersonic flows about bodies, the reduced equations are often referred to as the full viscous shock layer equations in contrast to the thin viscous shock layer model which is formulated with additional assumptions on the orders of magnitude of the inviscid terms.

The use of reduced equations is most expedient in the simulation of stationary flows. For stationary supersonic flows, the initial-boundary value problem is well-posed for the above equations and, therefore, a space-marching calculation procedure with respect to the streamwise coordinate applies. To calculate in this manner subsonic/supersonic flows, one has to suppress the elliptic solution behaviour in the subsonic regions. This can be achieved by means of one or another alteration of the gas dynamic fluxes in the streamwise direction, which excludes the upstream disturbance propagation. Because of well-posedness of the initial-boundary value problem, the reduced stationary Navier–Stokes equations are referred to as the parabolized equations. If the upstream influence has to be taken into account, the solution of the stationary reduced Navier–Stokes equations is obtained using global relaxation or time-asymptotic techniques. When investigating non-stationary flows, the reduced equations are integrated by the same methods as the full Navier–Stokes and non-stationary boundary layer equations.

2.1.1 Full viscous shock layer equations

This model of viscous shock layer flow can be derived using basic conceptions of the conventional boundary layer theory. The thickness of the near-wall region, where the viscosity and heat conductivity manifest themselves, and the normal velocity component in this region are supposed to be $O(\delta)$, where $\delta = 1/\sqrt{Re}$. The reduced equations omit the viscous terms which are, to an order of magnitude, less than δ or less than unity throughout the shock layer. The appropriate sets of reduced equations for supersonic blunt body flows were suggested by Tolstykh (1966) and Golovachov and Popov (1972a). We shall denote them by numbers I and II, respectively. In the case of an axisymmetric flow, set I of the full viscous shock layer equations is given by (1.1) – (1.4), in which the momentum and energy balance equations omit the terms written on the right of the double vertical bars. Set II lacks the terms of equations (1.2) – (1.4) written on the right of the single vertical bars. Davis and Flugge–Lotz (1964) and Davis (1970a) suggested an intermediate model, in which the equation for the momentum component normal to the body surface contained only inviscid terms and did not differ from that of set II, whereas the tangent momentum and energy balance equations did include some viscous terms equal, to an order of magnitude, to δ.

All above sets of the viscous shock layer equations contain all terms of the inviscid flow and boundary layer equations. Set II, which includes only the most viscous terms with the second derivatives and squares of the first derivatives in n, is sometimes referred to as the thin layer model meaning that it applies to high Reynolds number flows with a thin boundary layer adjacent to the body surface.

The initial conditions as well as the wall and bow shock boundary conditions for the reduced equations are similar to those for the full Navier–Stokes equations. Since set II applies to high Reynolds number flows, one may ignore the velocity slip and temperature jump effects on the body surface and impose the usual shock relationships on the outer boundary of the shock layer in this model. The total order of set II with respect to the transversal coordinate is a unity smaller than those of set I and full Navier–Stokes equations. So, the bow shock and wall boundary conditions are sufficient for calculation of the shock layer flow field and location of the bow shock. Using set I or the full Navier–Stokes equations, one has to formulate a supplementary relationship to close the problem; see subsection 1.1.2.

2.1.2 Range of applicability of the reduced Navier–Stokes equations

In accordance with evaluation of the viscous terms, one ought to expect the reduced Navier–Stokes equations to apply at fairly high Reynolds numbers. The quantitative data on the applicability of these equations can be obtained from comparison of their solutions with calculations using more general mathematical models.

Such comparisons were presented by Golovachov *et al* (1973), Golovachov and Leontieva (1987) for stationary supersonic flows about a sphere. Numerical solutions of the above sets of full viscous shock layer equations were obtained with the implicit unidirectional method by Golovachov and Popov (1972a) with the same boundary conditions and the same computational grids. The formulation of the problem was presented in subsection 1.1.2. The velocity slip and temperature jump on the body surface were neglected. The computational domain was bounded by the bow shock.

The results presented in figures 2.1 and 2.2 pertain to the flows of a perfect gas with specific heat ratio $\gamma = 1.4$, viscosity $\mu \propto \sqrt{T}$, Prandtl number $Pr = 0.7$ at free-stream Mach number $M_\infty = 6$. The full curves depict solutions of the full (A) and reduced I, II (B, C) equations, respectively. The dashed curve in figure 2.1 shows the wall pressure distribution calculated for the inviscid gas model. The dashed curves in figure 2.2 display the boundary layer calculations carried out with the finite-difference scheme of Buirkin and Schennikov (1970). The wall pressure is normalized by its stagnation point value.

It is seen that at $Re_\infty = 100$, set I and the full Navier–Stokes equations give the same results which differ noticeably from the solution of set II. The values of the skin friction coefficient provided by set II are close to the results of the boundary layer calculation. With increasing Reynolds number, the discrepancies between the solutions of all the sets diminish. At $Re_\infty = 3550$, both full and reduced Navier–Stokes equations provide the results which almost coincide with those of inviscid flow and boundary layer calculations. The above comparison enables one to estimate the range of validity of the reduced equations with respect to the Reynolds number. To extend the conclusions to flows with other Mach numbers M_∞, one should use, instead of Re_∞, the Reynolds number Re_s related to Re_∞ and M_∞ by formula (1.13).

The considerations exploited in the reduction of the Navier–Stokes equations are founded on the conventional boundary layer theory for non-separated flows. Nevertheless, the reduced equations are often used to calculate the flows which contain the recirculation zones. The validity of the reduced equations for simulating such flows is discussed below with reference to a stationary axisymmetric flow about a sphere immersed in a non-uniform wake-type supersonic stream (Golovachov and Leontieva 1987).

The far wake of an axisymmetric body at zero incidence is considered as a

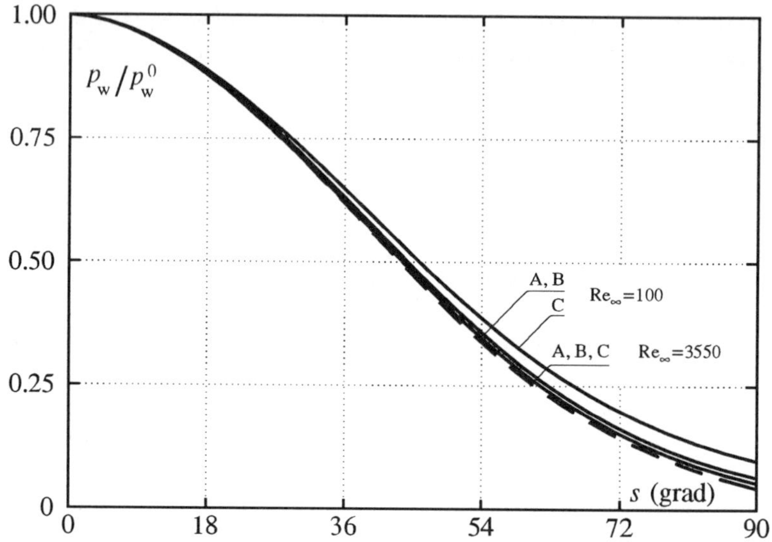

Figure 2.1 Pressure distributions over the sphere surface calculated from the viscous (full lines) and inviscid (dashed line) shock layer equations. $M_\infty = 6$, $\gamma = 1.4$, $T_w = 0.35T^*$.

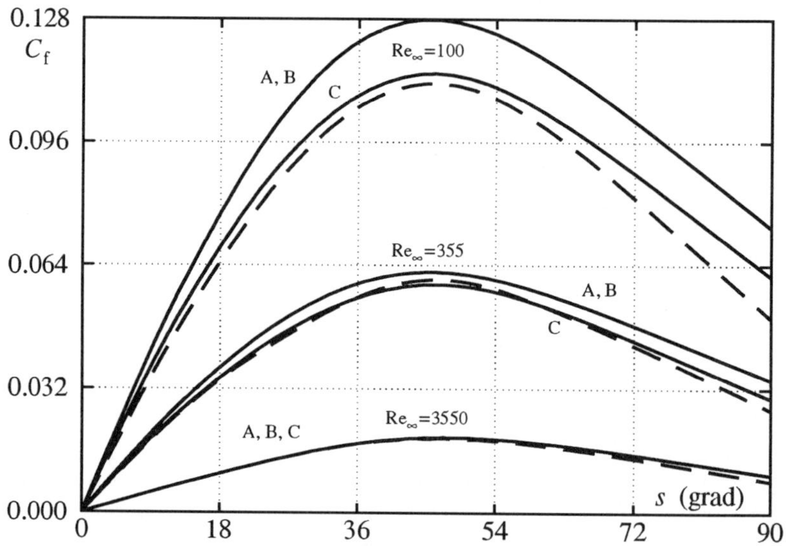

Figure 2.2 Variation of skin friction coefficient over the sphere surface calculated from the viscous shock layer (full lines) and boundary layer (dashed line) equations. $M_\infty = 6$, $\gamma = 1.4$, $T_w = 0.35T^*$.

parallel axisymmetric stream with constant static pressure and an axial velocity minimum and temperature maximum. The velocity and temperature distributions across the wake are described with the formulae suggested by Lin *et al* (1977)

$$V(r) = V(\infty)\left[1 - a \cdot \exp\left(-br^2\right)\right],$$
$$T(r) = T(0)\left\{1 + c\left[1 - \frac{V^2(r)}{V^2(0)}\right]\right\} \tag{2.1}$$

where r is the distance from the wake axis normalized by the sphere radius; V is the modulus of the velocity vector; a, b, c are the parameters which have to be determined from the wake calculations or experimental data.

The results presented below were obtained with $a = 0.1$, $b = 7.2$, $c = 3.0$. The sphere surface was supposed to be adiabatic, the Mach number on the wake axis, $M_0 = 6$, the Reynolds number varied within the range $177 \leqslant Re_0 \leqslant 5000$. The Reynolds number Re_0 is based on the sphere radius and the gas parameters on the wake axis. Its value was varied by means of the free stream density variation. Most calculations were carried out using an equally spaced computational grid $M \times N = 51 \times 25$, where M is the number of mesh points on the lines $s = const$; N is the number of lines.

A distinctive feature of the flow under study at fairly high oncoming stream non-uniformity is the formation of recirculating flows within the shock layer, which influences substantially the shape of the bow shock and the distribution of the drag and heat transfer parameters over the sphere surface. Numerical solutions of the problem with the use of full Navier–Stokes equations will be discussed in detail in subsection 3.1.2. Let us dwell here on a comparison of the results calculated with the full Navier–Stokes equations and reduced equations II. The full and reduced equations were solved subject to modified and conventional Rankine–Hugoniot relationships, respectively.

Calculations show that solutions of the full and reduced equations do not differ qualitatively. At the above parameters of the oncoming stream non-uniformity and sufficiently high Reynolds numbers, both sets provide the results demonstrating an existence of two vortices within the shock layer. The reduced equations predict a rise of the secondary vortex at smaller Reynolds numbers.

Figures 2.3 through 2.5 depict some calculation results. Here all distances are normalized by the sphere radius R, the pressure is normalized by $\rho_0 V_0^2$. The skin friction coefficient is defined as the ratio of the shear stress on the body surface to $\rho_0 V_0^2$. The subscript $\circ$ denotes the gas parameters on the wake axis. To estimate the dependence of the numerical results on mesh spacing, some calculations were made with the grids $M \times N = 26 \times 12$ and 101×50. Solutions of the full Navier–Stokes equations on these grids are plotted by the triangles Δ and ∇, respectively.

Figure 2.3 shows the shift of the outer boundary of the main recirculating flow with Reynolds number. One can see a good agreement between the results provided by the full and reduced equations even at the smallest value of the Reynolds number. A good agreement has been ascertained also for the bow shock distance from the sphere surface.

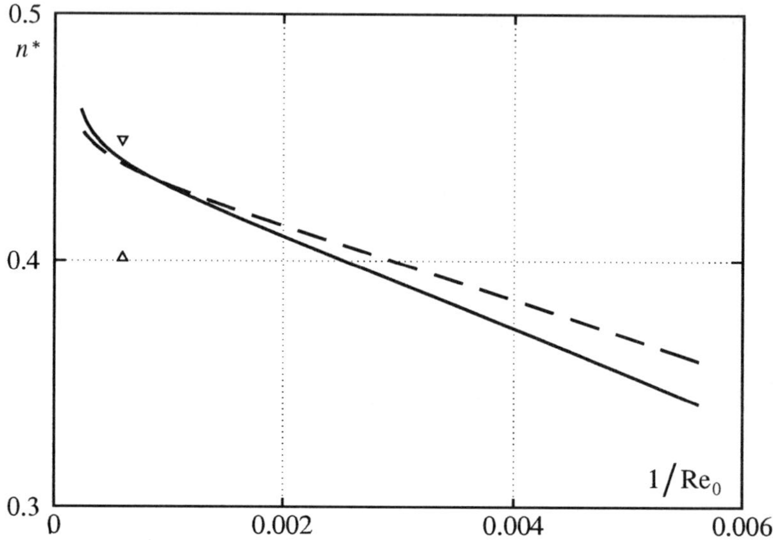

Figure 2.3 Extent of the recirculating flow on the symmetry axis ahead of a sphere in a wake-type oncoming stream. $M_0 = 6$, $\gamma = 1.4$. Full curve, full Navier–Stokes equations; dashed curve, reduced equations; Δ, $M \times N = 26 \times 12$; ∇, $M \times N = 101 \times 50$.

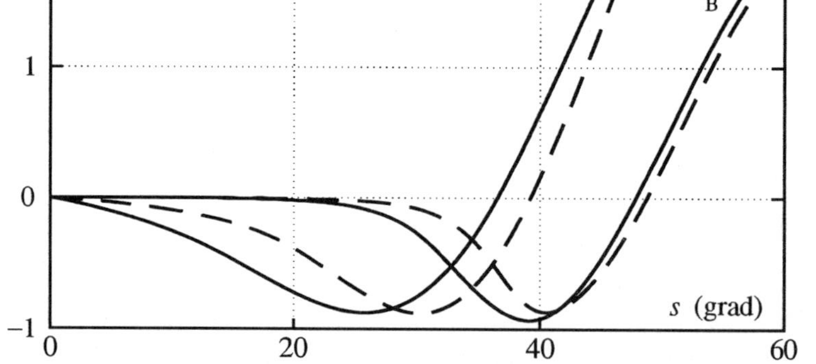

Figure 2.4 Variation of the skin friction coefficient over a sphere surface in a wake-type oncoming stream. $M_0 = 6$, $\gamma = 1.4$. Full curves, full Navier–Stokes equations; dashed curves, reduced equations; A, $Re_0 = 177$; B, $Re_0 = 1770$.

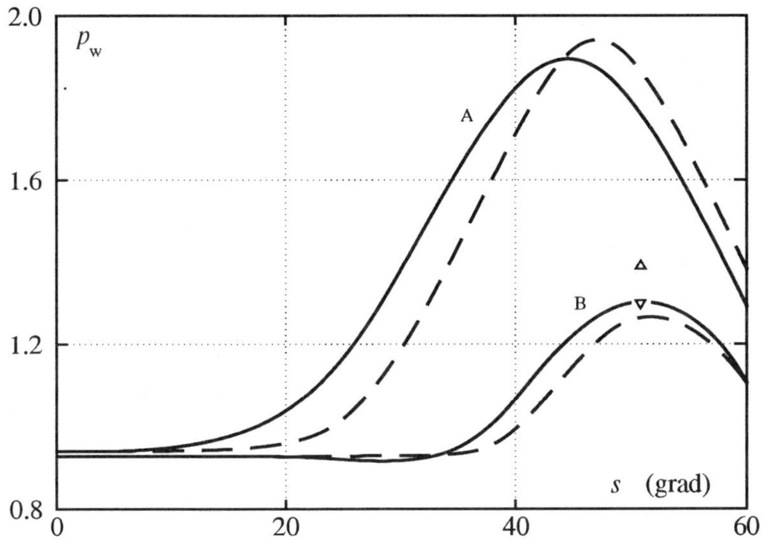

Figure 2.5 Pressure distribution over a sphere surface in a wake-type oncoming stream. $M_0 = 6$, $\gamma = 1.4$. Full curves, full Navier–Stokes equations; dashed curves, reduced equations; A, $Re_0 = 177$; B, $Re_0 = 1770$; Δ, $M \times N = 26 \times 12$; ∇, $M \times N = 101 \times 50$.

The next figures present the distributions of the skin friction coefficient and wall pressure over the sphere surface. It is seen that the discrepancy between the solutions of the full and reduced Navier–Stokes equations occurs mainly within the recirculation zone and diminishes with the Reynolds number increase.

The above comparison shows that the reduced Navier–Stokes equations apply to simulating flows with recirculation zones, though a marked difference between the results of the full and reduced equations within these zones remains up to fairly high Reynolds numbers. This conclusion is substantiated by calculations of Degani and Steger (1983) and Meleshko and Chernii (1985) who compared numerical solutions of the full and reduced Navier–Stokes equations for the turbulent separated flow in the compression corner and the shock wave interaction with the laminar boundary layer.

2.1.3 Formulation of the Cauchy problem for stationary reduced Navier–Stokes equations

The reduced equations do not contain the second derivatives with respect to the streamwise coordinate. However, the upstream influence in subsonic regions inheres these equations due to the inviscid mechanism which is put into effect through the streamwise component of the pressure gradient (Belotserkovskii *et al* 1967). In that view, one has, generally speaking, to formulate the boundary value problem for stationary subsonic and subsonic/supersonic flows.

To elucidate the problem, and especially the role of the streamwise component of the pressure gradient, let us consider set II of the reduced Navier–Stokes equations. For simplicity, we shall write these equations for the Cartesian coordinate system $(x,\ y)$ in which the x-axis is aligned with the main flow direction

$$\begin{aligned}
&\frac{\partial}{\partial x}(\rho u) + \frac{\partial}{\partial y}(\rho v) = 0 \\
&\rho u \frac{\partial u}{\partial x} + \rho v \frac{\partial u}{\partial y} + \frac{\partial p}{\partial x} = \frac{\partial}{\partial y}\left(\mu \frac{\partial u}{\partial y}\right) \\
&\rho u \frac{\partial v}{\partial x} + \rho v \frac{\partial v}{\partial y} + \frac{\partial p}{\partial y} = 0 \\
&\rho u \frac{\partial h}{\partial x} + \rho v \frac{\partial h}{\partial y} = u \frac{\partial p}{\partial x} + v \frac{\partial p}{\partial y} + \frac{\partial}{\partial y}\left(\lambda \frac{\partial T}{\partial y}\right) + \mu \left(\frac{\partial u}{\partial y}\right)^2
\end{aligned} \tag{2.2}$$

Following Lin and Rubin (1979) and making use of the equations of state, continuity and energy balance, let us eliminate the derivative $\partial p/\partial x$ from the second equation of set (2.2). This yields

$$\begin{aligned}
\rho u \frac{M^2-1}{M^2}\frac{\partial u}{\partial x} &= \frac{\partial}{\partial y}\left(\mu \frac{\partial u}{\partial y}\right) - \rho v \frac{\partial u}{\partial y} \\
&- \frac{1}{u}\left\{(\gamma-1)\left[\frac{\partial}{\partial y}\left(\lambda\frac{\partial T}{\partial y}\right) + \mu\left(\frac{\partial u}{\partial y}\right)^2\right] - v\frac{\partial p}{\partial y} - c^2 \rho \frac{\partial v}{\partial y}\right\}
\end{aligned} \tag{2.3}$$

where $c = \sqrt{\gamma p/\rho}$ is the speed of sound, $M = u/c$ is the local Mach number for the streamwise velocity component. One can see from (2.3) that if $u > 0$, then at $M > 1$, the coefficient at $\partial u/\partial x$ determining the sign of the effective viscosity is positive. At $M < 1$, the effective viscosity becomes negative. In the latter case, with $u > 0$, integration in the positive direction of the x-axis means solving an ill-posed problem. It is evident that one can obviate the above difficulty when the streamwise component of the pressure gradient is prescribed a priori, as it is in the conventional boundary layer theory. In this case, the Cauchy problem for set (2.2) is well-posed at any value of the local Mach number.

Consider the formulation of the Cauchy problem for the set of reduced Navier–Stokes equations in the limits of inviscid and fully viscous flows. By virtue of the superposition principle, this analysis allows one to establish the conditions for well-posedness of the Cauchy problem for the model set of equations with locally constant ('frozen') coefficients, which includes both inviscid and viscous terms.

Following Vigneron *et al* (1978), let us modify the set of reduced Navier–Stokes equations (2.2) replacing the second equation by

$$\rho u \frac{\partial u}{\partial x} + \rho v \frac{\partial u}{\partial y} + \omega \frac{\partial p}{\partial x} = \frac{\partial}{\partial y}\left(\mu \frac{\partial u}{\partial y}\right) \tag{2.4}$$

In the inviscid limit ($\mu = \lambda = 0$), the modified set of equations is written as

$$\frac{\partial \boldsymbol{X}}{\partial x} + \mathbf{A}\frac{\partial \boldsymbol{X}}{\partial y} = 0 \tag{2.5}$$

where

$$\boldsymbol{X} = \begin{Bmatrix} \rho \\ u \\ v \\ p \end{Bmatrix} \qquad \mathbf{A} = \frac{1}{u^2 - \omega c^2} \begin{Bmatrix} \frac{v}{u}(u^2 - \omega c^2) & -\rho v & \rho u & \omega \frac{v}{u} \\ 0 & uv & -\omega c^2 & -\omega \frac{v}{\rho} \\ 0 & 0 & \frac{v}{u}(u^2 - \omega c^2) & \frac{u^2 - \omega c^2}{\rho u} \\ 0 & -\rho v c^2 & \rho u c^2 & uv \end{Bmatrix}$$

The Cauchy problem for set (2.5) is well-posed provided it is hyperbolic in x, i.e. if all eigenvalues of matrix $\mathbf{A}$ are real. The eigenvalues of $\mathbf{A}$ are

$$\lambda_{1,2} = \frac{v}{u} \qquad \lambda_{3,4} = \frac{uv \pm c\sqrt{u^2 + \omega(v^2 - c^2)}}{u^2 - \omega c^2} \tag{2.6}$$

It is evident from (2.6) that when $\omega = 1$, set (2.5) is hyperbolic in x provided that $u^2 + v^2 \geqslant c^2$. This restriction is removed with $\omega = 0$.

Considering the other limit, we shall ignore the terms containing the first derivatives with respect to y. For the same vector $\boldsymbol{X}$, the set of equations in the fully viscous limit is written as

$$\frac{\partial \boldsymbol{X}}{\partial x} = \mathbf{B} \frac{\partial^2 \boldsymbol{X}}{\partial y^2} \tag{2.7}$$

where

$$\mathbf{B} = \frac{\mu}{u^2 - \omega c^2} \begin{Bmatrix} -\frac{\omega c^2}{u \rho Pr} & -1 & 0 & \frac{\omega \gamma}{\rho u Pr} \\ \frac{\omega c^2}{\rho^2 Pr} & \frac{u}{\rho} & 0 & -\frac{\omega \gamma}{\rho^2 Pr} \\ 0 & 0 & 0 & 0 \\ -\frac{u c^2}{\rho Pr} & -c^2 & 0 & \frac{\gamma u}{\rho Pr} \end{Bmatrix}$$

The problem with initial data in x for set (2.7) is well-posed if the real part of any eigenvalue of matrix $\mathbf{B}$ is not negative. It is easy to show that two of these eigenvalues are equal to zero ($\lambda_{1,2} = 0$), and two others can be found from the algebraic equation (assuming $u \neq 0$)

$$\begin{aligned} \lambda_{3,4}^2 - \frac{\mu}{\rho u}\left[\frac{1}{Pr} + \left(1 + \frac{\gamma - 1}{Pr}\right)\frac{u^2}{u^2 - \omega c^2}\right]\lambda_{3,4} \\ + \left(\frac{\mu}{\rho u}\right)^2 \frac{\gamma}{Pr}\frac{u^2}{u^2 - \omega c^2} = 0 \end{aligned} \tag{2.8}$$

Examination of the coefficients of (2.8) reveals that when $\omega = 1$ and $u > 0$ the roots $\lambda_{3,4}$ have the positive real parts, provided that $u > c$. If $u < c$, the roots are real and one of them is negative. When $\omega = 0$, the condition $u > 0$ ensures well-posedness of the Cauchy problem both in supersonic and subsonic flow regions.

Thus, the Cauchy problem for the set of reduced Navier–Stokes equations (2.2) is well-posed in both limits in the flow regions where $u > c$. If $u < c$, the Cauchy problem is well-posed for the modified set of equations with $\omega = 0$.

A more detailed analysis (Vigneron *et al* 1978) shows that well-posedness of the Cauchy problem in the subsonic region is guaranteed in both limits with

$$\omega < \frac{\gamma M^2}{1 + (\gamma - 1)M^2} = f(M) \tag{2.9}$$

where M is the local Mach number for the streamwise velocity component. Function $f(M)$ increases monotonously from zero to unity with the Mach number variation within the segment [0,1]. On the basis of the above analysis, Vigneron *et al* (1978) have formulated two conditions ensuring well-posedness of the Cauchy problem for the set of reduced (parabolized) Navier–Stokes equations. The first condition is given by inequality (2.9), which applies only in the subsonic regions. Function $f(M)$ determines the upper limit for the fraction of the streamwise pressure gradient term, $\omega \partial p/\partial x$, which can be treated as a derivative of an unknown function. Approaching the body surface, the Mach number $M \rightarrow 0$; therefore, the above fraction of the pressure gradient term approaches zero. The second condition is given by the inequality

$$u > 0 \tag{2.10}$$

which prohibits reverse flows.

2.2 PARABOLIZED EQUATIONS

The above analysis of the Cauchy problem founds the space-marching numerical methods for solving parabolized Navier–Stokes equations. In this approach, one can solve the problem with initial data in the streamwise coordinate instead of the boundary value problem, which saves significantly an amount of computations. Since a viscous shock layer flow contains a subsonic region near the body surface, one has to regularize the problem with initial data in the marching coordinate. The regularization causes an error which is usually admissible only for the flows containing thin subsonic regions, for which the effects of an upstream disturbance propagation are negligible.

2.2.1 Regularization of the initial-boundary value problem in subsonic regions

The wide-spread regularization procedures employed in the viscous shock layer calculations by the space-marching methods are the following.

(a) The simplest way to ensure well-posedness of the problem with initial data in the marching coordinate is to omit the streamwise pressure gradient terms of the governing equations. This method has been found to be admissible in calculating supersonic flows about a sharp leading edge and about slender conical bodies; see for example (Rudman and Rubin 1968).

(b) Another way to regularize the viscous shock layer problem in subsonic regions (Kovenya and Chernii 1979) is to use the following modified equation for the streamwise momentum component

$$\rho u(1+\chi)\frac{\partial u}{\partial x} + \rho v\frac{\partial u}{\partial y} + (1+\psi)\frac{\partial p}{\partial x} = \frac{\partial}{\partial y}\left(\mu\frac{\partial u}{\partial y}\right) \tag{2.11}$$

which includes the auxiliary terms $\chi\rho u(\partial u/\partial x)$ and $\psi(\partial p/\partial x)$. Elimination of the pressure derivative from (2.11) results in the following equation

$$(1+\chi)\rho u \frac{M^2 - \frac{1+\psi}{1+\chi}}{M^2} \frac{\partial u}{\partial x} = \frac{\partial}{\partial y}\left(\mu \frac{\partial u}{\partial y}\right) - \rho v \frac{\partial u}{\partial y} - \frac{1+\psi}{u} \times \left\{(\gamma - 1)\left[\frac{\partial}{\partial y}\left(\lambda \frac{\partial T}{\partial y}\right) + \mu\left(\frac{\partial u}{\partial y}\right)^2\right] - v\frac{\partial p}{\partial y} - c^2 \rho \frac{\partial v}{\partial y}\right\} \tag{2.12}$$

The positive sign of the coefficient at the derivative $\partial u/\partial x$ is ensured by functions χ and ψ which are introduced so that the above auxiliary terms do not exceed the neglected terms of the full Navier–Stokes equations in the supersonic region and provide well-posedness of the Cauchy problem in the subsonic region. Hence, the following regularization functions have been suggested

$$\chi(\boldsymbol{X}'_x) = \beta\sigma^\alpha \qquad \psi(\boldsymbol{X}'_x) = -\left(\frac{\beta\sigma}{1+\sigma}\right)^\nu \tag{2.13}$$

where $\boldsymbol{X}$ is the column-vector of unknown functions with k components, $\sigma = \sum_{(k)} |(X_k)'_x|$, $\beta \sim \mathrm{O}\left(1/\sqrt{Re}\right)$. The parameters α, β, ν are determined from the calculations. In the supersonic region, $|(X_k)'_x| \sim \mathrm{O}(1)$ and the functions χ and ψ are equal, to an order of magnitude, to β and β^ν, respectively. The Cauchy problem is well-posed and equation (2.11) only slightly differs from the original one. In the subsonic region, the instability resulting from ill-posedness of the Cauchy problem for the original set of reduced Navier–Stokes equations leads to an abrupt increase of the derivatives $(X_k)'_x$. However, this does not violate stability of the marching procedure since

$$\lim_{\sigma\to\infty} \frac{1+\psi}{1+\chi} = 0$$

and, as it is seen from (2.12), the Cauchy problem for the set of viscous shock layer equations with the modified momentum balance equation (2.11) remains well-posed.

(c) Applying the equations which include the streamwise component of the pressure gradient, one can regularize the problem in the subsonic region using backward difference formulae for the derivative $\partial p/\partial x$ in the momentum equation or the derivative $\partial u/\partial x$ in the continuity equation (Lubard and Helliwell 1974, Voronkin 1974). The simplest formula of this kind reads

$$\left(\frac{\partial f}{\partial x}\right)_{i+1,j} \sim \frac{f_{i,j} - f_{i-1,j}}{\Delta x} \tag{2.14}$$

where the indices i, j relate to the x, y-coordinates, respectively. The right-hand side of (2.14) contains the function values which have been calculated at the previous steps in the marching coordinate.

Li (1981) suggested a more complicated formula to evaluate the streamwise component of the pressure gradient in the near-wall region:

$$\left(\frac{\partial p}{\partial x}\right)_{i+1,j} = p_{i,j}\left(\frac{d\ln p}{dx}\right)_{i+1} \tag{2.15}$$

The derivative $(d\ln p/dx)_{i+1}$ was approximated by formula (2.14) using the pressure values averaged over the grid points located within the viscous region of the shock layer flow.

(d) Wide-spread regularization technique for the viscous shock layer calculations was suggested by Lin and Rubin (1975). In this method, which is referred to as the 'sublayer approximation', the pressure derivative with respect to the marching coordinate in the subsonic region is fetched down from the supersonic part of the flow. This approximation is based on the fact that the subsonic region of the viscous shock layer is usually a part of the near-wall boundary layer in which the transverse variation of pressure is negligible.

(e) The method suggested by Vigneron *et al* (1978) is based on the above conditions ensuring the Cauchy problem to be well-posed in the limits of inviscid and fully viscous flows. The pressure derivative term of the balance equation for the streamwise momentum component is written as the sum

$$\frac{\partial p}{\partial x} = \omega\frac{\partial p}{\partial x} + (1-\omega)\left(\frac{\partial p}{\partial x}\right)^{*} \tag{2.16}$$

where $\omega = \sigma f(M)$, function $f(M)$ is given by formula (2.9), $\sigma \lesssim 1$ is the safety factor. Going over to the finite-difference representation, the first term of (2.16) is approximated in the same manner as the derivatives of other functions. The term denoted with an asterisk is treated as the known quantity. If retained, it is evaluated using the backward difference formula or the 'sublayer approximation'. Note that supposing ω to be the step function

$$\omega = \begin{cases} 1 & \text{at } M > 1 \\ 0 & \text{otherwise} \end{cases}$$

the method considered reduces exactly to methods (c) or (d).

(f) To regularize the initial-boundary value problem in subsonic regions, some authors use the non-stationary equations at each new step in the marching coordinate. The time-asymptotic solution of these equations is supposed to give the searched function values at grid points of the marching section considered. This approach was first used in the viscous shock layer calculations by Lin and Rubin (1979). We will consider it in more detail following Golovachov (1984).

Let equations (2.2) complemented with the time-derivative terms be approximated by the implicit finite-difference scheme using the following difference formulae for the derivatives with respect to x and t

$$\left(\frac{\partial f}{\partial x}\right)^{n}_{i+1,j} \sim \frac{f^{n}_{i+1,j} - f_{i,j}}{\Delta x} \qquad \left(\frac{\partial f}{\partial t}\right)^{n}_{i+1,j} \sim \frac{f^{n}_{i+1,j} - f^{n-1}_{i+1,j}}{\Delta t} \tag{2.17}$$

Here f is any unknown function, n is the number of the new time level, indices i, j relate to the spatial coordinates of the grid points. Adding to and subtracting from the numerator of the second formula of (2.17) the quantity $f_{i,j}$, one obtains

$$\left(\frac{\partial f}{\partial t}\right)^n_{i+1,j} \sim \frac{\Delta x}{\Delta t}\left[\left(\frac{\partial f}{\partial x}\right)^n - \left(\frac{\partial f}{\partial x}\right)^{n-1}\right]_{i+1,j} \tag{2.18}$$

Relationship (2.18) is satisfied within the accuracy of the finite-difference approximation.

Let us approximate the non-stationary equations at the nodes of the computational grid with the coordinates $x = x_{i+1}$, $y = y_j$, $t = t^n$ and then eliminate the time derivatives using relationship (2.18). Further, exploiting the equations of state, continuity and energy balance, let us eliminate the derivative $(\partial p/\partial x)^n_{i+1,j}$ from the streamwise momentum balance equation. As a result, the latter takes the form

$$\begin{aligned}
&\rho\left(\frac{\Delta x}{\Delta t} + u\right)\frac{(\Delta x/c\Delta t + M)^2 - 1}{(\Delta x/c\Delta t + M)^2}\frac{\partial u}{\partial x} = \\
&\frac{\partial}{\partial y}\left(\mu\frac{\partial u}{\partial y}\right) - \rho v\frac{\partial u}{\partial y} - \frac{1}{\Delta x/\Delta t + u} \\
&\times\left\{(\gamma - 1)\left[\frac{\partial}{\partial y}\left(\lambda\frac{\partial T}{\partial y}\right) + \mu\left(\frac{\partial u}{\partial y}\right)^2\right] - v\frac{\partial p}{\partial y} - c^2\rho\frac{\partial v}{\partial y}\right\} \\
&+\frac{\Delta x}{\Delta t}\left[\rho\left(\frac{\partial u}{\partial x}\right)^{n-1} - \frac{1}{\Delta x/\Delta t + u}\left(\frac{\partial p}{\partial x}\right)^{n-1}\right]
\end{aligned} \tag{2.19}$$

where the low indices $i+1, j$ are omitted, the upper index is retained only for the quantities calculated at the preceding time level. Comparing (2.19) with (2.3), one can see that within the method discussed the effective viscosity coefficient at $M < 1$ can be made positive by a proper choice of the increments Δx and Δt. The conditions ensuring convergence of the time-asymptotic procedure at the current step in the marching coordinate are still open to question. Relevant information has been obtained from calculations which are discussed in the following subsection.

All above methods for regularization of the initial-boundary value problem in the subsonic regions can be divided into two groups.

The methods (a) and (b) are based on the modified differential equations which are written either without the pressure derivative term or without a part of this term or with the auxiliary terms. Another group includes the methods (c), (d) and (f) in which well-posedness of the initial-boundary value problem is ensured for the difference equations due to a truncation error. The method (e) can be attributed to one or another group depending on determination of the quantity denoted by an asterisk in formula (2.16).

To show a possibility to regularize the difference problem by a truncation error, let us investigate the stability of the simplest implicit T-scheme for the equation

$$\frac{\partial f}{\partial x} + b\frac{\partial f}{\partial y} = \frac{\partial}{\partial y}\left(\varepsilon\frac{\partial f}{\partial y}\right) \tag{2.20}$$

in which the coefficient ε at the second derivative may take negative values. Equation (2.20) can be considered as a model one for the streamwise momentum balance equation (2.3). Von Neumann stability analysis yields the condition

$$\varepsilon > -\Delta x \cdot \min\left[2\frac{\varepsilon^2}{(\Delta y)^2}, \frac{b^2}{2}\right] \tag{2.21}$$

Supposing the subsonic region to be located within the near-wall boundary layer, one can assume $a \approx \Delta y \sim \mathrm{O}\left(\sqrt{|\varepsilon|}\right)$. With such estimates, one concludes from (2.21) that the stability of the marching procedure can be ensured even at $\varepsilon < 0$ provided the mesh spacing in the marching coordinate Δx is sufficiently large. When $\Delta x \to 0$, the marching procedure appears to be stable only at $\varepsilon > 0$, which agrees with the condition for well-posedness of the differential problem.

The above result shows that the stability of marching methods in which the initial-boundary value problem is regularized by the truncation error, is restricted by the following condition on the allowable mesh spacing in the marching coordinate

$$\Delta x > (\Delta x)_{\mathrm{min}} \tag{2.22}$$

Condition (2.22) has been obtained by Lubard and Helliwell (1974) from the von Neumann stability analysis of the marching finite-difference scheme, in which the regularization method (c) was applied. According to the estimates by Rubin and Lin (1980), the value of $(\Delta x)_{\mathrm{min}}$ is equal, to an order of magnitude, to the length of the upstream disturbance propagation and is proportional to the thickness of the subsonic region of the shock layer.

Because of restriction (2.22), the marching methods discussed apply only to the calculation of flows with relatively thin subsonic regions for which $(\Delta x)_{\mathrm{min}}$ is less than the value of Δx required by the accuracy. It is worth noting also that a higher-order accuracy of the difference approximation with respect to the marching coordinate at fixed spacing Δx would deteriorate the stability of these methods. This conclusion is substantiated by the results of Reddy and Rubin (1984). The stability of marching methods based on the modified differential equations is not restricted by condition (2.22). However, the range of applicability of these methods is approximately the same as stated above because of the accuracy requirements.

2.2.2 Space-marching methods

The space-marching methods are considered below with reference to the reduced Navier–Stokes equations in which only the largest viscous terms with the second derivatives and the squares of the first derivatives in the transversal coordinate are retained.

(a) In the marching method suggested by Golovachov and Fursenko (1981b) for axisymmetric flows past the lateral part of blunt bodies, the governing equations are written using the cylindrical coordinate system (z, r) where z is the flow symmetry axis conforming approximately to the main flow direction. The set of governing equations looks as follows

$$\begin{gathered}
\frac{\partial}{\partial z}(\rho u)+\frac{\partial}{\partial r}(\rho v)+\frac{\rho v}{r}=0 \\
\frac{\partial}{\partial z}(\rho u v)+\frac{\partial}{\partial r}(\rho v^{2}+p)+\frac{\rho v^{2}}{r}=0 \\
\frac{\partial}{\partial z}(\rho u^{2}+p)+\frac{\partial}{\partial r}(\rho u v)+\frac{\rho u v}{r}-\frac{\partial}{\partial r}\left(\mu \frac{\partial u}{\partial r}\right)=0 \\
\frac{\partial}{\partial z}(\rho e u)+\frac{\partial}{\partial r}(\rho e v)+p\left(\frac{\partial u}{\partial z}+\frac{\partial v}{\partial r}\right) \\
+\frac{(\rho e+p) v}{r}-\frac{\partial}{\partial r}\left(\lambda \frac{\partial T}{\partial r}\right)-\mu\left(\frac{\partial u}{\partial r}\right)^{2}=0
\end{gathered} \tag{2.23}$$

Here ρe is the internal energy per unit volume, other notations are generally adopted.

The function values at the initial section of the shock layer $z = z_0$ are taken from a calculation of the flow about the body nose. The boundary conditions on the bow shock are given by the Rankine–Hugoniot relationships. On the body surface, both velocity components are assumed to be equal to zero, the gas temperature $T = T_w$.

With the independent variable transformation

$$\xi=\frac{r-r_{w}}{r_{s}-r_{w}} \tag{2.24}$$

the flow region considered is mapped onto a semi-infinite strip of unit width. Here $r = r_w(z)$, $r = r_s(z)$ are the equations of the body surface and bow shock, respectively. The computational grid is introduced by the formulae

$$\begin{aligned}
&z_{i}=z_{0}+i \Delta z, & & & i=0,1, \ldots; \\
&\xi_{j}=j \Delta \xi, & \Delta \xi &=1 / M, & j=0,1, \ldots, M
\end{aligned} \tag{2.25}$$

Having been written for the (z, ξ)-coordinates, equations (2.23) are approximated on grid (2.25) with the implicit finite-difference scheme.

Assuming the function values at $z = z_i$ to be known, let us describe the way to calculate the solution at $z = z_{i+1}$. Regularization of the problem in the subsonic region is carried out with method (f) of the previous subsection, which implies the use of non-stationary equations at each new value of the marching coordinate. The bow shock equation is rewritten as $r = r_s(z, t)$. The derivatives in the ξ-coordinate at $z = z_{i+1}$ are approximated using the function values at the new time level n. By analogy with the unidirectional method of subsection 1.2.2, the second-order equations in ξ are approximated at the grid points $\xi = \xi_j$, while the first-order equations in ξ are approximated at the grid points $\xi = \xi_{j\pm1/2}$, see figure 1.2. The difference formulae for the second-order and first-order equations are

$$\begin{aligned}
\frac{\partial f}{\partial t} &\sim \frac{1}{\Delta t}\left(f^n - f^{n-1}\right)_{i+1,j} \\
\frac{\partial f}{\partial z} &\sim \frac{1}{\Delta z}\left(f^n_{i+1} - f_i\right)_j \\
\frac{\partial f}{\partial \xi} &\sim \frac{1}{2\Delta \xi}\left(f_{j+1} - f_{j-1}\right)^n_{i+1} \\
\frac{\partial}{\partial \xi}\left(\gamma \frac{\partial f}{\partial \xi}\right) &\sim \frac{1}{(\Delta \xi)^2}\left[\gamma_{j+1/2}(f_{j+1} - f_j) - \gamma_{j-1/2}(f_j - f_{j-1})\right]^n_{i+1} \\
j &= 1, 2, \ldots, M-1
\end{aligned} \tag{2.26}$$

and

$$\begin{aligned}
\frac{\partial f}{\partial t} &\sim \frac{1}{2\Delta t}\left[(f_j + f_{j-1})^n - (f_j + f_{j-1})^{n-1}\right]_{i+1} \\
\frac{\partial f}{\partial \xi} &\sim \frac{1}{\Delta \xi}\left(f_j - f_{j-1}\right)^n_{i+1} \\
\frac{\partial f}{\partial z} &\sim \frac{1}{2\Delta z}\left[(f_j + f_{j-1})^n_{i+1} - (f_j + f_{j-1})_i\right] \\
j &= 1, 2, \ldots, M
\end{aligned} \tag{2.27}$$

The initial data for integrating in time at $z = z_{i+1}$ are provided by the solution at the preceding step in the marching coordinate. The function values $f^n_{i+1,j}$ at each new time level are evaluated using $f^{n-1}_{i+1,j}$ by means of the Newton iterative process. In each iteration, a linear set of difference equations with a block tridiagonal matrix of coefficients similar to (1.28) and (1.29) is obtained for the unknown function vector $\boldsymbol{X} = \{\rho\ \rho v\ \rho u\ p\}^T$. The method applied to solve this set of difference equations and to specify the bow shock distance from the body surface is quite similar to that described in subsection 1.2.2.

The above finite-difference scheme is first-order accurate in the z-coordinate and second-order accurate in the ξ-coordinate. Its stability can be estimated from the von Neumann analysis for the model equation

$$\frac{\partial f}{\partial t} + a\frac{\partial f}{\partial x} + b\frac{\partial f}{\partial y} = \frac{\partial}{\partial y}\left(\varepsilon \frac{\partial f}{\partial y}\right) \qquad a > 0, \quad |\varepsilon| \ll 1 \tag{2.28}$$

with constant coefficients. Investigating behaviour of the harmonics $f = f_0\, exp[\omega x + \mathrm{i}(py + lt)]$, it is easy to prove unconditional stability of the finite-difference scheme for both stationary and non-stationary equations provided $\varepsilon \geqslant 0$. At $\varepsilon < 0$, equation (2.28) simulates the problem for the subsonic region. In this case, the stability condition is

$$4\frac{\varepsilon^2}{a}\frac{\Delta x}{(\Delta y)^4}\sin^4\frac{\varphi}{2} - 2|\varepsilon|\left(1+\frac{2\Delta x}{a\Delta t}\sin^2\frac{\psi}{2}\right)\frac{\sin^2\varphi/2}{(\Delta y)^2}$$
$$+\left(1+\frac{\Delta x}{a\Delta t}\sin^2\frac{\psi}{2}\right)\frac{\sin^2\psi/2}{\Delta t}+\frac{\Delta x}{4a}\left(\frac{\sin\psi}{\Delta t}+\frac{b}{\Delta y}\sin\varphi\right)^2 \geqslant 0 \tag{2.29}$$

where $\varphi = p\Delta y$, $\psi = l\Delta t$ are the phase angles. Analogous inequality for the stationary equation differs from (2.29) by lack of the terms containing the time increment Δt. It reduces to (2.21) when $a = 1$. The above result shows that inclusion of the time derivative into the model equation can improve the stability.

More definite data on stability of the finite-difference scheme has been obtained from computations. They agree with the analysis of well-posedness of the Cauchy problem and the von Neumann stability analysis for model equation (2.28) demonstrating stability improvement with increasing mesh spacing in the marching coordinate and decreasing time increment. To accelerate convergence to the stationary solution, one ought to use the maximum time increment Δt whose allowable value has been found to vary nearly proportional to the mesh spacing Δz. The stationary solutions were achieved provided $\Delta z > (\Delta z)_{\min}$. For the most part, this restriction is not onerous because the value of $(\Delta z)_{\min}$ is less than that prescribed by the accuracy requirements.

(b) The marching method by Schiff and Steger (1980) is considered below with reference to a plane flow. The curvilinear coordinate system

$$\xi = \xi(x) \qquad \zeta = \zeta(x, z)$$

is used where x, z are the Cartesian coordinates, the x-axis aligns to the streamwise direction, the plane $\zeta = 0$ presents the body surface in the computational domain. The transformation $\zeta = \zeta(x, z)$ ensures the grid point clustering to the body surface. The Jacobian of the coordinate transformation and the metric coefficients are calculated in accordance with the formulae

$$J^{-1} = x_\xi z_\zeta, \quad \xi_x = Jz_\zeta, \quad \xi_z = 0, \quad \zeta_x = -Jz_\xi, \quad \zeta_z = Jx_\xi \tag{2.30}$$

For the (ξ, ζ)-coordinates the governing equations are written in the divergence form

$$\frac{\partial \boldsymbol{E}}{\partial \xi} + \frac{\partial \boldsymbol{G}}{\partial \zeta} = \frac{1}{Re}\frac{\partial \boldsymbol{S}}{\partial \zeta} \tag{2.31}$$

where the column-vector of unknown functions and the flux vectors are defined as

$$\boldsymbol{X} = \frac{1}{J}\begin{Bmatrix} \rho \\ \rho u \\ \rho w \\ \rho E \end{Bmatrix} \qquad \boldsymbol{E} = \frac{1}{J}\begin{Bmatrix} \rho U \\ \rho U u + \xi_x p \\ \rho U w \\ (\rho E + p)U \end{Bmatrix} \qquad \boldsymbol{G} = \frac{1}{J}\begin{Bmatrix} \rho W \\ \rho W u + \zeta_x p \\ \rho W w + \zeta_z p \\ (\rho E + p)W \end{Bmatrix}$$

$$S = \frac{1}{J}\left\{\begin{array}{c} 0 \\ \mu\left(\zeta_x^2+\zeta_z^2\right)u_\zeta + \mu/3\left(\zeta_x u_\zeta + \zeta_z w_\zeta\right)\zeta_x \\ \mu\left(\zeta_x^2+\zeta_z^2\right)w_\zeta + \mu/3\left(\zeta_x u_\zeta + \zeta_z w_\zeta\right)\zeta_z \\ \left(\zeta_x^2+\zeta_z^2\right)\left[\frac{\mu}{2}(u^2+w^2)_\zeta + \frac{\lambda}{Pr}\frac{1}{\gamma-1}(c^2)_\zeta\right] \\ +\frac{\mu}{3}(\zeta_x u + \zeta_z w)(\zeta_x u_\zeta + \zeta_z w_\zeta) \end{array}\right\}$$

Here u, w are the velocity vector components in the Cartesian coordinate system; $U = \xi_x u$, $W = \zeta_x u + \zeta_z w$ are the contravariant velocity components in the (ξ, ζ)-coordinate system; ρE is the total gas energy per unit volume; c is the speed of sound; other notations have been commented on before. Equations (2.31) are written for the non-dimensional quantities. The velocity components are normalized by c_∞, the density by ρ_∞, the energy per unit volume and gas pressure by $\rho_\infty c_\infty^2$. The Reynolds and Prandtl numbers are evaluated using free stream parameters.

The conditions ensuring well-posedness of the Cauchy problem for a model linearized set of equations with 'frozen' coefficients are surely satisfied in the case $u > 0$, provided that all eigenvalues of the Jacobian matrix $\mathbf{A} = \partial \boldsymbol{E}/\partial \boldsymbol{X}$ are positive. It is easy to show that one of them, namely, $\xi_x(u-c)$ is negative in the subsonic region. The initial-boundary value problem is regularized by means of a 'sublayer approximation' of the streamwise pressure gradient terms, which is applied in both the momentum and energy balance equations. The flux vector $\boldsymbol{E}$ of the governing equations is replaced by the modified vector

$$\boldsymbol{E}^* = \frac{1}{J}\left\{\begin{array}{c} \rho U \\ \rho U u + \xi_x p^* \\ \rho U w \\ U(\rho E + p^*) \end{array}\right\} \tag{2.32}$$

in which p^* is determined by the equation of state (1.5) in the supersonic region $(u > c+\delta,\ 0 < \delta \ll 1)$ and by the condition

$$\frac{\partial p^*}{\partial \zeta} = 0$$

in the region where $u < c+\delta$. Hence, p^* does not depend on the unknown function vector $\boldsymbol{X}$ within the subsonic region of the shock layer. As a result, all eigenvalues of the Jacobian matrix $\mathbf{A}^* = \partial \boldsymbol{E}^*/\partial \boldsymbol{X}$ are positive in both supersonic and subsonic regions provided $u > 0$:

$$\begin{cases} \sigma_{1,2,3,4} = U,\ U,\ U,\ U \pm \xi_x c & \text{when } u > c+\delta \\ \sigma_{1,2,3,4} = U,\ U,\ U,\ U & \text{otherwise} \end{cases}$$

Numerical solution of equations (2.31) with the modified flux vector $\boldsymbol{E}^*$ is obtained using the implicit non-iterative factored finite-difference scheme analogous to that suggested by Beam and Warming (1978) to solve the non-stationary Navier–Stokes equations. The derivatives with respect to the marching coordinate are replaced by the formula

$$\frac{\partial \boldsymbol{E}^*}{\partial \xi} = \frac{(\boldsymbol{E}_{i+1}^* - \boldsymbol{E}_i^*) - \alpha(\boldsymbol{E}_i^* - \boldsymbol{E}_{i-1}^*)}{(1-\alpha)\Delta\xi} \qquad \alpha = 0, 1/3 \tag{2.33}$$

Difference approximation (2.33) is first-order accurate when $\alpha = 0$ and second-order accurate when $\alpha = 1/3$. The derivatives with respect to ζ are approximated by the central second-order accurate difference formulae

$$\frac{\partial f}{\partial \zeta} \sim \delta_\zeta f = \frac{f_{j+1} - f_{j-1}}{2\Delta\zeta}$$

$$\frac{\partial}{\partial \zeta}\left(\gamma \frac{\partial f}{\partial \zeta}\right) \sim \overline{\delta}_\zeta \gamma \overline{\delta}_\zeta f \tag{2.34}$$

$$= \frac{(\gamma_{j+1} + \gamma_j)(f_{j+1} - f_j) - (\gamma_j + \gamma_{j-1})(f_j - f_{j-1})}{2\Delta\zeta^2}$$

where γ, f are any grid functions. Formulae (2.33) and (2.34) result in the following finite-difference scheme

$$\frac{(\boldsymbol{E}^*_{i+1} - \boldsymbol{E}^*_i) - \alpha(\boldsymbol{E}^*_i - \boldsymbol{E}^*_{i-1})}{(1-\alpha)\Delta\xi} + \delta_\zeta \boldsymbol{G}_{i+1} - \frac{1}{Re}\overline{\delta}_\zeta \boldsymbol{S}_{i+1} = 0 \tag{2.35}$$

which is unconditionally stable according to the linear stability analysis.

The flux vectors are the non-linear functions of the conservative variable vector $\boldsymbol{X}$. To avoid iterations, Taylor series expansion of the flux vectors is applied at each step in the marching coordinate with a truncation error $O(\Delta\xi^2)$. Keeping in mind a homogeneity of the flux vectors $\boldsymbol{E}, \boldsymbol{G}$ as functions of vector $\boldsymbol{X}$, the relevant formulae are written as follows

$$\begin{aligned} \boldsymbol{E}^*_{i+1} &= \tilde{\mathbf{A}}^*_i \boldsymbol{X}_{i+1} + \left(\frac{\xi_x}{J}\right)_{i+1} \boldsymbol{E}_{p,i} \\ \boldsymbol{E}_{p,i} &= \begin{cases} 0 & \text{at } u > c + \delta \\ p^*_i \, \{0,\ 1,\ 0,\ u\}^T & \text{otherwise} \end{cases} \\ \boldsymbol{G}_{i+1} &= \tilde{\mathbf{C}}_i \boldsymbol{X}_{i+1} \\ \boldsymbol{S}_{i+1} &= \tilde{\boldsymbol{S}}_i + \tilde{\mathbf{M}}_i \left(\boldsymbol{X}_{i+1} - \boldsymbol{X}_i\right) \end{aligned} \tag{2.36}$$

where $\mathbf{A}^* = \partial \boldsymbol{E}^*/\partial \boldsymbol{X}$, $\mathbf{C} = \partial \boldsymbol{G}/\partial \boldsymbol{X}$, $\mathbf{M} = \partial \boldsymbol{S}/\partial \boldsymbol{X}$ are the Jacobian matrices for the flux vectors. In the calculation of the quantities marked by a tilde, the $\boldsymbol{X}$-vector components at the i-th marching layer and the metric coefficients at the $(i+1)$-th layer are used. In order to retain the second-order accuracy in ξ at $\alpha = 1/3$ and the conservation property of the finite-difference scheme, the first formula of (2.36) is applied to both $\boldsymbol{E}^*_{i+1}$ and $\boldsymbol{E}^*_i$. Substitution of (2.36) into (2.35) results in the following set of linear difference equations for an increment of the unknown function vector $\Delta\boldsymbol{X} = \boldsymbol{X}_{i+1} - \boldsymbol{X}_i$

$$\begin{aligned} &\left[\tilde{\mathbf{A}}^*_i + (1-\alpha)\Delta\xi \left(\delta_\zeta \tilde{\mathbf{C}}_i - Re^{-1}\overline{\delta}_\zeta \tilde{\mathbf{M}}_i\right)\right] \Delta\boldsymbol{X}_i = \\ &\quad - \left(\tilde{\mathbf{A}}^*_i - \tilde{\mathbf{A}}^*_{i-1}\right) \boldsymbol{X}_i + \alpha \left(\boldsymbol{E}^*_i - \boldsymbol{E}^*_{i-1}\right) \\ &\quad - (1-\alpha)\Delta\xi \left[\left(\delta_\zeta \tilde{\mathbf{C}}_i - Re^{-1}\overline{\delta}_\zeta \tilde{\mathbf{M}}_i\right) \boldsymbol{X}_i - Re^{-1}\overline{\delta}_\zeta \tilde{\boldsymbol{S}}_i\right] \\ &\quad - \left[\left(\frac{\xi_x}{J}\right)_{i+1} \boldsymbol{E}_{p,i} - \left(\frac{\xi_x}{J}\right)_i \boldsymbol{E}_{p,i-1}\right] + \mathbf{D}\boldsymbol{X}_i \end{aligned} \tag{2.37}$$

The term $\mathbf{D}\boldsymbol{X}_i$ of the right-hand side of equation (2.37) represents the fourth-order artificial dissipation which is introduced to suppress the high-frequency oscillations of the numerical solution

$$\mathbf{D}\boldsymbol{X}_i = Q\tilde{\mathbf{A}}_i^*(J^{-1})_i(\nabla_\zeta\Delta_\zeta)^2(J\boldsymbol{X})_i \tag{2.38}$$

where

$$(\nabla_\zeta\Delta_\zeta)^2 f_{i,j} = (f_{j+2} - 4f_{j+1} + 6f_j - 4f_{j-1} + f_{j-2})_i$$

For the scheme of first-order accuracy in ξ ($\alpha = 0$), the linear stability analysis yields the restriction $Q < 1/8$.

The set of difference equations (2.37) is supplemented with the boundary conditions on the body surface $\zeta = 0$ and the outer boundary of the computational domain $\zeta = \zeta_{\max}$. The wall boundary conditions are given by the conventional relationships for the velocity components and internal energy (temperature) on the body surface. The outer boundary of the computational domain is located in the free stream. Because of the use of formulae (2.34), the continuity equation turns to be a second-order difference equation in ζ. Therefore, to close the set of difference equations one needs an additional relationship. It is given by the condition

$$\left.\frac{\partial p}{\partial \zeta}\right|_{\zeta=0} = 0$$

which is consistent with the 'sublayer approximation' applied to regularize the initial-boundary value problem in the subsonic region. The closed set of the difference equations at each marching step has a block tridiagonal matrix of coefficients and is solved by the block 'progonka' algorithm.

(c) To elucidate the main ideas of the marching method suggested by Kovenya and Chernii (1983), let us consider a model set of differential equations

$$\mathbf{A}\frac{\partial \boldsymbol{X}}{\partial x} + \mathbf{B}\frac{\partial \boldsymbol{X}}{\partial y} = \mathbf{C}\frac{\partial^2 \boldsymbol{X}}{\partial y^2} \tag{2.39}$$

where $\boldsymbol{X} = \{\rho\ u\ v\ p\}^T$ is the column-vector of unknown functions and the matrices of coefficients are

$$\mathbf{A} = \begin{Bmatrix} u & \rho & 0 & 0 \\ 0 & u & 0 & 1/\rho \\ 0 & 0 & u & 0 \\ 0 & \gamma p & 0 & u \end{Bmatrix} \quad \mathbf{B} = \begin{Bmatrix} v & 0 & \rho & 0 \\ 0 & v & 0 & 0 \\ 0 & 0 & v & 1/\rho \\ 0 & 0 & \gamma p & v \end{Bmatrix}$$

$$\mathbf{C} = \begin{Bmatrix} 0 & 0 & 0 & 0 \\ 0 & m & 0 & 0 \\ 0 & 0 & m & 0 \\ 0 & 0 & 0 & \gamma m/Pr \end{Bmatrix} \qquad m = \mu/\rho Re$$

Set (2.39) is solved subject to the initial data prescribed on the line $x = x_0$.

A computational grid with mesh spacing Δx, Δy and the difference operators Λ_2, Λ_{22} for the derivatives in the y-coordinate are introduced. The central difference operator Λ_{22} approximates the derivative $\partial^2/\partial y^2$ on a three-point stencil with second-order accuracy. As for operator Λ_2, one can choose the central Λ_2^0 or non-central Λ_2^k operator approximating the derivative $\partial/\partial y$ with the second or k-th order accuracy, respectively. Equations (2.39) are replaced by the following finite-difference scheme

$$\mathbf{A}\frac{\boldsymbol{X}_{i+1,j} - \boldsymbol{X}_{i,j}}{\Delta x} + (\mathbf{B}\Lambda_2 - \mathbf{C}\Lambda_{22})\left[\sigma\boldsymbol{X}_{i+1,j} + (1-\sigma)\boldsymbol{X}_{i,j}\right] = 0 \tag{2.40}$$

Matrices $\mathbf{A}$, $\mathbf{B}$, $\mathbf{C}$ are evaluated at the grid points $[\sigma(i+1) + (1-\sigma)i, j]$. Below, the subscript j will be omitted. Scheme (2.40) is second-order accurate in x when $\sigma = 0.5$ and first-order accurate otherwise.

When $\sigma \neq 0$, set (2.40) is non-linear with respect to the function values at the $(i+1)$-th marching layer. Its solution is obtained using the following iteration procedure

$$\begin{aligned}\frac{\boldsymbol{X}_{i+1}^{\nu+1} - \boldsymbol{X}_{i+1}^{\nu}}{\tau} + \mathbf{A}^{\nu}\frac{\boldsymbol{X}_{i+1}^{\nu+1} - \boldsymbol{X}_i}{\Delta x} + (\mathbf{B}^{\nu}\Lambda_2 - \mathbf{C}^{\nu}\Lambda_{22})\\ \times\left[\sigma\boldsymbol{X}_{i+1}^{\nu+1} + (1-\sigma)\boldsymbol{X}_i\right] = 0\end{aligned} \tag{2.41}$$

Here τ is the iteration parameter, ν is the iteration number, matrices $\mathbf{A}^{\nu}$, $\mathbf{B}^{\nu}$, $\mathbf{C}^{\nu}$ are evaluated in accordance with the formula

$$\mathbf{D}^{\nu} = \sigma\mathbf{D}_{i+1}^{\nu} + (1-\sigma)\mathbf{D}_i$$

where $\mathbf{D}$ is any of matrices $\mathbf{A}$, $\mathbf{B}$ or $\mathbf{C}$. Set (2.41) is obtained from (2.40) by adding the term

$$\frac{\boldsymbol{X}_{i+1}^{\nu+1} - \boldsymbol{X}_{i+1}^{\nu}}{\tau}$$

which is equivalent to introducing the time derivatives. Both schemes, (2.41) and (2.40), approximate equation (2.39) with the same order of accuracy provided the iterations have converged.

Omitting the subscript $(i+1)$, let us rewrite the linear set of difference equations (2.41) in the following conventional form

$$\mathbf{K}^{\nu}\frac{\boldsymbol{X}^{\nu+1} - \boldsymbol{X}^{\nu}}{\tau} = -\Omega^{\nu} \tag{2.42}$$

where

$$\mathbf{K}^{\nu} = \mathbf{I} + \frac{\tau}{\Delta x}\mathbf{A}^{\nu} + \tau\sigma(\mathbf{B}^{\nu}\Lambda_2 - \mathbf{C}^{\nu}\Lambda_{22})$$

$$\Omega^{\nu} = \mathbf{A}^{\nu}\frac{\boldsymbol{X}^{\nu} - \boldsymbol{X}_i}{\Delta x} + (\mathbf{B}^{\nu}\Lambda_2 - \mathbf{C}^{\nu}\Lambda_{22})\left[\sigma\boldsymbol{X}^{\nu} + (1-\sigma)\boldsymbol{X}_i\right]$$

$\mathbf{I}$ is the identity matrix.

Set (2.42) is solved using operator splitting in the physical processes. The difference operator $\mathbf{K}^\nu$ is replaced by the factored operator $\mathbf{T}^\nu$

$$\mathbf{K}^\nu \approx \mathbf{T}^\nu = \mathbf{T}_1^\nu \mathbf{T}_2^\nu \mathbf{T}_3^\nu \tag{2.43}$$

where

$$\mathbf{T}_1^\nu = \mathbf{I} + \frac{\tau}{\Delta x}\mathbf{A}^\nu \quad \mathbf{T}_2^\nu = \mathbf{I} + \tau\sigma\left(\mathbf{B}_1^\nu \Lambda_2^k - \mathbf{C}^\nu \Lambda_{22}\right) \quad \mathbf{T}_3^\nu = \mathbf{I} + \tau\sigma \mathbf{B}_2^\nu \Lambda_2^0$$

$$\mathbf{B}_1^\nu = \left\{\begin{matrix} v & 0 & 0 & 0 \\ 0 & v & 0 & 0 \\ 0 & 0 & v & 0 \\ 0 & 0 & 0 & v \end{matrix}\right\} \qquad \mathbf{B}_2^\nu = \left\{\begin{matrix} 0 & 0 & \rho & 0 \\ 0 & 0 & 0 & 0 \\ 0 & 0 & 0 & 1/\rho \\ 0 & 0 & \gamma p & 0 \end{matrix}\right\}$$

matrices $\mathbf{A}^\nu$, $\mathbf{C}^\nu$ are the same as in (2.39). Formulae (2.43) answer the following splitting: operator $\mathbf{A}^\nu$ takes into account the transfer in the x-direction; operators $\mathbf{B}_1^\nu \Lambda_2^k$, $\mathbf{C}^\nu \Lambda_{22}$ and $\mathbf{B}_2^\nu \Lambda_2^0$ present the convective and molecular transport and pressure terms in the y-direction.

With approximate factorization (2.43), the finite-difference scheme (2.42) is implemented through three fractional steps

$$\begin{aligned} \mathbf{T}_1^\nu (\Delta \boldsymbol{X})^{1/3} &= -\tau\Omega^\nu, \qquad \mathbf{T}_2^\nu (\Delta \boldsymbol{X})^{2/3} = (\Delta \boldsymbol{X})^{1/3} \\ \mathbf{T}_3^\nu (\Delta \boldsymbol{X}) &= (\Delta \boldsymbol{X})^{2/3} \end{aligned} \tag{2.44}$$

followed by evaluation of the function values at the $(\nu+1)$-th iteration by formula $\boldsymbol{X}^{\nu+1} = \boldsymbol{X}^\nu + \Delta \boldsymbol{X}$.

The set of difference equations at the first fractional step is solved explicitly, because the operator $\mathbf{T}_1^\nu$ contains no derivatives and its inversion is equivalent to the inversion of matrix 4×4. The difference equations of the second fractional step are solved successively, one after another, using a scalar version of the 'progonka' algorithm in accordance with the diagonal structure of the operator $\mathbf{T}_2^\nu$. The equations of the third fractional step are

$$\begin{aligned} \Delta\rho + \tau\sigma\rho^\nu \Lambda_2^0 \Delta v &= (\Delta\rho)^{2/3} \qquad & \Delta u &= (\Delta u)^{2/3} \\ \Delta v + \tfrac{\tau\sigma}{\rho^\nu}\Lambda_2^0 \Delta p &= (\Delta v)^{2/3} \qquad & \Delta p + \tau\sigma\gamma p^\nu \Lambda_2^0 \Delta v &= (\Delta p)^{2/3} \end{aligned} \tag{2.45}$$

They are solved as follows. Using the third equation of (2.45) to express Δv through Δp and substituting the result into the fourth equation, one obtains the following scalar equation for Δp

$$\Delta p - (\tau\sigma)^2 \gamma p^\nu \Lambda_2^0 \left(\frac{1}{\rho^\nu}\Lambda_2^0 \Delta p\right) = (\Delta p)^{2/3} - \tau\sigma\gamma p^\nu \Lambda_2^0 (\Delta v)^{2/3} \tag{2.46}$$

which is solved by the 'progonka' algorithm. The increment Δp being known, one obtains the increment Δv from the third equation of (2.45) and then calculates the increment $\Delta\rho$ from the first equation. The second equation of (2.45) yields the increment Δu.

In this way, the solution of the set of difference equations in each iteration reduces to a sequence of the scalar 'progonka' procedures and the calculations by explicit formulae. The iterations are continued until the condition

$$\frac{||\boldsymbol{X}^{\nu+1} - \boldsymbol{X}^{\nu}||}{\tau} \leqslant \delta$$

is satisfied, where $||\cdot||$ is an appropriate norm of the vector grid functions, $\delta \sim O(h^2)$, $h = \max(\Delta x,\ \Delta y)$. The solution thus obtained provides the initial data for the next marching step calculations.

The well-posedness of the initial-boundary value problem in the subsonic region is ensured by regularization functions which are introduced into the momentum balance equation, see method (b) of the previous subsection. Convergence of iterations and stability of the marching procedure have been substantiated by the von Neumann analysis.

2.2.3 Examples

The validity of the above approach to the numerical simulation of viscous shock layer flows has been ascertained by comparison of calculation results with solutions of the boundary value problem and experimental data.

Figures 2.6 and 2.7 show the calculations of Schiff and Steger (1980) pertaining to plain supersonic flow about the symmetric parabolic arc airfoil with a relative thickness of 10 per cent. The results of the space-marching method are plotted by the circles. The full lines depict the time-asymptotic solution of the full Navier–Stokes equations. Both calculations have been carried out for the turbulent flow regime using the algebraic turbulence model by Baldwin and Lomax (1978). The Reynolds number Re_∞ is evaluated using the airfoil chord c. The airfoil surface is supposed to be adiabatic. The initial data for the marching calculation were taken from the time-asymptotic solution. The marching calculation was carried out from $x/c = 0.1$ to $x/c = 1.0$. In both cases, the same computational grid was used with the nodes concentrated towards the body surface.

The next figures present calculations by Golovachov and Fursenko (1981a,b) of axisymmetric laminar flows about spherically blunted cones. Here, the gas specific heat ratio $\gamma = 1.4$, viscosity $\mu \propto \sqrt{T}$, the Prandtl number $Pr = 0.7$. The computational grid contained from 26 to 51 nodes on each line z =const. The grid points were clustered towards the body surface by transformation (1.43). Most calculations were made with the time increment $\Delta t = 0.1\ R/V_\infty$, where R is the body nose radius. At each step in the z-coordinate, the equations were integrated in time until the relative values of all time derivatives became less than 5×10^{-3}. A further approach to the stationary solution was found to have a negligible effect on the results. The number of time steps required was equal only to a few on the smooth parts of the body surface but increased up to some dozens in the vicinities of the discontinuities of the body shape variation.

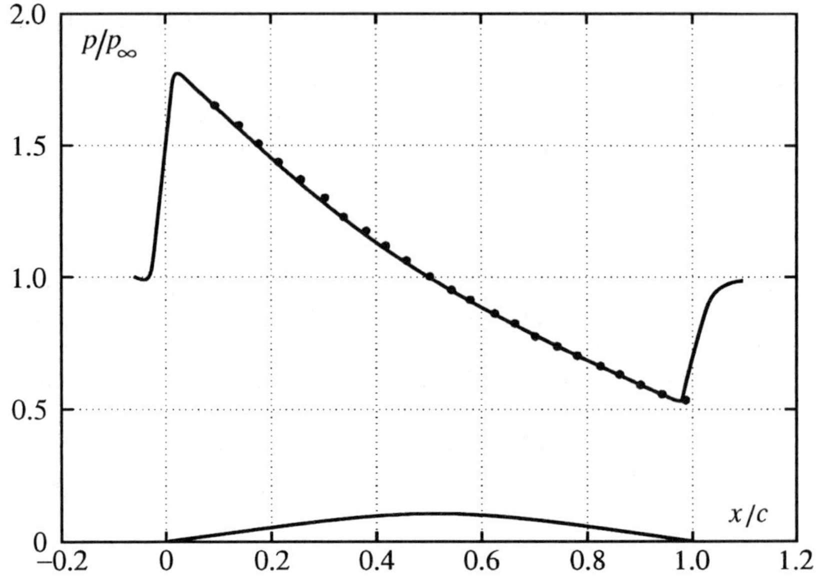

Figure 2.6 Pressure distribution over the airfoil. $M_\infty = 2$, $Re_\infty = 1.85 \times 10^6$. Line, full Navier–Stokes equations; circles, parabolized Navier–Stokes equations.

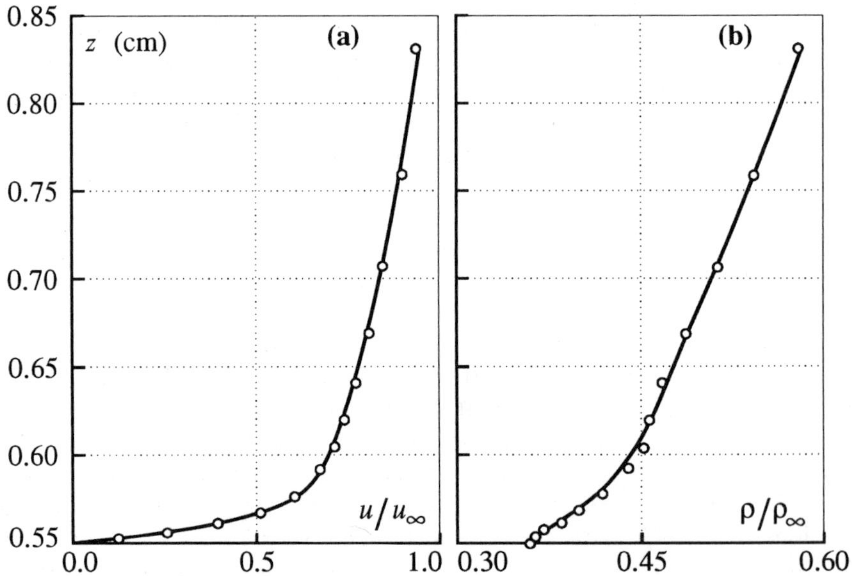

Figure 2.7 Streamwise velocity component and density profiles in the shock layer on the airfoil at $x/c = 0.9$. $M_\infty = 2$, $Re_\infty = 1.85 \times 10^6$, $c = 1m$. Lines, full Navier–Stokes equations; circles, parabolized Navier–Stokes equations.

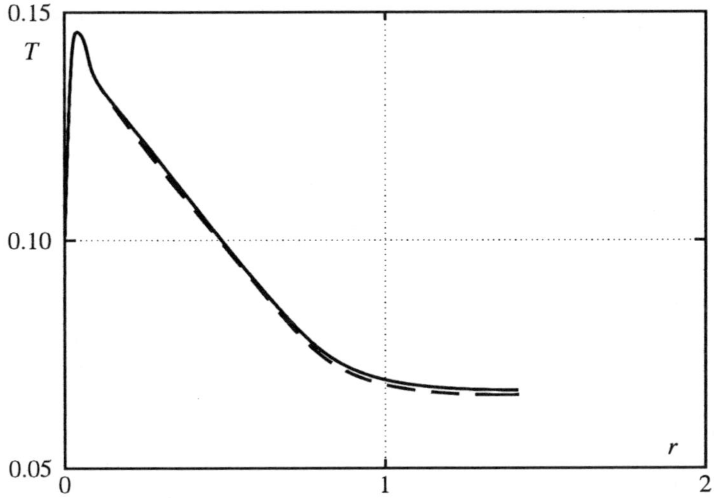

Figure 2.8 Temperature profiles in the shock layer of a spherically blunted cone with half-angle $\theta_c = 10^o$. $M_\infty = 6$, $Re_\infty = 35500$, $T_w = 0.35T^*$, $z = 4.09$. Full line, parabolized Navier–Stokes equations; dashed line, full Navier–Stokes equations.

Figure 2.8 shows the temperature profile in the shock layer. Here z is the distance from the stagnation point along the body axis, r is the distance from the body surface. Both distances are normalized by the body nose radius R. The marching calculation started from $z = 1.17$. The dashed curve depicts the temperature profile provided by the solution of the boundary value problem using the full Navier–Stokes equations. One can see a good agreement between both calculations. The shock layer thickness and wall pressure agree within 1 per cent. The greatest discrepancies, about 5 per cent, occur for the heat flux and the skin friction coefficient.

Figure 2.9 compares the calculated heat flux distribution over the cone with experimental data of Pappas and Lee (1970). Here s is the distance from the stagnation point along the cone generatrix normalized by the body nose radius. The marching method was applied to solve the viscous shock layer equations beginning from $s = 1.8$. There is a good agreement between the calculated and measured heat flux distributions in spite of a relatively low value of the Reynolds number.

In figure 2.10, the calculated values of the wall pressure and heat transfer coefficients are compared with experimental data presented by Griffith and Lewis (1964). For convenience, the similarity variables (Chernii 1959) are used

$$\tilde{z} = \frac{\theta}{\sqrt{\frac{\gamma-1}{\gamma+1} C_D^o}} \frac{z}{d} \qquad \tilde{c}_p = \frac{p_w - p_\infty}{\rho_\infty V_\infty^2 \theta^2}$$

$$\tilde{c}_H = \frac{\left(\frac{\gamma-1}{\gamma+1} C_D^o\right)^{1/4} C_H \sqrt{2Re_\infty}}{\sqrt{C_*}\left(M_\infty^2 \theta^2 + \frac{1}{\gamma \tilde{c}_p}\right)}$$

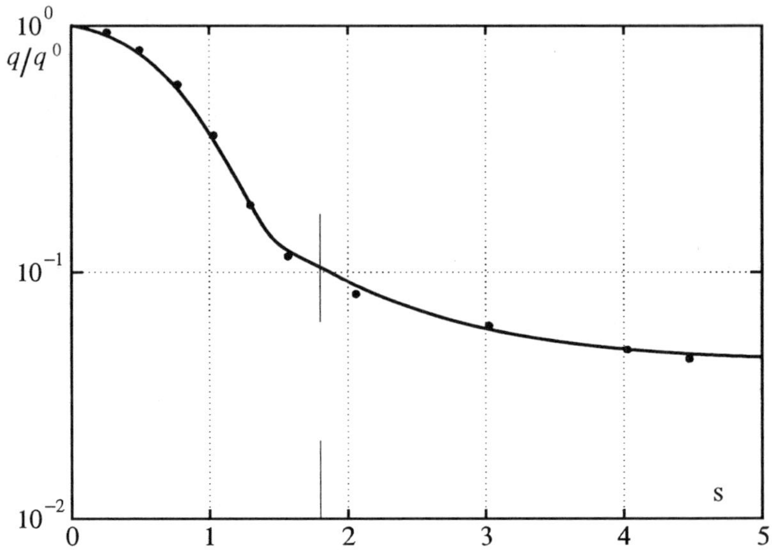

Figure 2.9 Heat flux distribution over a spherically blunted cone with half-angle $\theta_c = 7.5^o$. $M_\infty = 13.4$, $Re_\infty = 1515$, $T_w = 0.073T^*$. Line, calculation of Golovachov and Fursenko (1981a); circles, measurements of Pappas and Lee (1970).

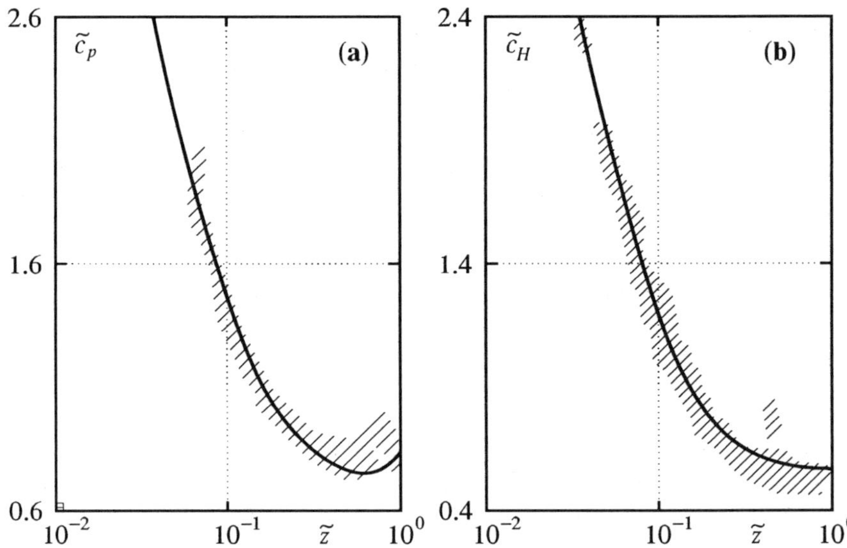

Figure 2.10 Wall pressure (a) and heat transfer (b) coefficient variation over a spherically blunted cone with half-angle $\theta_c = 9^o$. $M_\infty = 18$, $Re_\infty = 2240$, $T_w = 0.1T^*$. Lines, calculations of Golovachov and Fursenko (1981b); shading, the experimental data of Griffith and Lewis (1964).

Here C_D^o is the drag coefficient of the body nose, d is the diameter of the bluntness, $C_H = q/\rho_\infty V_\infty(h^* - h_w)$ is the heat transfer coefficient, $C_* = (\mu_*/\mu_\infty)(T_\infty/T_*)$ is the Chapman–Rubesin parameter, $T_* = \frac{1}{6}(T^* + 3T_w)$. The experiments were performed at $M_\infty = 14 \div 20$, $Re_\infty = (1 \div 6.75) \times 10^3$. The lines are the calculation results for $M_\infty = 18$, $Re_\infty = 2240$, $T_w = 0.1T^*$.

In conclusion, it is worth noting that the marching procedure can be generalized to a block version in which several layers in the marching coordinate are calculated simultaneously instead of one layer in each marching step (Afonina *et al* 1983, Belova *et al* 1985). In this case, one can take into account the upstream disturbance propagation and remove the restriction mentioned in subsection 2.2.1. With an increasing number of simultaneously calculated layers, the block version of the marching method approaches the conventional iterative procedure for solving the boundary value problem.

2.3 MULTISWEEP METHODS

Investigating shock layer flows in which the upstream disturbance propagation is important, one ought to use the methods ensuring information transfer throughout the computational domain. The methods to be discussed, which are also referred to as global iteration methods, satisfy the above requirement, but unlike the relaxation methods of a general type, like the time-asymptotic procedure, they use the previous iteration data to evaluate only those terms of the governing equations which are responsible for upstream disturbance propagation. Each global iteration is implemented with a space-marching procedure.

Davis (1970a) was the first who applied such calculation procedure to solve the problems of supersonic flow around a body. He used the set of full viscous shock layer equations mentioned in subsection 2.1.1. In each global iteration, the solution for the stagnation streamline was first obtained and then the parabolic set of equations was integrated in the downstream direction. The 'elliptic' terms of the governing equations and the shape of the bow shock were determined with the previous global iteration data. The initial distribution of gas dynamic functions was calculated using the thin viscous shock layer model which ignored the effects of upstream disturbance propagation. Calculations of hypersonic axisymmetric flows about hyperboloids have demonstrated a rapid convergence of the above multisweep relaxation method. However, for some other cases, in particular for the flows around spherically blunted cones, the global iterations proved to be divergent. To remove this drawback, some modifications have been suggested (Waskiewich *et al* 1978, Srivastava *et al* 1978, 1979, Vasilyevskii and Tirskii 1979, 1987) concerning generation of the initial data, evaluation of the bow shock shape, solving the set of difference equations and accounting for the discontinuities of the body generatrix curvature.

2.3.1 Multisweep relaxation of the streamwise pressure gradient term

In the full viscous shock layer equations, the upstream disturbance propagation manifests itself through the bow shock shape and the streamwise component of the pressure gradient. It will be shown below that the bow shock shape can be determined in each global iteration without the previous iteration data. The streamwise component of the pressure gradient is approximated by the forward difference formulae with the use of the previous global iteration data for pressure values at the downstream grid points.

One of the most popular difference formulae reads

$$\left.\frac{\partial p}{\partial x}\right|_i \sim \omega \frac{p_i - p_{i-1}}{\Delta x} + (1-\omega)\frac{p_{i+1} - p_i}{\Delta x} \qquad 0 \leqslant \omega < 1 \tag{2.47}$$

where x is the streamwise coordinate; subscript i numbers the grid points in this direction.

It is evident from (2.47) that in the method discussed, the boundary condition for the pressure on the downstream boundary of the computational domain is required. This boundary is located in the flow region where the upstream disturbance propagation is negligible and so the approximate conditions

$$\frac{\partial p}{\partial x} = 0 \qquad \text{or} \qquad \frac{\partial^2 p}{\partial x^2} = 0$$

apply.

The stability and convergence rate of global iterations are influenced by the value of parameter ω. If it is assumed to be constant, the best results are obtained with $\omega = 0$ (Rubin 1981). In accordance with analysis of the Cauchy problem for reduced Navier–Stokes equations, the value of ω is expected to depend on the local Mach number based on the streamwise velocity component. The upstream influence is taken into account by the 'elliptic' part of the streamwise pressure gradient term

$$(1-\omega)\frac{\partial p}{\partial x}$$

which is approximated by the forward difference formula using the previous global iteration data. The stability condition for the marching procedure in each sweep is satisfied if

$$0 \leqslant \omega \leqslant f(M) \tag{2.48}$$

within the subsonic region. Function $f(M)$ is defined by relationship (2.9). Ramakrishnan and Rubin (1987) suggested the formulae

$$\omega = \begin{cases} \alpha f(M), & \text{if } M < 1 \\ \alpha, & \text{otherwise} \end{cases} \tag{2.49}$$

with $0 \leqslant \alpha \leqslant 1$. Their calculations revealed a noticeable influence of parameter α on the convergence of global and internal iterations. This parameter was

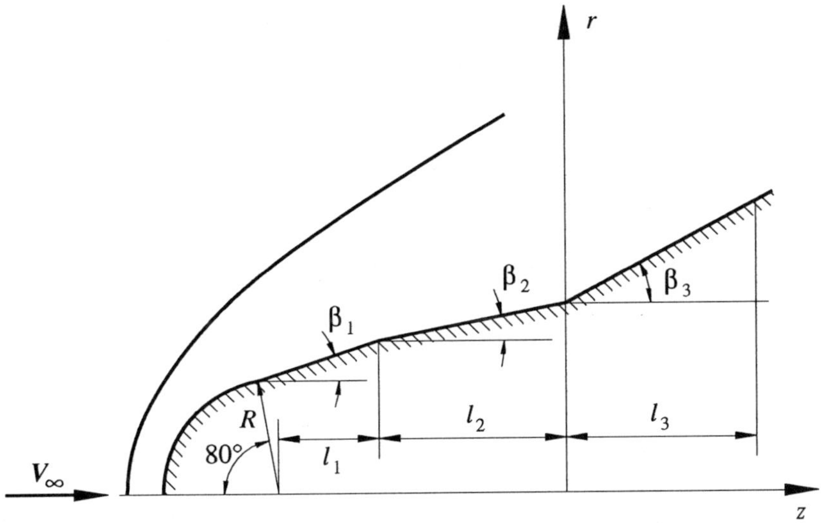

Figure 2.11 Meridional section of the combined body of revolution.

found to be especially important in the shock-capturing procedures because of its significant effect on the location of intensive shocks and their smearing over the cells of the computational grid. The stability and convergence rate of the methods considered can be improved by the use of damping procedures in evaluation of the pressure field and the streamwise pressure gradient term (Barnett and Davis 1986, Golovachov and Timofeyev 1988a).

2.3.2 Examples

Application of the multisweep global iteration methods to the viscous shock layer problems was discussed by Golovachov and Timofeyev (1988a) with reference to an axisymmetric flow about the lateral part of a combined blunted body of revolution shown in figure 2.11.

The shock layer flow was calculated with reduced Navier–Stokes equations (2.23). The boundary conditions on the bow shock and the body surface were the same as those of subsection 2.2.2a. On the downstream boundary, the condition

$$\frac{\partial^2 p}{\partial z^2} = 0. \tag{2.50}$$

was imposed. The profiles of gas dynamic functions at the initial value of the streamwise coordinate were obtained using the time-asymptotic method by Golovachov and Popov (1972a) or the space-marching method by Golovachov and Fursenko (1981b). The marching procedure applied in each sweep (global iteration) differed from that described in subsection 2.2.2a only by the use of stationary equations and the bow shock slope to the z-axis instead of the bow shock velocity.

The term $\partial p/\partial z$ responsible for upstream disturbance propagation was approximated by the following difference formulae

$$\left.\frac{\partial p}{\partial z}\right|_{i+1} \sim \frac{p_{i+2,j}^{\nu-1} - p_{i+1,j}^{\nu}}{\Delta z} \tag{2.51a}$$

$$\left.\frac{\partial p}{\partial z}\right|_{i+1} \sim \frac{1}{\Delta z}\left(p_{i+2,j}^{\nu-1} - p_{i+1,j}^{\nu-1}\right) \tag{2.51b}$$

$$\left.\frac{\partial p}{\partial z}\right|_{i+1} \sim \frac{1}{\Delta z}\Big[\omega_i\left(p_{i+1,j}^{\nu} - p_{i,j}^{\nu}\right) + (1-\omega_i)\left(\left(p_{i+2,j}^{\nu-1}\right)^* - p_{i+1,j}^{\nu}\right)\Big] \tag{2.51c}$$

$$\left.\frac{\partial p}{\partial z}\right|_{i+1} \sim \frac{1}{\Delta z}\Big[\omega_i\left(p_{i+1,j}^{\nu} - p_{i,j}^{\nu}\right) + (1-\omega_{i+1})\left(\left(p_{i+2,j}^{\nu-1}\right)^* - p_{i+1,j}^{\nu}\right)\Big] \tag{2.51d}$$

where ω is given by formula (2.49),

$$\left(p_{i+2,j}^{\nu-1}\right)^* = \begin{cases} p_{i+2,j}^{\nu-1} & \text{or} \\ \left(p_{i+2,j+1}^{\nu-1} + p_{i+2,j-1}^{\nu-1}\right)/2 \end{cases}$$

superscript ν denotes the number of the current sweep.

The marching finite-difference scheme exploited in each sweep is not monotonic. Hence, in calculating flows with inner shocks, local smoothing procedures were applied for gas dynamic fluxes in the transversal coordinate.

The convergence of the multisweep calculation procedure was checked by the pressure field. For the most part, the global iterations were carried out until 0.5 per cent accuracy was achieved. To accelerate the convergence, the following damping was applied

$$p_{i,j}^{\nu} = \beta\overline{p}_{i,j}^{\nu} + (1-\beta)p_{i,j}^{\nu-1}$$

where $\overline{p}_{i,j}^{\nu}$ is the preliminary value of pressure obtained in the ν-th sweep; β is the parameter which was usually assumed to be equal to 0.4.

When calculating flows with inner shocks which intersected the downstream boundary of the computational domain, convergence of the pressure field in the vicinity of the intersection points was not achieved, which was evidently caused by an inappropriate downstream boundary condition (2.50).

The calculations were carried out for the flows around a smooth blunted cone ($l_1 \gg 1$) and a combined body ($l_1 = 0.17$, $l_2 = 1.1$, $l_3 = 1.0$, $\beta_1 = 10°$, $\beta_2 = 0$, $\beta_3 = 10 \div 25°$) at $M_\infty = 3$, $Re_\infty = 3 \times 10^4$, $T_w = 0.476T^*$, $\gamma = 1.4$, $\mu \propto \sqrt{T}$, $Pr = 0.7$. The number of grid points across the shock layer varied within the range $26 \div 51$. The grid points were concentrated towards the body surface.

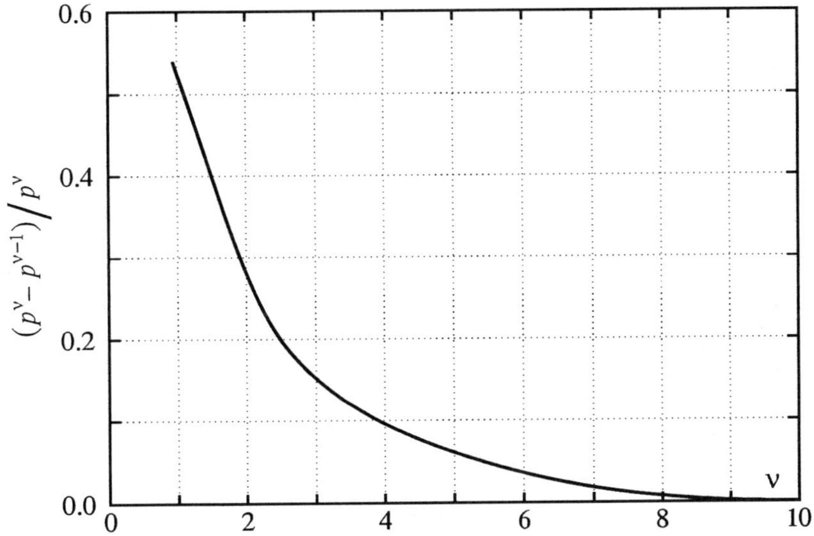

Figure 2.12 Variation of the maximum pressure increment in multisweep calculation of shock layer flow about the combined body. $M_\infty = 3$, $Re_\infty = 3 \times 10^4$, $T_w = 0.476T^*$, $\beta_3 = 15°$.

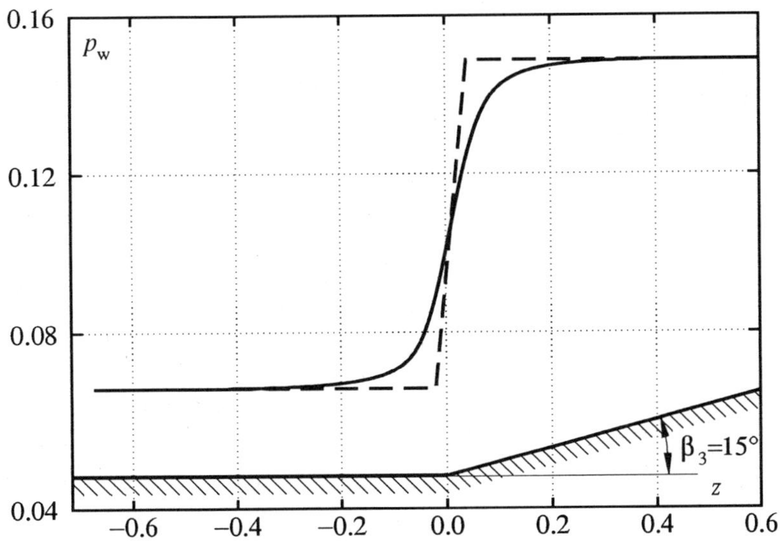

Figure 2.13 Wall pressure distributions calculated with the multisweep method (full line) and marching method (dashed line). $M_\infty = 3$, $Re_\infty = 3 \times 10^4$, $T_w = 0.476T^*$, $\beta_3 = 15°$.

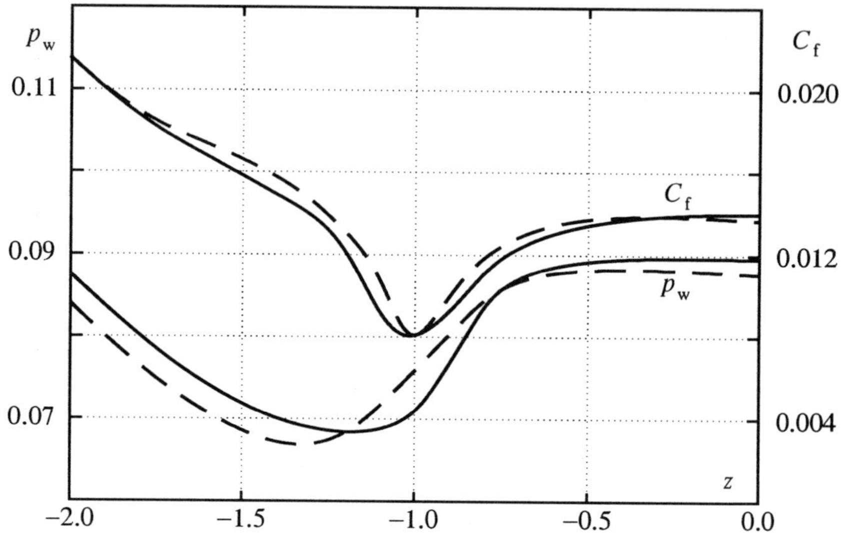

Figure 2.14 Variation of the wall pressure and skin friction coefficient in the vicinity of the body contour slope discontinuity. $M_\infty = 10$, $Re_\infty = 5000$, $T_w = 0.286T^*$, $l_1 = l_2 = 1.0$, $\beta_2 = \beta_3 = 20°$. Full lines, solution of the reduced Navier–Stokes equations by the multisweep method; dashed lines, solution of the full Navier–Stokes equations by the time-asymptotic method.

The results demonstrate the convergence rate of global iterations and the effects of various difference approximations for the streamwise pressure gradient term, initial pressure distribution in the shock layer and downstream boundary condition. The conclusions are formulated as follows:

(a) The use of multisweep relaxation removes the lower limit for the allowable mesh spacing in the marching coordinate.

(b) The initial pressure distribution can be calculated with the marching scheme under the assumption $\partial p/\partial z = 0$. In serial calculations it is advisable to take the initial pressure distribution from available data for the similar flight conditions.

(c) The downstream boundary condition influences the solution within a region which does not extend over several mesh cells adjacent to the boundary.

(d) Formula (2.51-c) with $\alpha = 0.95$ was found to be the best one among tested difference approximations for the streamwise pressure gradient term. With this formula, the solution can be obtained provided that at least several nodes are within the subsonic region of the boundary layer. Otherwise, approximations (2.51-a,b) are preferable, the first providing better results as compared with the fully explicit formula (2.51-b).

(e) The total number of global iterations (sweeps) did not exceed 30. It depended substantially on the slope angle β_3 and the initial pressure field. With mesh refinement in the marching coordinate, the required number of sweeps increased because of deceleration of the information transport in the upstream direction.

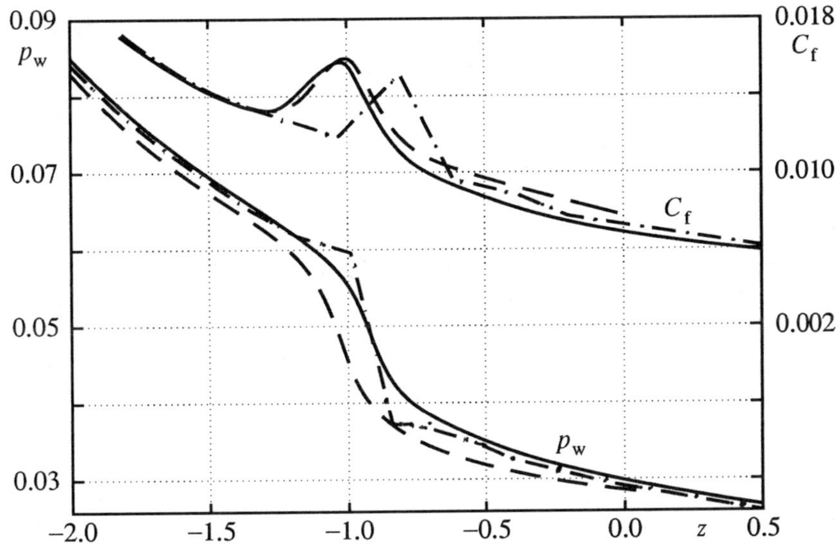

Figure 2.15 Variation of the wall pressure and skin friction coefficient in the vicinity of the body contour slope discontinuity. $M_\infty = 10$, $Re_\infty = 5000$, $T_w = 0.286T^*$, $l_1 = l_2 = 1.0$, $\beta_2 = \beta_3 = 0$. Full and dash-dotted lines, solutions of the reduced Navier–Stokes equations by the multisweep and marching methods, respectively; dashed lines, solution of the full Navier–Stokes equations by the time-asymptotic method.

Figure 2.12 displays the convergence rate of global iterations for a flow about the combined body. Here, mesh spacing in the marching coordinate $\Delta z = 0.1$, the initial pressure field was calculated using the marching method by Golovachov and Fursenko (1981b). The number of inner iterations at each marching step varied from two to six.

Figure 2.13 compares the results of the multisweep and marching calculations. Here, the wall pressure distribution is plotted in the vicinity of the body generatrix slope discontinuity. It is seen that the above solutions differ substantially only within a relatively small region near the discontinuity point.

The next two figures present a comparison with the time-asymptotic solution of the full Navier–Stokes equations calculated by Chernii (1983). Here the gas viscosity $\mu \propto T^{0.75}$. Figures 2.14 and 2.15 show the comparisons for compression and expansion flows, respectively. One can see a satisfactory agreement of the results. The discrepancy in the wall pressure values is caused by the difference of input data on the upstream boundary of the computational domain. In the case of expansion flow (figure 2.15), the dash-dotted curves depict the calculations by the marching method of Golovachov and Fursenko (1981b). This method fails in the case of compression flow.

2.3.3 Calculation of flows with extensive subsonic and recirculation zones

The effects of upstream disturbance propagation are by far more important in flows containing extensive subsonic and recirculation regions. To calculate such flows, more sophisticated versions of the global iteration method were elaborated.

For subsonic flows, Rubin (1981) suggested a global iteration method in which the pressure field was evaluated in each global iteration from the Poisson equation after determination of the other gas dynamic functions. This version of the global iteration method is similar to the procedure which is widely used in calculations of incompressible flows.

Calculating flows with recirculation zones, the convective terms of the governing equations are evaluated within the regions of reverse flow as follows:

(a) the convective terms are assumed to be equal to zero (FLARE-approximation by Reyhner and Flugge–Lotz (1968));

(b) the longitudinal velocity component $u = 0$ or $u = |u|$, see Barnett and Davis (1986);

(c) the difference approximation of the derivatives with respect to the marching coordinate is reversed, namely, the convective terms are approximated by the forward difference formula using the previous global iteration data, while the pressure gradient term is approximated by the backward difference formula (Meleshko and Chernii 1985).

The downstream boundary is usually placed outside the recirculation zone. Therefore, there is no need in the downstream boundary conditions for the velocity components. The initial velocity field can be obtained by means of marching calculation of the flow under study omitting the streamwise component of the pressure gradient.

Calculations of separated flows have proved that method (c) provides the most exact results. The methods (a) and (b) are acceptable provided the reverse flow velocity is small within the recirculation region. Barnett and Davis (1986), Ramakrishnan and Rubin (1987) stressed the significant influence of mesh spacing and difference representation of the longitudinal convective term on the feasibility and stability of the numerical methods for separated flows. The use of FLARE-approximation and a coarse computational mesh can provide similar results inconsistent with the real flow structure. For instance, the use of upwind difference formulae instead of FLARE-approximation allowed to reveal a secondary vortex within the separation bubble on the airfoil (Davis and Carter 1984). The results of the above authors indicate that the appearance of the recirculation flow within the computational domain deteriorates convergence of the global and inner iterations, which can be crucially affected by parameters of the difference approximation of the streamwise pressure gradient term and convective transport. Below, some numerical results are discussed which concern application of the global iteration method to the calculation of laminar separated flow resulting from the shock wave-boundary layer interaction (Meleshko and Chernii 1985) and separated flow about a compression ramp (Barnett and Davis 1986).

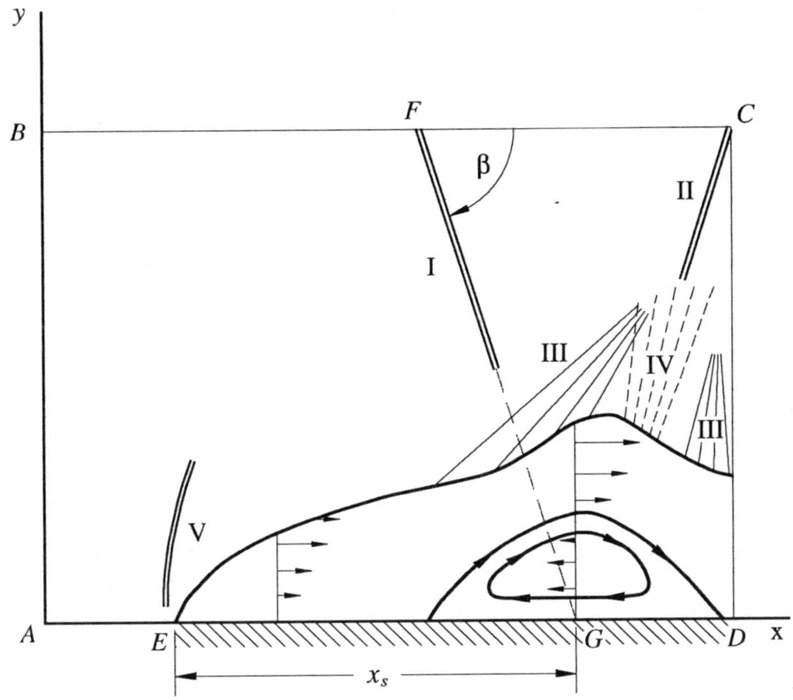

Figure 2.16 Outline of the shock wave-boundary layer interaction.

A flow induced by oblique shock interaction with the boundary layer on a flat plate immersed in a supersonic stream is schematically shown in figure 2.16. If the intensity of the incident shock I is large enough, the pressure increase results in boundary layer separation. Pressure perturbation caused by the incident shock spreads upstream through a subsonic region of the boundary layer. Therefore, the separation point is located on the left of the shock incidence point. The shift is larger for a laminar flow which is characterized by a relatively thicker subsonic region of the boundary layer. Separation and subsequent attachment of the boundary layer leads to the formation of the regions with concave streamlines in which the compression waves III are generated making the reflected shock II. The compression regions are separated by expansion flow IV. Bow shock V arising near the leading edge of the plate is also shown in figure 2.16.

Figures 2.17 and 2.18 demonstrate distributions of the pressure and skin friction coefficient over the plate. The Reynolds number Re_∞ is evaluated with the free stream parameters and the distance x_s of the shock incidence point from the leading edge. Solution of reduced equations II obtained by the global iteration method is plotted with the dashed lines. The full lines show the solution of the boundary value problem with the full Navier–Stokes equations. Detailed discussion of numerical solutions of the full Navier–Stokes equations for the problem under study will be presented in subsection 3.3.1.

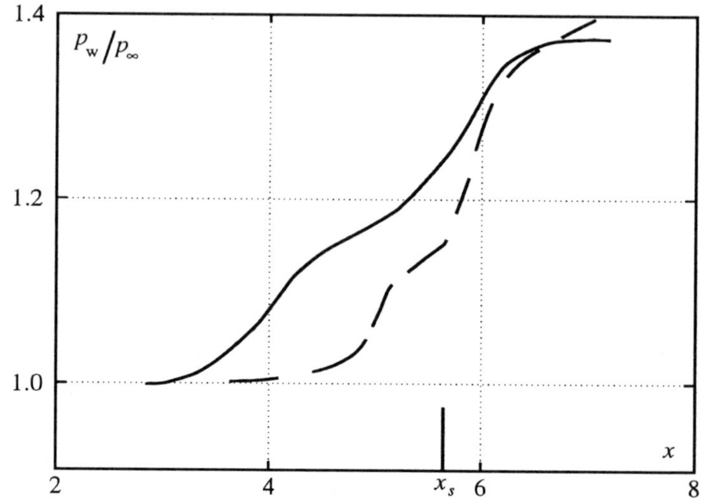

Figure 2.17 Pressure distribution over a plate in the vicinity of the oblique shock incidence point. $M_\infty = 2$, $Re_\infty = 2.96 \times 10^5$, $\beta = 33°$. Full line, solution of the full Navier–Stokes equations by the time-asymptotic method; dashed line, solution of the reduced equations by the multisweep relaxation method.

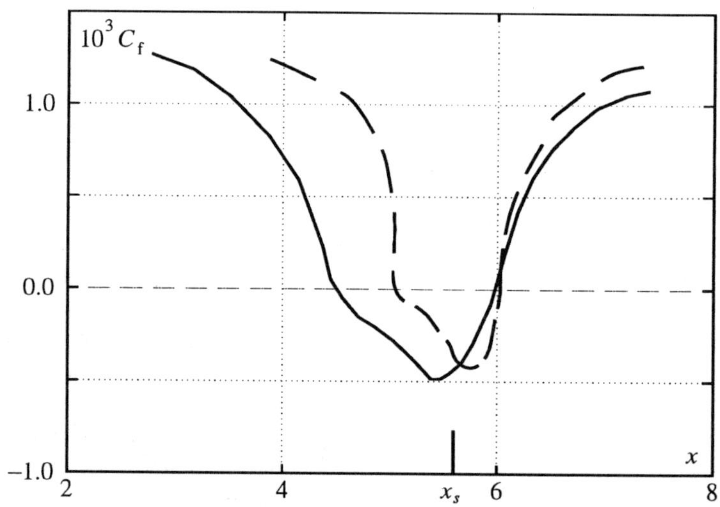

Figure 2.18 Variation of the skin friction coefficient over a plate in the vicinity of the oblique shock incidence point. $M_\infty = 2$, $Re_\infty = 2.96 \times 10^5$, $\beta = 33°$. Full line, solution of the full Navier–Stokes equations by the time-asymptotic method; dashed line, solution of the reduced equations by the multisweep relaxation method.

Figure 2.19 presents calculation results for plane laminar separated flow in the compression corner. Here x is the distance from the leading edge, the Reynolds

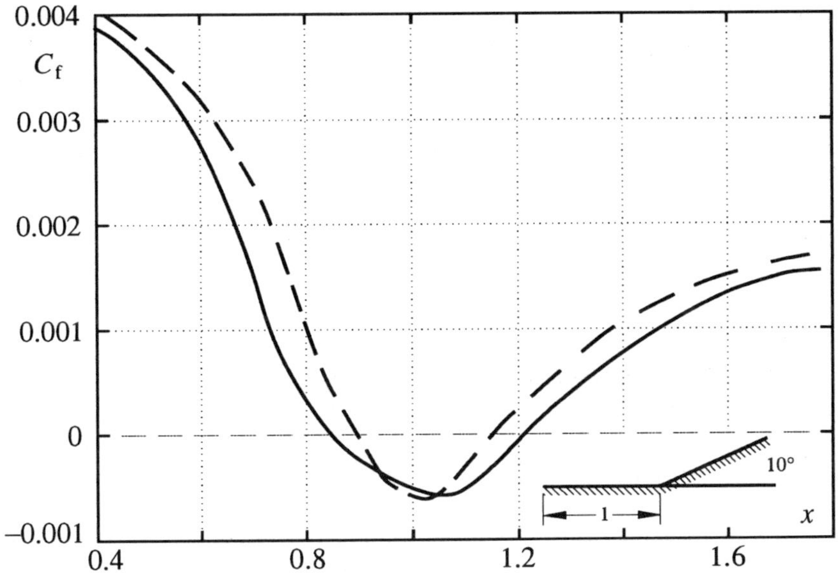

Figure 2.19 Variation of skin friction coefficient in the vicinity of compression corner apex. $M_\infty = 3$, $Re_\infty = 1.68 \times 10^4$. Full line, solution of the full Navier–Stokes equations by the time-asymptotic method; dashed line, solution of the reduced equations by the multisweep relaxation method.

number is based on the distance between the leading edge and the corner apex. In this case, about three hundred sweeps were needed to obtain the solution.

In the above examples one can see noticeable discrepancies between the solutions of the full and reduced Navier–Stokes equations. This is not surprising since the separated flows under study are characterized by a length scale equal, to an order of magnitude, to the boundary layer thickness at the beginning of the free interaction zone. Under the conditions considered, the Reynolds number based on this characteristic length does not exceed some hundreds.

The convergence rate of the global iterations is limited by the speed of the upstream information transfer over the computational grid. It can be increased by means of the use of difference schemes of higher-order accuracy with respect to the longitudinal coordinate written with the extended stencil, by the use of multigrid algorithms and by means of correction of the pressure field after each global iteration. The correction can be made in the form of a two-step calculation procedure for each global iteration which includes the consecutive marching sweeps in the forward and backward directions. Such global iteration procedures were suggested and applied by Barnett and Davis (1986), Karatayev and Koterov (1990).

As compared with the methods for solving the boundary value problem using the full Navier–Stokes equations and the space-marching methods for parabolized equations, the multisweep global iteration methods take an intermediate position with respect to both the accuracy of the flow prediction and the amount of computations. The average number of sweeps required to calculate a non-separated flow with a thin subsonic region varies from some ones to some tens. When calculating flows with a strong viscous-inviscid interaction and separation, the convergence of the global iterations comes slower and their number increases up to some hundreds. In this case, it may become necessary to store not only the pressure but the velocity field as well. For such flows, the global iteration methods lose their advantage over relaxation methods of the general type elaborated to solve the elliptic problems. Besides that, one has to keep in mind the error caused by the reduction of the Navier–Stokes equations.

2.4 THIN VISCOUS SHOCK LAYER MODEL

At hypersonic flight speed, the blunt body problem includes a small parameter δ which is the density ratio at the bow shock wave. Its value is approximately equal, for a body of revolution, to the shock layer thickness on the stagnation streamline normalized by the body nose radius. The model of a thin viscous shock layer suggested by Cheng (1961, 1963) is based on the asymptotic analysis of the problem at $\delta \to 0$, $Re \to \infty$, $\delta \cdot Re \sim \mathrm{O}(1)$.

2.4.1 Asymptotic approach to Navier–Stokes equations for hypersonic flows

Since the above papers by Cheng, an asymptotic analysis of a hypersonic viscous flow about a blunted body was carried out by many authors; see for instance Bush (1964), Magomedov (1970), Markov (1974). Here, we shall follow the papers by Cheng (1963) and Magomedov (1970) who considered the two-layer model which included the shock layer proper, i.e. the region bounded by the rear of the bow shock wave and the body surface and the region of the bow shock wave. The thickness of both regions is supposed to be small as compared with the body nose radius.

With the above assumption, the solution of the Navier–Stokes equations is searched in the form

$$\begin{aligned}
n &= \delta n_0 & u &= u_0 + \delta u_1 + \dots \\
v &= \delta(v_0 + \delta v_1 + \dots) & p &= p_0 + \delta p_1 + \dots \\
\rho &= \delta^{-1}(\rho_0 + \delta \rho_1 + \dots) & h &= h_0 + \delta h_1 + \dots
\end{aligned} \qquad (2.52)$$

Here $\delta = \rho_\infty / \rho_s$ is the density ratio at the bow shock wave, n is the distance from the body surface; u, v are the velocity components for the body-fitted curvilinear orthogonal coordinate system $(s,\ n)$; ρ, p, h are the density, pressure and specific enthalpy.

Retaining only the main terms of series expansions (2.52), one can derive the following equations for the plane ($j = 0$) and axisymmetric ($j = 1$) shock layer flows in the above limits with respect to the parameters δ and Re:

$$\begin{gathered}
\frac{\partial}{\partial s}(\rho u r^j) + \frac{\partial}{\partial n}(\rho v r^j) = 0 \\
\rho u \frac{\partial u}{\partial s} + \rho v \frac{\partial u}{\partial n} + \frac{\partial p}{\partial s} = \frac{\partial}{\partial n}\left(\mu \frac{\partial u}{\partial n}\right) \\
\frac{\partial p}{\partial n} = \frac{\rho u^2}{R} \\
\rho u \frac{\partial H}{\partial s} + \rho v \frac{\partial H}{\partial n} = \frac{1}{Pr}\frac{\partial}{\partial n}\left(\mu \frac{\partial h}{\partial n}\right) + \frac{\partial}{\partial n}\left(\mu u \frac{\partial u}{\partial n}\right)
\end{gathered} \tag{2.53}$$

Here r is the distance from the axis or plane of symmetry to the body surface, R is the body generatrix radius, $H = h + u^2/2$. The second equation of set (2.53) includes a higher-order term $\partial p/\partial s$ which becomes important in the near-wall region at high Reynolds numbers.

The thin viscous shock layer equations contain the terms of the boundary layer equations and the inviscid hypersonic shock layer flow. They present a parabolic set of partial differential equations which ought to be integrated from the stagnation line in the domain bounded by the body surface and the rear of the bow shock wave.

The function values at $s = 0$ are provided by a self-similar solution for the stagnation streamline; see subsection 2.5.2. The wall boundary conditions are formulated for the velocity components and gas temperature. They do not differ from those considered in subsection 1.1.1 for the full Navier–Stokes equations. The only remark that should be made here concerns the velocity slip and temperature jump. It seems inexpedient to take these effects into account in the thin viscous shock layer model since it does not ensure adequate description of the flow ahead of a blunted body at low Reynolds numbers.

The boundary conditions on the bow shock wave are given by the modified Rankine–Hugoniot relationships which are simplified in accordance with a hypersonic speed of the oncoming stream. Using series expansions (2.52) and assuming the whole perturbed flow region to be thin, one can obtain from (1.9)

$$\begin{gathered}
\rho v_* = -\rho_\infty V_\infty \sin\sigma \\
p = \rho_\infty V_\infty^2 \sin^2\sigma \\
\rho v_*(u - V_\infty \cos\sigma) = \mu \frac{\partial u}{\partial n} \\
\rho v_*(H - H_\infty) = \frac{\mu}{Pr}\frac{\partial h}{\partial n} + \mu u \frac{\partial u}{\partial n}
\end{gathered} \tag{2.54}$$

Here $H_\infty = h_\infty + V_\infty^2/2 \approx V_\infty^2/2$ is the total specific enthalpy of the oncoming stream, $v_* = v - \varepsilon_s' \cdot u$ is the projection of the gas velocity vector on the normal to the bow shock, $\varepsilon(s)$ is the shock layer thickness, V_∞ is the speed of the oncoming stream, σ is the slope of the bow shock to the oncoming stream direction. At high Reynolds numbers, the shock relationships neglecting the molecular transport processes can be used.

The total order of set (2.53) with respect to the n-coordinate is equal to six. Therefore, three wall boundary conditions along with four relationships on the bow shock wave are sufficient for integration across the shock layer and the location of the outer boundary of the shock layer flow.

2.4.2 Methods for integrating thin viscous shock layer equations

The problem formulated in subsection 2.4.1 can be solved by methods analogous to those which are applied to boundary layer flows. The main distinction consists in implementation of the boundary conditions on the outer boundary of the computational domain. As an example, let us consider the method suggested by Bruikina (1978, 1979), which is based on the Picard iterative procedure applied earlier by Kovach and Tirskii (1970) to solve the boundary layer equations.

Let us transform the thin viscous shock layer equations and the boundary conditions to new independent variables of Dorodnitsin's type

$$\xi = \frac{s}{R_0} \qquad \eta = \frac{1}{\Delta R_0 \rho_\infty} \int_0^n \rho dn \tag{2.55}$$

where R_0 is the body generatrix radius at the stagnation point,

$$\Delta = \frac{1}{R_0 \rho_\infty} \int_0^\varepsilon \rho dn.$$

Introduce the non-dimensional variables

$$\begin{aligned} u' &= \frac{u}{V_\infty \cos\sigma} & v' &= \frac{v}{-V_\infty \sin\sigma} & \rho' &= \frac{\rho}{\rho_\infty} \\ g &= \frac{H - H_w}{H_\infty - H_w} & \mu' &= \frac{\mu}{\mu^*} & R' &= \frac{R}{R_0} \\ r' &= \frac{r}{R_0} & p' &= \frac{p}{\rho_\infty V_\infty^2} & & \end{aligned} \tag{2.56}$$

and the reduced stream function $f(\xi, \eta)$ related to the conventional stream function ψ by the formula

$$\psi = \rho_\infty V_\infty \cos\sigma r_w^j R_0^{j+1} \Delta f(\xi, \eta)$$

The reference value of the viscosity μ^* can be evaluated at the free stream stagnation temperature T^*. Further, the primes of the non-dimensional quantities will be omitted.

Set (2.53) is written for the new variables (2.55) and (2.56) as follows

$$
\begin{aligned}
&\frac{\partial f}{\partial \eta} = u, \qquad \beta f + \xi \frac{\partial f}{\partial \xi} = -\rho(k_1 u + k_2 v),\\
&\beta_1 u^2 + \xi u \frac{\partial u}{\partial \xi} - \left(\beta f + \xi \frac{\partial f}{\partial \xi}\right)\frac{\partial u}{\partial \eta} = -\frac{\xi}{\rho \cos^2\sigma}\frac{\partial p}{\partial \xi}\\
&\qquad + \frac{\partial}{\partial \eta}\left(\frac{l\xi}{Re\Delta^2 \cos\sigma}\frac{\partial u}{\partial \eta}\right),\\
&\frac{\partial p}{\partial \eta} = \frac{\Delta \cos^2\sigma}{R} u^2,\\
&\beta_2 u(1-g) + \xi u \frac{\partial g}{\partial \xi} - \left(\beta f + \xi \frac{\partial f}{\partial \xi}\right)\frac{\partial g}{\partial \eta}\\
&= \frac{\partial}{\partial \eta}\left[\frac{l\xi}{PrRe\Delta^2 \cos\sigma}\frac{\partial}{\partial \eta}\left(g - \frac{1-Pr}{1-g_w}u^2\cos^2\sigma\right)\right]
\end{aligned}
\tag{2.57}
$$

where

$$
\begin{aligned}
&\beta = \frac{d\ln\Delta}{d\ln\xi} + \beta_w \qquad \beta_w = \xi\left(\frac{\tan\sigma}{R} + \frac{j\sin\sigma}{r_w}\right) \qquad \beta_1 = \frac{\xi\tan\sigma}{R}\\
&\beta_2 = -\frac{d\ln(1-g_w)}{d\ln\xi} \qquad g_w = \frac{h_w}{H_\infty} \qquad k_1 = \frac{s}{\rho}\frac{d\eta}{ds}\\
&k_2 = \frac{-\xi\tan\sigma}{\Delta} \qquad l = \rho\mu \qquad Re = \frac{\rho_\infty V_\infty R_0}{\mu^*}.
\end{aligned}
$$

The boundary conditions take the form:
bow shock wave

$$
\begin{aligned}
&u(\xi,1) = 1 - \frac{l}{Re\Delta\sin\sigma}\frac{\partial u}{\partial \eta}\\
&g(\xi,1) = 1 - \frac{l}{RePr\Delta\sin\sigma}\frac{\partial}{\partial \eta}\left(g - \frac{1-Pr}{1-g_w}\cos^2\sigma u^2\right)\\
&p(\xi,1) = \sin^2\sigma\\
&f(\xi,1) = \frac{r_w}{(1+j)\Delta\cos\sigma}
\end{aligned}
\tag{2.58}
$$

body surface (without mass injection)

$$
f(\xi,0) = 0, \qquad u(\xi,0) = 0, \qquad g(\xi,0) = 0 \tag{2.59}
$$

Let us integrate twice the ξ-projection of the momentum balance equation and the energy balance equation with respect to η (the first time from η to 1 and the second time from 0 to η). Taking into account the boundary conditions, it results in two integro-differential equations for the functions $u(\xi, \eta)$ and $g(\xi, \eta)$:

$$
\begin{aligned}
u &= \Delta^2 \frac{\cos\sigma}{\xi} \left[\beta_w A(u) + \beta_1 B(u) + C(u)\right] + \Delta \frac{d\Delta}{d\xi} \cos\sigma A(u), \\
g &- \frac{1 - Pr}{1 - g_w} \cos^2\sigma u^2 = \Delta^2 \frac{\cos\sigma}{\xi} \\
&\times \left[\beta_w A(g) + \beta_2 B(g) + C(g)\right] + \Delta \frac{d\Delta}{d\xi} \cos\sigma A(g),
\end{aligned}
\tag{2.60}
$$

where

$$
\begin{aligned}
A(u) &= Re \int_0^\eta \left[(1-u)f + \theta(u)\right] l^{-1} d\eta \\
B(u) &= Re \int_0^\eta \theta_p l^{-1} d\eta \\
C(u) &= Re\xi \int_0^\eta \left[(1-u)\frac{\partial f}{\partial \xi} + \frac{\partial}{\partial \xi}\theta(u)\right] l^{-1} d\eta \\
A(g) &= Re \int_0^\eta \left[(1-g)f + \theta(g)\right] l^{-1} Pr d\eta \\
B(g) &= Re \int_0^\eta \theta(g) l^{-1} Pr d\eta \\
C(g) &= Re\xi \int_0^\eta \left[(1-g)\frac{\partial f}{\partial \xi} + \frac{\partial}{\partial \xi}\theta(g)\right] l^{-1} Pr d\eta \\
\theta(u) &= \int_\eta^1 u(1-u) d\eta \\
\theta(g) &= \int_\eta^1 u(1-g) d\eta \\
\theta_p &= -\int_\eta^1 \left(u^2 + \frac{\xi}{\beta_1 \cos^2\sigma} \frac{1}{\rho} \frac{\partial p}{\partial \xi}\right) d\eta \\
f &= \int_0^\eta u d\eta \\
\Delta &= \frac{r_w}{(1+j)\cos\sigma f(\xi, 1)}.
\end{aligned}
$$

The simplest way for evaluating the functions $u(\xi, \eta)$ and $g(\xi, \eta)$ is the use of formulae (2.60) whose right-hand sides are calculated using the previous iteration data. To increase the convergence rate, Bruikina (1978) suggested another iteration procedure ensuring a more accurate implementation of the bow shock boundary conditions in the current iteration. It looks as follows

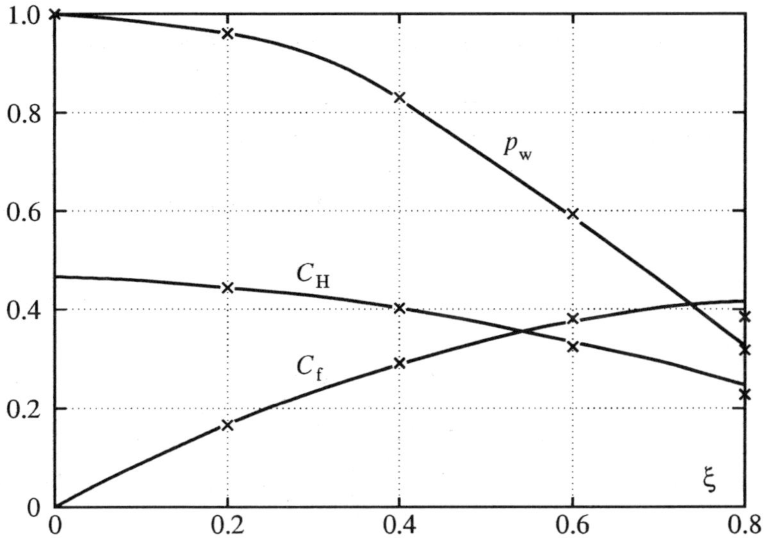

Figure 2.20 Variation of pressure, skin friction and heat transfer coefficients over the sphere surface in a hypersonic stream. $Re = 20$, $T_w = 0.015T^*$.

$$\begin{aligned} u^{\nu+1} &= \left(\Delta_u^{\nu+1}\right)^2 \frac{\cos\sigma^\nu}{\xi} \left[\beta_w A(u^\nu) + \beta_1 B(u^\nu) + C(u^\nu)\right] \\ &\quad + \Delta_u^{\nu+1} \frac{d\Delta_u^{\nu+1}}{d\xi} \cos\sigma^\nu A(u^\nu), \\ g^{\nu+1} &- \frac{1-Pr}{1-g_w} \cos^2\sigma^\nu \left(u^{\nu+1}\right)^2 = \left(\Delta_H^{\nu+1}\right)^2 \frac{\cos\sigma^\nu}{\xi} \\ &\times \left[\beta_w A(g^\nu) + \beta_2 B(g^\nu) + C(g^\nu)\right] + \Delta_H^{\nu+1} \frac{d\Delta_H^{\nu+1}}{d\xi} \cos\sigma^\nu A(g^\nu) \end{aligned} \tag{2.61}$$

where ν is the iteration number. The functions $\Delta_u^{\nu+1}(\xi)$ and $\Delta_H^{\nu+1}(\xi)$ are the solutions of the ordinary differential equations resulting from substitution of $u^{\nu+1}$ and $g^{\nu+1}$ into the first two relationships of (2.58).

The functions $u^{\nu+1}$, $g^{\nu+1}$ being known, the functions $f^{\nu+1}(\xi,\eta)$, $\Delta^{\nu+1}(\xi)$, $p^{\nu+1}(\xi,\eta)$, $\rho^{\nu+1}(\xi,\eta)$ and $l^{\nu+1}(\xi,\eta)$ can be easily evaluated. The pressure is determined by the formula

$$p = \sin^2\sigma - \Delta \frac{cos^2\sigma}{R} \int_\eta^1 u^2 d\eta \tag{2.62}$$

The density is obtained from the equation of state (1.5). Later on, the shape of the bow shock is specified and one can proceed to the next iteration.

The shock layer calculation starts from the linear profiles of unknown functions which satisfy the boundary conditions on the bow shock and the body surface. With such initial data, the first iteration can be obtained in an analytical form.

The calculations show that for Reynolds numbers $Re \lesssim 100$ the subsequent iterations make the solution more accurate by only a few per cent.

Figure 2.20 presents some calculation results pertaining to hypersonic flow of a perfect monatomic gas ($\gamma = 5/3$, $Pr = 0.75$) about a sphere. The skin friction coefficient is defined here as the shear stress-to-$\rho_\infty V_\infty^2/2$ ratio. The lines and the crosses show the results obtained by the above method and the finite-difference scheme of Ankudinov (1970), respectively.

2.4.3 Comments on the validity and accuracy of the thin viscous shock layer approach

When formulating the thin viscous shock layer model, one assumes the thickness of the perturbed flow ahead of the body to be small as compared to the body characteristic length. This assumption becomes invalid with decreasing Mach and Reynolds numbers. To clear up the limit of applicability of the thin viscous shock layer model to the calculation of rarefied gas flows, Jain and Adimurthy (1974) compared the solutions of the thin viscous shock layer equations and the full Navier–Stokes equations for a hypersonic stagnation line flow ahead of a sphere. Both sets of equations were solved by the shock-capturing finite-difference method subject to the velocity slip and temperature jump wall boundary conditions. Numerical solutions were compared also with the results provided by kinetic models and measurements. Comparisons have shown that the range of validity of the thin viscous shock layer model is limited by the Reynolds numbers $Re_s \approx 50$. At $Re_s \lesssim 50$, this model leads to a significant error which is especially large at high wall temperature.

The above conclusion is illustrated in figure 2.21 by numerical results of Jain and Adimurthy (1974). The figure shows variation of pressure at the stagnation point of a heat-insulated sphere with Reynolds number based on the free stream parameters, sphere radius and gas viscosity at the free stream stagnation temperature. Here the gas specific heat ratio $\gamma = 1.67$, the Prandtl number $Pr = 0.75$.

In spite of a significant error in prediction of the flow field structure at low Reynolds numbers, the thin viscous shock layer model provides at the same time quite accurate heat flux values on the body surface. Figure 2.22 shows variation of the heat transfer coefficient at the stagnation point of a sphere with a rarefaction parameter K^2 suggested by Cheng (1963). The full and dashed lines depict solutions of the full Navier–Stokes equations and thin viscous shock layer equations, respectively. Here

$$C_H = \frac{q}{\rho_\infty V_\infty (h^* - h_w)} \qquad K^2 = \sqrt{\frac{\gamma - 1}{\gamma + 1}} Re_* \left(\frac{T_*}{T^*}\right),$$

$$Re_* = \frac{\rho_\infty V_\infty R_0}{\mu_*}, \qquad \mu_* = \mu(T_*)$$

$$T_* = \frac{1}{2}(T^* + T_w)$$

The value $C_H = 1$ corresponds to a free-molecular limit with the body surface accommodation coefficient equal to unity. The numerical solutions were obtained

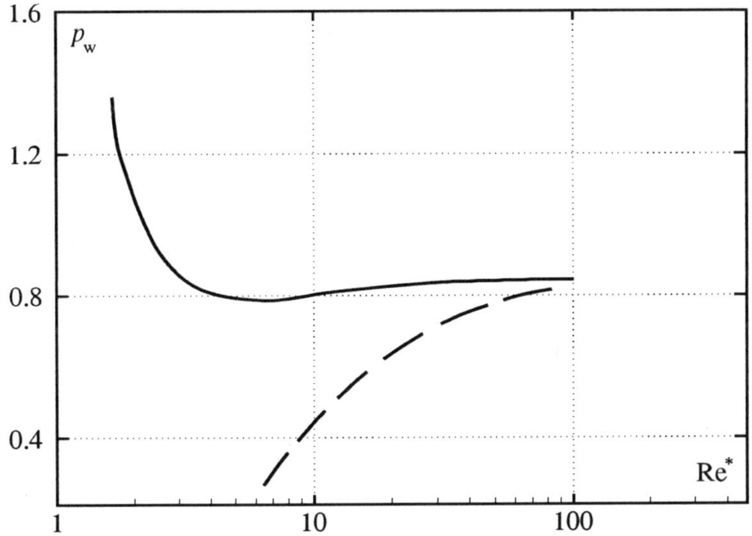

Figure 2.21 Variation of gas pressure at the stagnation point of a heat-insulated sphere with Reynolds number. $M_\infty = 10$. Full line, the full Navier–Stokes equations; dashed line, the thin viscous shock layer equations.

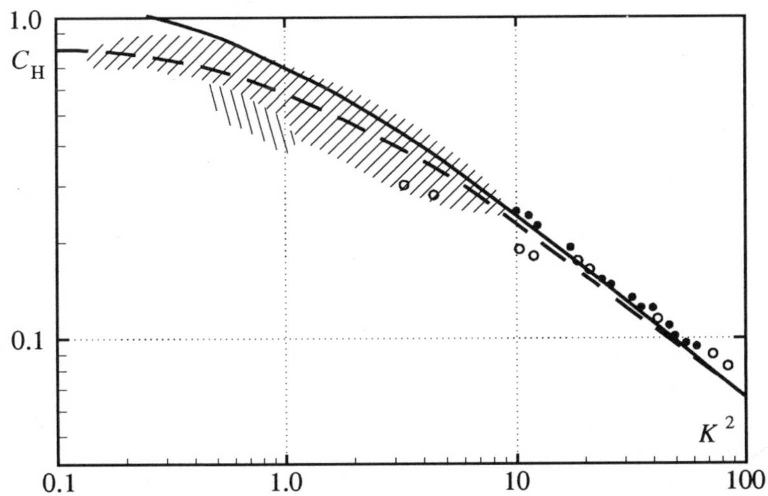

Figure 2.22 Variation of the heat transfer coefficient at the stagnation point of a sphere with rarefaction parameter.

at $M_\infty = 10$, $\gamma = 1.33$, $T_w = 0.1T^*$. With increasing rarefaction parameter K^2, the function $C_H(K^2)$ turns to be linear in accordance with the boundary layer theory. The hatched areas present the experimental data of Vidal and Wittliff (1963) and Boylan (1971). The bright and filled circles are the measurements by Wilson

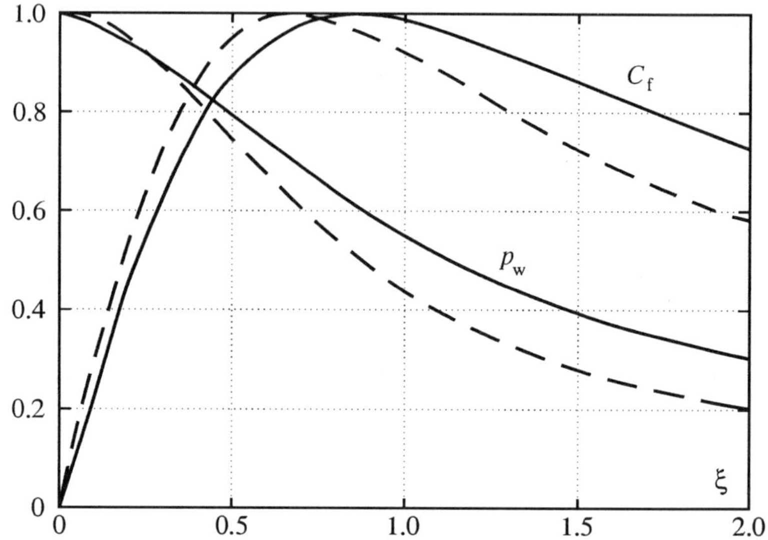

Figure 2.23 Distribution of wall pressure and skin friction coefficient over a paraboloid of revolution. $M_\infty = 20$, $Re = 10^3$, $T_w = 0.03T^*$. Full lines, full Navier–Stokes equations; dashed lines, thin viscous shock layer model.

and Wittliff (1962), Ferry and Zakkay (1962). The measurements were carried out at Mach numbers $8 \leqslant M_\infty \leqslant 24$ and wall temperature $0.1 \leqslant T_w/T^* \leqslant 0.46$.

An essential drawback of the thin viscous shock layer model concerns the allowable body shape. It results from the use of a simplified balance equation for the normal momentum component, in which pressure variation is determined only by the centrifugal force. For this reason, applying the thin viscous shock layer model to the flow about a sphere one obtains at $s \approx R$ unreal results, namely zero wall pressure and separation of the shock layer from the body surface. For bodies with smoothly diminishing generatrix curvature, such as a hyperboloid or paraboloid, this model works at a large distance from the stagnation point.

Figure 2.23 compares the solutions of the thin viscous shock layer and the full Navier–Stokes equations for a hypersonic flow of a perfect mono-atomic gas about a paraboloid of revolution. The results of the thin viscous shock layer model are borrowed from (Ankudinov 1973), the solution of the full Navier–Stokes equations was obtained by the method presented in subsection 1.2.2. The pressure and skin friction coefficient are normalized by their maximums. The Reynolds number is evaluated using gas viscosity at the free stream stagnation temperature. It is seen that the thin viscous shock layer model results in a large error for the distribution of the flow characteristics outside the vicinity of the stagnation point. The error can be diminished by the use of exact formulae for the Lamé coefficients (Peigin and Tirskii 1988).

2.5 SELF-SIMILAR SOLUTIONS

This section deals with shock layer flows with inherent symmetry and similarity properties which allow one to reduce the number of the independent variables.

2.5.1 Solution of Navier–Stokes equations on the stagnation streamline

To investigate the flow in the vicinity of the stagnation streamline, unknown functions are presented by series expansions with respect to the parameter characterizing the distance from this line. Substituting the expansions into the Navier–Stokes equations, one can obtain, for a stationary flow, the sets of ordinary differential equations for the expansion coefficients. Because of the elliptic nature of the problem, each set contains the next expansion coefficients. A procedure for series truncation has been suggested by Kao (1964). Following this paper, let us use the trigonometric expansions

$$\begin{array}{ll|l}
u(s,n) = u_1(n)\sin(s/R) & +u_2(n)\sin^3(s/R) + \dots \\
v(s,n) = v_0(n)\cos(s/R) & +v_2(n)\cos(s/R)\sin^2(s/R) + \dots \\
T(s,n) = T_0(n) & +T_2(n)\sin^2(s/R) + \dots \\
\mu(s,n) = \mu_0(n) & +\mu_2(n)\sin^2(s/R) + \dots \\
\rho(s,n) = \rho_0(n) & +\rho_2(n)\sin^2(s/R) + \dots \\
p(s,n) = p_0(n) + p_2(n)\sin^2(s/R) & +p_4(n)\sin^4(s/R) + \dots
\end{array} \tag{2.63}$$

where R is the body nose radius. Substituting (2.63) into the Navier–Stokes equations (1.1)–(1.4) and retaining only the terms written in (2.63) on the left of the vertical bar, one obtains the following set of equations for the expansion coefficients

$$(\rho v)' + 2\rho W = 0 \tag{2.64}$$

$$\rho v v' + p' - \frac{4}{3}(\mu v')' - \frac{2}{3}\frac{\mu}{r+n}(u' + 4v' - 7W) + \frac{4}{3}W\mu' = 0 \tag{2.65}$$

$$\begin{aligned} \rho v u' + \rho u W + 2\frac{p_2}{r+n} - \mu\left[u'' + \frac{1}{r+n}\left(2u' - \frac{8}{3}W - \frac{1}{3}v'\right)\right] \\ - (u' - W)\mu' = 0 \end{aligned} \tag{2.66}$$

$$\rho v h' - v p' - \frac{4}{3}\mu(v' - W)^2 - (\lambda T')' - 2\frac{\lambda}{r+n}T' = 0 \tag{2.67}$$

$$p' + p_2' - \rho u W = 0 \tag{2.68}$$

Here $W = (u+v)/(R+n)$, all expansion coefficients are written without subscripts, except for p_2. Equation (2.68) is derived from the balance equation for the normal momentum component (1.2) by equating the terms proportional to $\sin^2(s/R)$.

Estimating the order of magnitude of the terms responsible for the molecular momentum and energy transfer with respect to parameter $\delta = 1/\sqrt{Re}$ (see subsection 2.1.1), one can obtain from (2.64)–(2.68) the reduced Navier–Stokes equations for high Reynolds number flows. Assuming the density ratio at the bow shock to be large and the extent of the perturbed flow to be small, one obtains the reduced equations of a hypersonic flow (Levinsky and Yoshihara 1962). For hypersonic flow about a spherical bluntness, variation of the gas dynamic functions along the shock layer is well described by series expansions (2.63) up to $s/R \approx 25°$.

The boundary conditions for the full and reduced equations are formulated as follows. The wall boundary conditions are given by the conventional relationships for the velocity components and gas temperature. The boundary conditions on the outer edge of the computational domain located in the free stream are

$$u = V_\infty, \quad v = -V_\infty, \quad T = T_\infty, \quad p = p_\infty, \quad p_2 = 0 \tag{2.69}$$

where V_∞ is the modulus of the free stream velocity vector. If the computational domain is bounded by the bow shock, the outer boundary conditions can be obtained by substitution of expansions (2.63) into modified Rankine–Hugoniot relationships (1.9). As a result, one obtains

$$\begin{aligned}
\rho v &= \rho_\infty v_\infty \\
\rho v(v - v_\infty) + p - p_\infty &= \frac{4}{3}\mu v' \\
\rho v(u - u_\infty) &= \mu u' \\
\rho v\left(h + \frac{v^2}{2} - h_\infty - \frac{v_\infty^2}{2}\right) &= \lambda T' + \frac{4}{3}\mu v v' \\
p + p_2 &= p_\infty
\end{aligned} \tag{2.70}$$

When using boundary conditions (2.70), one has to know a priori the bow shock shape which determines the value of u_∞ (the first expansion coefficient for the free stream velocity). For hypersonic flows about smooth blunt bodies, the bow shock near the stagnation streamline can be assumed to be concentric with the body, i.e. $u_\infty = V_\infty$. This becomes inappropriate for strongly blunted, for instance, flat-nosed bodies, or at small supersonic speeds of the oncoming stream.

Numerical solutions of the above boundary value problems have been obtained by many authors (Levinsky and Yoshihara 1962, Kao 1964, Jain and Adimurthy 1974, Gershbein and Kolesnikov 1979, Golovachov 1979c). At first, let us consider the results of Jain and Adimurthy (1974) who obtained solutions of the full Navier–Stokes equations including the region of the bow shock wave. The calculation results for heat transfer at the stagnation point of a sphere are shown by the full curve in figure 2.22. At $K^2 \lesssim 0.3$, the heat transfer coefficient provided by the Navier–Stokes equations slightly exceeds the free-molecular limit $C_H = 1$. This is not surprising if one takes into account that these flow regimes are beyond the limits of validity of the continuous medium approach, see subsection 1.3.2. Figure 2.24 presents the temperature profile on the stagnation streamline ahead of a sphere, which is compared with the experimental data of Ahouse and Bogdonoff (1969). The comparison demonstrates a good agreement between the self-similar

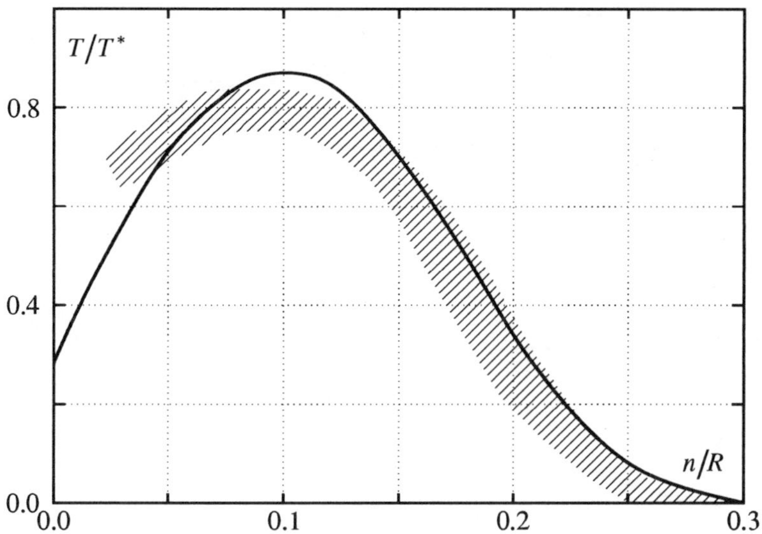

Figure 2.24 Temperature profile on the stagnation streamline ahead of a sphere. $M_\infty = 20.8$, $\gamma = 1.4$, $Re_\infty = 2782$, $T_w = 0.19T^*$. Full line, self-similar solution of the Navier–Stokes equations (Jain and Adimurthy 1974); hatched area, experimental data of Ahouse and Bogdonoff (1969).

solution of the Navier–Stokes equations and measurements at hypersonic flight speed.

Golovachov (1979c) compared the self-similar solutions with the solutions of the two-dimensional problem for a flow about the whole front of a sphere ($s \leqslant \pi/2$). In both cases the computational domain was bounded by the bow shock. Some results are presented in figures 2.25 and 2.26. They pertain to a flow of diatomic gas with specific heat ratio $\gamma = 1.4$, viscosity $\mu \propto \sqrt{T}$, Prandtl number $Pr = 0.71$. The full lines are the profiles provided by solution of the two-dimensional problem. The dashed and dash-dotted curves depict the self-similar solutions for the stagnation streamline. The dashed curves show the results obtained with the exact value of u_∞ borrowed from the solution of the two-dimensional problem; the dash-dotted curves show the results calculated with $u_\infty = V_\infty$, which means the bow shock concentricity with the body surface.

It is seen that the error of a self-similar solution manifests itself mainly in underestimating the shock layer thickness. With decreasing Mach number, the error increases. It is also seen that the accuracy of a self-similar solution can be improved by using the exact data on the bow shock shape. The calculations were carried out for Reynolds numbers $Re_\infty \geqslant 10^2$. The accuracy of the self-similar solutions was found to be only slightly dependent on this parameter. Table 2.1 presents the non-dimensional heat flux values at the stagnation point of a sphere. The rows 1 through 3 contain the results obtained from a solution of the two-dimensional problem, from a self-similar solution with an exact value of u_∞ and

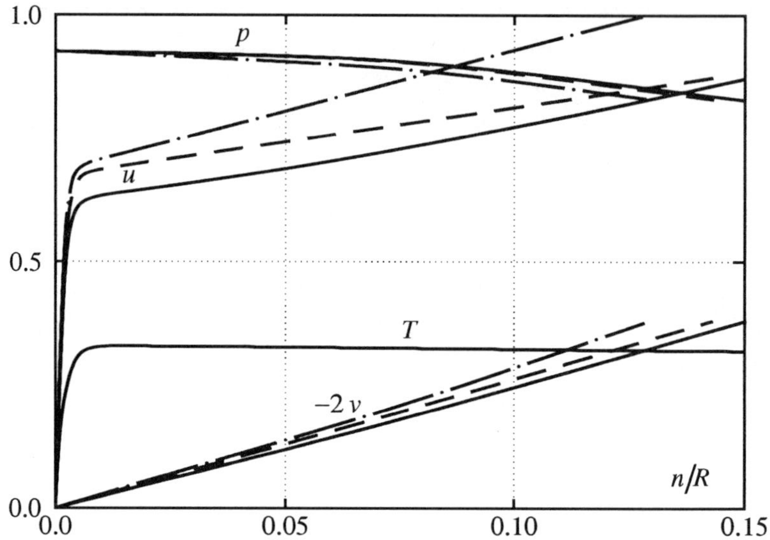

Figure 2.25 Profiles of gas dynamic functions on the stagnation streamline ahead of a sphere. $M_\infty = 6$, $Re_\infty = 71000$, $T_w = 0.35T^*$. Full lines, solution of two-dimensional problem; dashed and dash-dotted lines, self-similar solutions.

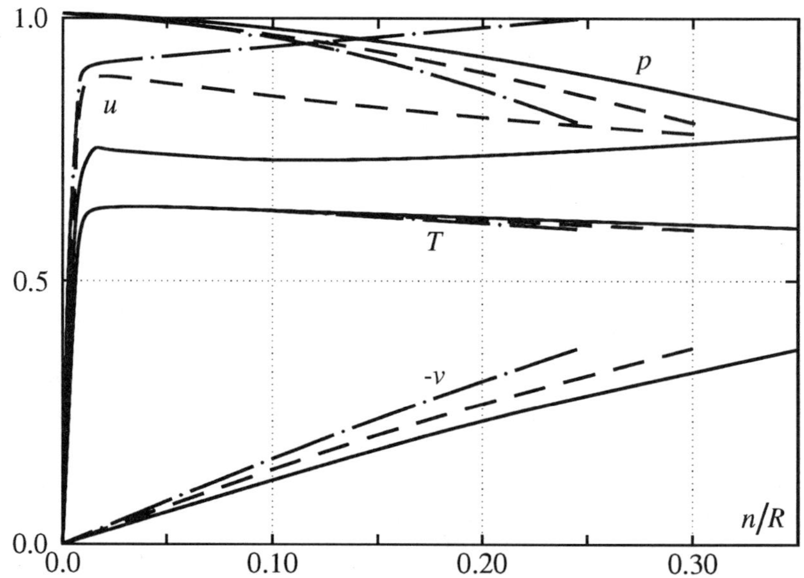

Figure 2.26 Profiles of gas dynamic functions on the stagnation streamline ahead of a sphere. $M_\infty = 2$, $Re_\infty = 11830$, $T_w = 0.35T^*$. Full lines, solution of two-dimensional problem; dashed and dash-dotted lines, self-similar solutions.

Table 2.1

	$M_\infty = 6, Re_\infty = 71000$	$M_\infty = 6, Re_\infty = 100$	$M_\infty = 2, Re_\infty = 11830$
1	0.00447	0.140	0.0120
2	0.00458	0.146	0.0132
3	0.00463	0.156	0.0132

from a self-similar solution with $u_\infty = V_\infty$, respectively.

2.5.2 Solution of thin viscous shock layer equations on the stagnation streamline

Using the two-layer model of Cheng (1963) to calculate a hypersonic stagnation line flow, one can neglect the viscous energy dissipation and assume the pressure to be constant along this streamline. With these simplifications, the set of governing equations for an axisymmetric flow is written as follows

$$\begin{aligned} \frac{\partial}{\partial s}(\rho u s) + \frac{\partial}{\partial n}(\rho v s) &= 0 \\ \rho u \frac{\partial u}{\partial s} + \rho v \frac{\partial u}{\partial n} + \frac{dp}{ds} &= \frac{\partial}{\partial n}\left(\mu \frac{\partial u}{\partial n}\right) \\ \rho u \frac{\partial h}{\partial s} + \rho v \frac{\partial h}{\partial n} &= \frac{\partial}{\partial n}\left(\lambda \frac{\partial T}{\partial n}\right) \end{aligned} \tag{2.71}$$

The bow shock is taken to be concentric with the body surface, the wall pressure distribution is approximated by the formula

$$p(s) = \rho_\infty V_\infty^2 \left(1 - \frac{\rho_\infty}{\rho_s}\right)(1 - s^2/R^2) \tag{2.72}$$

Introducing the stream function ψ in accordance with the continuity equation and using Dorodnitsin's independent variables, one can rewrite the momentum and energy balance equations as follows

$$\begin{aligned} (lf'')' + ff'' - \frac{1}{2}(f')^2 + \frac{\rho_\infty}{\rho}\left(1 - \frac{\rho_\infty}{\rho_s}\right) &= 0 \\ \left(\frac{l}{Pr} g'\right)' + fg' &= 0 \end{aligned} \tag{2.73}$$

where

$$f = \frac{\psi}{\sqrt{2\xi}} \qquad f' = \frac{u}{u_s} \qquad l = \frac{\rho\mu}{\rho_w \mu_w}$$

$$g = \frac{h}{h_s} \qquad u_s = \frac{sV_\infty}{R} \qquad \xi = \int_0^s \rho_w \mu_w u_s s^2 ds$$

$$\eta = \frac{s u_s}{\sqrt{2\xi}} \int_0^n \rho dn$$

the prime denotes differentiation with respect to η, subscripts s, w denote the quantities on the bow shock and body surface, respectively.

Normalizing the η-coordinate by η_s, denoting $\varphi = f'$ and exploiting the wall impermeability condition, one can present equations (2.73) in the form

$$\begin{gathered} \frac{1}{\eta_s^2}(l\varphi')' + f\varphi' - \frac{\varphi^2}{2} + \frac{\rho_\infty}{\rho}\left(1 - \frac{\rho_\infty}{\rho_s}\right) = 0 \\ \frac{1}{\eta_s^2}\left(\frac{l}{Pr}g'\right)' + fg' = 0 \end{gathered} \tag{2.74}$$

where $f = \int_0^\zeta \varphi d\zeta$; the prime denotes differentiation with respect to $\zeta = \eta/\eta_s$.

The boundary conditions on the body surface read

$$\varphi(0) = 0, \qquad g(0) = g_w \tag{2.75}$$

The modified Rankine–Hugoniot relationships take the form

$$l\varphi' + \eta_s^2 f(1)(\varphi - 1) = 0, \qquad \frac{l}{Pr}g' + \eta_s^2 f(1)(g - 1) = 0 \tag{2.76}$$

where

$$\eta_s = \frac{\rho_\infty}{f(1)}\sqrt{\frac{V_\infty R}{2\rho_w \mu_w}}$$

Numerical solution of the above boundary value problem can be obtained by the following finite-difference method (Golovachov 1972):

Equations (2.74) and boundary conditions (2.75) and (2.76) are approximated on a uniform computational grid

$$\zeta_i = i\Delta\zeta, \qquad i = 0,\ 1, \ldots,\ M; \qquad \Delta\zeta = 1/M$$

with the second-order accurate difference formulae

$$\begin{aligned} (mz')'_i &\sim \frac{m_{i+1/2}(z_{i+1} - z_i) - m_{i-1/2}(z_i - z_{i-1})}{(\Delta\zeta)^2} \\ (z')_i &\sim \frac{z_{i+1} - z_{i-1}}{2\Delta\zeta} \\ (z')_M &\sim \frac{3z_M - 4z_{M-1} + z_{M-2}}{2\Delta\zeta} \end{aligned}$$

where $m = l,\ l/Pr$; $z = \varphi, g$. The set of difference equations is solved by iterations. Evaluating the quantities $\rho,\ l,\ f,\ \eta_s$ from the previous iteration data, one obtains the set of linear difference equations for each of the functions $\varphi(\zeta)$ and $g(\zeta)$ in the current iteration. These sets have tridiagonal matrices of coefficients and are solved by the 'progonka' algorithm. The above method has been successfully applied to the calculation of hypersonic viscous shock layer flows in a wide range of flight conditions.

Assuming additional simplifications, one can obtain the solution of the thin viscous shock layer equations in an analytical form. Following Magomedov (1970), let us consider hypersonic flow about a sphere at low and moderate Reynolds numbers when the pressure gradient term of the momentum balance equation (the second equation of (2.71)) can be neglected. The governing equations and boundary conditions are written for non-dimensional variables with the use of R, V_∞, ρ_∞, $\rho_\infty V_\infty^2$, V_∞^2 and μ_s as the reference values of length, velocity, density, pressure, enthalpy and viscosity. Using Dorodnitsin's independent variable $\eta = \int_0^n \rho dn$ and the following presentation of the unknown functions

$$u = u_1(\eta)\sin s \qquad \rho v + u\eta'_s = -v_0(\eta)\cos s$$
$$p = p_0(\eta) + p_2(\eta)\sin^2 s \qquad h = h_0(\eta) + h_2(\eta)\sin^2 s$$
$$l = \mu\rho = l_0(\eta)\cos s$$

one obtains a set of ordinary differential equations

$$\begin{aligned} &2u_1 - v_0' = 0 \qquad u_1^2 - v_0 u_1' = m(l_0 u_1')' \qquad p_0' = 0 \\ &p_2' = u_1^2 \qquad -v_0 h_0' = \frac{m}{Pr}(l_0 h_0')' \end{aligned} \tag{2.77}$$

Here $m = 1/\delta Re$, $\delta = \rho_\infty/\rho_s$, $Re = \rho_\infty V_\infty R/\mu_s$.

The boundary conditions are

$$\eta = 0: \qquad u_1 = v_0 = 0 \qquad h_0 = h_w \tag{2.78}$$

$$\begin{aligned} \eta = \eta_s: \qquad & u_1 + m l_0 u_1' = 1 \qquad p_0 = 1 \\ & h_0 + m l_0 Pr^{-1} h_0' = H_\infty \qquad p_2 = -1 \\ & v_0 = 1 \end{aligned} \tag{2.79}$$

where H_∞ is total enthalpy of the free stream.

Then one assumes $l_0 = l_0(h_*) = const$ with $h_* = 1/2(h_s + h_w)$. This assumption is well appropriate for the adiabatic wall case. In the case of a cooled wall, one can try to improve the solution by means of another choice of h_*.

With the above assumptions, the solution of problem (2.77)–(2.79) can be written in the following analytical form

$$\begin{aligned} &u_1 = \eta/\Delta^2 \qquad v_0 = \eta^2/\Delta^2 \qquad p_0 = 1 \\ &p_2 = -1 - \frac{1}{3\Delta}\left(1 - \frac{\eta^3}{\Delta^3}\right) \\ &\frac{h_0 - h_w}{H_\infty - h_w} = \frac{\gamma^{1/3}(t)}{\left[\gamma^{1/3}(\omega) + \omega^{-2/3}e^{-\omega}\right]}, \\ &\Delta = \eta_s = \frac{1}{2}\left(1 + \sqrt{1 + 4ml_0}\right) \end{aligned} \tag{2.80}$$

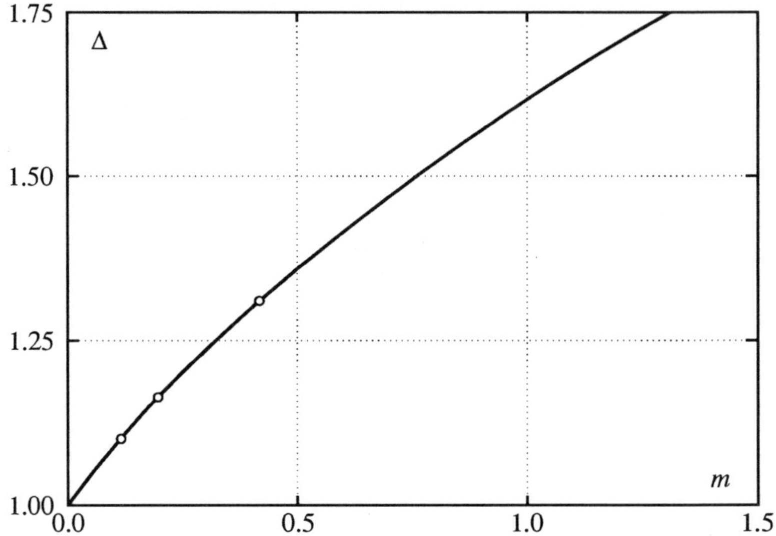

Figure 2.27 Variation of the bow shock coordinate on the stagnation streamline ahead of an adiabatic sphere with rarefaction parameter. Line, analytical solution of the thin viscous shock layer equations (Magomedov 1970); circles, numerical solution of the full viscous shock layer equations (Belotserkovskii *et al* 1967).

Here $t = \omega(\eta/\Delta)^3$, $\omega = Pr/3ml_0$, $\gamma^{1/3}(t)$ is the incomplete γ-function of order 1/3.

The accuracy of the analytical solution is illustrated in figure 2.27 which presents a comparison with the numerical solution of the full viscous shock layer equations (Belotserkovskii *et al* 1967). One can see an excellent agreement of the results in spite of a fairly large value of the parameter $\delta = 0.17$.

The bow shock distance from the cooled wall ought to be specified in accordance with the formula

$$\varepsilon = \frac{\Delta}{H_\infty}\left[h_w + (H_\infty - h_w)f_\varepsilon\right] \tag{2.81}$$

where the bow shock distance ε is normalized by the product δR,

$$f_\varepsilon = \frac{3\omega I_2(\omega)e^\omega}{1 + 3\omega I_1(\omega)e^\omega}$$

$$I_1(\omega) = \frac{\gamma^{1/3}(\omega)}{3\omega^{1/3}} = \sum_{n=0}^{\infty}\frac{(-1)^n}{n!}\frac{\omega^n}{1+3n}$$

$$I_2(\omega) = \frac{1}{3}\left[\frac{\gamma^{1/3}(\omega)}{\omega^{1/3}} - \frac{\gamma^{2/3}(\omega)}{\omega^{2/3}}\right] = \sum_{n=0}^{\infty}\frac{(-1)^n}{n!}\frac{\omega^n}{(1+3n)(2+3n)}$$

To evaluate the functions $I_1(\omega)$ and $I_2(\omega)$, Magomedov (1970) suggested the approximate formulae

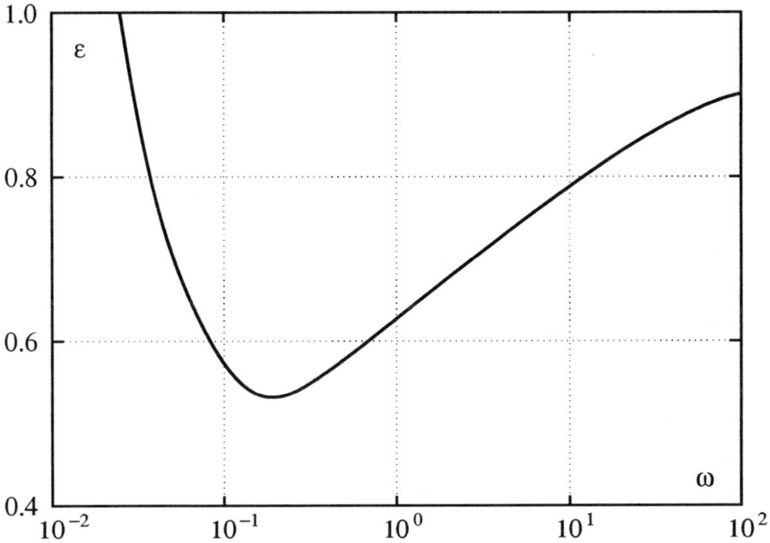

Figure 2.28 Variation of the shock layer thickness on the stagnation streamline ahead of a cooled sphere with the rarefaction parameter. Analytical solution of the thin viscous shock layer equations (Magomedov 1970).

$$\left.\begin{aligned} I_1(\omega) &= 1 - \frac{1}{4}\omega + \frac{1}{14}\omega^2 - \frac{1}{60}\omega^3 + \frac{1}{312}\omega^4 - \frac{1}{1920}\omega^5 + \dots \\ I_2(\omega) &= \frac{1}{2} - \frac{1}{20}\omega + \frac{1}{112}\omega^2 - \frac{1}{660}\omega^3 + \frac{1}{4508}\omega^4 - \dots \end{aligned}\right\} \omega < 2$$

$$\left.\begin{aligned} I_1(\omega) &= \frac{1}{3}\omega^{-1/3}\left(2.68 - \omega^{-2/3}e^{-\omega}\right) \\ I_2(\omega) &= \frac{1}{3}\omega^{-1/3}\left(2.68 - 1.35\omega^{-2/3}\right) \end{aligned}\right\} \omega \geqslant 2$$

Figure 2.28 presents the dependence of the bow shock coordinate on the parameter ω at $h_w = 0.05h^*$. Non-monotonous variation of the shock layer thickness with the rarefaction parameter agrees with numerical solutions of the full Navier–Stokes equations at low and moderate Reynolds numbers (Belotserkovskii *et al* 1974).

The above analytical solution yields the following formulae for the pressure, stagnation point skin friction and heat transfer coefficients

$$\begin{aligned} C_p &= \frac{p_w - p_\infty}{1/2\rho_\infty V_\infty^2} = 2 - \left(2 + \frac{2}{3\Delta}\right)\sin^2 s \\ C_f &= \frac{\tau_w}{1/2\rho_\infty V_\infty^2} = \frac{2ml_0}{\Delta^2} \\ C_H &= \frac{q}{\rho_\infty V_\infty (H_\infty - h_w)} = \frac{e^\omega}{1 + 3\omega I_1(\omega)e^{-\omega}} \end{aligned} \tag{2.82}$$

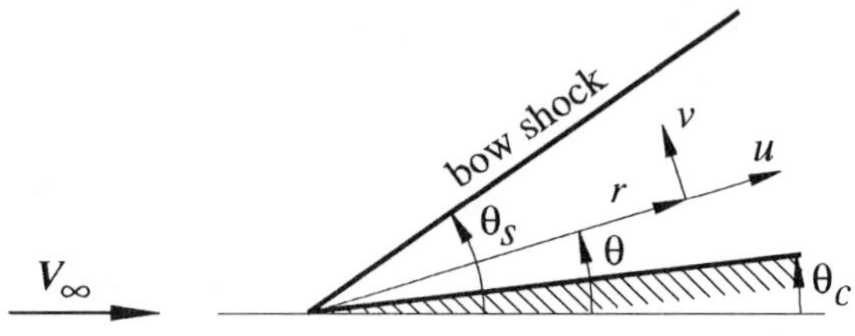

Figure 2.29 Coordinate system for a supersonic flow about a pointed circular cone.

In the limit $m \to \infty$, $\omega \to 0$, formulae (2.82) provide the values consistent with those for the free molecular flow regime. Certainly, it should not be considered as an evidence for applicability of the thin viscous shock layer model to the calculation of free molecular flows since the range of validity of the Navier–Stokes equations is bounded by the Reynolds number $Re_s \approx 10$; see subsection 1.3.2.

2.5.3 Local conical approximation

This approach is applied in calculations of stationary supersonic flows about pointed bodies with an attached bow shock wave. Within the framework of local conical approximation, all gas dynamic functions are supposed to remain constant along the rays drawn from the apex of the body between its surface and the bow shock wave. In the case of an axisymmetric flow, the gas parameters are supposed to be invariable on the conic surfaces co-axial with the body. This property which ensures an existence of the self-similar solutions inheres in the inviscid gas flows around sharp cones referred to as the conical flows.

In the flows around bodies with a curvilinear generatrix, for instance ogival bodies, the slope of the bow shock changes with a distance from the body apex. It violates conical behaviour of the shock layer flow. In this case, the local self-similar solutions apply which are obtained assuming the conical flow behaviour in accordance with the local slope of the bow shock or its local distance from the body surface.

The viscous shock layer flows are not the conical ones even for the conical bodies. The conical behaviour is voilated since the boundary layer thickness increases with a distance from the body apex in a non-linear way. Deviation from the linear law is especially large for the laminar boundary layer whose thickness is proportional to the square root of the distance from the cone apex. Therefore, the concept of a conical behaviour for the viscous shock layer flows is an approximation of which the accuracy depends on the boundary layer flow regime determined by the local value of the Reynolds number.

It is often convenient to obtain the local conical solutions using the spherical coordinate system with the pole located at the body apex. This coordinate system is shown in figure 2.29 for a flow about a circular cone.

Assuming the shock layer flow to be conical, one ought to omit all the terms of the governing equations which contain derivatives with respect to the radial

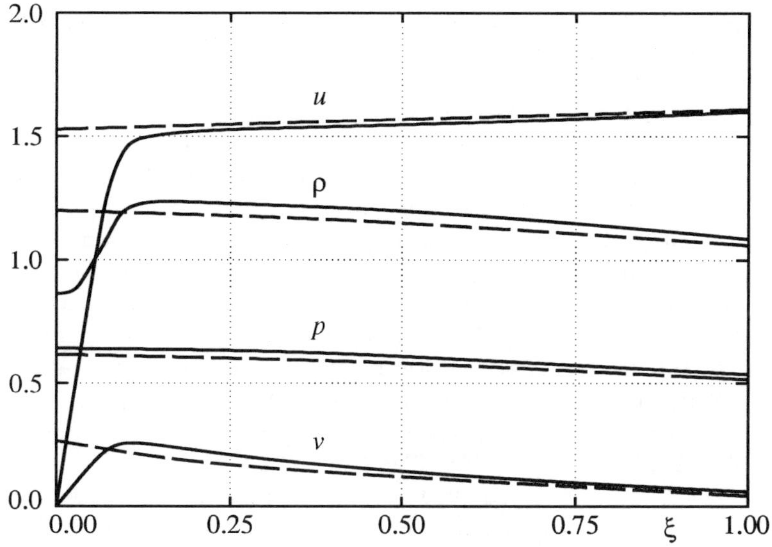

Figure 2.30 Profiles of gas dynamic functions in the shock layer of a pointed circular cone. $M_\infty = 2$, $Re_\infty = 3778$, $\theta_c = 10°$, $T_w = 0.55T^*$. Full lines, Navier–Stokes equations; dashed lines, inviscid flow model.

coordinate. As a result, the full Navier–Stokes equations are reduced to the following set of ordinary differential equations (the prime denotes differentiation in the θ-coordinate):

$$
\begin{aligned}
& v\rho' + \rho(v' + 2u + v\cot\theta) = 0 \\
& \rho v(u' - v) - \frac{1}{r}\left[\mu(u' - v)\right]' \\
& \qquad + \frac{\mu}{r}\left(\frac{4}{3}v' + \frac{8}{3}u + \frac{7}{3}v\cot\theta - u'\cot\theta\right) = 0 \\
& \rho v(v' + u) + p' - \frac{2}{r}\left\{\frac{1}{3}\left[\mu(2v' + u - v\cot\theta)\right]'\right. \\
& \qquad \left. + \mu\left[u' - v + (v' - v\cot\theta)\cot\theta\right]\right\} = 0 \\
& \rho v h' - v p' - \frac{\mu}{r}\left\{2\left[(v' + u)^2 + (u + v\cot\theta)^2\right]\right. \\
& \qquad \left. + (v - u')^2 - \frac{2}{3}(2u + v' + v\cot\theta)^2\right\} \\
& \qquad - \frac{1}{r}\left[(\lambda T')' + \lambda T'\cot\theta\right] = 0
\end{aligned}
\tag{2.83}
$$

When calculating high Reynolds number flows, one can use the reduced equations which can be obtained from (2.83) by estimating the order of magnitude of the viscous terms with respect to parameter $\delta = 1/\sqrt{Re}$; see subsection 2.1.1.

In the local conical approach, the equations for an axisymmetric viscous shock layer flow are solved within the segment $\theta_c \leqslant \theta \leqslant \theta_s$, where θ_c is the cone half-angle, θ_s is the slope of the bow shock. The conventional boundary conditions are imposed on the body surface and the bow shock wave.

Numerical solution of the above non-linear boundary value problem does not meet any difficulties. It can be obtained by various methods including, in particular, the space-marching schemes of subsection 2.2.2. In this case, the mesh spacing with respect to the marching coordinate is used as an iteration parameter.

Figure 2.30 illustrates an application of the local conical approximation to the calculation of supersonic axisymmetric flow about a sharp circular cone. Here the gas dynamic function profiles are plotted which were obtained using the full Navier–Stokes equations. The results are compared with the inviscid flow calculations of Babenko *et al* (1964). The calculations were carried out using the cylindrical coordinate system aligned with the cone axis. In the figure, ξ is the radial coordinate normalized by the shock layer thickness, u and v are the axial and radial velocity components which are normalized by the critical sound speed c_*, the pressure and density are normalized by $\rho_\infty c_*^2$ and ρ_∞, respectively. The Reynolds number Re_∞ is evaluated using the free stream parameters and the distance from the cone apex.

E. N. S. H. ..I. G.
BIBLIOTHÈQUE

3

Applications to plane and axisymmetric flows

The examples of calculations in the preceding chapters were largely aimed at demonstrating the accuracy and range of validity of the viscous shock layer models. Most of them pertained to stationary flows about the bodies of revolution placed in a uniform supersonic stream. This chapter deals with simulation of more complicated viscous shock layer flows.

3.1 FLOW AROUND A BLUNTED BODY IN THE AERODYNAMIC WAKE

A supersonic flow about a couple of bodies moving one after another is accompanied by the formation of a bow shock wave ahead of the rear body provided it is far enough from the front one. Available data (Eremeitsev *et al* 1988) indicate a possibility of two types of shock layer flows ahead of the rear body: flow with recirculation zones and non-separated flow. The flows of the first type have to be investigated using the full Navier–Stokes equations while non-separated flows can be investigated within the framework of more simple viscous shock layer models.

3.1.1 Parameters of the oncoming stream

The supersonic far wake of a body of revolution at zero incidence can be considered as a parallel axisymmetric stream with constant static pressure, an axial velocity minimum and a temperature maximum. Velocity and temperature variations across the wake are approximated by formulae (2.1) which involve three parameters characterizing the above axial extrema and relative width of the wake. These parameters can be evaluated using the formulae

$$a = 2.5C_x^{1/3}l^{-2/3}, \qquad b = 5.027C_x^{-2/3}l^{-2/3}\left(\frac{d_2}{d_1}\right)^2$$
$$c = \frac{(1-a)^2(\gamma-1)PrM_\infty^2}{(2-a)\left[1+(\gamma-1)PrM_\infty^2 a\right]} \tag{3.1}$$

L.I.G.H.E.G. BIBLIOTHÈQUE

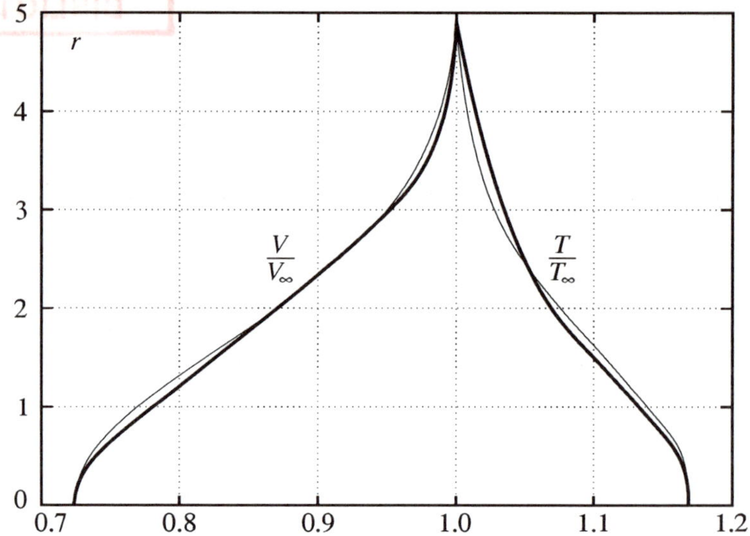

Figure 3.1 Velocity and temperature profiles in the far wake behind a spherically blunted cylinder. $M_\infty = 3$, $\gamma = 1.4$, $Re_\infty = 200$, $T_w = 0.257T^*$. Full lines, numerical solution; hair lines, formulae (2.1).

derived by Eremeitsev *et al* (1988) from asymptotic analysis of a supersonic far wake. Here C_x is the drag coefficient of the front body, l is the distance between the bodies; d_1, d_2 are the diameters of the middle cross-sections of the front and rear bodies; M_∞ is the Mach number of the uniform oncoming stream, other notations are generally adopted.

The parameters of formulae (2.1) can be obtained as well from comparison with available calculations or experimental data on the wake flows. As an example, figure 3.1 demonstrates the velocity and temperature profiles behind a spherically blunted cylinder with $L = 3.6R_c$, where L, R_c are the length and radius of the cylinder. The profiles pertain to the wake cross-section located at a distance $l = 50R_c$ from the cylinder base. The Reynolds number Re_∞ is evaluated using the cylinder radius. The full lines depict the results of wake flow calculation (Kokoshinskaya *et al* 1980), the hair lines are plotted with the use of formulae (2.1) with parameters determined from a comparison with the above numerical solution: $a = 0.277$, $b = 0.199$, $c = 0.163$.

3.1.2 Flows with recirculation zones

Below, we discuss numerical solutions for the stationary axisymmetric flows about a sphere whose centre is located on the wake axis (Golovachov and Leontieva 1985, Golovachov 1991, Golovachov and Zemlyakov 1993b). The gas properties are the following: the specific heat ratio $\gamma = 1.4$, the viscosity $\mu \propto \sqrt{T}$, the Prandtl number $Pr = 0.7$.

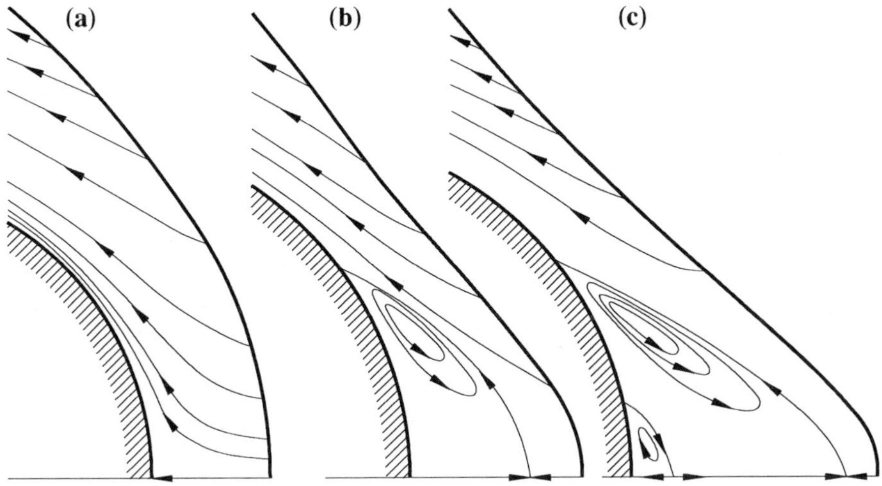

Figure 3.2 Supersonic flow about a sphere in the wake of a blunted cylinder. $M_\infty = 3$, $\gamma = 1.4$, $Re_\infty = 200$, $T_w = 0.257T^*$; (a) $l = 70$, $R = 2.5$; (b) $l = 70$, $R = 10$; (c) $l = 30$, $R = 10$.

The laminar flows were calculated with the full Navier–Stokes equations using the unidirectional method presented in subsection 1.2.2. Most calculations were carried out using the uniform computational grid with mesh spacing $\Delta s = 2.5°$, $\Delta\xi = 0.02$ where ξ is the distance from the sphere surface normalized by the shock layer thickness. To test the accuracy, some calculations were carried out with double and half the number of grid points in each spatial coordinate. The mass conservation law was also exploited to check the accuracy of numerical solutions. The total mass disbalance was found to be less than one per cent. The results presented below demonstrate the dependence of the shock layer structure, drag and heat transfer parameters on the geometric factors, Reynolds number and wall temperature.

Figures 3.2 and 3.3 show the calculations of laminar flows ahead of a heat-insulated sphere located in the wake of a spherically blunted cylinder. The free stream parameters and cylinder dimensions are the same as those in subsection 3.1.1. The parameters of formulae (2.1) providing the distributions of gas dynamic functions in the wake were determined from a comparison with numerical solutions by Kokoshinskaya *et al* (1980).

Figure 3.2 demonstrates a dependence of the flow structure on the geometric parameters of the couple of bodies. It depicts the bow shock and the streamlines within the shock layer for several sphere radii R and distances between the bodies l normalized by the cylinder radius R_c. At $l = 70$ and $R = 2.5$, the shock layer flow is non-separated. With increasing sphere radius, a recirculating flow arises. The same happens when the sphere approaches the cylinder. At $l = 30$, $R = 10$ one can see a secondary vortex.

Figure 3.3 shows variations of the stagnation point pressure p_w, bow shock

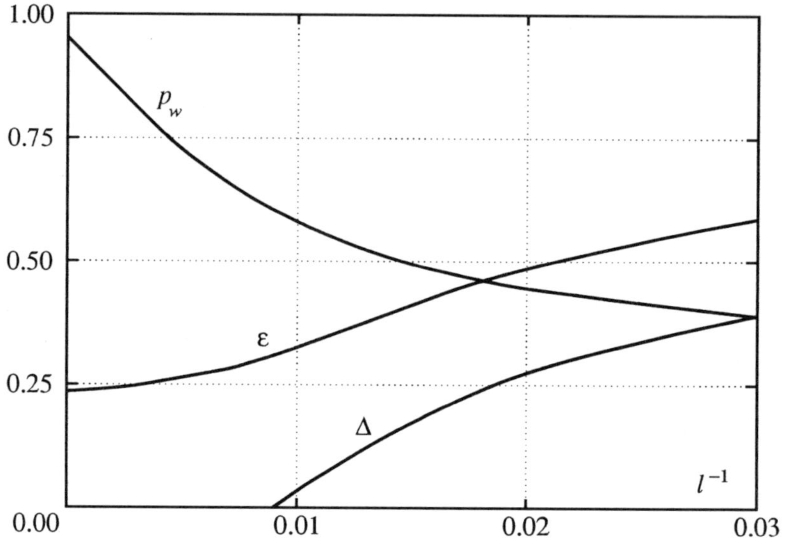

Figure 3.3 Variation of characteristics of supersonic flow ahead of a sphere with a distance from the cylinder. $M_\infty = 3$, $\gamma = 1.4$, $Re_\infty = 200$, $T_w = 0.257T^*$, $R = 5$.

detachment ε and thickness Δ of the recirculation zone on the symmetry axis with a distance between the bodies. The shock layer and recirculation zone thicknesses are normalized by the sphere radius, the wall pressure is normalized by $\rho_\infty V_\infty^2$. In this case, the recirculating flow arises at $l \approx 110$ and the secondary vortex does not appear.

The next figures pertain to the flow about a sphere located in the wake-type stream with parameters $b = 7.2$, $c = 3.0$ at Mach number on the wake axis $M_0 = 6$. The density and velocity values on the wake axis are chosen as the reference ones when normalizing the calculation results. The Reynolds number Re_0 is based on the sphere radius. It is varied by means of gas density variation.

Figure 3.4 demonstrates the Reynolds number effect on distribution of the heat transfer parameter over the sphere in a uniform ($a = 0$) and non-uniform ($a = 0.04$) oncoming stream. In the latter case, the Reynolds number increase results in a drastic reduction of the heat transfer parameter in the vicinity of the stagnation point, which is caused by the formation of a recirculating flow within the shock layer. The rise of recirculating flows can also be stimulated by cooling the body surface.

Figure 3.5 displays the Reynolds number and body surface cooling effects on the wall pressure distribution. The full lines depict the results for an adiabatic wall, the hair line pertains to the case of a fixed wall temperature. It is seen that the maximum pressure is achieved downstream from the stagnation point. With increasing Reynolds number, the wall pressure decreases and the maximum point shifts downstream. The body cooling effects the wall pressure distribution in an opposite manner.

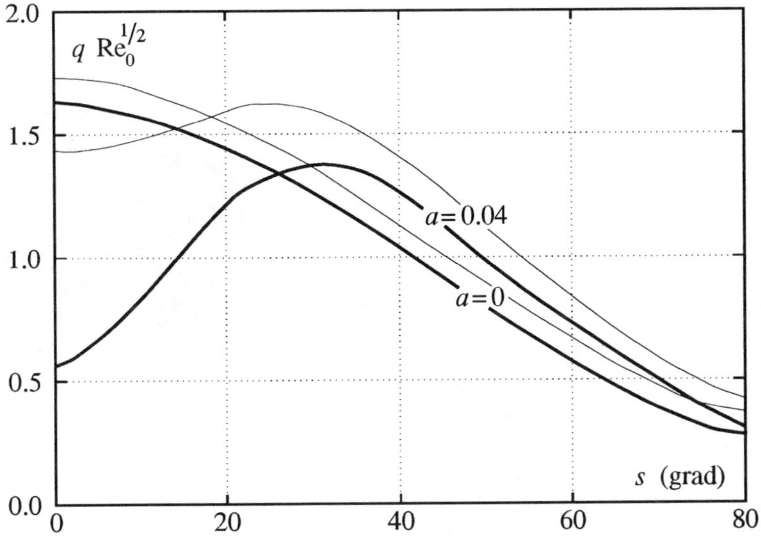

Figure 3.4 Variation of the heat transfer parameter over a sphere surface. $M_0 = 6$, $T_w = 0.15T_0^*$, $b = 7.2$, $c = 3.0$. Full lines, $Re_0 = 3550$; hair lines, $Re_0 = 177$.

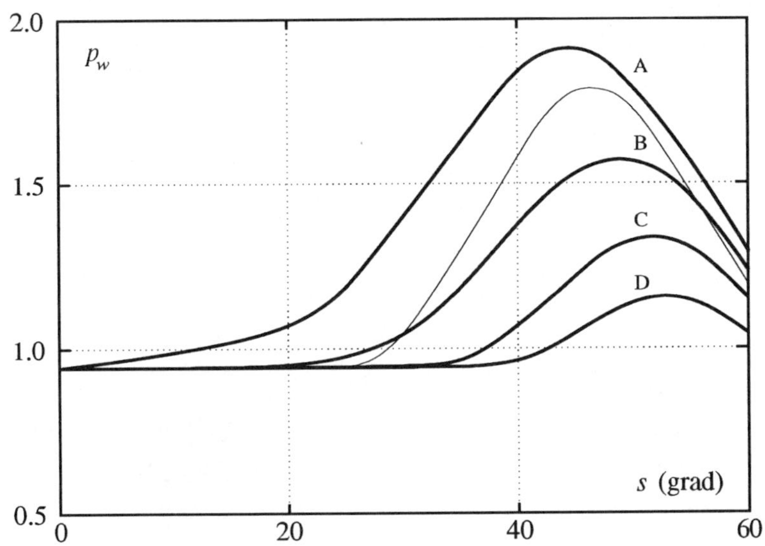

Figure 3.5 Pressure distribution over a sphere surface. $M_0 = 6$, $a = 0.1$, $b = 7.2$, $c = 3.0$. Curves A–D, $Re_0 = 177$, 533, 1770 and 5000, respectively; full lines, adiabatic wall; hair line B, $T_w = 0.3T_0^*$.

In the calculation of turbulent shock layer flows, the distributions of gas dynamic functions of the oncoming stream were obtained with formulae (2.1) and (3.1).

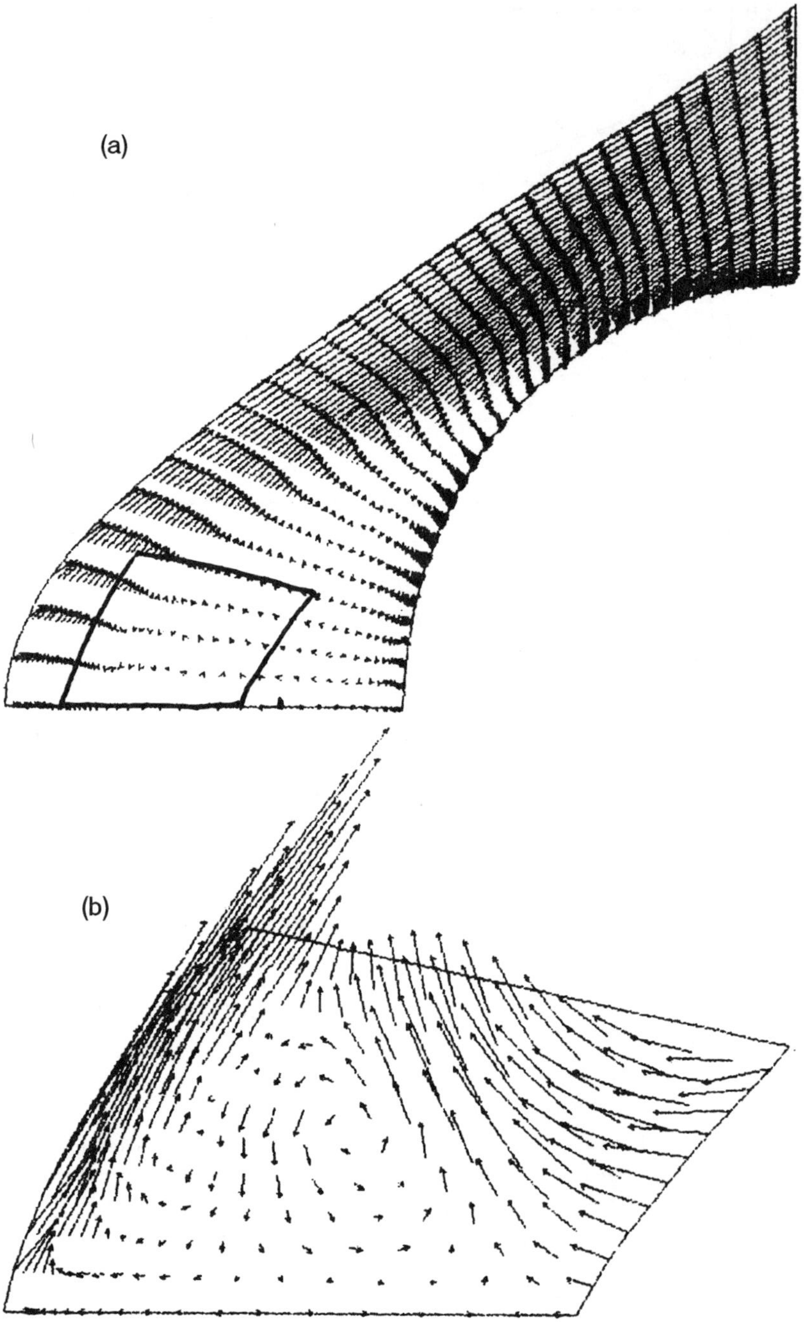

Figure 3.6 Velocity map of a turbulent shock layer flow ahead of a sphere in supersonic wake. $M_\infty = 3$, $Re_\infty = 7.6 \times 10^5$, $l = 20$.

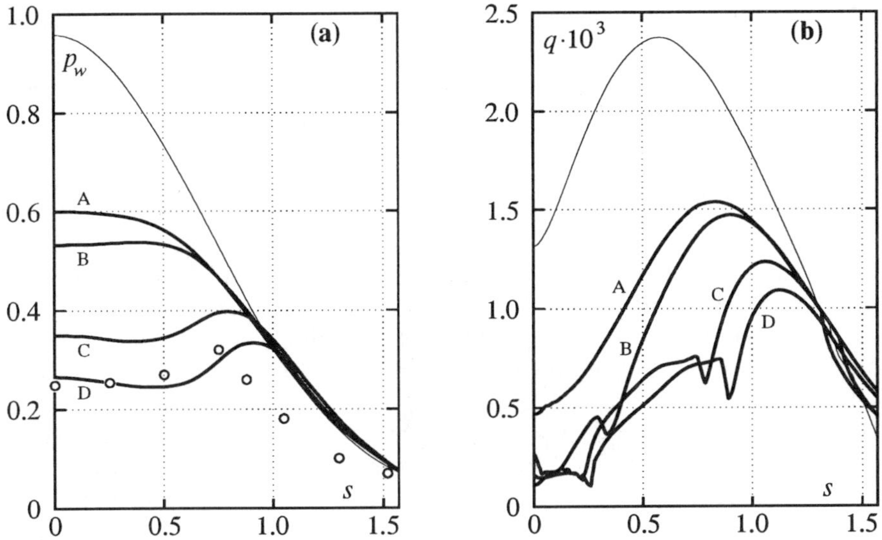

Figure 3.7 Pressure (a) and heat flux (b) distributions over a sphere in supersonic wake at various distances between the bodies. $M_\infty = 3$, $Re_\infty = 7.6 \times 10^5$. A, $l = 100$; B, $l = 50$; C, $l = 30$; D, $l = 20$; hair lines, uniform oncoming stream.

The calculations were carried out with the reduced Reynolds equations which differed from set II of subsection 2.1.1 only by the use of the effective transport coefficients. The gas viscosity was evaluated with Sutherland's formula for air. The eddy viscosity was obtained from the two-layer algebraic turbulence model of Baldwin and Lomax (1978) using the coordinates of the global maximum of vorticity function on the lines $s = const$ as an appropriate length scale. The intermittancy factor was evaluated with the formulae of Chen and Thyson (1971) supposing the laminar-turbulent transition to start from the stagnation streamline. The solutions were obtained using the unidirectional implicit finite-difference scheme with exponential approximation of the second-order equations with respect to the transversal coordinate (Belousov *et al* 1991, Golovachov and Zemlyakov 1994). Calculations were carried out with the adaptive grids generated using the equidistribution law (1.50); see subsection 1.2.3. An example of a computational grid was presented in figure 1.7.

The results presented in figures 3.6 and 3.7 pertain to turbulent flow about the front of a sphere located in the wake of another sphere. The Reynolds number Re_∞ is based on the radius of the rear body R_2, its wall temperature $T_w = 0.4T^*$, the sphere radius ratio $R_2/R_1 = 3.2$. Figure 3.6 displays the shock layer velocity map (a) and a fragment (b) at $l = 20$. The distance between the bodies l is normalized by radius R_2. One can see a rather complicated flow structure which contains several recirculation zones.

Figure 3.7 shows pressure and heat flux distributions over the rear sphere at various distances between the bodies. The above quantities are normalized by

$\rho_\infty V_\infty^2$ and $10^{-3} \times \rho_\infty V_\infty^3$, respectively. Variation of the above distributions with a distance l correlates with the variation of the shock layer structure and with the data of wind tunnel measurements presented by Eremeitsev *et al* (1988) for $l = 5.6$ which are shown in figure 3.7(a) by the circles.

3.1.3 Non-separated flows

The oncoming stream non-uniformity noticeably affects the shock layer flow field in a non-separated flow regime as well. To calculate such flows, one can use simpler gas dynamic models instead of the full Navier–Stokes equations. The results discussed below have been obtained by Pilugin and Talipov (1986) and Eremeitsev *et al* (1987) within the framework of the thin viscous shock layer model.

At first, consider an analytical solution for the vicinity of the stagnation point of a blunt body, which has been obtained by analogy with the self-similar solution of Magomedov (1970), see subsection 2.5.2. Restricting the analysis to the flows at low and moderate Reynolds numbers, the momentum balance equation of set (2.71) is written without the pressure gradient term. The product $l = \rho\mu$ is supposed to be constant.

With the use of the non-dimensional functions and Dorodnitsin's variables, the solution is searched in the form of a power series expansion with respect to the distance from the stagnation point. Substitution into the governing equations and equating the terms with the same power index results in a set of ordinary differential equations (2.77). The boundary conditions on the body surface are given by formulae (2.78). Taking into account the oncoming stream non-uniformity, the boundary conditions on the bow shock ($\eta = \eta_s$) are written as follows:

$$\begin{aligned} v_0 &= B(1-a), \qquad u_1 = 1 - a - \frac{ml_0}{B(1-a)}u_1', \\ p_0 &= B(1-a)^2, \qquad p_2 = B(1-a)^2\left[\frac{2ab(1+c)}{1-a} - 1\right], \\ h_0 &= H - \frac{ml_0}{B(1-a)Pr}h_0' \end{aligned} \tag{3.2}$$

Here

$$B = 1 + c\left[1 - (1-a)^{-2}\right] \qquad H = \frac{h_\infty}{B} + \frac{(1-a)^2}{2}$$

a, b, c are the non-uniformity parameters of formulae (2.1), H is the non-dimensional total gas enthalpy on the oncoming stream axis, other notations and reference values of the gas dynamic functions are the same as those in subsection (2.5.2). Subscript ∞ denotes the oncoming stream parameters at $r \to \infty$.

To solve the problem, one has to prescribe the bow shock shape near the stagnation streamline. Boundary conditions (3.2) are written assuming the bow shock to be concentric with the body surface, which is quite admissible at hypersonic flight speed, see subsection 2.5.1. It seems even more appropriate in the case of a non-uniform wake-type oncoming stream since the non-uniformity increases the bow shock curvature in the vicinity of the stagnation streamline.

Solution of set (2.77) subject to boundary conditions (2.78) and (3.2) is given by the following formulae

$$
\begin{gathered}
v_0 = \frac{B(1-a)}{\Delta^2}\eta^2, \quad u_1 = \frac{B(1-a)}{\Delta^2}\eta, \quad p_0 = B(1-a)^2 \\
p_2 = \frac{B^2(1-a)^2}{3\Delta}\left(\frac{\eta^3}{\Delta^3} - 1\right) + B(1-a)^2\left[\frac{2ab(1+c)}{1-a} - 1\right] \\
\frac{h_0 - h_w}{h_\infty/B + \frac{(1-a)^2}{2} - h_w} = \frac{\gamma^{1/3}(t)}{\gamma^{1/3}(\omega) + \omega^{-2/3}e^{-\omega}} \\
\Delta = \eta_s = \frac{B}{2}\left(1 + \sqrt{1 + \frac{4ml_0}{B^2(1-a)}}\right)
\end{gathered}
\tag{3.3}
$$

where

$$
t = \omega\eta^3/\Delta^3, \quad \omega = \frac{Pr \cdot B(1-a)}{3ml_0}, \quad \gamma^{1/3}(t) = \sum_{n=0}^{\infty} \frac{(-1)^n}{n!}\frac{t^{n+1/3}}{n+1/3}
$$

Using the above analytical solution, Pilugin and Talipov (1986) derived the following formulae for the shock layer thickness on the flow symmetry axis, the heat transfer and skin friction coefficients at the stagnation point of a sphere and the wall pressure coefficient:

$$
\varepsilon/R = \int_0^\Delta \frac{d\eta}{\rho} = \frac{2\Delta}{p_0}\left[h_w + (H - h_w)f_\varepsilon\right] \tag{3.4}
$$

$$
C_H = \frac{2q}{\rho_\infty V_\infty^3} = \frac{2(H - h_w)e^\omega B(1-a)}{1 + 3\omega I_1(\omega)e^{-\omega}} \tag{3.5}
$$

$$
C_f = \frac{2\tau_w}{\rho_\infty V_\infty^2} = \frac{2ml_0 B(1-a)}{\Delta^2} \tag{3.6}
$$

$$
C_p = \frac{2p_w}{\rho_\infty V_\infty^2} = 2B(1-a)^2\left\{1 - \left[\frac{B}{3\Delta} + 1 - \frac{2ab(1+c)}{1-a}\right]\sin^2 s\right\} \tag{3.7}
$$

Functions $f_\varepsilon(\omega)$, $I_1(\omega)$ are presented in subsection 2.5.2 along with the approximate formulae which apply at small and large argument values. For a uniform oncoming stream ($a = 0$, $B = 1$), formulae (3.3)–(3.7) turn into those obtained by Magomedov (1970).

Formulae (3.4) and (3.5) indicate an increase of the shock layer thickness and a decrease of the heat transfer coefficient with increasing oncoming stream non-uniformity, which agrees with available numerical and experimental data.

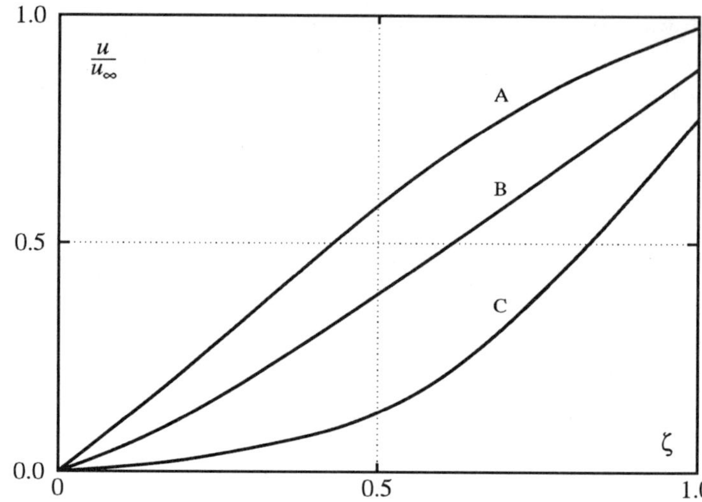

Figure 3.8 Velocity profiles on the stagnation streamline in a hypersonic wake flow about a sphere. $Re = 100$, $b = 2.6$, $c = 4.0$; A, $a = 0.0$; B, $a = 0.04$; C, $a = 0.07$.

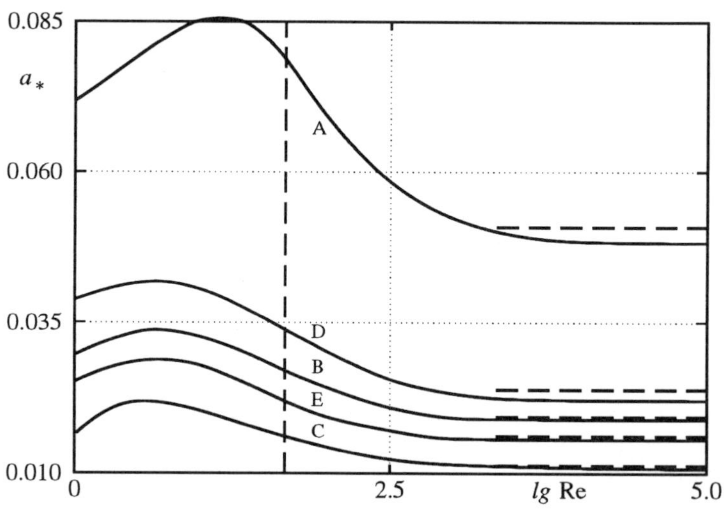

Figure 3.9 Variation of the critical value of non-uniformity parameter with Reynolds number. Curves A,B,C, $b = 2.6,\ 7.2,\ 11.8$ at $c = 4.0$; curves D,E, $c = 3.0,\ 5.0$ at $b = 7.2$.

It has been noted in subsection 2.4.3 that the thin viscous shock layer model results in an appearance of a zero pressure point on the sphere surface (Newtonian separation point). This point can be located from the condition $C_p = 0$ using (3.7), which yields

$$s^s = \arcsin\left[\frac{B}{3\Delta} + 1 - \frac{2ab(1+c)}{1-a}\right]^{-1/2}$$

The above formula displays the dependence of the separation point coordinate on the oncoming stream rarefaction and non-uniformity.

When investigating the flow about the whole bluntness, the thin viscous shock layer equations and the boundary conditions were also written for Dorodnitsin's variables. The continuity equation was satisfied using the stream function. The term $\partial p/\partial \xi$ in the tangent momentum balance equation was determined from the first-order ordinary differential equation resulting from a differentiation of the normal momentum balance equation (the third equation of set (2.53)) with respect to ξ. The boundary condition for this equation was imposed on the bow shock wave. It was obtained by differentiation of the boundary condition for the gas pressure.

Starting from the stagnation streamline, the above parabolic problem was solved by the space-marching finite-difference scheme with iterations in each marching step. In each iteration, the equations for gas pressure and its derivative $\partial p/\partial \xi$ were first integrated from the bow shock to the body surface with the use of Simpson quadrature formula. Then, the equations for the stream function and gas temperature were solved. They were approximated by the implicit finite-difference scheme of Petukhov (1964) of the fourth-order accuracy in the η-coordinate and second-order accuracy in the ξ-coordinate. The resultant sets of difference equations were solved, one after another, by means of the 'progonka' algorithm. Later on, the gas density, viscosity and bow shock distance from the body surface were specified and the next iteration was started.

Calculations were carried out for hypersonic flow of a perfect gas with specific heat ratio $\gamma = 1.4$, viscosity $\mu \propto \sqrt{T}$, Prandtl number $Pr = 0.7$. The body surface temperature was assumed to be $T_w = 0.15T^*_\infty$. Figure 3.8 shows the velocity profiles on the stagnation streamline. Here the Reynolds number is defined as $Re = \rho_\infty V_\infty R/\mu(T^*_\infty)$. It is seen that with increasing non-uniformity parameter a, the velocity profiles become less filled, their slope near the body surface decreases. At some value $a = a_*$, the derivative $\partial u/\partial \zeta$ at the body surface becomes equal to zero. A further increase of the above parameter gives rise to a recirculating flow within the shock layer. The critical value a_* depends on the Reynolds number, see figure 3.9. The values of parameter a which relate to a non-separated flow regime are below the curves $a_* = a_*(\lg Re)$. An increase of the non-uniformity parameter b (decrease of the relative wake width) at fixed values of other parameters results in a shrinkage of the range of a-values which correspond to the non-separated flow regime. The same is caused by an increase of the parameter c (curves D,B,E) characterizing the temperature non-uniformity within the wake. The dashed horizontal lines in figure 3.9 depict the values a^∞_* determined from the inviscid flow model by a change of the pressure gradient sign at the stagnation point (Eremeitsev and Pilugin 1986):

$$a^\infty_* = \frac{4/3 - \delta}{7/6 - \delta + 2b(1-\delta)(1+c)}, \quad \delta = \frac{\gamma - 1}{2\gamma} \tag{3.8}$$

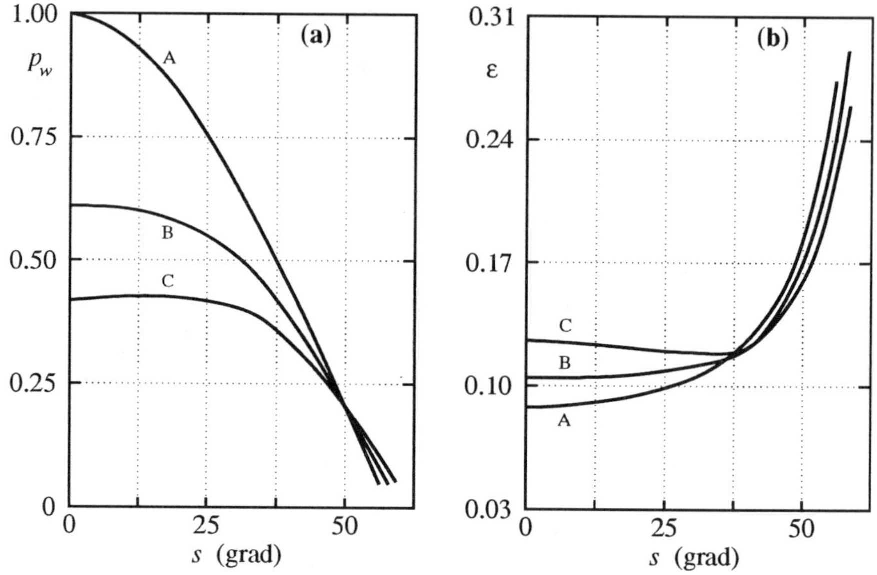

Figure 3.10 Variation of wall pressure shock layer thickness in a hypersonic wake flow about a sphere. $Re = 100$, $b = 2.6$, $c = 4.0$. A, $a = 0.0$; B, $a = 0.04$; C, $a = 0.06$.

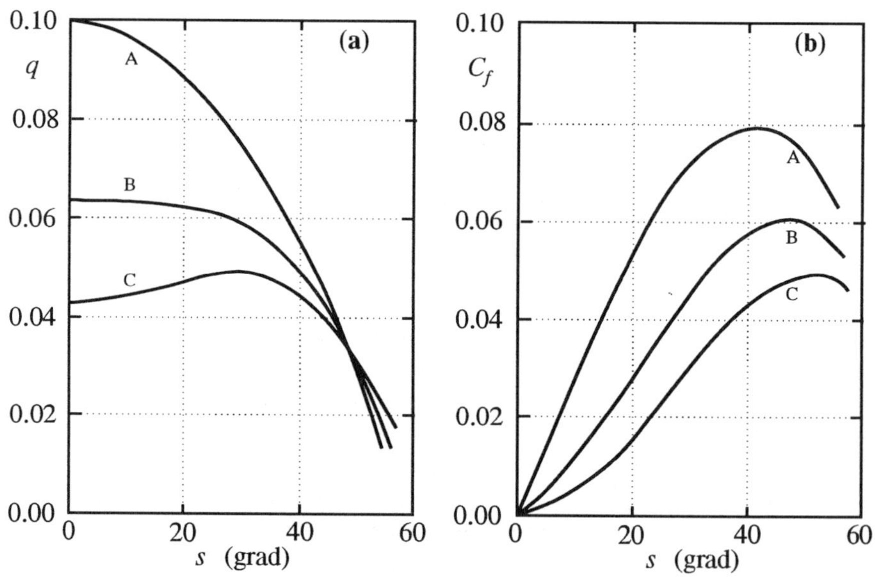

Figure 3.11 Variation of the heat flux and skin friction coefficient over a sphere surface in a hypersonic wake flow. $Re = 100$, $b = 2.6$, $c = 4.0$. A, $a = 0.0$; B, $a = 0.04$; C, $a = 0.06$.

Figure 3.9 shows that a_*^∞ presents approximately the lower boundary of the range of the a_*-values at which the recirculating flow can appear within the shock layer. At $Re \lesssim 50$, the functions $a_*(Re)$ become non-monotonous; however, one ought to keep in mind that these Reynolds numbers are beyond the scope of the thin viscous shock layer model, whose bound is shown in figure 3.9 by the dashed vertical line.

The next two figures show the distributions of the shock layer thickness, wall pressure, heat flux and skin friction coefficient over the sphere surface. The heat flux is normalized here by $\rho_\infty V_\infty^3/2$, the reference values of other functions are usually adopted. The variation of these distributions with increasing oncoming stream non-uniformity agrees with the solutions of the full Navier–Stokes equations. It concerns non-monotonous variation of the shock layer thickness, formation of the downstream wall pressure and heat flux maxima, diminution of the maximum skin friction coefficient and its shift downstream.

3.2 FLOW FROM A SUPERSONIC SOURCE PAST A SPHERICAL BLUNTNESS

The flow from a supersonic source presents the simplest mathematical model for a supersonic under-expanded jet. Such flows are employed in a number of experimental and technical facilities. Below, we consider the results of viscous shock layer calculations which pertain to source flows about a sphere at Reynolds numbers $Re_s \gtrsim 50$ (Golovachov and Leontieva 1983). The analysis of the numerical solutions reveals a similarity in behaviour of the flow characteristics, which is also discussed.

3.2.1 Formulation of the problem

The flow under study is schematically shown in figure 3.12. The centres of the body bluntness and the spherical source are separated from each other by a distance l. The pressure p_*, density ρ_* and gas velocity $V_* = \sqrt{\gamma p_*/\rho_*}$ are prescribed on the source surface. The gas parameters in front of the bow shock wave are obtained from an analytical solution for a stationary supersonic radial inviscid gas flow. Subject to the above boundary conditions on the source surface, it gives the following formulae for gas parameters at the point located at a distance r from the source centre

$$\begin{aligned}
W &= \left(\frac{r_*}{r}\right)^2 \left(\frac{2}{\gamma+1}\right)^{1/(\gamma-1)} \left(1 - \frac{\gamma-1}{\gamma+1}W^2\right)^{-1/(\gamma-1)} \\
\frac{p}{p_*} &= \left(\frac{r_*}{r}\right)^2 \frac{\gamma+1}{2W} \left(1 - \frac{\gamma-1}{\gamma+1}W^2\right) \\
\frac{\rho}{\rho_*} &= \left(\frac{r_*}{r}\right)^2 \frac{1}{W}
\end{aligned} \qquad (3.9)$$

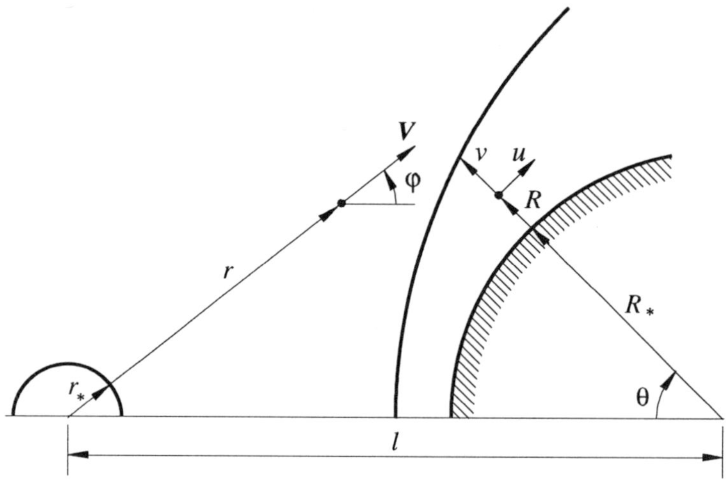

Figure 3.12 Outline of the flow from a supersonic source about a blunt body.

Here $W = V/V_*$, V is the modulus of the velocity vector. The distance r and the slope of the velocity vector to the symmetry axis φ are expressed through the spherical coordinates (R, θ) with a pole, located at the centre of the spherical bluntness

$$r = \left[(l - R\cos\theta)^2 + R^2 \sin^2\theta\right]^{1/2}$$
$$\varphi = \arctan \frac{R\sin\theta}{l - R\cos\theta} \tag{3.10}$$

The pressure p_* and density ρ_* on the source surface and the distance l are prescribed so that the required values of Mach and Reynolds numbers would be ensured on the symmetry axis ahead of the bow shock wave. In this case, the oncoming stream non-uniformity is determined only by the sphere-to-source radius ratio $d = R_*/r_*$, $d = 0$ corresponding to the uniform oncoming stream.

The flow between the bow shock wave and the body surface is calculated using the Navier–Stokes equations. The gas parameters behind the bow shock are determined by modified Rankine-Hugoniot relationships with the use of formulae (3.9) and (3.10). The body surface is supposed to have a fixed temperature T_w. Both velocity components on the body surface are assumed to be equal to zero. The symmetry relationships and approximate boundary conditions (1.12) are imposed on two other boundaries of the computational domain. The calculations were carried out with the implicit unidirectional method described in subsection 1.2.2.

3.2.2 Effects of non-uniformity and rarefaction of the oncoming stream

The effects of non-uniformity and rarefaction of the oncoming stream are demonstrated in figures 3.13 through 3.16 which pertain to the flows about a sphere at

Figure 3.13 Bow shocks and sonic lines ahead of a sphere in a non-uniform source-type oncoming stream. $M_0 = 6$, $Re_0 = 35500$; A, $d = 0.0$; B, $d = 4.0$; C, $d = 8.0$.

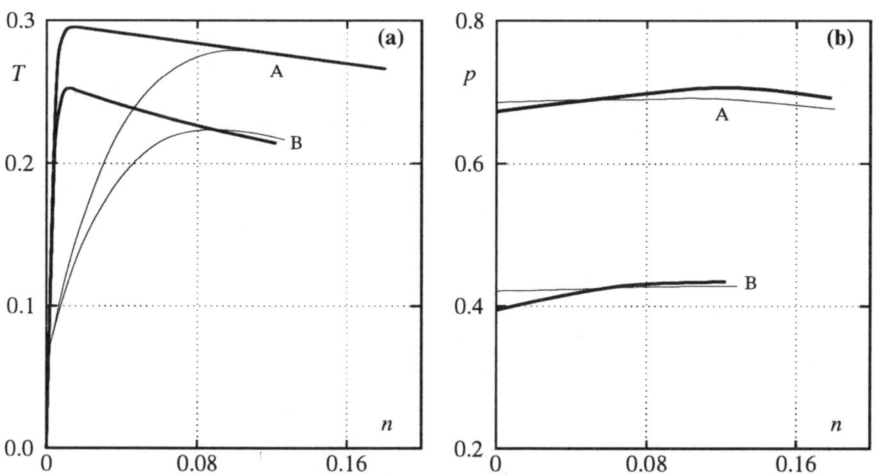

Figure 3.14 Temperature and pressure profiles on the line $\theta = 30°$. $M_0 = 6$. A, $d = 0.0$; B, $d = 4.0$; full lines, $Re_0 = 35500$; hair lines, $Re_0 = 355$.

Mach number $M_0 = 6$ and wall temperature $T_w = 0.15T^*$. The gas specific heat ratio $\gamma = 1.4$, viscosity $\mu \propto \sqrt{T}$, Prandtl number $Pr = 0.7$. Reynolds number Re_0 is evaluated using the sphere radius. Here the subscript 0 denotes the gas parameters on the flow symmetry axis in front of the bow shock wave. The results are presented in a non-dimensional form using the quantities $V_0^2/2R^*$, $\rho_0 V_0^2$, $\rho_0 V_0^3$ as the reference values of temperature, pressure and heat flux, respectively. The distance from the sphere surface is normalized by radius R_*.

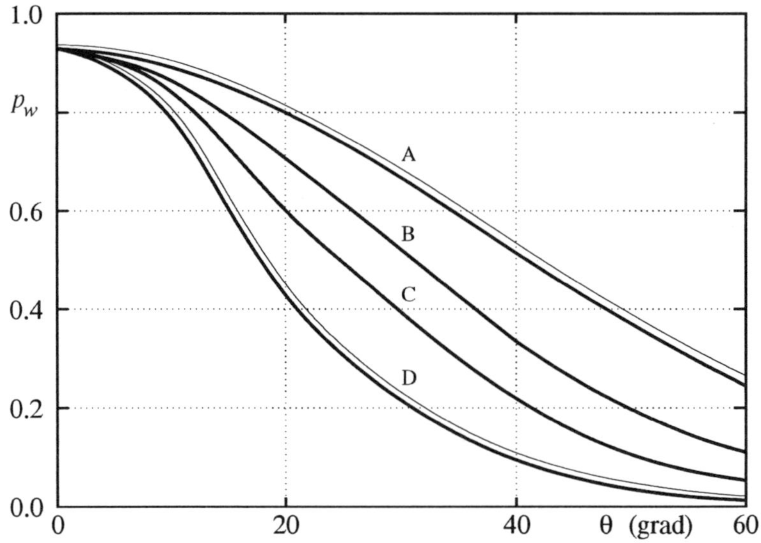

Figure 3.15 Wall pressure distribution over a sphere in a non-uniform source-type oncoming stream . $M_0 = 6$. Curves A–D, $d = 0.0, 2.0, 4.0, 8.0$, respectively; full lines, $Re_0 = 35500$; hair lines, $Re_0 = 355$.

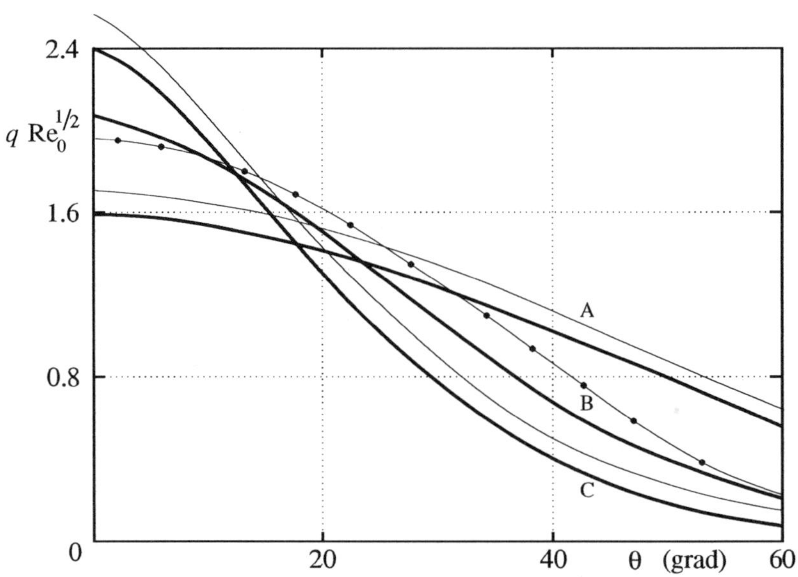

Figure 3.16 Heat flux distribution over a sphere in a non-unifirm source-type oncoming stream. $M_0 = 6$. A, $d = 0.0$; B, $d = 4.0$; C, $d = 8.0$; full lines, $Re_0 = 35500$; hair lines, $Re_0 = 355$; hair-dotted line, formula (3.11).

Figure 3.13 depicts the bow shock and sonic lines ahead of a sphere. It is seen that with an increasing non-uniformity parameter d, the bow shock straightens itself and approaches the body in the vicinity of the symmetry axis. The subsonic part of the shock layer shrinks. Figure 3.14 shows the gas temperature and pressure profiles in the shock layer in the cases of a uniform and non-uniform oncoming stream at two Reynolds numbers.

Figures 3.15 and 3.16 show the wall pressure and heat flux distributions over the sphere surface. The wall pressure at the stagnation point of a sphere is not influenced by the oncoming stream non-uniformity. At fairly high Reynolds numbers considered, it is nearly independent of this parameter as well and is determined only by the gas specific heat ratio γ and Mach number M_0. The more non-uniform the oncoming stream, the more rapid the diminution of the wall pressure over a sphere. An increase of the non-uniformity results in a significant increase of the heat flux to the stagnation point of a sphere, which agrees with the increasing gas velocity at the outer edge of the boundary layer. Similar to the wall pressure distribution, the heat flux variation over the sphere surface is more rapid in the case of a non-uniform oncoming stream. The dash-dotted curve in figure 3.16 depicts the distribution of the heat transfer parameter at $Re_0 = 35500$ calculated with the formula

$$q(\theta) = q_u(0)\sqrt{1+k}(\cos\theta)^{4/3+(5k+8)k/3} \tag{3.11}$$

which has been suggested by Eremeitsev and Pilugin (1981) using the thin inviscid shock layer and the boundary layer models with wall pressure distribution prescribed by the Newton formula. Here $k = R_*/(l - R_*)$, $q_u(0)$ is the heat flux at the stagnation point in the case of a uniform oncoming stream.

Comparison of the calculations for two Reynolds numbers shows that rarefaction manifests itself in the supersonic divergent flow about a blunt body in the same manner as in the case of a uniform oncoming stream. Variation of the Reynolds number significantly influences the profiles of the temperature and tangent velocity component while its effect of the shock layer thickness, pressure and normal velocity component is by far less. The heat transfer and wall shear stress coefficients on the whole front of a sphere slightly increase with decreasing Reynolds number.

3.2.3 Prediction of the local aerodynamic characteristics

Calculations of the flow under study within the framework of an inviscid gas model (Lunyov and Khramov 1970, Shapiro 1975) revealed a similarity in variation of the shock layer parameters with the angular coordinate scaled by the value dependent on the free-stream non-uniformity. Being extended to the viscous shock layer flows, this property allows us to suggest universal relationships for the evaluation of the shock layer thickness, wall pressure, shear stress and heat flux on the spherical bluntness (Golovachov 1985).

The above relationships are founded on numerical solutions of the inviscid gas, boundary layer and viscous shock layer equations. The angular coordinate is normalized by its value at the sonic point on the sphere surface in an inviscid

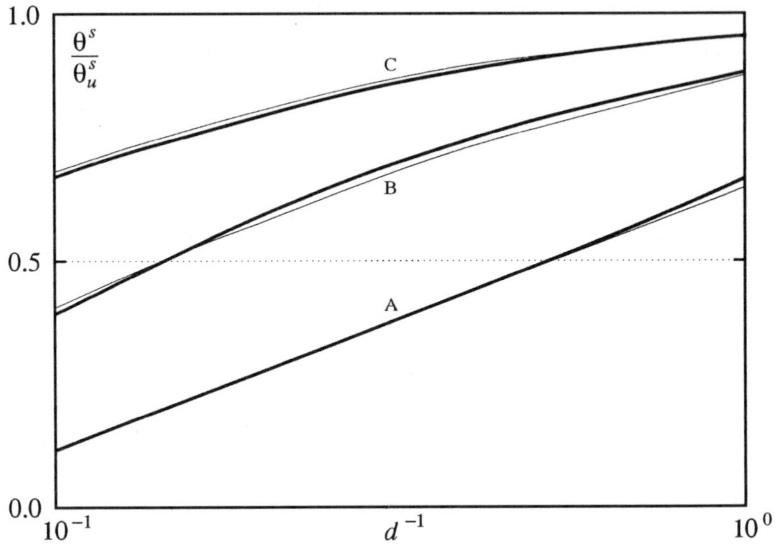

Figure 3.17 Variation of the sonic point coordinate on a sphere surface with oncoming stream non-uniformity. A, $M_0 = 3$; B, $M_0 = 6$; C, $M_0 = 10$; full lines, numerical solutions; hair lines, formula (3.12).

gas flow. Figure 3.17 shows the dependence of the sonic point coordinate on the non-uniformity parameter d. The figure depicts the results of numerical solutions and the results provided by the approximate formula of Shapiro (1975):

$$\theta^s/\theta^s_u = \frac{1}{2}\left[1 + \frac{1}{K} - \sqrt{\left(1 - \frac{1}{K}\right)^2 + \frac{4}{Kl}}\right], \quad K = \varepsilon^0_u/(1+\varepsilon^0_u) \tag{3.12}$$

Here ε is the bow shock distance from the body surface, the subscript u relates to the uniform oncoming stream, the superscript o relates to the parameters on the stagnation streamline. The shock layer thickness and the distance between the source and sphere are normalized by the sphere radius. It is seen that the value θ^s_u being known, formula (3.12) allows a quite accurate prediction of the sonic point coordinate θ^s. The inviscid flow field data for the case of a uniform oncoming stream can be taken from tables (Belotserkovskii *et al* 1967, Lubimov and Rusanov 1970, Belotserkovskii *et al* 1974).

The next two figures present distributions of the shock layer thickness, wall pressure, heat flux and skin friction coefficient over a sphere surface with the similarity variable $\eta = \theta/\theta^s$. Most of these results were obtained from the inviscid gas and boundary layer equations. The bright circles depict solutions of the Navier–Stokes equations, k denotes the temperature ratio $k = T_w/T^*$. One can see that the distributions of all above quantities for a non-uniform oncoming stream are well described up to $\eta \approx 1.5$ by the full curves which depict the results for a uniform stream. At greater values of the η-coordinate there is a deviation from the

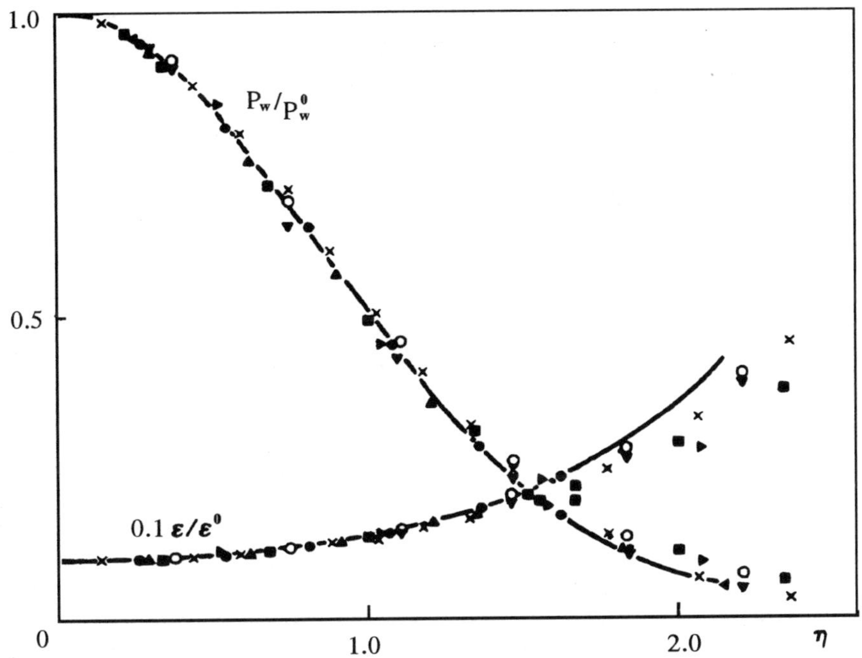

Figure 3.18 Variation of the shock layer thickness and wall pressure over a sphere in a non-uniform source-type oncoming stream. Line, $d = 0$, $M_\infty = 6$, $k = 0.15$; •, $d = 1$, ▲, $d = 2$, ▼, $d = 4$, ▶, $d = 8$, $M_0 = 6$, $k = 0.15$; ■, $d = 4$, $M_0 = 3$, $k = 0.22$; ×, $d = 4$, $M_0 = 10$, $k = 0.06$; ○, $d = 4$, $M_0 = 6$, $Re_0 = 177$, $k = 0.15$.

full curves and separation of the results with respect to the Mach number. The rarefaction effect does not exceed the dispersion caused by the oncoming stream non-uniformity.

To make use of the above universal distributions, one ought to know the shock layer thickness on the symmetry axis, the wall pressure and heat flux at the stagnation point and the maximum skin friction coefficient.

At fixed Mach number on the symmetry axis in front of the bow shock wave, the pressure at the stagnation point of a sphere is the same as in the case of a uniform oncoming stream with the same value of the Mach number. The dependence of others of the above quantities on the oncoming stream non-uniformity is shown in figure 3.20. The full lines depict the results of inviscid flow and boundary layer calculations. The circles show the solution of the Navier–Stokes equations at $M_0 = 6$, $Re_0 = 355$. It is seen that the rarefaction influence on the normalized values of the shock layer thickness, heat flux and skin friction coefficient is negligible at the Reynolds numbers considered. It is also worth noting a closing of the curves with increasing Mach number. The hair lines in figure 3.20 depict the results for Mach numbers $M_0 = 6$ and 10 obtained with the approximate formulae suggested by Shapiro (1975) and Eremeitsev and Pilugin (1981)

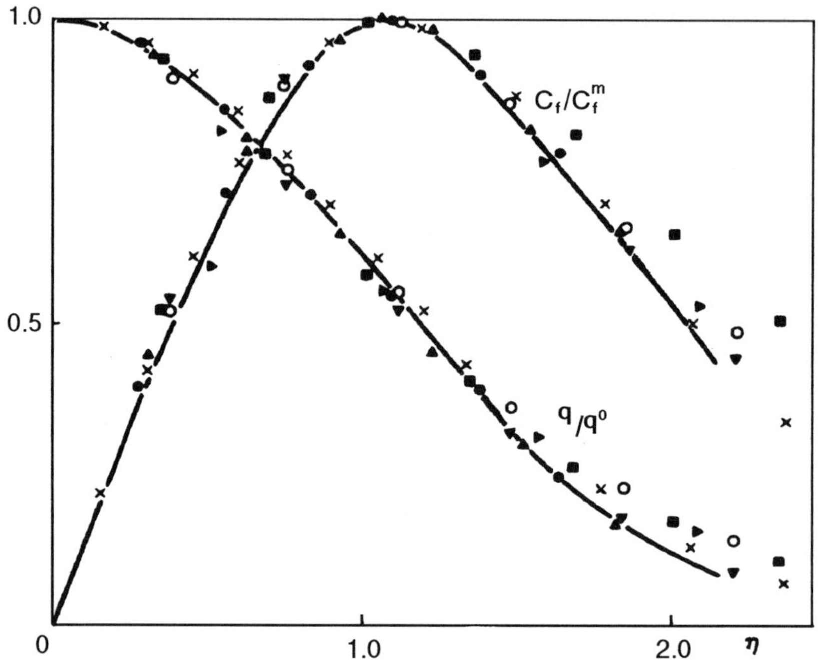

Figure 3.19 Distributions of the heat flux and skin friction coefficient over a surface in a non-uniform source-type oncoming stream. Line, $d = 0$, $M_\infty = 6$, $k = 0.15$; •, $d = 1$, ▲, $d = 2$, ▼, $d = 4$, ▶, $d = 8$, $M_0 = 6$, $k = 0.15$; ■, $d = 4$, $M_0 = 3$, $k = 0.22$; ×, $d = 4$, $M_0 = 10$, $k = 0.06$; ○, $d = 4$, $M_0 = 6$, $Re_0 = 177$, $k = 0.15$.

$$\frac{\varepsilon^0}{\varepsilon_u^0} = \frac{\theta^s}{\theta_u^s}\frac{1}{1+\varepsilon_u^0(1-\theta^s/\theta_u^s)}, \qquad \frac{q^0}{q_u^0} = \sqrt{\frac{l}{l-1}} \tag{3.13}$$

The second formula was derived for hypersonic oncoming stream velocities. However, it results in a fairly large error even in this case.

The above relationships allow one to evaluate the parameters of a non-uniform flow about a sphere through their values in the case of a uniform oncoming stream. In the latter case within the Reynolds number range considered, the oncoming stream rarefaction alters the stagnation point pressure by no more than several per cent, see for example (Belotserkovskii *et al* 1974). Within this accuracy, the pressure can be estimated with the formula

$$p_w^0 = p_0\left(\frac{2\gamma}{1+\gamma}M_0^2 - \frac{\gamma-1}{\gamma+1}\right)\left[\frac{2\gamma+(\gamma-1)^2/2}{2\gamma M_0^2-(\gamma-1)}\right]^{\frac{\gamma}{\gamma-1}} M_0^{\frac{2\gamma}{\gamma-1}} \tag{3.14}$$

The results for the shock layer thickness, heat flux and shear stress in the case of a uniform oncoming stream are presented in figures 3.21 and 3.22. Here Re_s is the Reynolds number based on the gas parameters behind a normal shock wave. It relates to Re_∞ by formula (1.13).

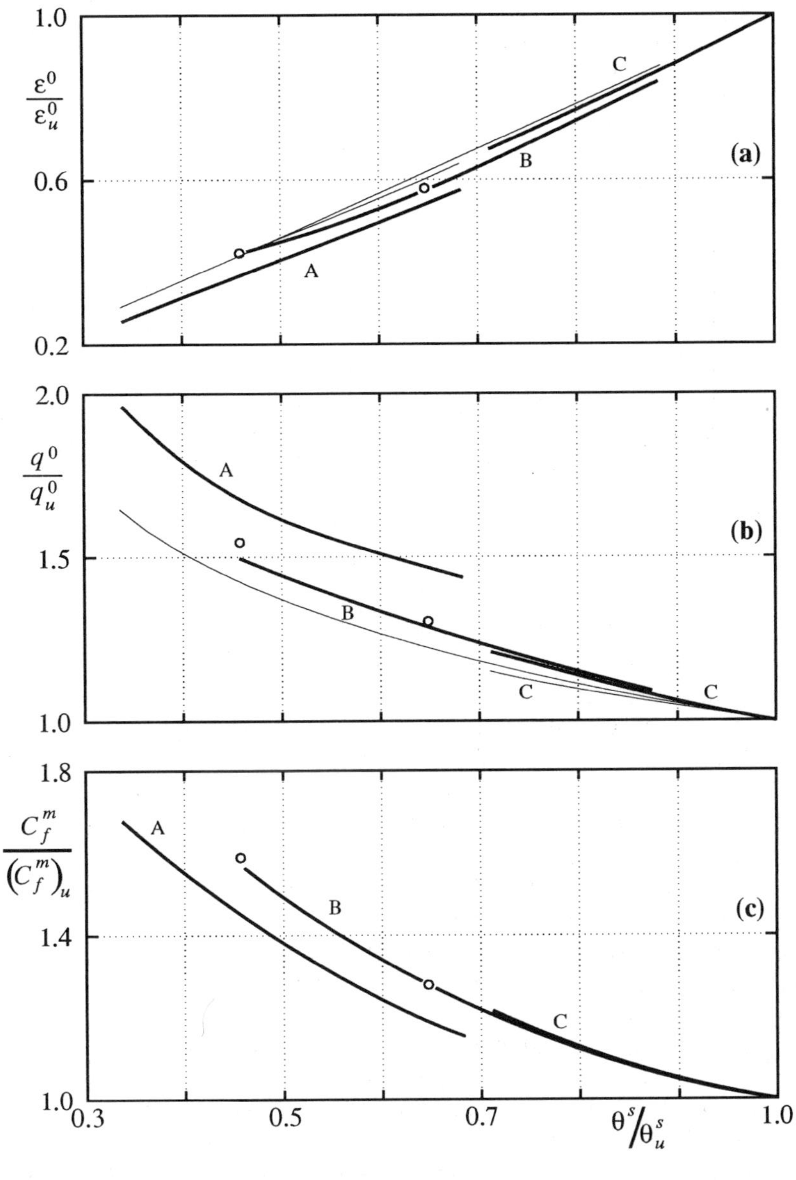

Figure 3.20 Dependence of the flow characteristics on the oncoming stream non-uniformity. A, $M_0 = 3$; B, $M_0 = 6$, C, $M_0 = 10$; ○, $M_0 = 6$, $Re_0 = 355$; full lines, numerical solutions; hair lines, approximate formulae (3.13).

In figure 3.21 the full curves depict a variation of the shock layer thickness on the symmetry axis at several values of the wall temperature factor. The hair curve corresponds to the adiabatic wall case. The shock layer thickness is normalized by its value in an inviscid gas flow. The results represent solutions of the Navier–

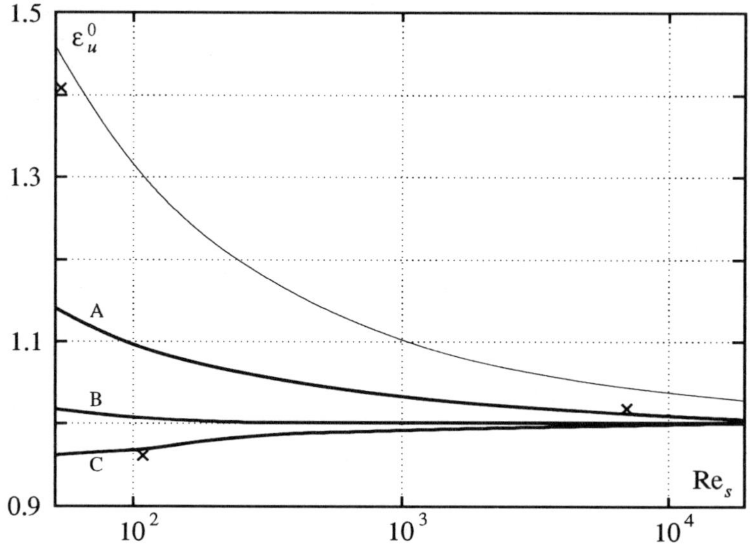

Figure 3.21 Dependence of the shock layer thickness on the flow symmetry axis on the Reynolds number. A, $k = 0.35$; B, $k = 0.15$; C, $k = 0.075$; hair line, adiabatic wall.

Stokes equations at Mach number $M_\infty = 6$. They demonstrate a substantial effect of the wall temperature on the shock layer thickness at small Reynolds numbers. At $Re_s \gtrsim 10^3$, the shock layer thickness is close to that provided by the inviscid gas model. To demonstrate a universality of the criterion Re_s, the results for $M_\infty = 13.4$, $k = 0.073$; $M_\infty = 2$, $k = 0.35$ and $M_\infty = 3.74$, adiabatic wall are also plotted in this figure by means of crosses.

The dependences of the stagnation point heat transfer parameter C_H and of the maximum skin friction coefficient on the sphere surface C_f^m on the Reynolds number Re_s are close to the linear boundary layer ones. At $k \lesssim 0.3$, they are only slightly affected by the wall temperature in the whole range of Reynolds numbers considered. The full lines present, accurate to within some per cent, calculations for $M_\infty = 6$, $0.075 \leqslant k \leqslant 0.35$. The experimental data for heat transfer at Mach numbers $2 \leqslant M_\infty \leqslant 18$ are plotted in figure 3.22 as well. The crosses show the skin friction coefficient values calculated for several Mach numbers from the above range. The hair line depicts the variation of the maximum of skin friction coefficient on an adiabatic wall. In this case, the skin friction coefficient is higher than on a cooled wall, which agrees with a larger gas viscosity on the body surface. A slower increase of the skin friction coefficient on the adiabatic wall with decreasing Reynolds number is explained by the different rarefaction effects on the shock layer thickness in the above cases; see figure 3.21.

The above formulae and graphs allow one to determine approximately the shock layer thickness, wall pressure, heat flux and skin friction coefficient at any point on the front of a sphere located in a non-uniform divergent stream. Their application requires only the data on the inviscid flow characteristics in the case of a uniform

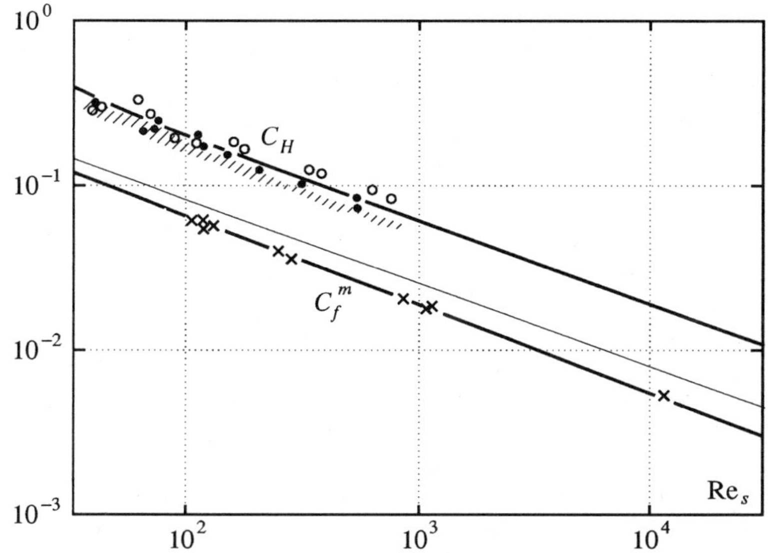

Figure 3.22 Dependence of the heat transfer and skin friction coefficients on the Reynolds number. The lines, calculations at $M_0 = 6$; ///, ○, •, experimental data for heat transfer by Hickman and Giedt (1963), Wilson and Wittliff (1962), Gusev and Nikolskii (1971), respectively; × calculated C_f^m values on a cooled sphere at $2 \leqslant M_\infty \leqslant 18$; hair line, C_f^m on an adiabatic sphere.

free stream. The above results are based on calculations carried out for a perfect gas with the specific heat ratio $\gamma = 1.4$. Available data on supersonic blunt body flows of other gases and gas mixtures (Shapiro 1975, Golovachov and Timofeyev 1988b) indicate that the similarity properties inhere in these flows as well.

3.3 VISCOUS SHOCKED FLOWS

This section deals with the viscous gas flows about bodies located in a shocked supersonic stream. The shocks can be generated by, for instance, interaction of a supersonic flow with the elements of an aircraft or an experimental facility. Interaction of an incident shock with a shock layer flow may give rise to separation and appearance of the shear layers, jets and inner shock waves. Numerical simulation of such flows is carried out with the full Navier-Stokes equations.

3.3.1 Shock wave–boundary layer interaction

The flow under study is induced by an oblique shock incidence on a flat plate located in a supersonic high Reynolds number stream. An outline of the flow has been presented in figure 2.16. With the use of the full Navier–Stokes equations, the problem is formulated as follows. The computational domain is the rectangle

$ABCD$ in the plane of Cartesian coordinates (x, y). A supersonic stream with fixed values of gas dynamic functions flows into the computational domain through the boundary AB. On the segment AE of the lower boundary in front of the plate, the symmetry relationships are imposed. On the plate surface ED, both velocity components are assumed to be equal to zero and the wall temperature is prescribed. The upper boundary conditions take into account the oblique shock FG. Accordingly, the free stream parameters are prescribed on the segment BF while the gas parameters along the segment FC are determined from the oblique shock relationships. On the downstream boundary CD, the approximate conditions (1.12) are imposed.

First, let us dwell on calculations of the laminar flow regime. The problem under study is often used as a benchmark in estimation of the numerical methods for solving the Navier–Stokes equations (Mac Cormack 1971, Beam and Warming 1978, Belotserkovskii *et al* 1982, Savelyev and Tolstykh 1987). The results are compared with experimental data of Hakkinen *et al* (1959) for the following conditions: slope of the incident shock $\varphi = 32.6°$; distance from the leading edge of the plate to the shock incidence point $x_s = 0.047m$; free stream Mach number $M_\infty = 2$; Reynolds number $Re = \rho_\infty V \infty x_s / \mu_\infty = 0.296 \times 10^6$. The above calculations agree well with the experimental data demonstrating a recirculation zone near the plate surface. Figure 3.23 shows the wall pressure and skin friction coefficient distributions calculated by Savelyev and Tolstykh (1987). Here the pressure and shear stress are normalized by p_∞ and $10^{-3} \times \rho_\infty V_\infty^2$, respectively. The wall temperature was assumed to be $T_w = 1.7T_\infty$. The calculations were carried out using a compact third-order accurate finite-difference scheme on the grid with 32×39 nodes clustering near the plate surface.

Hussani *et al* (1980) compared numerical solutions of the Navier–Stokes equations with the results of the asymptotic triple-deck theory by Neyland (1969), Stewartson and Williams (1969). The Navier–Stokes equations were integrated using the method of Mac Cormack (1978). The calculations were carried out on two grids with 32×32 and 128×32 nodes which were clustered towards the plate surface. Figure 3.24 shows the wall pressure and skin friction coefficient distributions. The results are presented in a non-dimensional form using the reference values of the triple-deck theory. The abscissa X presents the non-dimensional distance from the separation point

$$X = [(x - x^s)/x^s] \left(\frac{Re}{C}\right)^{3/8} (M_\infty^2 - 1)^{3/8} \lambda^{5/4} (T_\infty/T_w)^{3/2}$$

where x is the distance from the leading edge of the plate, x^s is the separation point coordinate, C is the Chapman–Rubesin parameter, $\lambda = 0.3321$ is the slope of the gas velocity profile on the plate in an undisturbed boundary layer. Figure 3.24-a displays the variation of the non-dimensional pressure

$$P = (M_\infty^2 - 1)^{1/4} [p(x) - \tilde{p}_w] \Big/ \left(Re^{1/4} \gamma C^{1/4} M_\infty^2 \tilde{p}_w\right)$$

where $\tilde{p}_w$ is the wall pressure in front of the free interaction zone. The skin friction coefficient is normalized by its value in the Blasius problem of the boundary layer

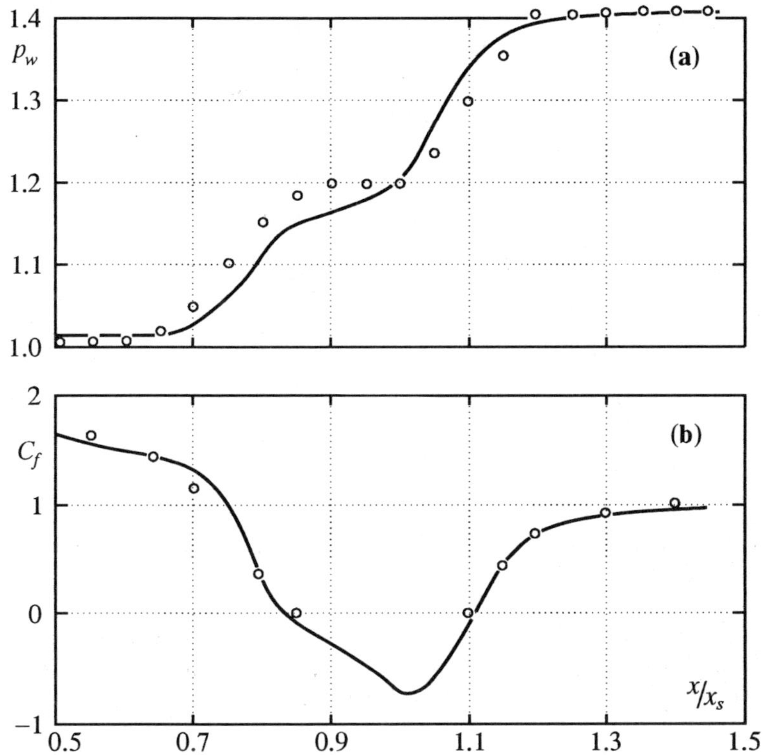

Figure 3.23 Variation of the pressure and skin friction coefficient on a plate in the vicinity of the oblique shock incidence point. $M_\infty = 2$, $Re = 0.296 \times 10^6$. Lines, calculations by Savelyev and Tolstykh (1987); circles, experiment by Hakkinen *et al* (1959).

on a plate at constant pressure. The full lines present the results of the asymptotic theory at $Re \to \infty$, the hair lines show numerical solutions of the Navier–Stokes equations at two values of the Reynolds number $Re = \rho_\infty V_\infty x^s/\mu_\infty$. With increasing Reynolds number, numerical solutions approach the asymptotic limits. Nevertheless, one can see a noticeable discrepancy even at higher Reynolds number. This is not surprising since the boundary layer thickness δ at the beginning of the free interaction zone ought to be considered as a proper characteristic length of the separated flow under study instead of the distance x^s. The Reynolds numbers Re_δ based on this characteristic length are within the range $10^2 \div 10^3$.

Figure 3.25 demonstrates the dependence of the incident shock intensity, required for the boundary layer separation, on the Reynolds number. Here I denotes the quantity

$$I = Re^{1/4}\left(\frac{p - p_\infty}{p_\infty}\right)$$

where p_∞, p are the pressure values in front and behind the incident shock, re-

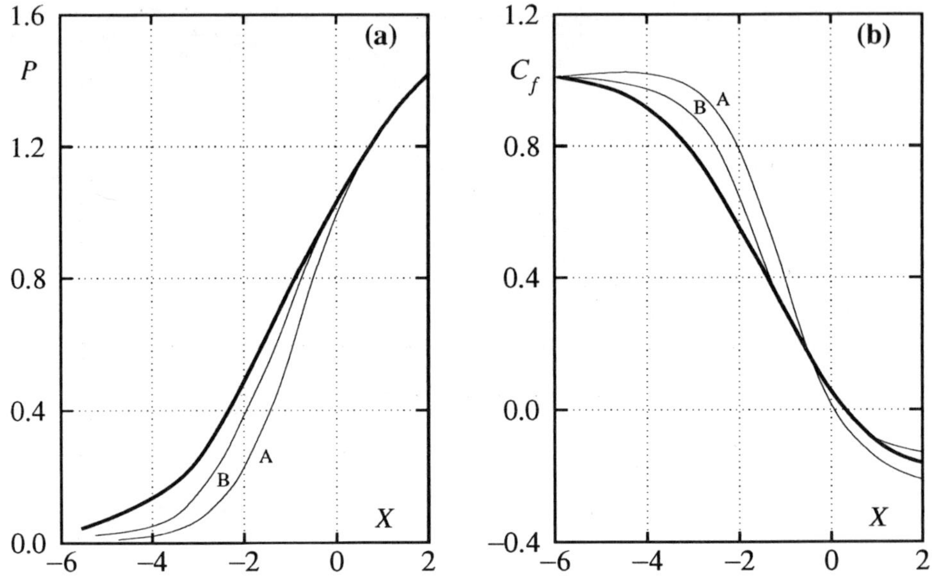

Figure 3.24 Variation of the pressure and skin friction coefficient on a plate in the vicinity of the oblique shock incidence point. $M_\infty = 2$. Full lines, triple-deck theory; hair lines, numerical solutions of Navier–Stokes equations; A, $Re = 0.4 \times 10^5$; B, $Re = 4 \times 10^5$.

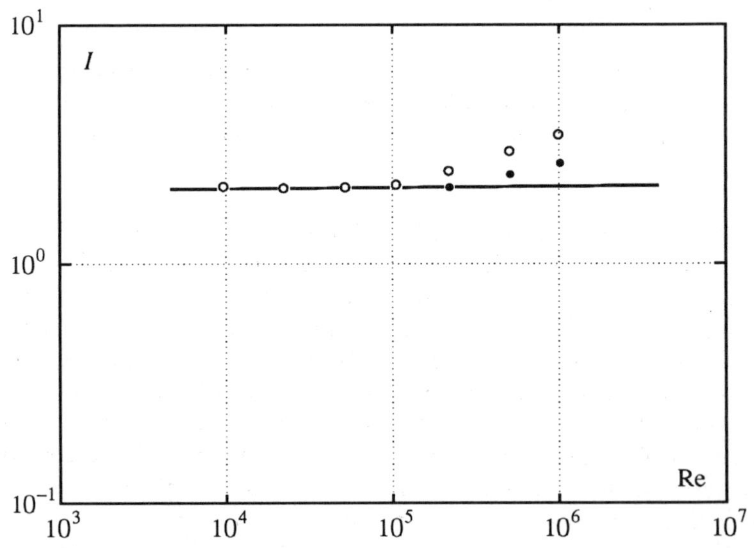

Figure 3.25 Comparison of numerical solutions of the Navier–Stokes equations (circles) with the asymptotic triple-deck theory. ○, grid 32 × 32; •, grid 128 × 32.

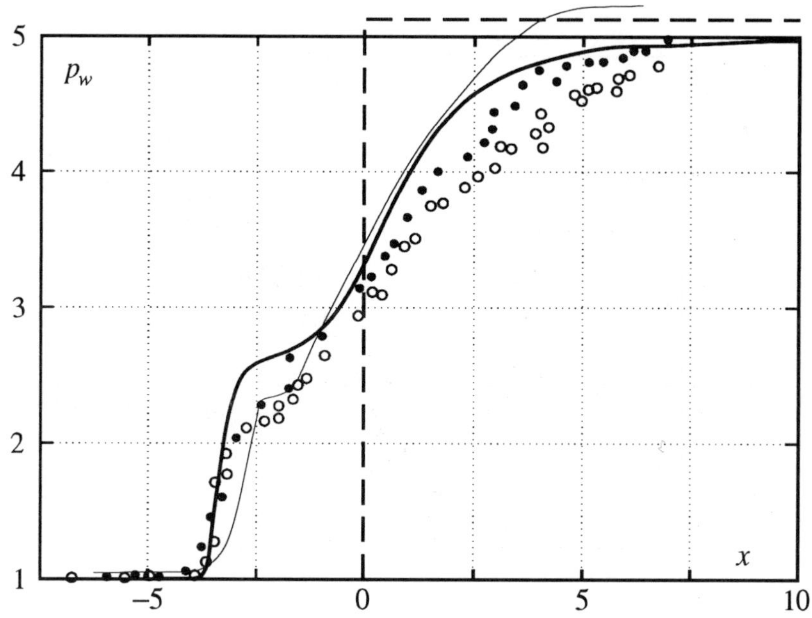

Figure 3.26 Pressure variation over a plate in the vicinity of the shock wave-turbulent boundary layer interaction. $M_\infty = 2.96$, $Re_\delta = 10^6$; full line, calculations of Savelyev and Tolstykh (1987); hair line, calculations of Baldwin and Mac Cormack (1976); circles, experimental data of Reda and Murphy (1972), Murthy and Rose (1978).

spectively. The straight line corresponds to the triple-deck theory which gives:

$$\frac{p - p_\infty}{p_\infty} \sim Re^{-1/4}$$

The bright circles depict the calculations on a grid with 32 × 32 nodes, the filled circles depict the calculations on the grid with a fourfold number of grid points along the plate. Deviation of the numerical results from the asymptotic limit allows one to estimate the range of Reynolds numbers in which an adequate numerical simulation of the flow can be carried out with the above method and computational grids.

The study of shock wave interaction with a turbulent boundary layer presents a more difficult problem. This being the case, an asymptotic analysis does not provide such comprehensive information as for the laminar flow while a numerical method has to meet stricter resolution requirements. Besides, the calculation results depend on the turbulence model. Some authors have noted a non-stationary behaviour of the mean flow within the separation zone (Murthy and Rose 1978, Adamson and Messiter 1980).

Some results of numerical and experimental investigations are presented in figures 3.26 and 3.27 which display wall pressure and skin friction distributions. The above quantities and the distance x are normalized by the free stream pressure

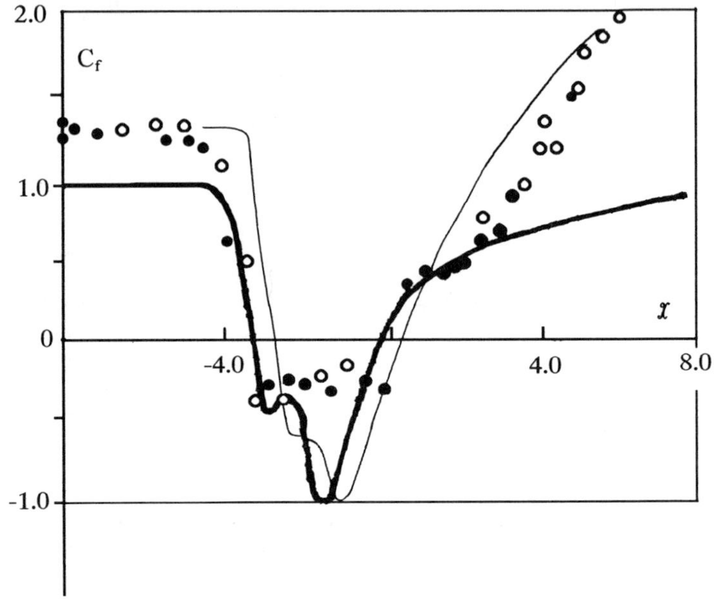

Figure 3.27 Variation of the skin friction coefficient over a plate in the vicinity of the shock wave-turbulent boundary layer interaction. The flow conditions and notations are the same as those in figure 3.26.

p_∞, $10^{-3} \times \rho_\infty V_\infty^2$ and the thickness of the boundary layer δ in front of the interaction zone, respectively. The full curves show calculations of Savelyev and Tolstykh (1987) for the free stream Mach number $M_\infty = 2.96$; the angle of the velocity vector turn at the incident shock $\theta = 12.75°$; the Reynolds number $Re_\delta = 10^6$. The wall temperature was assumed to be equal to the free stream stagnation temperature. The calculations were carried out using the differential turbulence model of Saffman and Wilcox (1974) and a third-order accurate compact finite-difference scheme. The boundary conditions on the plate surface were implemented using the wall functions (Launder and Spalding 1974). The computational grid contained 40×25 nodes which were clustered to the plate surface and to the shock incidence point. The dashed curve in figure 3.26 depicts the pressure distribution in an inviscid gas model. In both figures, the circles show the results of wind tunnel measurements presented by Reda and Murphy (1972), Murthy and Rose (1978). The hair curves depict calculations of Baldwin and Mac Cormack (1976) carried out with the algebraic turbulence model. Figure 3.26 demonstrates a good agreement of the pressure distributions calculated with the above turbulence models with each other and the experimental data. The measured values of the skin friction coefficient shown in figure 3.27 demonstrate an increase of the shear stress behind the separation zone as compared with the flow in front of this zone and agree better with calculations by Baldwin and Mac Cormack (1976). Within the separation zone, both numerical solutions provide close results which differ substantially from the experimental data.

3.3.2 Viscous shock layer interaction with an impinging shock

Numerical investigation of this problem for a stationary plane flow near the front of a circular cylinder was carried out by Tannehill *et al* (1976). The computational domain was bounded by the bow shock wave. The impinging shock was taken into account by the discontinuous distribution of the oncoming stream parameters in the bow shock boundary conditions. The stationary solution of the Navier–Stokes equations was obtained by the time-asymptotic method with the use of the explicit second-order accurate finite-difference scheme (Mac Cormack 1969, Mac Cormack and Baldwin 1975). Integration in time was started from the flow field which had been calculated for the case of a uniform supersonic free stream. The computational grid contained 31 nodes in the transversal and 51 nodes in the streamwise direction. The grid points were clustered to the cylinder surface, the mesh spacing in the streamwise coordinate was chosen to be constant.

The gas dynamic effects arising from the shock-viscous shock layer interaction depend, to a great extent, on geometric factors and the intensity of the bow and impinging shocks. The above authors considered the following conditions: $M_\infty = 4.6$, $Re_\infty = 10^4$, $\gamma = 1.4$, $Pr = 0.72$, $T_w = 0.636T^*_\infty$. The Reynolds number is evaluated using the cylinder diameter. The gas viscosity was determined by Sutherland's formula (1.8). The parameters of the plane impinging shock were chosen to be the following:

$$\begin{array}{llll} \text{(a)} & \beta = 20.9^\circ, & \theta = 36^\circ, & \overline{p} = 2.98; \\ \text{(b)} & \beta = 16.1^\circ, & \theta = \ \ 9^\circ, & \overline{p} = 1.73; \\ \text{(c)} & \beta = 20.9^\circ, & \theta = \ \ 9^\circ, & \overline{p} = 2.98; \end{array}$$

Here β is the angle between the impinging shock plane and the oncoming stream velocity vector; θ is the angular coordinate of the intersection point of the impinging and bow shocks (ignoring their interaction) measured from the symmetry plane of the undisturbed flow; $\overline{p}$ is the pressure ratio at the impinging shock.

Figures 3.28a-c show the Mach number contours in the shock layer for the above parameters of the impinging shock. The isolines are plotted with spacing $\Delta\ M = 0.05$. In case (a), the intersection point of the impinging and bow shocks is located just under the sonic line of the undisturbed flow field. The shear layer arising at the shock intersection point reaches the downstream boundary of the computational domain without interaction with the cylinder surface. In case (b), the above shear layer is tangential to the cylinder surface. In case (c), there is an additional discontinuity of the bow shock slope, besides the one that happens at the shock intersection point. Accordingly, there are two shear layers. The gas velocity near the bow shock between the discontinuity points exceeds the speed of sound. The region of supersonic flow is closed by the shock originating at the

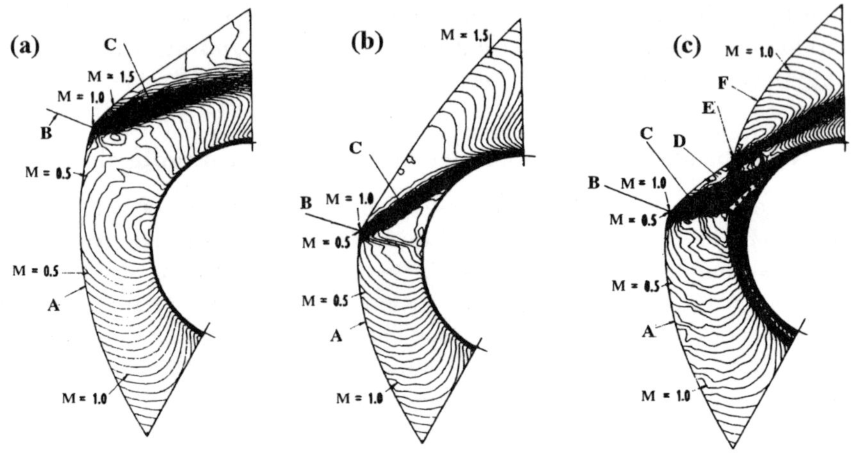

Figure 3.28 Mach number contours in the shock layer ahead of a circular cylinder with an impinging shock. A, bow shock; B, impinging shock; C, shear layer; D, supersonic region; E, imbedded shock wave; F, displaced bow shock.

upper discontinuity point. This shock crosses the shear layer which issues from the incidence point of the impinging shock before this layer reaches the body surface. It should be noted here that the numerical solutions do not reproduce all details of the complicated flows under study because of a rather coarse computational grid.

Figures 3.29 and 3.30 show the wall pressure and heat flux distributions over the cylinder surface for the above conditions of the shock layer-impinging shock interaction. The hair lines correspond to the case of a uniform free stream. The pressure and heat flux are normalized by their values at the stagnation point of a cylinder placed in a uniform free stream. One can see that the impinging shock affects the wall pressure and heat flux distributions to the greatest extent in case (c) when the shear layer produced by the interaction of the bow and impinging shocks reaches the body surface. In this case, there is a small region of the cylinder surface where the wall pressure and heat flux exceed more than twice their maximum values for the case of a uniform free stream.

The calculation results were compared with experimental data of Edney (1968) for three-dimensional flow ahead of an axisymmetric blunt body with an impinging plane shock. The comparison revealed a qualitative agreement with numerical results for the plane flow.

Moon and Holt (1989) carried out a numerical investigation of the problem for the conditions which corresponded to maximum pressure and heat flux values on the body surface. They calculated both laminar and turbulent flows. In the latter case, the Reynolds equations complemented with the turbulence model of

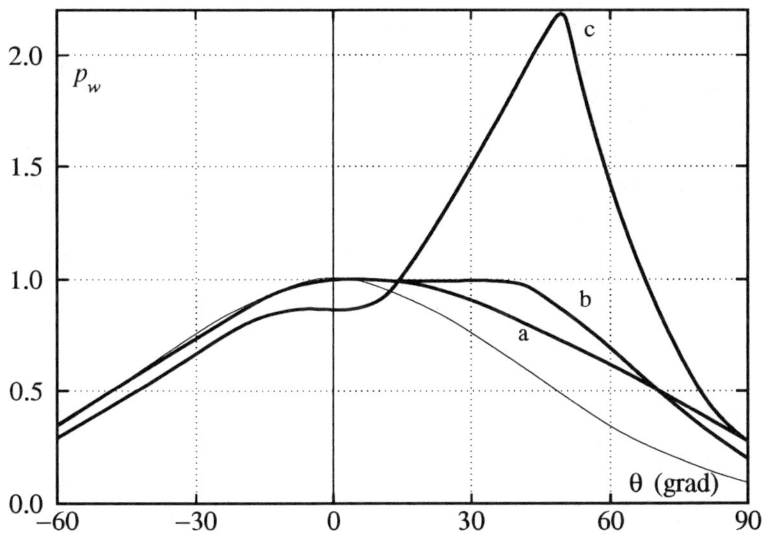

Figure 3.29 Wall pressure distribution over a circular cylinder with an impinging shock; hair line, uniform free stream.

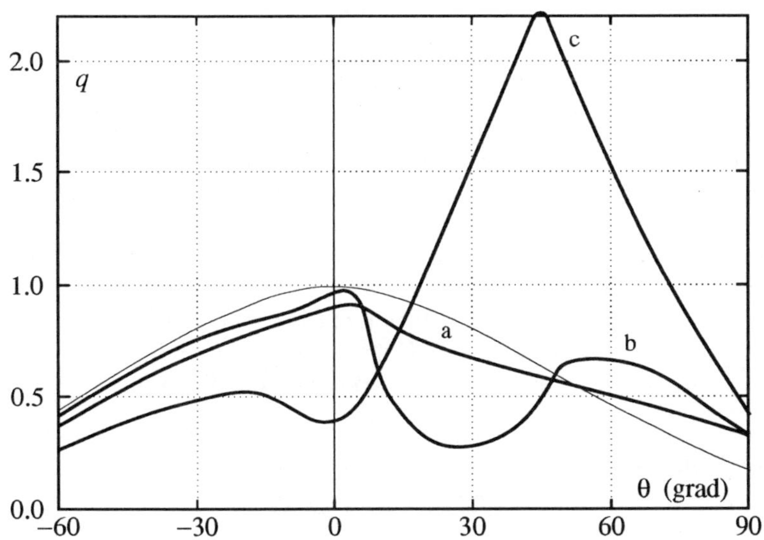

Figure 3.30 Heat flux distribution over a circular cylinder with an impinging shock; hair line, uniform free stream.

Baldwin and Lomax (1978) were used. The numerical method used a finite-volume notation of the governing equations with flux and coordinate splitting. Integration in time was performed with the implicit Euler scheme. Calculations were carried out with a non-uniform grid which contained 151×81 nodes. The comparison with the measurements of Wieting (1987) showed that the calculations reproduced the real flow structure including such details as the triple shock configurations, shear layers, jets and separation of the boundary layer. Nevertheless, the calculated wall pressure and heat flux distributions over the cylinder surface demonstrated only qualitative agreement with the experimental data. The authors attributed the discrepancies to the use of a coarse computational grid.

3.4 FLOWS WITH MASS INJECTION

Mass blowing is an effective way to reduce aerodynamic heating at high supersonic flight speed and to control, to some extent, the aerodynamic characteristics. In the section which follows, we assume the foreign gas to be injected into the shock layer through the permeable body surface and its properties to be identical with those of the free stream gas.

3.4.1 Flow on the stagnation streamline

Flow in the vicinity of the stagnation streamline ahead of a blunt body can be investigated with the use of self-similar solutions based on series expansions (2.63). An applicability of the self-similar solutions to shock layer flows with mass injection was investigated by Gershbein (1974). As for the case of an impermeable body surface, the study of such flows can generally be carried out retaining only the largest terms of (2.63).

Consider, at first, the numerical solution of the problem in the domain including the bow shock wave (Gershbein and Kolesnikov 1979). Navier–Stokes equations for the stagnation streamline flow are written using Dorodnitsin's independent variable

$$\eta = (1 + \nu) \int_1^r \rho r^\nu \, dr \tag{3.15}$$

where r is the distance from the centre of the bluntness normalized by its radius; $\nu = 0$ and 1 for plane and axisymmetric flows, respectively; the density is normalized by the free stream value.

Introducing the non-dimensional variables and the stream function $f = -r^{1+\nu} \rho v$, one obtains the following set of ordinary differential equations:

$$
\begin{aligned}
&\left[\frac{1+\nu}{Re_\infty}\mu\rho r^{1+2\nu}\left(u'-V\right)+fu\right]' \\
&\quad+\frac{\mu r^\nu}{Re_\infty}\left(u'-V\right)-u^2-\frac{1}{1+\nu}u(u+v) \\
&\quad-\frac{2}{1+\nu}\frac{p_2}{\rho}+\frac{2}{3}\frac{\mu r^\nu}{Re_\infty}\left[v'-(2-\nu)V\right]=0 \\
&\left[\frac{1+\nu}{Re_\infty}\mu\rho r^{1+2\nu}\left(-\frac{4}{3}v'+\frac{2}{3}\frac{u+v}{\rho r^{1+\nu}}\right)-fv+r^{1+\nu}p\right]' \\
&\quad+uv-\frac{p}{\rho}-\frac{(1+\nu)\mu r^\nu}{Re_\infty}\left(u'-V\right) \\
&\quad+\frac{2}{3}\frac{1+\nu}{Re_\infty}\mu r^\nu\left(-v'+(2-\nu)V\right)=0 \\
&\left\{\frac{1+\nu}{Re_\infty}\frac{\mu\rho r^{1+2\nu}}{Pr}\left[h'-Pr\,v\left(-\frac{4}{3}v'+\frac{2}{3}\frac{u+v}{\rho r^{1+\nu}}\right)\right]+f\left(h+\frac{v^2}{2}\right)\right\}' \\
&\quad-u\left(h+\frac{v^2}{2}\right)+\frac{1+\nu}{Re_\infty}\mu r^\nu \\
&\quad\times\left[\frac{2}{3}\left(-v'+(2-\nu)V\right)u+\left(u'-V\right)v\right]=0 \\
&p_2'=-p'+\rho uV \qquad u=f'
\end{aligned}
\tag{3.16}
$$

Here $V=\frac{u+v}{(1+\nu)\rho r^{1+\nu}}$, the subscript indicating the number of the expansion term is retained only for the second coefficient of the pressure expansion; the prime denotes the derivatives with respect to η; the velocity components, specific enthalpy, pressure and temperature are normalized by V_∞, V_∞^2, $\rho_\infty V_\infty^2$, $(\gamma-1)M_\infty^2 T_\infty$, respectively. The Reynolds number is defined using the free stream parameters and the body nose radius.

The geometric coordinate is determined from the differential equation

$$
\left(r^{1+\nu}\right)'=1/\rho \tag{3.17}
$$

subject to the boundary condition

$$
r(0)=1 \tag{3.18}
$$

The boundary conditions for set (3.16) are imposed in the free stream and on the body surface. The free stream boundary conditions are given by relationships (2.69). The wall boundary conditions are

$$
u=0 \qquad v=v_w \qquad h=h_w \tag{3.19}
$$

The mass injection rate $G=-f_w=\rho_w v_w$ can be prescribed on the body surface instead of the normal velocity component v_w.

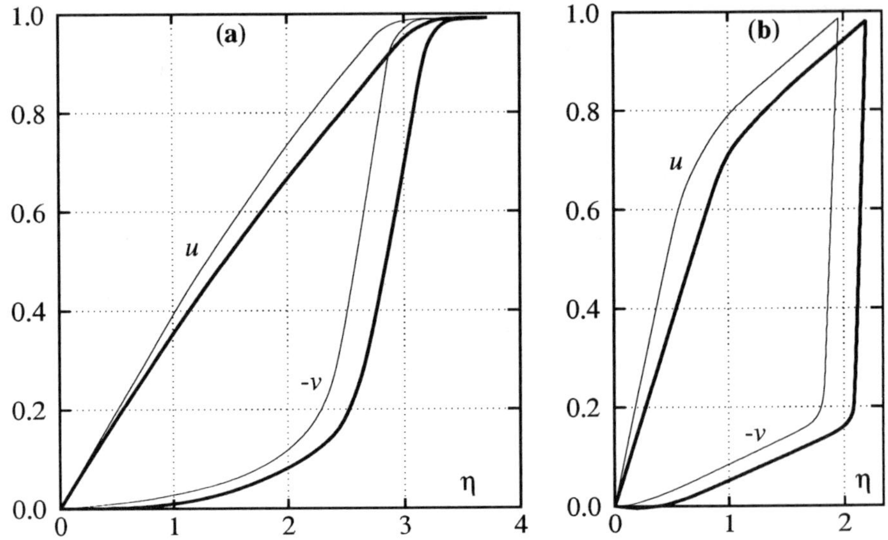

Figure 3.31 Profiles of gas velocity components on the stagnation streamline with and without mass injection. (*a*), $Re_\infty = 100$; (*b*), $Re_\infty = 1100$; full lines, $G = 0.1$; hair lines, $G = 0$.

In a general case, set (3.16) has two node-type singular points, one being located in the free stream and another at the point where $f = 0$. The second singular point is within the shock layer in the flows with mass injection and disappears in the flows with mass suction through the body surface. For an impermeable body, it is the stagnation point on the body surface.

After some simplifications consistent with the hypersonic flight speed, the governing equations were approximated by the implicit fourth-order accurate finite-difference scheme of Petukhov (1964). The non-linear set of difference equations was solved by iterations. In each iteration the first three equations of (3.16) were solved one after another using the 'progonka' algorithm. Then, the function p_2 was calculated with the Simpson quadrature formula.

The results illustrating the mass blowing effect on the profiles of gas dynamic functions are presented in figure 3.31. They pertain to flows of a perfect gas with specific heat ratio $\gamma = 1.4$, viscosity $\mu \propto \sqrt{T}$, Prandtl number$Pr = 0.75$ at free stream Mach number $M_\infty = 10$ and wall temperature $T_w = 0.3T^*$. The full and hair lines depict calculations of the flows with and without mass injection, respectively. It is seen that at fixed mass injection rate, the decrease of the Reynolds number results in a diminution of the injection effect. At $Re_\infty = 1100$, the bow shock wave becomes nearly a discontinuity and the boundary layer appears. Mass injection makes the boundary layer thicker with simultaneous increase of the bow shock detachment from the body surface.

The next figures demonstrate the dependence of the flow characteristics on the mass injection rate. The calculations were carried out with the full Navier–Stokes

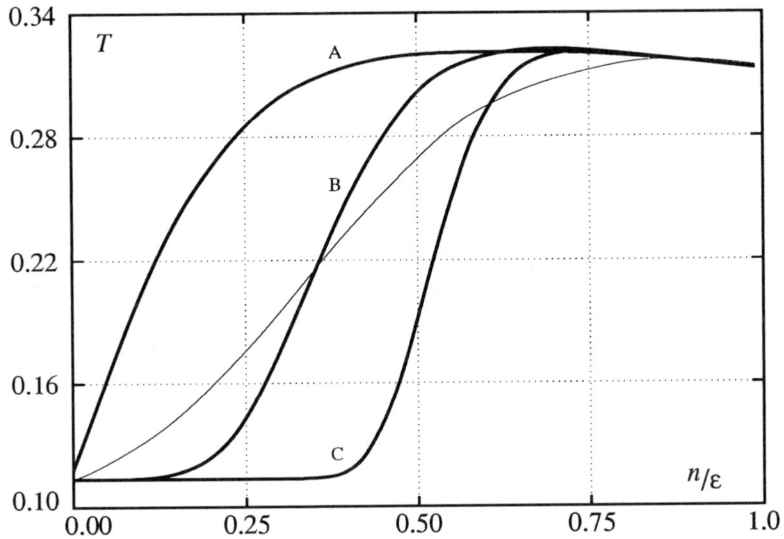

Figure 3.32 Temperature profiles on the stagnation streamline. Full lines, $Re_\infty = 710$; hair line, $Re_\infty = 177$; A, $G = 0.0$; B, $G = 0.50$; C, $G = 1.11$.

equations (2.64)–(2.68) in the domain bounded by the bow shock wave. The solutions were obtained by a one-dimensional version of the numerical method described in subsection 1.2.2; see also subsection 2.5.1.

The calculations presented in figures 3.32 and 3.33 pertain to flows of a perfect gas with specific heat ratio $\gamma = 1.4$; viscosity $\mu \propto \sqrt{T}$, Prandtl number $Pr = 0.7$ at the free stream Mach number $M_\infty = 6$ and wall temperature $T_w = 0.35T^*$. Figure 3.32 shows the temperature profiles on the stagnation streamline at two Reynolds numbers. It is seen that at $Re_\infty \gtrsim 700$ and $G \gtrsim 0.5$, the shock layer flow consists of two inviscid regions divided by a shear layer. Figure 3.33 displays the variation of the wall pressure and heat flux on the sphere surface with mass injection rate. The heat flux values are normalized by those calculated with $G = 0$. The hair-dotted curve depicts the heat flux variation obtained from the boundary layer calculations (Anfimov 1966, Avduevskii and Glebov 1970). One can see a noticeable reduction of the mass injection efficiency in respect of heat protection of the body surface with a Reynolds number decrease. This is also seen from the temperature profiles plotted in figure 3.32. Higher mass injection rate results in a decreasing wall pressure. The same effect has been revealed by Gilinskii and Lebedev (1977) in inviscid flow calculations.

Figure 3.34 shows the variation of the distances from the body surface to the stagnation point Δ and to the bow shock ε with the injection parameter $\Lambda = \rho_w v_w^2/\rho_\infty V_\infty^2$. Both distances are normalized by the sphere radius. Here the gas specific heat ratio $\gamma = 1.4$, viscosity $\mu \propto T^{3/4}$. The calculations show a nearly linear variation of the above distances with $\sqrt{\Lambda}$. The circles depict the experimental data by Katzen and Kaattari (1965).

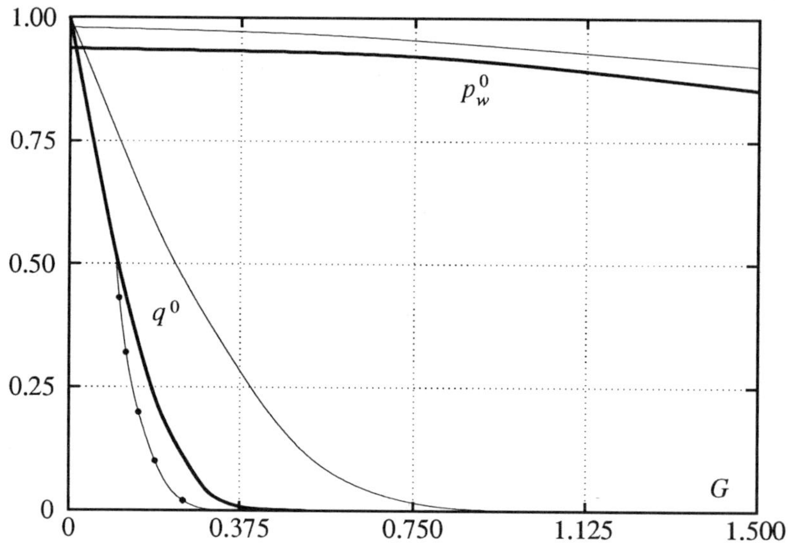

Figure 3.33 Dependence of the heat flux and wall pressure on mass injection rate. Full lines, $Re_\infty = 710$; hair lines, $Re_\infty = 177$; hair-dotted line, boundary layer theory.

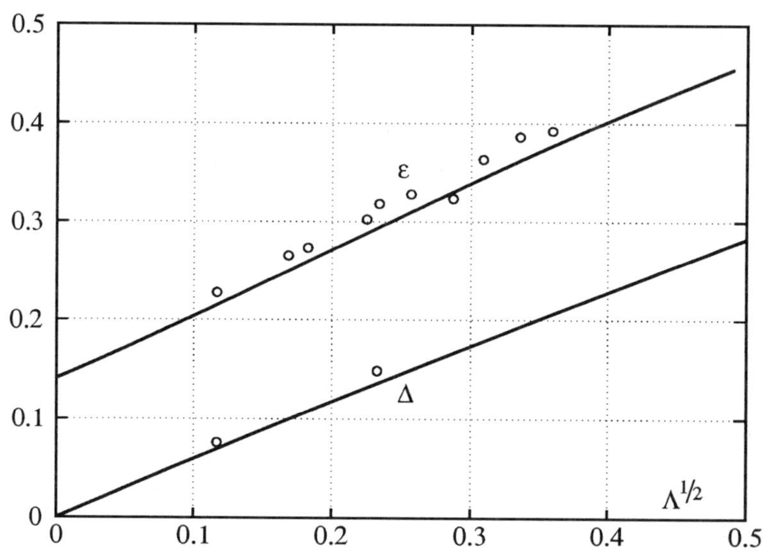

Figure 3.34 Variation of the distances from the body surface to the stagnation point and to the bow shock on the stagnation streamline with mass injection rate. $M_\infty = 10$, $Re_\infty = 10^4$, $T_w = 0.6T^*$. Lines, self-similar solution of the Navier–Stokes equations; circles, experimental data of Katzen and Kaattari (1965).

3.4.2 Flow about the whole front and lateral parts of a blunt body

Let us consider numerical solutions of some two-dimensional problems beginning with calculations of the flow about a bluntness. Since mass injection is applied generally to reduce the heat flux to the body surface, the optimum distribution of the mass injection rate in this respect is of interest. To investigate various means of mass blowing, Krasilov *et al* (1986 a,b) calculated heat transfer on the sphere surface prescribing the mass injection rate distribution in accordance with the formula

$$G = G_0 \exp\left[-(s/s_*)^2\right] \tag{3.20}$$

where the distance s is measured from the stagnation line along the body generatrix. At large values of s_*, $G(s) \to G_0$, i.e. the mass injection rate is constant over the body surface. At small values of s_*, the mass injection is localized within a stagnation region. The case $G_0 = 0$ corresponds to an impermeable body surface.

Since the total mass injection rate

$$G_\Sigma = 2\pi \int_0^S G(s) r(s) ds$$

is usually limited, the optimum distribution is searched for at fixed G_Σ. The value of s_* being prescribed, the condition $G_\Sigma = const$ is satisfied by determination of G_0 with the formula

$$G_0 = G_\Sigma \Big/ 2\pi \int_0^S \exp\left[-(s/s_*)^2\right] r(s) ds \tag{3.21}$$

Calculations were carried out within the framework of the thin viscous shock layer model whose applicability to hypersonic shock layer flows at sufficiently small values of the mass injection parameter $\Lambda = (\rho_w v_w^2)/(\rho_\infty V_\infty^2)$ has been shown by Gershbein (1974).

Figure 3.35 shows the heat flux distributions over a sphere surface calculated at several values of parameter s_*. Here the hypersonic similarity criterion $\delta = (\gamma - 1)/2\gamma = 0.1$, the viscosity $\mu \propto \sqrt{T}$, the Prandtl number $Pr = 0.71$, the total non-dimensional mass injection rate $G_\Sigma = 0.063$, the Reynolds number $Re = (\rho_\infty V_\infty R)/\mu(T^*) = 500$, and the wall temperature $T_w = 0.1T^*$. The hair line depicts the heat flux distribution over an impermeable body. The smaller s_*, the more intensive the mass injection in the stagnation region. As a result, the dependence $q(s)$ becomes non-monotonous. At $s_* = 0.2$, the heat flux diminishes to zero within a part of the body surface bounded by the line $s \approx 0.15$. The results of figure 3.35 demonstrate a non-monotonous variation of the maximum heat flux with parameter s_*. So, there is an optimum mass injection rate distribution (3.20) with fixed total rate providing the lowest value of q_{max}. The optimum distribution

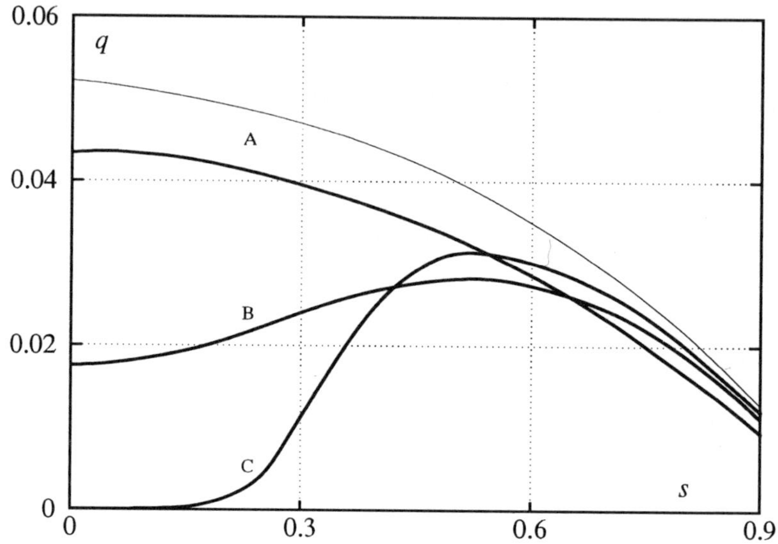

Figure 3.35 Heat flux variation over a sphere surface at various mass injection rate distributions. A, $s_* = 1.2$; B, $s_* = 0.4$; C, $s_* = 0.2$; hair line, $G = 0$.

of the mass injection rate ensures a more effective heat protection as compared with the uniform one with the same value of the total rate G_Σ.

The next figures present calculations of the flow past a spherically blunted cone with mass injection on its spherical part. Asymptotic analysis of such flows at high Reynolds and Mach numbers and intensive blowing ($v_w/V_\infty \gg 1/\sqrt{Re}$) was curried out by Ermack (1983). Numerical results discussed below were obtained by Belousov and Golovachov (1992) using the reduced Navier–Stokes equations (set II of subsection 2.1.1). The gas specific heat ratio $\gamma = 1.4$, the viscosity $\mu \propto \sqrt{T}$, the Prandtl number $Pr = 0.7$, and the cone half-angle $\theta_c = 10°$. The calculations were carried out by the implicit unidirectional method of subsection 1.2.2 with adaptive computational grids ensuring resolution of both the internal shear layer in the vicinity of bluntness and the boundary layer adjacent to the lateral part of the body surface. The adaptive grid was generated using the equidistribution law; see subsection 1.2.3.

Figure 3.36 shows the temperature profiles in several cross-sections of the shock layer. The dimensionless blowing velocity on the spherical part of the body ($s \leqslant 1.396$) is assumed to be $v_w = 0.02$ while the conical part is assumed to be impermeable. It is seen that, similar to the previous problem, the shock layer flow in the vicinity of the body bluntness consists of two inviscid regions divided by a mixing layer. It is worth noting that because of expansion about the spherical bluntness, the injected gas temperature falls to a smaller value as compared with the wall temperature T_w. This effect gradually disappears near the conical part of the body as the injected gas is swallowed up by the shear layer.

Figure 3.37 shows the variation of the heat flux to the conical part of the

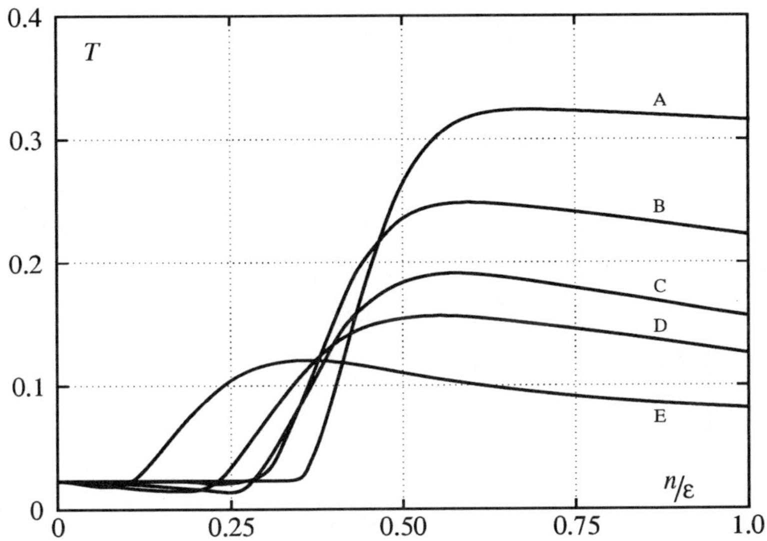

Figure 3.36 Temperature profiles in the shock layer on a spherically blunted cone with mass injection on its spherical part. $M_\infty = 6$, $Re_\infty = 10^3$, $T_w = 0.075T^*$; A–E, $s = 0, 0.87, 1.40, 1.92, 4.01$.

body surface when the normal velocity component on the spherical bluntness is prescribed by the formula

$$v_w = 0.01 \cos s \qquad s \leqslant 1.396$$

The heat flux is normalized by its value at the stagnation point of the impermeable body. The hair line depicts the heat flux variation over a cone without mass injection.

The next figures demonstrate the effects of mass blowing through slots located on the lateral parts of a circular cylinder with the flat butt and a flat plate (Kulnev and Paskinov 1983). The full Navier–Stokes equations were solved by the implicit finite-difference scheme of Cheraneva (1980) in the domain including the bow shock wave. Calculations were carried out for the slot $d = 0.25$ wide, whose midst was located at a distance $c = 1.975$ from the front butt of the body. Here, all distances are normalized by the plate thickness or cylinder diameter.

Figure 3.38 displays the flow structure near the plate at Mach number at the slot brink $M_w = 1.04$. The hatched areas show the bow shock wave and the inner shock arising from the deceleration of a supersonic flow by the jet. The above shocks are noticeably smeared because of a fairly low Reynolds number. The figure depicts also the streamlines. The streamline $\psi = 0$ represents the jet edge. The behaviour of the streamline $\psi = 0.01$ near the plate surface upstream of the slot demonstrates an ejection effect. One can also see a small recirculation zone in front of the slot. The mass injection strongly influences the sonic lines shown by the dotted curves. At the injection rate considered, the supersonic regions of

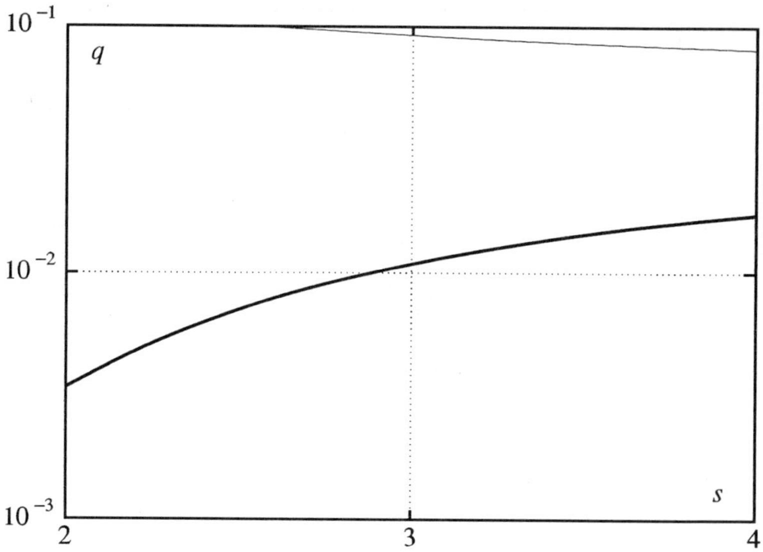

Figure 3.37 Heat flux distribution over the lateral part of a spherically blunted cone with (full line) and without (hair line) mass injection on its spherical part. $M_\infty = 6$, $Re_\infty = 10^3$, $T_w = 0.075T^*$.

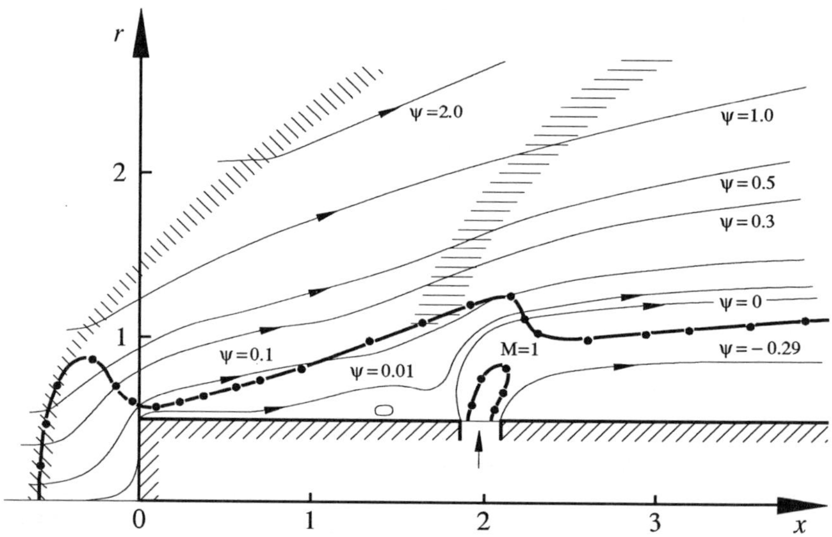

Figure 3.38 Flow near a plate in a supersonic stream with mass injection through the slot. $M_\infty = 4$, $M_w = 1.04$, $Re_\infty = 200$, $T_w = 0.714T^*$.

the jet and the main flow do not merge and the subsonic region of the shock layer remains unbroken.

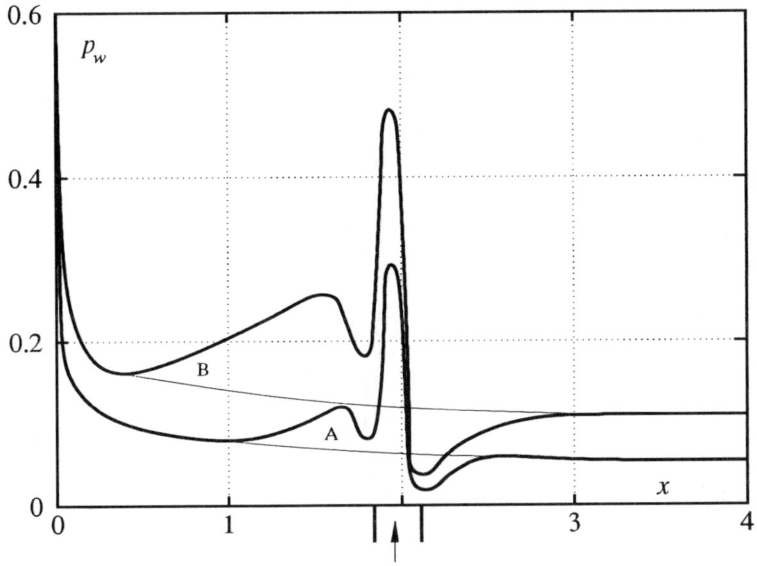

Figure 3.39 Wall pressure distribution on the cylinder (A) and plate (B). $M_\infty = 4$, $M_w = 1.04$, $Re_\infty = 200$, $T_w = 0.714T^*$.

Figure 3.39 shows the wall pressure distributions over the lateral parts of the cylinder and the plate. The hair lines depict the results calculated without mass injection. Wall pressure variations on the plate and the cylinder are similar, but in a plane flow, the pressure is higher and the region influenced by mass injection is by far more extensive both upstream and downstream. It is seen that in both cases there is a region of higher pressure in front of the slot. It is followed by a pressure minimum. An absolute pressure maximum is achieved in the injection region. Just behind the slot, the wall pressure is far less than on the impermeable surface and gradually approaches this level with increasing distance from the slot. The same behaviour of the wall pressure, except the minimum in front of the slot, has been revealed by Speid and Zukoski (1968) for flows at Reynolds numbers $10^5 \div 10^6$.

Figures 3.40 and 3.41 demonstrate the effect of the Mach number M_w on the heat flux distribution and sonic lines in the flow past a cylinder. With increasing M_w, the mass injection rate increases. For the above Mach numbers M_w, the injection parameter $G = (\rho v)_w / \rho_\infty V_\infty$ takes the following values: 0.0, 0.15, 1.0, 2.2.

One can see in figure 3.40 that deceleration of the main flow by the jet results in a rise of the gas temperature ahead of the slot and a greater heat flux to the body surface. With increasing injection rate, the heat flux in this region monotonously increases.

The graphs in figure 3.41 show that at a relatively small injection velocity ($M_w = 0.35$), the sonic line is only slightly distorted. With increasing injection

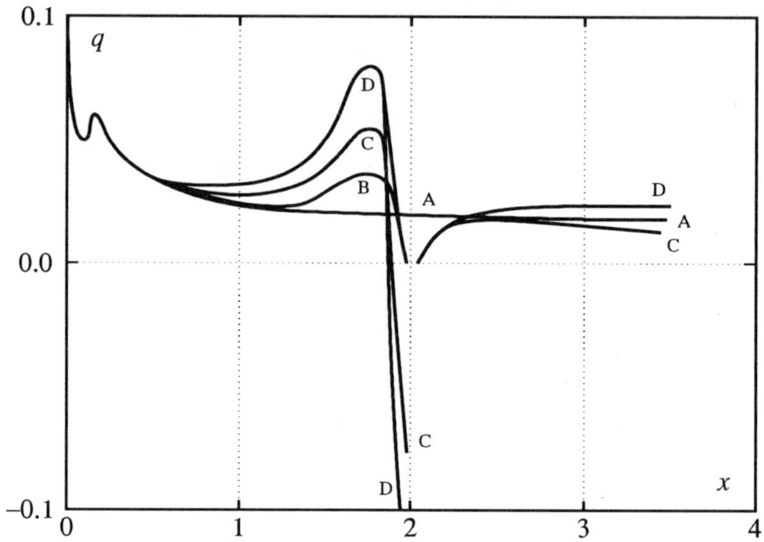

Figure 3.40 Dependence of the heat flux distribution over a cylinder on the injection Mach number M_w. $M_\infty = 4$, $Re_\infty = 200$, $T_w = 0.714T^*$; A, $M_w = 0$; B, $M_w = 0.35$; C, $M_w = 1.04$; D, $M_w = 1.4$.

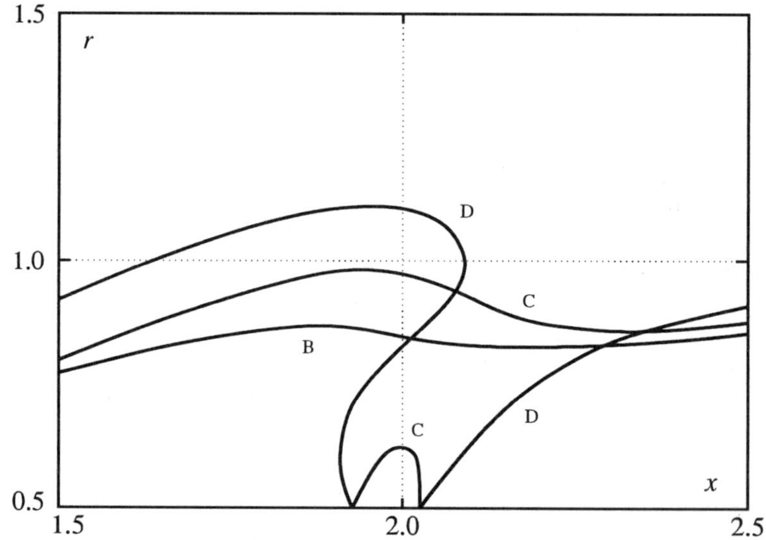

Figure 3.41 Effect of injection Mach number on the sonic line in the flow past a cylinder. $M_\infty = 4$, $Re_\infty = 200$, $T_w = 0.714T^*$; B, $M_w = 0.35$; C, $M_w = 1.04$; D, $M_w = 1.4$.

rate, the sonic line moves away from the slot. At $M_w = 1.04$, the small supersonic region appears just over the slot, which expands with increasing M_w. At $M_w = 1.4$, it divides the subsonic region of the shock layer into two parts located upstream and downstream from the slot.

A more detailed discussion of the numerical results including, in particular, the effects of Reynolds number, location of the slot and injection velocity distribution on the flow field structure, and the drag and heat transfer parameters can be found in a paper by Kulnev and Paskonov (1983).

3.5 UNSTEADY FLOWS

Most available calculation data on unsteady supersonic flows about bodies have been obtained within the inviscid gas model; see for instance the book of Belotserkovskii *et al* (1974) and the papers of Turchak (1976), Lipnitskii and Lyakhov (1976), Tugazakov (1980), Kamenetskii and Turchak (1984), Goloviznin *et al* (1987). The inviscid gas model provides, in particular, the wall pressure distribution, which enables one to solve the boundary layer problem and to determine the heat flux and shear stress on the body surface. However, with increasing Strouhal number, the boundary layer theory fails even for high Reynolds number flows, which is caused by a rise and development of intensive wave processes within the shock layer. So, it seems expedient to calculate essentially unsteady flows using the viscous shock layer approach. Such calculations were carried out for a plane shock interaction with a shock layer on a circular cylinder (Scala and Gordon 1968), the flow about an accelerating body (Belousov *et al* 1988), flows with variable mass injection rate (Markov 1988), flows around bodies moving in non-homogeneous media (Golovachov and Leontieva 1989a–c). The calculation results for some of the above problems are discussed below.

3.5.1 Supersonic flow about an accelerating blunt body

Consider an axisymmetric flow past a blunt body moving during the time $t_1 \leqslant t \leqslant t_2$ at a speed increasing from V_1 to V_2. The flight velocities before and past the above time interval, V_1 and V_2, are supposed to be constant and correspond to the free stream Mach numbers M_1 and M_2, respectively. The shock layer flow is described by the reduced Navier–Stokes equations (set II, see subsection 2.1.1) which are written for the body-fitted non-inertial coordinate system. The continuity and energy balance equations do not differ from (1.1) and (1.4) while the momentum balance equations include the components of vector $\rho \boldsymbol{W}$ accounting for the coordinate system acceleration $\boldsymbol{W}(t)$. The characteristic time of the unsteady flow is by far greater than the time of gas flow through the bow shock wave. Accordingly, the conventional Rankine-Hugoniot relationships are used. Other boundary conditions are also similar to those applied in the calculations of stationary flows. The initial flow field data are obtained from solutions of the stationary problems.

The calculations were carried out by the implicit unidirectional finite-difference scheme of subsection 1.2.2 on the grids which, for the most part, contained 16 equally spaced lines $s = const$ with 51 nodes on each line. The grid points were

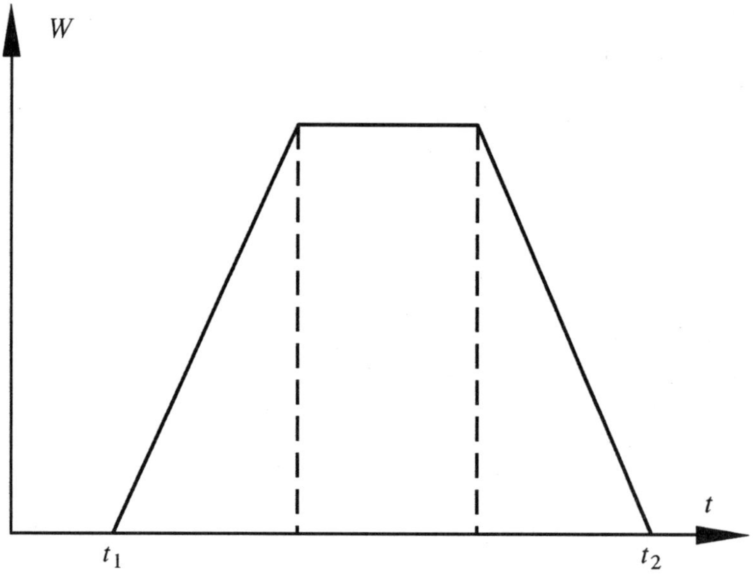

Figure 3.42 Time dependence of body acceleration.

distributed along the lines by means of either coordinate transformation (1.43) or the equidistribution principle of subsection 1.2.3.

Consider some calculation results which pertain to supersonic flow past a sphere accelerating as shown in figure 3.42. The free stream Mach number varies from $M_1 = 4$ to $M_2 = 6$; the gas specific heat ratio $\gamma = 1.4$, the viscosity $\mu \propto \sqrt{T}$, the Prandtl number $Pr = 0.7$ and wall temperature $T_w = 0.293T_1^*$ where T_1^* is the stagnation temperature of the oncoming stream with Mach number M_1.

The results are presented in a non-dimensional form using the following reference values of time, pressure and heat flux: $\Delta t = t_2 - t_1$ $(t_1 = 0)$, $\rho_\infty V_1^2$, $\rho_\infty V_1^3$. The shock layer thickness is normalized by the sphere radius R.

Figures 3.43a–c show the variation of the shock layer thickness, pressure at the stagnation point of a sphere and pressure just behind the bow shock on the symmetry axis at two Strouhal numbers $Sh = R/(V_1 \Delta t)$. The results are compared with the quasi-stationary solution. Here, the Reynolds number is defined as $Re = \rho_\infty R V_1 / \mu(T_\infty)$. One can see an essentially non-stationary behaviour of the shock layer parameters under the flight conditions considered. Also note a delay in p_s^0 increase, which is equal approximately to the time of the compression wave propagation from the body to the bow shock.

The compression waves and their interaction with the bow shock and the body surface are clearly seen in figure 3.44 which presents pressure profiles on the stagnation streamline at Strouhal number $Sh = 5$.

Figures 3.45 and 3.46 show the variation of the heat transfer parameter at the stagnation point of a sphere and its distribution over the sphere surface. The discrepancy of the results for two Reynolds numbers as well as a large pressure

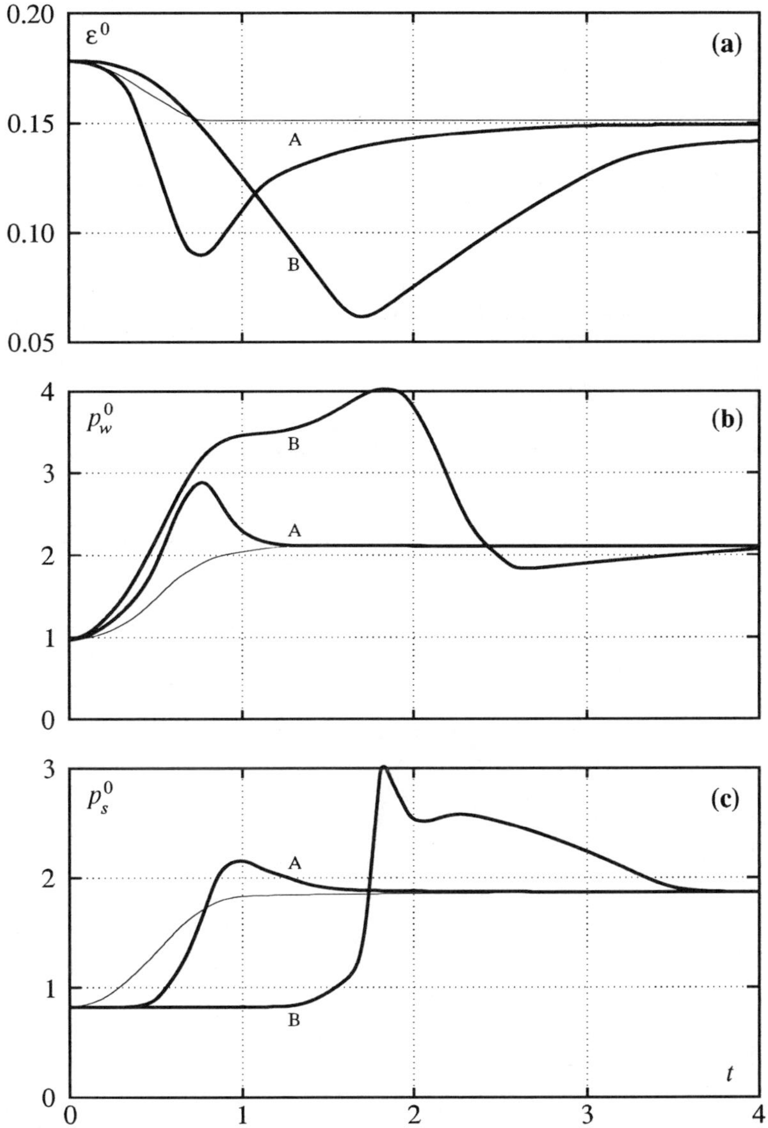

Figure 3.43 Time variation of the shock layer thickness, wall pressure and pressure just behind the bow shock on the symmetry axis ahead of an accelerating sphere. $Re = 2370$; A, $Sh = 1$; B, $Sh = 5$; hair lines, quasi-stationary solution.

gradient in the near-wall region of the shock layer (see figure 3.44) indicate an invalidity of the conventional boundary layer theory.

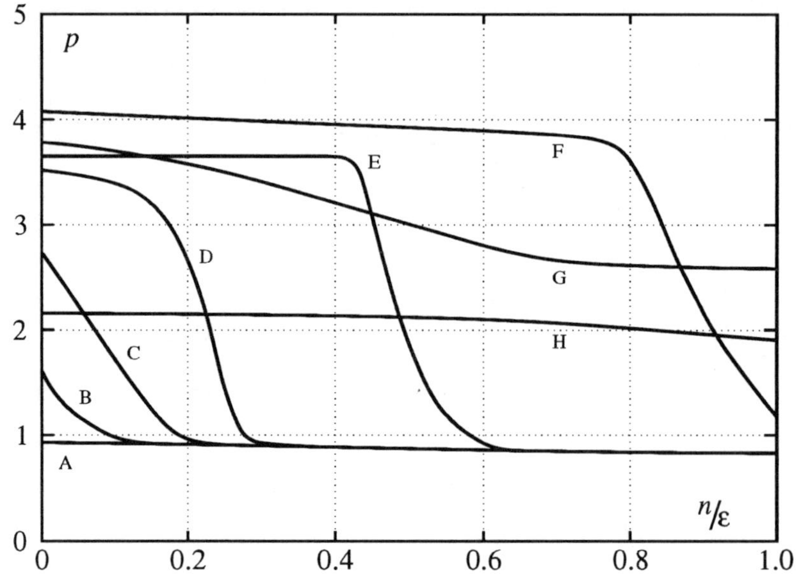

Figure 3.44 Pressure profiles on the stagnation streamline in front of an accelerating sphere. $Re = 2370$, $Sh = 5$; A–H, $t = 0, 1/3, 2/3, 1, 4/3, 5/3, 2, \infty$.

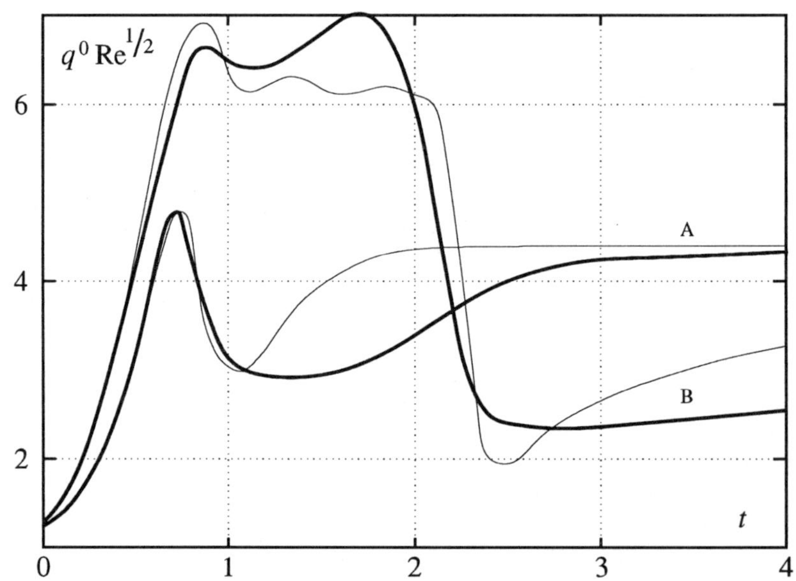

Figure 3.45 Time variation of the heat transfer parameter at the stagnation point of an accelerating sphere. Full lines, $Re = 2370$; hair lines, $Re = 237$; A, $Sh = 1$, B, $Sh = 5$.

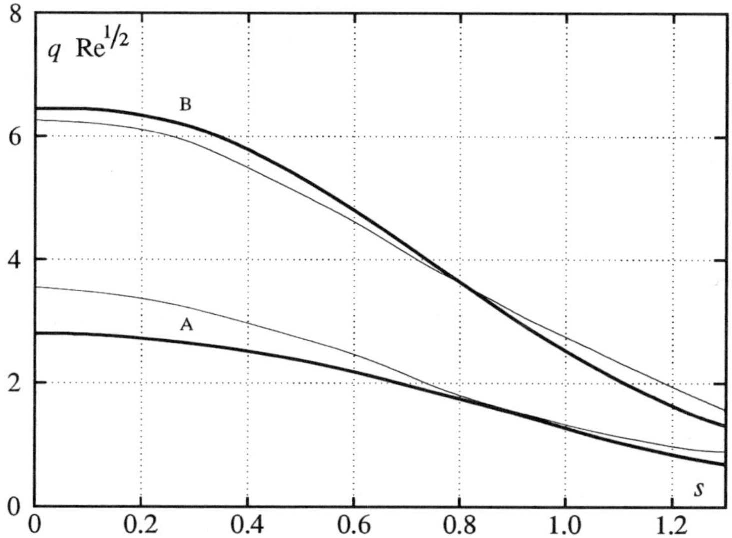

Figure 3.46 Distribution of the heat transfer parameter over an accelerating sphere at $t = 4/3$. Full lines, $Re = 2370$; hair lines, $Re = 237$; A, $Sh = 1$; B, $Sh = 5$.

3.5.2 Flow about a blunt body moving through thermal non-homogeneities

The non-homogeneities considered are infinite plane gas layers with constant pressure equal to its ambient value and variable temperature and density. The temperature distributions within the non-homogeneities are schematically shown in figure 3.47. They are described by the formula

$$T(x) = T_0 \left\{1 - a \exp\left[-(x/b)^2\right]\right\}^n \tag{3.22}$$

where the temperature overfall and non-homogeneity width are characterized by the parameters a and b. The distributions A and B are described by formula (3.22) with $n = \mp 1$, respectively, applying this formula only at $x \leqslant 0$. To describe case C, formula (3.22) with $n = -1$ is employed at $-\infty < x < \infty$.

The body moves in the direction which is normal to the non-homogeneity, so that the flow under study is axisymmetric. The flight speed is assumed to be constant and to exceed the local speed of sound.

The computational domain is bounded by the bow shock wave. The shock layer flow is calculated using the reduced Navier–Stokes equations (set II, see subsection 2.1.1). Their validity for the simulation of the flows under study has been proved by a comparison with solutions of the full Navier–Stokes equations. The bow shock boundary conditions are formulated using the time-dependent free stream parameters. The gas viscosity is calculated by Sutherland's formula (1.8), the Prandtl number $Pr = 0.7$, the gas specific heat ratio $\gamma = 1.4$ and wall temperature $T_w = T_0 = 300$ K. The numerical method is the same as that in the preceding problem.

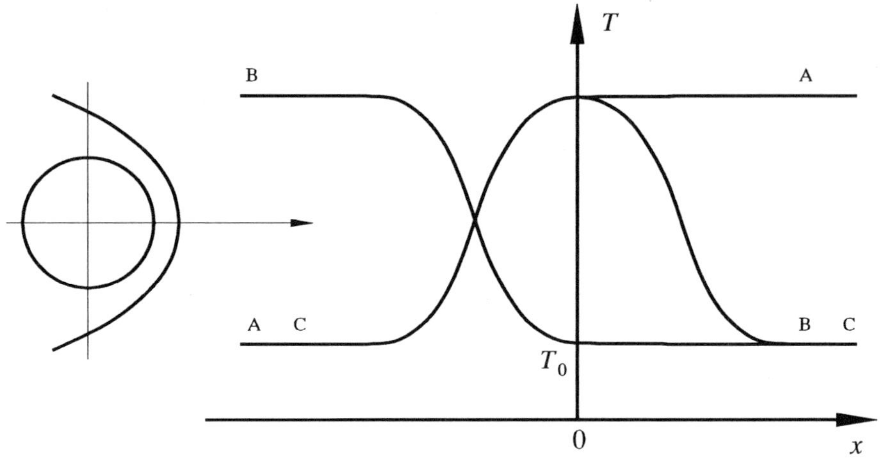

Figure 3.47 Gas temperature distributions in the thermal non-homogeneities.

Figures 3.48 through 3.50 present calculations of the flow past a sphere moving in a gas with increasing ($n = -1$) and decreasing ($n = +1$) temperature at $a = 0.9$, $b = 0.45R$, where R is the sphere radius. With the above non-homogeneity parameters, the gas temperature varies ten times on a length equal approximately to the body characteristic length. The stationary flow field obtained at $t \to \infty$ with $n = -1$ was used for the initial data in the calculation of the flow with $n = +1$. In both cases, the flight speed V was supposed to be the same. The Mach and Reynolds numbers evaluated using the flight speed V, body nose radius R and gas parameters in the region with the higher density, are the following: $M = 6$, $Re = 10^4$. The calculation results are presented in a non-dimensional form using the following reference values of time, pressure and heat flux: R/V, $\rho_* V^2$, $\rho_* V^3$, where ρ_* is the highest of the density values at $|x| \to \infty$.

Figure 3.48 demonstrates the variation of the bow shock shape and some distinctive features of the shock layer flow field. In this figure, the velocity maps in the shock layer are plotted at the moments when the middle plane of a sphere passes by the non-homogeneity region with maximum temperature gradient. The jet-type flows are seen which appear because of abrupt deceleration of the bow shock motion from the body surface in the vicinity of the stagnation streamline at $n = -1$ and because of compression of the shock layer in this region at $n = +1$.

Figure 3.49 shows the variation of the stagnation point heat flux. One can see clearly an essentially non-stationary nature of the heat transfer on the body surface. Because of the smaller shock layer thickness, the stationary flow regime is achieved sooner when the body moves in a gas with decreasing temperature. In this case the maximum heat flux exceeds by an order of magnitude the values corresponding to the stationary flow regimes.

Figure 3.50 displays the heat flux distributions over the sphere surface. The full lines correspond to the stationary flow regimes at $x \to \pm\infty$. The curve $a(b)$

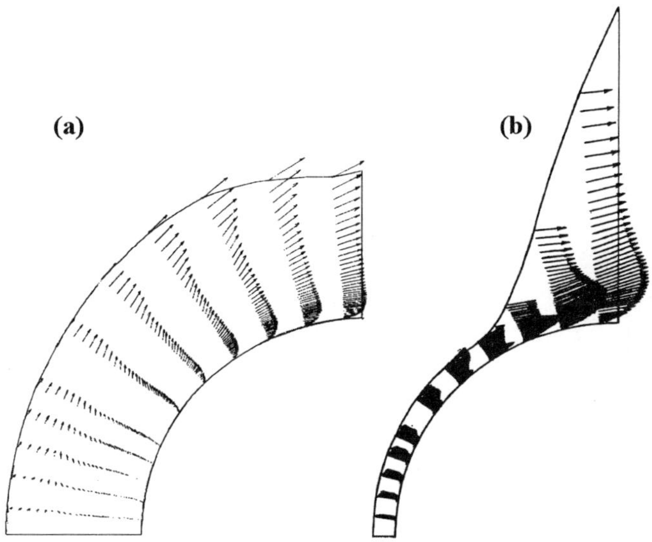

Figure 3.48 Velocity map in the shock layer ahead of a sphere moving in a gas with increasing (a, $t = 2.0$) and decreasing (b, $t = 2.4$) temperature.

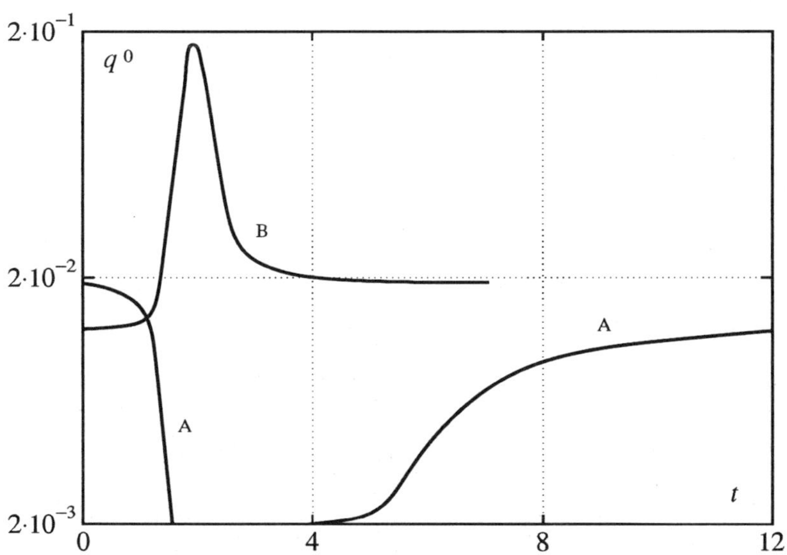

Figure 3.49 Variation of the heat flux at the stagnation point of a sphere moving in a gas with increasing (A) and decreasing (B) temperature.

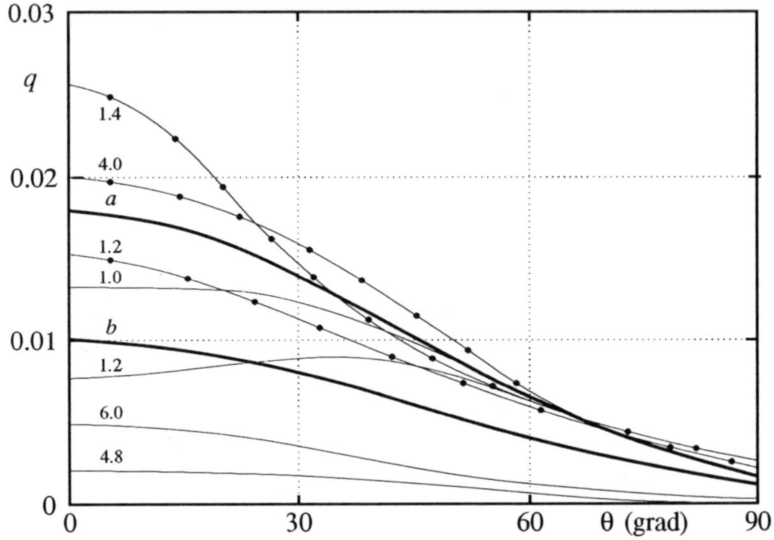

Figure 3.50 Heat flux distributions over a sphere moving in a gas with increasing (hair lines) and decreasing (hair-dotted lines) temperature. Full lines, stationary flows.

pertains to the initial moment in case $A(B)$ of the temperature distribution within the non-homogeneity and to the final moment in case $B(A)$. The heat flux distributions in the cases A and B are plotted by the hair and hair-dotted curves, respectively. Numerals above the curves indicate the non-dimensional time. It is worth noting that the maximum heat flux can be achieved outside the vicinity of the stagnation point.

The next figures present the results for case C. Here, the Mach and Reynolds numbers are evaluated using the gas parameters outside the non-homogeneity. Calculations show that in this case supersonic flow about a blunt body is also accompanied by the appearance of a high-temperature jet which spreads from the symmetry axis along the shock layer. This effect is most prominent at low supersonic speeds.

Figure 3.51 shows pressure contours in the shock layer. They are plotted with spacing $\Delta p = 0.05$ for several moments with $t = 0$ corresponding to the time when the oncoming stream temperature differs by 0.1 per cent from its value outside the non-homogeneity. The pressure contours demonstrate a rise and propagation of intensive expansion and compression waves within the shock layer.

Figure 3.52 depicts the variation of the heat transfer parameter at the stagnation point of a sphere. From the moment when the sphere centre crosses the symmetry plane of the non-homogeneity up to restoration of the stationary flow regime, the heat transfer parameter exceeds substantially its stationary value demonstrating a separation of the results with respect to the Reynolds number. With increasing Reynolds number, the maximum heat transfer parameter is achieved later and diminishes.

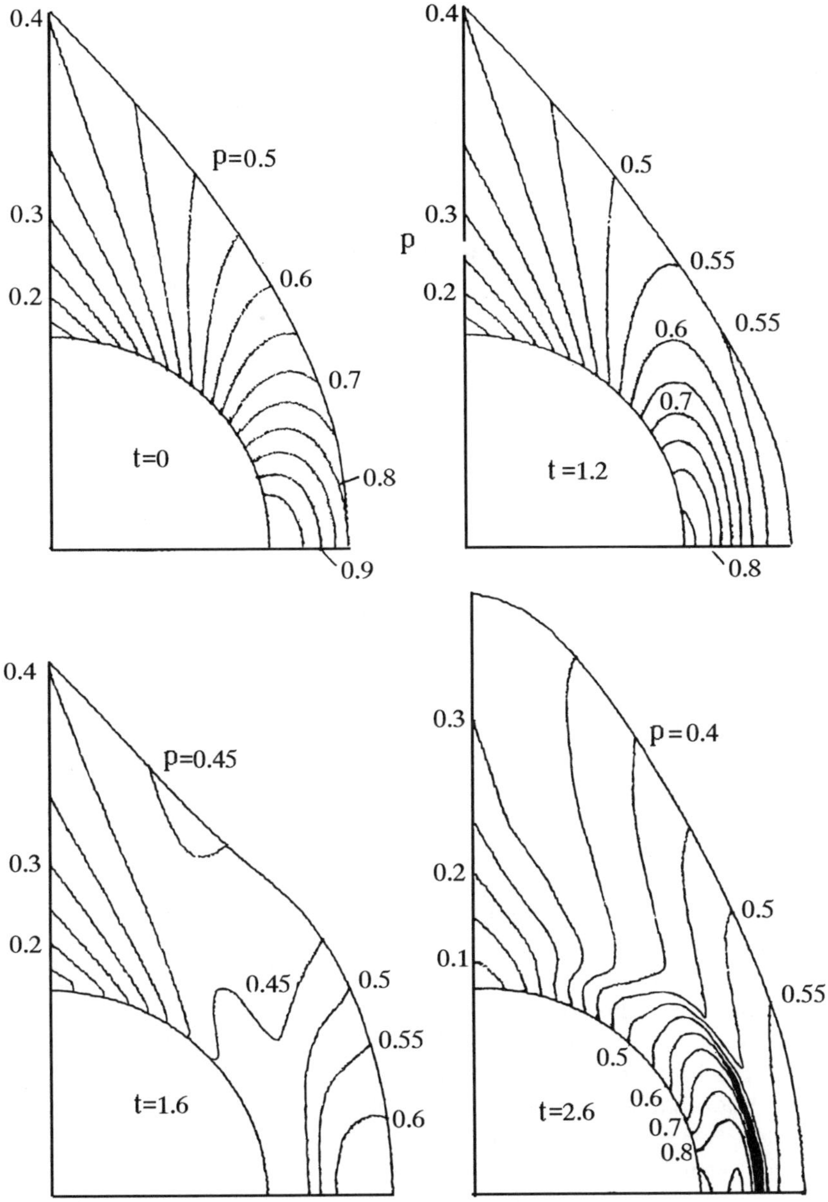

Figure 3.51 Pressure contours in the shock layer ahead of a sphere moving through the symmetric thermal non-homogeneity. $M = 2$, $Re = 10^4$. $a = 0.7$, $b = 0.45R$.

The hair curve in figure 3.52 shows the self-similar solution. To obtain it, one has to assume the bow shock shape in the vicinity of the stagnation streamline

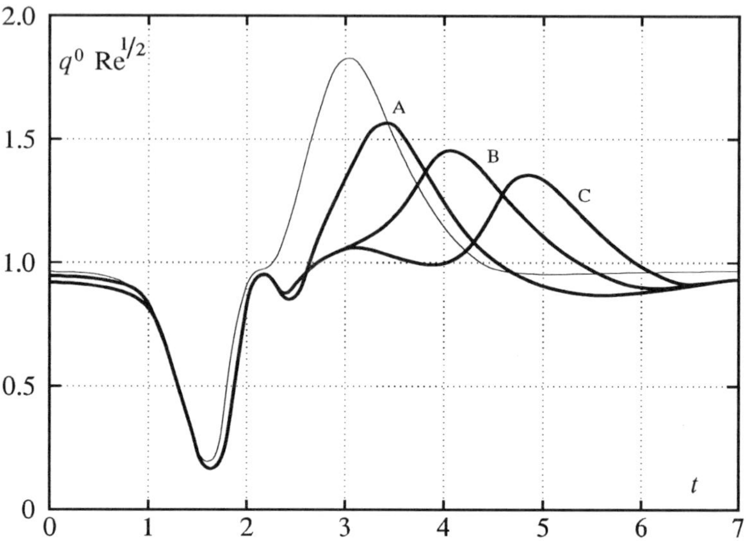

Figure 3.52 Time dependence of the heat transfer parameter at the stagnation point of a sphere moving through the symmetric thermal non-homogeneity. $M = 2$, $a = 0.5$, $b = 0.45R$; A, $Re = 10^3$; B, $Re = 10^4$; C, $Re = 10^5$.

during the flight through non-homogeneity to be the same as in the stationary flow. Accordingly, the bow shock boundary condition for the function u (see subsection 2.5.1) is borrowed from the solution of the stationary two-dimensional problem and does not change in calculating the non-stationary flow. The above quantity differs from the stationary value to the greatest extent when the bow shock wave crosses the symmetry plane of the non-homogeneity. Under the flight conditions considered, the maximum difference amounts to 20 per cent. It is seen in figure 3.52 that the above assumption results in a large error in the prediction of the stagnation point heat flux. The error diminishes with increasing Mach number and is only slightly influenced by the Reynolds number.

The dependence of the non-stationary effects on the length scale of the non-homogeneity is illustrated in figure 3.53 where the numerals depict the values of parameter b normalized by the sphere radius R. Here, the time is normalized by the quantity b/V which characterizes a duration of the body motion within the non-homogeneity. It is seen that the use of this reference time instead of R/V brings together the calculation results for $b \gtrsim R$. At small width of the non-homogeneity ($b \lesssim R$), the ratio R/V characterizing the time of disturbance propagation through the shock layer, seems a more appropriate time scale.

Figure 3.54 compares the calculations with experimental data by Goloviznin *et al* (1987). Here the length scale of non-homogeneity is approximately equal to the sphere diameter.

The above numerical results pertained to laminar flows. Below, some calculations of turbulent flows (Golovachov and Zemlyakov 1993a) are discussed. The

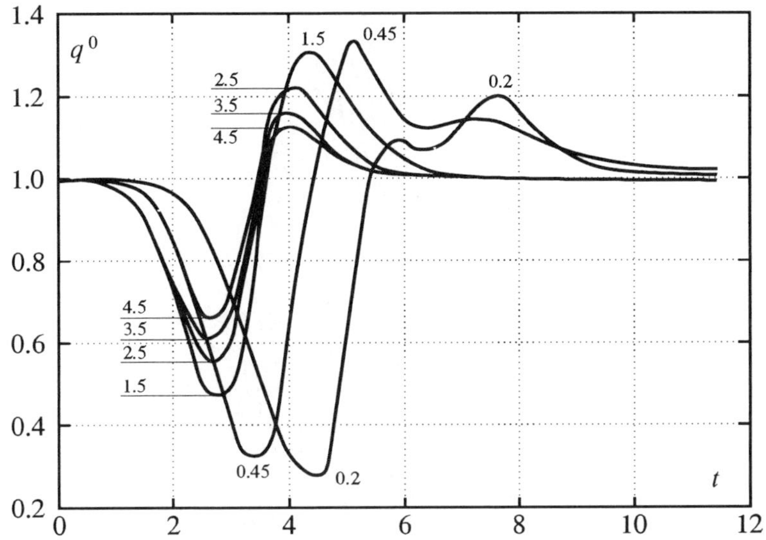

Figure 3.53 Time dependence of the stagnation point heat flux at various widths of the symmetric thermal non-homogeneity. $M = 6$, $Re = 10^4$, $a = 0.5$.

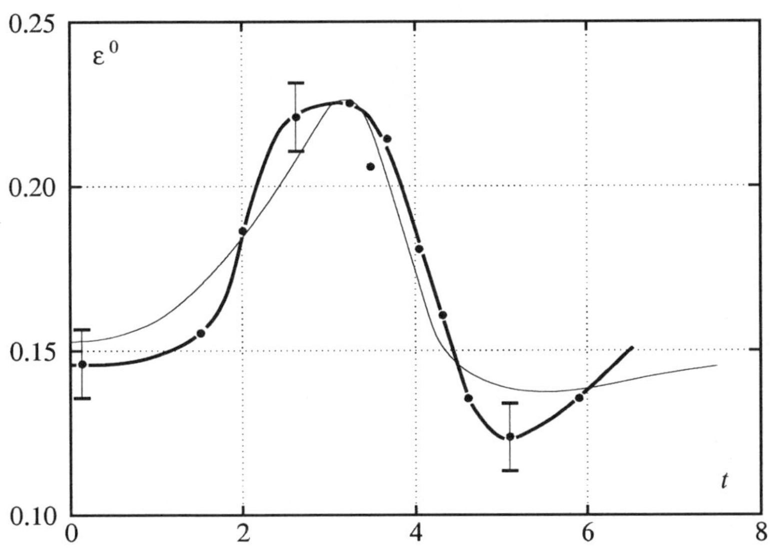

Figure 3.54 Variation of the shock layer thickness on the stagnation streamline ahead of a sphere moving through the symmetric thermal non-homogeneity. $V = 1.98$ km/s, $R = 1$ cm, $p = 13.27$ kPa, $T_0 = 343$ K, $T_{max} = 700$ K; full dotted line, experiment by Goloviznin *et al* (1987); hair line, calculations by Golovachov and Leontieva (1989).

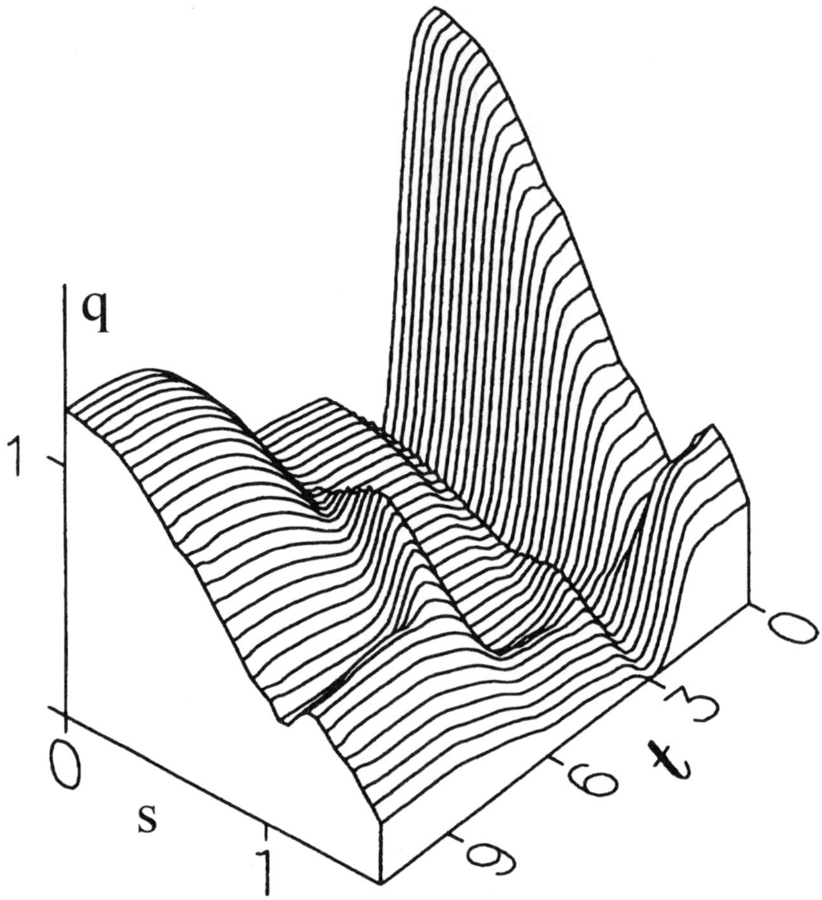

Figure 3.55 Heat flux variation in the laminar/turbulent flow over a sphere moving through the gas layer with increasing temperature. $M = 6$, $Re = 10^6$, $T_w = 0.12T^*$, $a = 0.7$, $b = 0.45R$.

calculations were carried out with the reduced Reynolds equations analogous to those employed in calculations of laminar flows and the algebraic turbulence model by Baldwin and Lomax (1978). The longitudinal intermittancy factor was evaluated with the formula suggested by Chen and Thyson (1971). The laminar-turbulent transition was located with the Reynolds number Re^* based on the characteristic length

$$L = 1.6 \times F_{\max} \times n_{\max}/u_{\max} \tag{3.23}$$

where $n_{\max}$, $u_{\max}$ are the distance from the body surface and longitudinal velocity component at the grid point with maximum vorticity function $F_{\max}$. The characteristic length (3.23) was obtained from analysis of the turbulence models of Baldwin and Lomax (1978) and Cebeci and Smith (1974). The applicability of the transition criterion has been substantiated by calculations of the stationary flows about a sphere located in a homogeneous supersonic stream. This being the case, the above criterion results in nearly the same location of the transition region as obtained with a conventional evaluation of Re^* with the momentum boundary layer thickness. The numerical solutions were obtained using the implicit exponential unidirectional method of Belousov *et al* (1991) and the adaptive grids.

The calculations proved the main features of the turbulent shock layers on the bodies moving through thermal non-homogeneities to remain similar to those of the laminar flow regime. The laminar-turbulent transition influences, to the greatest extent, the heat flux distribution over the body surface. As an example, figure 3.55 shows the heat flux variation over the sphere moving through the gas layer with increasing temperature. The heat flux values are normalized by $10^{-3}\rho_* V^3$ where ρ_* is the largest of the gas densities outside the non-homogeneity, V is the flight speed. The figure demonstrates a substantial increase of the heat flux within the transition region and a shift of this region over the sphere surface.

4

Simulation of three-dimensional flows

Numerical investigation of three-dimensional flows requires, in general, a large amount of computations, which is caused, to a great extent, by more severe restrictions on the allowable value of time increment and the lower convergence rate of the iterative procedures as compared with one- and two-dimensional problems. For this reason, the well-founded use of gas dynamic models and numerical methods appears to be especially important.

The chapter which follows deals with an extension of the viscous shock layer models and calculation procedures to the flows depending on three spatial coordinates. The examples of three-dimensional flow calculations are presented. Some approximate methods for evaluation of the aerodynamic characteristics are discussed as well.

4.1 NOTATION OF THE GOVERNING EQUATIONS

When calculating the flows about smooth bodies, the viscous shock layer equations are usually written in the differential form. Investigating the flows with inner shocks or other discontinuities, one can find the finite-volume notation to be more convenient. In the latter case, the governing equations are the integral relationships which express the mass, momentum and energy balance for the cells of the computational grid.

4.1.1 Navier–Stokes equations for a general curvilinear coordinate system

Using the differential form, it is convenient to write three-dimensional viscous shock layer equations for a general curvilinear coordinate system proceeding from tensor notation of the Navier–Stokes equations (Kovenya and Yanenko 1981, Shevelev 1986). For the moving curvilinear coordinates $q^i (i = 1, 2, 3)$ related to the Cartesian coordinates $x^\alpha (\alpha = 1, 2, 3)$ through a sufficiently smooth nondegenerated transformation, the Navier–Stokes equations are written as follows: the continuity equation

$$\frac{\partial}{\partial t}\left(J^{-1}\rho\right) + \frac{\partial}{\partial q^i}\left[J^{-1}\rho\left(u^i - \omega^i\right)\right] = 0 \tag{4.1}$$

the momentum balance equations

$$\begin{gathered}\frac{\partial}{\partial t}\left[J^{-1}\rho\left(u^i-\omega^i\right)\right]+\frac{\partial}{\partial q^i}\left[J^{-1}\rho\left(u^i-\omega^i\right)\left(u^j-\omega^j\right)+J^{-1}P^{ij}\right] \\ +J^{-1}\Gamma^i_{jk}\left[P^{jk}+\rho\left(u^j-\omega^j\right)\left(u^k-\omega^k\right)\right]=0 \\ i=1,2,3\end{gathered} \tag{4.2}$$

the energy balance equation

$$\frac{\partial}{\partial t}\left(J^{-1}\rho E\right)+\frac{\partial}{\partial q^i}\left[J^{-1}\rho E\left(u^i-\omega^i\right)+P^{ki}g_{kj}u^j-J^{-1}\lambda g^{ij}\frac{\partial T}{\partial q^j}\right]=0 \tag{4.3}$$

Here

$$J^{-1}=\det\left\{\frac{\partial x^\alpha}{\partial q^i}\right\},\qquad \omega^i=\frac{\partial q^i}{\partial t},$$

$$\Gamma^i_{jk}=\frac{1}{2}g^{li}\left(\frac{\partial g_{lj}}{\partial q^k}+\frac{\partial g_{lk}}{\partial q^j}-\frac{\partial g_{jk}}{\partial q^l}\right),$$

u^i are the contravariant velocity components; ρE is the total energy per unit volume; g_{ij}, g^{ij} are the covariant and contravariant metric tensor components; J is the Jacobian of the coordinate transformation; ω^i are the contravariant components of the coordinate system velocity; Γ^i_{jk} are the Christoffel symbols; P^{ij} are the stress tensor components. All Latin and Greek indices take the values: 1, 2, 3. The summation is supposed to be made with respect to the repeated indices.

The components of the metric and stress tensors are defined by the formulae

$$g_{ik}=\frac{\partial x^\alpha}{\partial q^i}\frac{\partial x^\alpha}{\partial q^k}\qquad g_{ik}g^{kj}=\delta^j_i=\begin{cases}1 & i=j\\ 0 & i\neq j\end{cases} \tag{4.4}$$

$$\begin{aligned}P^{ij}&=g^{ij}p+\frac{2}{3}\mu g^{ij}J\frac{\partial}{\partial q^k}\left(J^{-1}u^k\right)\\ &\quad-\mu\left[g^{ik}\frac{\partial u^j}{\partial q^k}+g^{jk}\frac{\partial u^i}{\partial q^k}+u^k\left(g^{il}\Gamma^j_{kl}+g^{jl}\Gamma^i_{kl}\right)\right]\end{aligned} \tag{4.5}$$

It is seen from the above that in contrast with the continuity and energy balance equations which are the scalar ones, the momentum balance equations are written in a weak conservation law form including the source terms which appear because of the velocity vector transformation to the curvilinear coordinate system. A strong conservation law form for these equations can be obtained provided that one retains the Cartesian velocity components as the dependent variables treating them as the scalar quantities invariant with respect to coordinate transformation.

4.1.2 Finite-volume notation of the Navier–Stokes equations

This notation of the governing equations is obtained proceeding from the mass, momentum and energy conservation laws written in the integral form for an arbitrary volume bounded by a closed surface. Write out these laws for the elementary

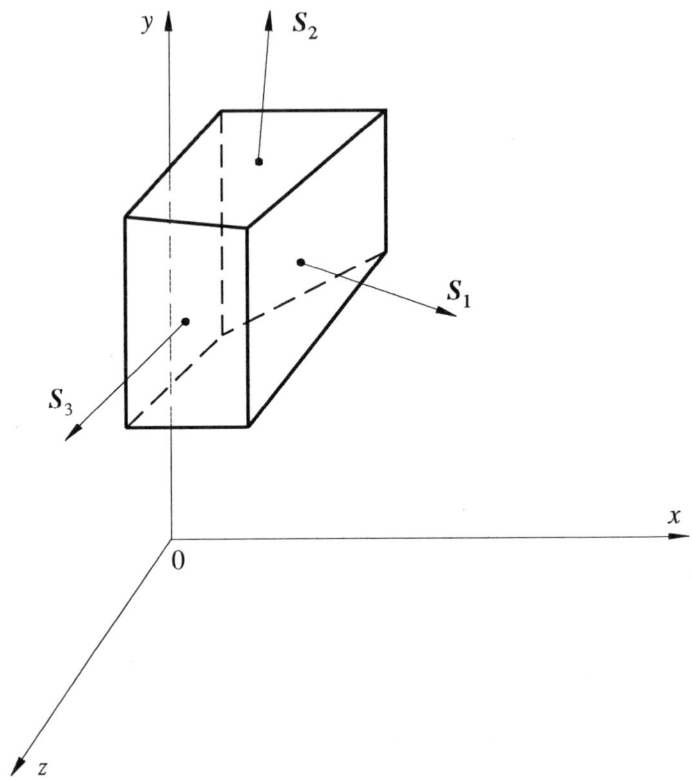

Figure 4.1 On finite-volume notation of the governing equations.

volume v shown in figure 4.1

$$\frac{\partial}{\partial t}(\boldsymbol{X}v) + \partial_i(\boldsymbol{S}_i \cdot \mathsf{F}) = 0 \qquad i = 1, 2, 3 \tag{4.6}$$

Here $\boldsymbol{X} = \{\rho \ \rho u_x \ \rho u_y \ \rho u_z \ \rho E\}^T$ is the column-vector of the conservative variables; $\boldsymbol{S}_i$ is the side vector of the elementary volume;

$$\mathsf{F} = (\boldsymbol{u} - \boldsymbol{\omega})\boldsymbol{X} + \mathsf{J} \qquad \mathsf{J} = \begin{Bmatrix} 0 \\ \mathsf{P} \\ \mathsf{P} \cdot \boldsymbol{u} + \boldsymbol{q} \end{Bmatrix}$$

$$\mathsf{P} = \mathsf{I}p - \frac{\mu}{v}\left[\boldsymbol{S}_j(\partial_j \boldsymbol{u}) + (\partial_j \boldsymbol{u})\boldsymbol{S}_j - \frac{2}{3}\mathsf{I}\boldsymbol{S}_j \cdot \partial_j \boldsymbol{u}\right]$$

$$\boldsymbol{q} = -\frac{\lambda}{v}\boldsymbol{S}_j \partial_j T$$

I is the unit tensor, $\boldsymbol{\omega} = \partial \boldsymbol{r}/\partial t$ is the velocity vector of the local moving curvilinear coordinate system whose axes are aligned with the vectors $\boldsymbol{S}_i$; $\boldsymbol{r}$ is the radius-vector of the point under consideration; ∂_i defines the difference between the function values on the volume sides corresponding to the i-th direction; summation is supposed to be made with respect to the repeated indices.

Employing index i (1,2,3) for the curvilinear coordinates and indices α, β (x, y, z) for the Cartesian coordinates, one can write the equations for the vector $\boldsymbol{X}$ components in the following conservation law form:
the continuity equation

$$\frac{\partial}{\partial t}(\rho v) + \partial_i \left[\boldsymbol{S}_i \cdot (\boldsymbol{u} - \boldsymbol{\omega})\rho\right] = 0 \tag{4.7}$$

the momentum balance equations

$$\frac{\partial}{\partial t}\left[(\rho u_\alpha)v\right] + \partial_i \left[\boldsymbol{S}_i \cdot (\boldsymbol{u} - \boldsymbol{\omega})\rho u_\alpha\right] + \partial_i \left[S_{i\beta} P_{\alpha\beta}\right] = 0 \qquad (\alpha = x, y, z) \tag{4.8}$$

the energy balance equation

$$\frac{\partial}{\partial t}\left[(\rho E)v\right] + \partial_i \left[\boldsymbol{S}_i \cdot (\boldsymbol{u} - \boldsymbol{\omega})\rho E\right] + \partial_i \left[S_{i\beta}\left(P_{\beta\alpha} u_\alpha - \frac{\lambda}{v} S_{j\beta} \partial_j T\right)\right] = 0 \tag{4.9}$$

The gas velocity components for the local coordinate system are expressed in terms of the Cartesian components by the formula

$$u_i = \hat{S}_{i\alpha} u_\alpha \tag{4.10}$$

where $\hat{S}_{i\alpha} = S_{i\alpha}/S_i$ are the components of the unit vector $\hat{\boldsymbol{S}}_i = \boldsymbol{S}_i/S_i$, S_i is the side area of the elementary volume. The side vectors and the volume of the elementary cell can be calculated through the apex coordinates using the formulae presented by Coakley (1981), Rizzi and Ericksson (1981), Kordulla and Vinokur (1983).

4.2 METHODS FOR SOLVING THE NON-STATIONARY EQUATIONS

The non-stationary equations are applied in calculations of both stationary and non-stationary flows. A wide-spread time-asymptotic technique for solving stationary boundary value problems is based on the time-marching numerical procedure whose efficiency, in this case, is determined by the rate of convergence to the stationary solution. From this standpoint, it seems preferable to use unconditionally stable (within the framework of linear analysis) implicit finite-difference schemes allowing larger values of the time increment Δt. When calculating non-stationary flows, the allowable time increment Δt is limited not only by the stability condition but also by the required accuracy with respect to time. This being the case, the explicit finite-difference schemes may be found to be more expedient.

4.2.1 Implicit factored scheme

This method for calculating three-dimensional viscous blunt body flows (Pulliam and Steger 1978, Kutler *et al* 1980) is based on the ideas of Beam and Warming (1978) concerning the implicit non-iterative factored finite-difference schemes for solving the Navier–Stokes equations.

Restricting themselves to the study of high Reynolds number flows, the authors use the reduced Navier–Stokes equations which include only the largest terms responsible for the molecular momentum and energy transport. The computational domain is bounded by the bow shock and the body surface. It is mapped onto a parallelepiped using the transformation

$$\tau = t, \qquad q^i = q^i(t,x,y,z), \qquad i = 1,2,3 \tag{4.11}$$

where x, y, z are the Cartesian coordinates; $q^3 = 0, 1$ correspond to the body surface and the bow shock, respectively. The governing equations are written for the coordinates (4.11) in a strong conservation law form retaining the Cartesian momentum components as the dependent variables:

$$\frac{\partial \boldsymbol{X}}{\partial \tau} + \frac{\partial \boldsymbol{E}}{\partial q^1} + \frac{\partial \boldsymbol{F}}{\partial q^2} + \frac{\partial \boldsymbol{G}}{\partial q^3} = \frac{1}{Re}\frac{\partial \boldsymbol{S}}{\partial q^3} \tag{4.12}$$

Here

$$\boldsymbol{X} = J^{-1}\left\{\begin{matrix} \rho \\ \rho u \\ \rho v \\ \rho w \\ \rho E \end{matrix}\right\} \qquad \boldsymbol{E} = J^{-1}\left\{\begin{matrix} \rho U \\ \rho u U + q^1_x p \\ \rho v U + q^1_y p \\ \rho w U + q^1_z p \\ (\rho E + p)U - q^1_t p \end{matrix}\right\}$$

$$\boldsymbol{F} = J^{-1}\left\{\begin{matrix} \rho V \\ \rho u V + q^2_x p \\ \rho v V + q^2_y p \\ \rho w V + q^2_z p \\ (\rho E + p)V - q^2_t p \end{matrix}\right\} \qquad \boldsymbol{G} = J^{-1}\left\{\begin{matrix} \rho W \\ \rho u W + q^3_x p \\ \rho v W + q^3_y p \\ \rho w W + q^3_z p \\ (\rho E + p)W - q^3_t p \end{matrix}\right\}$$

$$\boldsymbol{S} = J^{-1}\left\{\begin{matrix} 0 \\ \mu m \frac{\partial u}{\partial q^3} + \frac{\mu}{3} l q^3_x \\ \mu m \frac{\partial v}{\partial q^3} + \frac{\mu}{3} l q^3_y \\ \mu m \frac{\partial w}{\partial q^3} + \frac{\mu}{3} l q^3_z \\ m\left[0.5\mu \frac{\partial}{\partial q^3}(u^2+v^2+w^2) + \frac{\lambda}{Pr}\frac{1}{\gamma-1}\frac{\partial}{\partial q^3}(c^2)\right] \\ +\frac{\mu}{3} l\left(q^3_x u + q^3_y v + q^3_z w\right) \end{matrix}\right\}$$

$$U = q^1_t + q^1_x u + q^1_y v + q^1_z w \qquad V = q^2_t + q^2_x u + q^2_y v + q^2_z w$$

$$W = q^3_t + q^3_x u + q^3_y v + q^3_z w \qquad \rho E = \frac{p}{\gamma - 1} + \rho\frac{u^2+v^2+w^2}{2}$$

$$m = \left(q^3_x\right)^2 + \left(q^3_y\right)^2 + \left(q^3_z\right)^2 \qquad l = q^3_x\frac{\partial u}{\partial q^3} + q^3_y\frac{\partial v}{\partial q^3} + q^3_z\frac{\partial w}{\partial q^3}$$

u, v, w are the Cartesian velocity components in the x, y, z directions, respectively; c is the speed of sound; J is the Jacobian of transformation (4.11); other symbols are usually adopted. Equations (4.12) are written for non-dimensional quantities.

The Cartesian coordinates of the grid points being known, the metric coefficients can be evaluated by either analytical differentiation of relationships (4.11) or by numerical differentiation in accordance with the formulae

$$\begin{aligned}
q_x^1 &= J\left(\frac{\partial y}{\partial q^2}\frac{\partial z}{\partial q^3} - \frac{\partial y}{\partial q^3}\frac{\partial z}{\partial q^2}\right) & q_x^2 &= J\left(\frac{\partial z}{\partial q^1}\frac{\partial y}{\partial q^3} - \frac{\partial y}{\partial q^1}\frac{\partial z}{\partial q^3}\right)\\
q_y^1 &= J\left(\frac{\partial z}{\partial q^2}\frac{\partial x}{\partial q^3} - \frac{\partial x}{\partial q^2}\frac{\partial z}{\partial q^3}\right) & q_y^2 &= J\left(\frac{\partial x}{\partial q^1}\frac{\partial z}{\partial q^3} - \frac{\partial x}{\partial q^3}\frac{\partial z}{\partial q^1}\right)\\
q_z^1 &= J\left(\frac{\partial x}{\partial q^2}\frac{\partial y}{\partial q^3} - \frac{\partial y}{\partial q^2}\frac{\partial x}{\partial q^3}\right) & q_z^2 &= J\left(\frac{\partial y}{\partial q^1}\frac{\partial x}{\partial q^3} - \frac{\partial x}{\partial q^1}\frac{\partial y}{\partial q^3}\right)\\
q_x^3 &= J\left(\frac{\partial y}{\partial q^1}\frac{\partial z}{\partial q^2} - \frac{\partial z}{\partial q^1}\frac{\partial y}{\partial q^2}\right) & q_t^1 &= -\frac{\partial x}{\partial \tau}q_x^1 - \frac{\partial y}{\partial \tau}q_y^1 - \frac{\partial z}{\partial \tau}q_z^1\\
q_y^3 &= J\left(\frac{\partial x}{\partial q^2}\frac{\partial z}{\partial q^1} - \frac{\partial x}{\partial q^1}\frac{\partial z}{\partial q^2}\right) & q_t^2 &= -\frac{\partial x}{\partial \tau}q_x^2 - \frac{\partial y}{\partial \tau}q_y^2 - \frac{\partial z}{\partial \tau}q_z^2\\
q_z^3 &= J\left(\frac{\partial x}{\partial q^1}\frac{\partial y}{\partial q^2} - \frac{\partial y}{\partial q^1}\frac{\partial x}{\partial q^2}\right) & q_t^3 &= -\frac{\partial x}{\partial \tau}q_x^3 - \frac{\partial y}{\partial \tau}q_y^3 - \frac{\partial z}{\partial \tau}q_z^3\\
J^{-1} &= \frac{\partial x}{\partial q^1}\frac{\partial y}{\partial q^2}\frac{\partial z}{\partial q^3} + \frac{\partial x}{\partial q^3}\frac{\partial y}{\partial q^1}\frac{\partial z}{\partial q^2} + \frac{\partial x}{\partial q^2}\frac{\partial y}{\partial q^3}\frac{\partial z}{\partial q^1} - \frac{\partial x}{\partial q^1}\frac{\partial y}{\partial q^3}\frac{\partial z}{\partial q^2}\\
&\quad - \frac{\partial x}{\partial q^2}\frac{\partial y}{\partial q^1}\frac{\partial z}{\partial q^3} - \frac{\partial x}{\partial q^3}\frac{\partial y}{\partial q^2}\frac{\partial z}{\partial q^1}
\end{aligned} \tag{4.13}$$

Equations (4.12) are approximated by the implicit finite-difference scheme operating with the increments of function values at two time levels (δ-form of the implicit Euler scheme). Using Taylor series expansions for the flux vectors $\boldsymbol{E}$, $\boldsymbol{F}$, $\boldsymbol{G}$, $\boldsymbol{S}$ and approximate factorization of the three-dimensional difference operator, one can present the finite-difference scheme in the form

$$\begin{aligned}
&\left(\mathbf{I} + \Delta t\delta_1\mathbf{A}^n - \varepsilon_1 J^{-1}\nabla_1\Delta_1 J\right)\left(\mathbf{I} + \Delta t\delta_2\mathbf{B}^n - \varepsilon_2 J^{-1}\nabla_2\Delta_2 J\right)\\
&\times\left(\mathbf{I} + \Delta t\delta_3\mathbf{C}^n - \Delta t/Re\delta_3 J^{-1}\mathbf{M}^n - \varepsilon_3 J^{-1}\nabla_3\Delta_3 J\right)\Delta\boldsymbol{X} =\\
&-\Delta t\left(\delta_1\boldsymbol{E}^n + \delta_2\boldsymbol{F}^n + \delta_3\boldsymbol{G}^n - 1/Re\delta_3\boldsymbol{S}^n\right)\\
&-\varepsilon^* J^{-1}\left[(\nabla_1\Delta_1)^2 + (\nabla_2\Delta_2)^2 + (\nabla_3\Delta_3)^2\right] J\boldsymbol{X}^n
\end{aligned} \tag{4.14}$$

Here $\Delta\boldsymbol{X} = \boldsymbol{X}^{n+1} - \boldsymbol{X}^n$ is the increment of the unknown function vector; $\mathbf{I}$ is the identity matrix; $\mathbf{A} = \partial\boldsymbol{E}/\partial\boldsymbol{X}$, $\mathbf{B} = \partial\boldsymbol{F}/\partial\boldsymbol{X}$, $\mathbf{C} = \partial\boldsymbol{G}/\partial\boldsymbol{X}$, $\mathbf{M} = \partial\boldsymbol{S}/\partial\boldsymbol{X}$ are the Jacobi matrices of the flux vectors; δ_i, ∇_i, Δ_i are the central and one-sided difference operators with respect to the q^i coordinate ($i = 1, 2, 3$).

The terms factored by ε_i in the left-hand side of (4.14) represent a second-order implicit artificial dissipation which is introduced to improve the stability. These terms do not affect the stationary solution. The quantity factored by ε^* in the right-hand side of (4.14) represents a fourth-order explicit artificial dissipation required to suppress the high frequency perturbations of the gas dynamic field.

To close the set of difference equations including the unknown bow shock velocity, the following technique is applied. First, the gas pressure behind the bow shock is evaluated by the explicit formula

$$p^{n+1} = p^n + (p_\tau)^n \, \Delta\tau \tag{4.15}$$

The derivative p_τ^n is determined from the following combination of the continuity equation and the energy balance equation written without viscous terms which do not influence the flow near the bow shock at high Reynolds numbers

$$\begin{aligned}\frac{\partial p}{\partial \tau} = -\Bigg[& U\frac{\partial p}{\partial q^1} + V\frac{\partial p}{\partial q^2} + W\frac{\partial p}{\partial q^3} \\ & + \rho c^2 \left(q_x^1 \frac{\partial u}{\partial q^1} + q_y^1 \frac{\partial v}{\partial q^1} + q_z^1 \frac{\partial w}{\partial q^1} + q_x^2 \frac{\partial u}{\partial q^2} \right. \\ & \left. + q_y^2 \frac{\partial v}{\partial q^2} + q_z^2 \frac{\partial w}{\partial q^2} + q_x^3 \frac{\partial u}{\partial q^3} + q_y^3 \frac{\partial v}{\partial q^3} + q_z^3 \frac{\partial w}{\partial q^3} \right) \Bigg]\end{aligned} \tag{4.16}$$

The right-hand side of (4.16) is evaluated with the previous time level data. The derivatives with respect to q^1 and q^2 are replaced by the central difference formulae, the derivatives in q^3 are approximated by the one-sided formulae. The pressure behind the bow shock being known, the other gas dynamic functions and the bow shock velocity can be easily obtained from the Rankine–Hugoniot relationships. The bow shock velocity is used to locate the bow shock at the moment $\tau = \tau^{n+1}$. Relevant formulae are presented by Kutler *et al* (1980).

The boundary conditions on the body surface are given by the conventional relationships for the velocity components and gas temperature. At the downstream boundary, the linear extrapolation of the unknown functions from within the computational domain is applied, which is equivalent to the use of boundary conditions (1.12) with $k = 2$. Besides that, the pressure within the subsonic segment of this boundary was assumed to be equal to its value at the grid point located in the supersonic region, which is nearest to the wall. Numerical implementation of the above boundary conditions meets no difficulties.

Finite-difference scheme (4.14) complemented with the above relationships is implemented through three fractional steps. At each of them one obtains the set of difference equations with a block tridiagonal matrix of coefficients which is solved by a block Thomas ('progonka') algorithm.

4.2.2 Implicit unidirectional scheme

As an example of the unsplitted implicit finite-difference methods, let us consider the unidirectional method by Karyakin and Popov (1977). As in the previous method, the shock layer flow is calculated with the moving curvilinear coordinate system q^i $(i = 1, 2, 3)$ in which the surfaces $q^3 = 0$ and $q^3 = 1$ correspond to the body surface and the bow shock. Coordinates q^i are chosen on the basis of a spherical coordinate system (r, θ, φ)

$$q^1 = \theta, \qquad q^2 = \varphi, \qquad q^3 = \frac{r - r_w}{r_s - r_w} \tag{4.17}$$

Here $r = r_w(\theta, \varphi)$, $r = r_s(\theta, \varphi, t)$ are the equations of the body surface and the bow shock. The Cartesian coordinates x^α and contravariant components of the local coordinate system velocity ω^i are determined by the formulae

$$\begin{aligned} &x^1 = r\cos q^1 \quad && x^2 = r\sin q^1 \cos q^2 \quad && x^3 = r \sin q^1 \sin q^2 \\ &\omega^1 = 0 && \omega^2 = 0 && \omega^3 = q^3 \frac{\partial r_s}{\partial t} \Big/ (r_s - r_w) \end{aligned} \tag{4.18}$$

where

$$r = r_w(q^1, q^2) + \left[r_s(q^1, q^2, t) - r_w(q^1, q^2)\right] q^3$$

The problem is solved using the reduced Navier–Stokes equations analogous to those of the preceding subsection. Being written in the tensor form, the governing equations are

$$\begin{aligned} \frac{\partial \rho}{\partial t} + \left(u^j - \omega^j\right) \nabla_j \rho + \rho \nabla_j u^j = 0 \\ \rho\left(\frac{\partial u_i}{\partial t} - u_j \nabla_i \omega^j\right) + \rho(u^j - \omega^j)\nabla_j u_i + \nabla_i p - g^{33}\nabla_3\left(\mu \nabla_3 u_i\right) = 0 \\ (i = 1, 2) \\ \rho\left(\frac{\partial u^3}{\partial t} + u^j \nabla_j \omega^3\right) + \rho(u^j - \omega^j)\nabla_j u^3 + g^{3j}\nabla_j p = 0 \\ \rho \frac{\partial e}{\partial t} + \rho(u^j - \omega^j)\nabla_j e + p\nabla_j u^j - g^{33}\left(\nabla_3 \lambda \nabla_3 T + \mu \nabla_3 u_j \nabla_3 u^j\right) = 0 \end{aligned} \tag{4.19}$$

The momentum balance equations of (4.19) are written for the covariant components in the q^1, q^2-coordinates and for the contravariant component in the q^3-coordinate. In the energy balance equation, e is the specific internal gas energy. The other symbols are the same as previously adopted.

The co- and contravariant velocity components and the covariant tensor derivatives are calculated using the Cartesian components in accordance with the formulae (Kochin 1965)

$$\begin{aligned} &u_i = u_\alpha \frac{\partial x^\alpha}{\partial q^i} && \nabla_3(\mu \nabla_3 u_i) = \frac{\partial}{\partial q^3}\left(\mu \frac{\partial u_\alpha}{\partial q^3}\right)\frac{\partial x^\alpha}{\partial q^i} \\ &u^i = g^{ij} u_j && \frac{\partial u_i}{\partial t} - u_j \nabla_i \omega^j = \frac{\partial u_\alpha}{\partial t}\frac{\partial x^\alpha}{\partial q^i} \\ &\nabla_i u^j = g^{\alpha j} \nabla_i u_\alpha && \frac{\partial u^3}{\partial t} + u^j \nabla_j \omega^3 = g^{3j} \frac{\partial u_\alpha}{\partial t}\frac{\partial x^\alpha}{\partial q^j} \\ &\nabla_i u_j = \frac{\partial u_\alpha}{\partial q^i}\frac{\partial x_\alpha}{\partial q^j} \end{aligned} \tag{4.20}$$

Here u_α are the velocity components for the Cartesian coordinates x^α $(\alpha = 1, 2, 3)$. Covariant derivatives of the scalar quantities are the conventional partial derivatives.

The boundary conditions for set (4.19) are imposed on the body surface, bow shock and the surface $q^1 = \Theta$ which is the downstream boundary of the computational domain. If the flow under study has symmetry planes, the computational domain is bounded by these planes in the q^2-coordinate.

The boundary conditions on the body surface are written as follows

$$\begin{gathered} u_1 = u_2 = u^3 = 0 \\ \frac{\varkappa}{\sqrt{g^{33}}} g^{3i} \nabla_i T = (1 - \varkappa)(T - T_w) \end{gathered} \tag{4.21}$$

where $\varkappa = 0, 1$ correspond to the fixed wall temperature and to the adiabatic wall, respectively.

The tensor form of the modified Rankine–Hugoniot relationships reads

$$\begin{aligned} \rho(u^3 - D^3) &= \rho_\infty \left(u^3_\infty - D^3\right) \\ \rho(u^3 - D^3)u_i - \mu g^{33} \nabla_3 u_i &= \rho_\infty \left(u^3_\infty - D^3\right) u_{i\infty} \\ &\quad (i = 1, 2) \\ \rho(u^3 - D^3)^2 + p g^{33} - \frac{4}{3}\mu g^{33} \nabla_3 u^3 &= \rho_\infty \left(u^3_\infty - D^3\right)^2 + p_\infty g^{33} \\ \rho(u^3 - D^3)H - g^{33}\left[\lambda \nabla_3 T + \mu(u^i - D^i)\nabla_3 u_i\right] & \\ -\frac{4}{3}\mu(u^3 - D^3)\nabla_3 u^3 &= \rho_\infty \left(u^3_\infty - D^3\right) H_\infty \end{aligned} \tag{4.22}$$

Here $D^3 = \omega^3\big|_{q^3=1}$ is the bow shock velocity; $H = h + 1/2(u_i - D_i)(u^i - D^i)$ is the total specific enthalpy.

The approximate conditions (1.12) are imposed on the downstream boundary $q^1 = \Theta$. The boundary conditions in the circumferential q^2-coordinate take into account the symmetric or periodic behaviour of gas dynamic functions in this direction. If only the windward part of the shock layer flow is calculated, then approximate conditions (1.12) are imposed on the downstream boundary of the computational domain in the circumferential coordinate.

The finite-difference approximation of the governing equations and boundary conditions takes into account the singularity of the coordinate system on the line $q^1 = 0$. With that end in view, the computational grid is shifted by a half spacing with respect to this line. The difference approximation of spatial derivatives ensures an exact presentation of the simplest particular solutions of the problem, namely, the uniform and axisymmetric streams. It has been found that the above prevents oscillations of the numerical solution near the line $q^1 = 0$ without any artificial viscosity or smoothing procedures.

Let us consider the numerical method in more detail. The computational grid is introduced in the domain $0 \leqslant q^1 \leqslant \Theta$, $0 \leqslant q^2 \leqslant \Phi$, $0 \leqslant q^3 \leqslant 1$, $t \geqslant 0$ in accordance with the formulae

$$
\begin{array}{lll}
q_k^1 = (k - 1/2)\Delta q^1 & \Delta q^1 = \dfrac{\Theta}{K - 1/2} & k = 1, 2, \ldots, K; \\
q_m^2 = (m - 1/2)\Delta q^2 & \Delta q^2 = \dfrac{\Phi}{M - 1/2} & m = 1, 2, \ldots, M; \\
q_l^3 = l\Delta q^3 & \Delta q^3 = 1/L & l = 0, 1, \ldots, L; \\
t^n = n\Delta t & & n = 1, 2, \ldots
\end{array}
\tag{4.23}
$$

The velocity field is determined by the Cartesian components u_α ($\alpha = 1, 2, 3$). Equations (4.19) are approximated at the grid points $(k,\ m,\ l + 1/2)^n$ by the implicit finite-difference scheme which is written with a stencil including 12 grid points of the new time level $(k,\ m,\ l + 1/2 \pm 1/2)^n$,$(k,\ m,\ l + 1/2 \pm 3/2)^n$, $(k \pm 1,\ m,\ l + 1/2 \pm 1/2)^n$, $(k,\ m \pm 1,\ l + 1/2 \pm 1/2)^n$ and two grid points of the preceding time level $(k,\ m,\ l + 1/2 \pm 1/2)^{n-1}$. The first and second derivatives with respect to q^3 are replaced by the symmetric two- and four-point difference formulae, respectively. The difference formulae look as follows

$$
\begin{aligned}
\frac{\partial f}{\partial t} &\sim \frac{1}{\Delta t}\left(f^n - f^{n-1}\right)_{k,m,l+1/2} \\
\frac{\partial f}{\partial q^1} &\sim \frac{1}{2\sin\Delta q^1}\left(f_{k+1} - f_{k-1}\right)^n_{m,l+1/2} \\
\frac{\partial f}{\partial q^2} &\sim \frac{1}{2\sin\Delta q^2}\left(f_{m+1} - f_{m-1}\right)^n_{k,l+1/2} \\
\frac{\partial f}{\partial q^3} &\sim \frac{1}{\Delta q^3}\left(f_{l+1} - f_l\right)^n_{k,m} \\
\frac{\partial}{\partial q^3}\left(\gamma\frac{\partial f}{\partial q^3}\right) &\sim \frac{1}{2(\Delta q^3)^2}\left[\gamma_{l+1}\left(f_{l+2} - f_l\right) - \gamma_l\left(f_{l+1} - f_{l-1}\right)\right]^n_{k,m}
\end{aligned}
\tag{4.24}
$$

where f, γ are any of the grid functions,

$$
f^n_{kml} = f\left(q_k^1, q_m^2, q_l^3, t^n\right) \qquad f_{l+1/2} = \frac{1}{2}\left(f_{l+1} + f_l\right)
$$

The metric tensor and velocity vector components are obtained with formulae (4.4), (4.18) and (4.20) in which the derivatives $\partial r_i/\partial q^j$ ($i = w, s; j = 1, 2$) are calculated with formulae

$$
\begin{aligned}
\left(\frac{\partial r_i}{\partial q^1}\right)^n_{k,m} &\sim \left[(r_i)^n_{k,m}\right]^2 \frac{1/(r_i)^n_{k-1,m} - 1/(r_i)^n_{k+1,m}}{2\sin\Delta q^1} \\
\left(\frac{\partial r_i}{\partial q^2}\right)^n_{k,m} &\sim \left[(r_i)^n_{k,m}\right]^2 \frac{1/(r_i)^n_{k,m-1} - 1/(r_i)^n_{k,m+1}}{2\sin\Delta q^2}
\end{aligned}
\tag{4.25}
$$

When the function values at the nodes located outside the computational domain are required for difference approximation of the problem, they are evaluated taking into account the symmetry or continuity relationships.

The above finite-difference scheme approximates equations (4.19) with an order of accuracy $O[(\Delta q^1)^2 + (\Delta q^2)^2 + (\Delta q^3)^2 + \Delta t]$ and ensures an exact presentation of the uniform and axisymmetric streams.

As a result of the above difference approximation, one obtains a set of five algebraic equations at each grid point $(k, m, l + 1/2)^n$

$$(H_i)^n_{k,m,l+1/2} = 0 \qquad i = 1, \ldots, 5 \tag{4.26}$$

Since the second-order equations in the q^3-coordinate are approximated at the grid points with semi-whole values of the index l, set (4.26) complemented with approximations of the boundary conditions includes three surplus equations on each line q^3. They have been found to be the above difference equations written near the body surface with $l = 0$. These difference equations are excluded from set (4.26).

The function values at a new time level n are obtained by iterations in accordance with the formula

$$\boldsymbol{X}^{n(q+1)} = \boldsymbol{X}^{n(q)} + \omega \Delta \boldsymbol{X} \tag{4.27}$$

where q is the iteration number, $\boldsymbol{X}^{n(q)}$ is the unknown function vector at the time level n in the q-th iteration, $\Delta \boldsymbol{X}$ is the increment of vector $\boldsymbol{X}^{n(q)}$, ω is the damping factor.

According to the basic idea of the iteration methods, the residual vectors $\boldsymbol{H}^{n(q)}$ are evaluated using the values of $\boldsymbol{X}^{n(q)}$ and then the corrections $\Delta \boldsymbol{X}$ are obtained by solving the iterative set of difference equations. The coefficients of the iterative set are chosen to be the same as those of the difference equations approximating, to first-order accuracy, the terms of (4.19) which contain the derivatives with respect to q^3 and t. As a result, one obtains

$$\begin{aligned}
&\mathbf{P}^{n(q)}_{l+1/2} (\Delta \boldsymbol{X}_{l+1} + \Delta \boldsymbol{X}_l) + \frac{\Delta t}{\Delta q^3} \mathbf{R}^{n(q)}_{l+1/2} (\Delta \boldsymbol{X}_{l+1} - \Delta \boldsymbol{X}_l) \\
&- g^{33}_{l+1/2} \frac{\Delta t}{(\Delta q^3)^2} \Big[\mathbf{S}^{n(q)}_{l+1} (\Delta \boldsymbol{X}_{l+1} - \Delta \boldsymbol{X}_l) \\
&- \mathbf{S}^{n(q)}_{l} (\Delta \boldsymbol{X}_{l} - \Delta \boldsymbol{X}_{l-1}) \Big] + \boldsymbol{H}^{n(q)}_{l+1/2} = 0
\end{aligned} \tag{4.28}$$

Here the indices k, m are omitted. For the unknown function vector $\boldsymbol{X} = \{u_1 \; u_2 \; u^3 \; T \; \rho\}^T$, the coefficients of (4.28) are

$$\mathbf{P} = \begin{Bmatrix} \rho & 0 & 0 & 0 & 0 \\ 0 & \rho & 0 & 0 & 0 \\ 0 & 0 & \rho & 0 & 0 \\ 0 & 0 & 0 & \rho e'_T & 0 \\ 0 & 0 & 0 & 0 & 1 \end{Bmatrix} \qquad \mathbf{S} = \begin{Bmatrix} \mu & 0 & 0 & 0 & 0 \\ 0 & \mu & 0 & 0 & 0 \\ 0 & 0 & 0 & 0 & 0 \\ 0 & 0 & 0 & \lambda & 0 \\ 0 & 0 & 0 & 0 & 0 \end{Bmatrix}$$

$$\mathbf{R} = \begin{Bmatrix} \rho V^3 & 0 & 0 & 0 & 0 \\ 0 & \rho V^3 & 0 & 0 & 0 \\ 0 & 0 & \rho V^3 & g^{33} p'_T & g^{33} p'_\rho \\ 0 & 0 & p & \rho V^3 e'_T & 0 \\ 0 & 0 & \rho & 0 & V^3 \end{Bmatrix} \qquad \boldsymbol{H} = \begin{Bmatrix} H_1 \\ H_2 \\ H_3 \\ H_4 \\ H_5 \end{Bmatrix}$$

$$V^3 = U^3 - \omega^3$$

where $\boldsymbol{H}$ is the vector of the residuals of equations (4.19).

The stationary solution obtained by the above finite-difference scheme is second-order accurate with respect to all spatial coordinates.

Due to definition of the velocity vector by its covariant components with respect to q^1 and q^2, the first and second equations of (4.28) along with the relevant boundary conditions on each line q^3 represent independent scalar tridiagonal sets. They are solved one after another by the 'progonka' algorithm, which provides the covariant velocity vector components u_1 and u_2.

The other three equations are solved simultaneously by the block 'progonka' algorithm. In these calculations, the wall boundary condition for the normal velocity component u^3 is transferred to the bow shock wave where it is exploited along with Rankine–Hugoniot relationships to determine the vector $\Delta \boldsymbol{X}_L$ and the bow shock velocity $(D^3)^{n(q+1)}$. The procedure is quite analogous to that described in subsection 1.2.2 for an axisymmetric flow. Having finished the 'progonka' calculations, one evaluates the vector $\boldsymbol{X}_l^{n(q+1)}$ at all grid points on the line q^3. Later on, the Cartesian velocity components are determined and the bow shock distance from the body surface is specified. These calculations are carried out along all lines q^3 and then one can proceed to the next iteration or the next time step calculations.

4.3 METHODS FOR SOLVING THE PARABOLIZED EQUATIONS

Parabolized Navier–Stokes equations apply to the calculation of stationary flows in which the upstream influence can be neglected. They are integrated in the streamwise coordinate by means of space-marching methods. In a general case, the two-dimensional boundary value problem has to be solved at each step in the marching coordinate. So this approach to numerical simulation of stationary three-dimensional flows consumes approximately the same computer resources as the solution of a two-dimensional non-stationary problem. It can be extended to problems with elliptic properties using multisweep calculation procedures to specify the terms responsible for upstream disturbance propagation.

4.3.1 Coordinate system and grid generation

When applying the space-marching methods to solve the viscous shock layer problems, a computational grid is generated as the solution advances in the marching coordinate. In calculation of three-dimensional flows about the complex shapes, one has to give special attention to the choice of the coordinate system and grid generation. The matter is that the parabolized Navier–Stokes equations assumes the longitudinal components of the viscous stress tensor and the heat flux vector to be small as compared with the transversal ones. Accordingly, the second derivatives with respect to the marching coordinate are neglected. If the coordinate system is not aligned with the main flow direction, the parabolized Navier–Stokes equations ignore the terms which may be important for the flow under study. Further, the initial-boundary value problem is well-posed provided the velocity component in the marching coordinate exceeds the speed of sound in the inviscid part of the shock layer. Hence, for a reduction of the Navier–Stokes equations and for stability of the marching calculation procedure, the family of coordinate lines chosen as the marching one ought to be aligned with the main flow velocity vector and to be orthogonal with two other families in the regions where the molecular transport processes are important.

In the calculation of shock layer flows around simple shapes, the above requirements can be satisfied to some extent using a cylindrical coordinate system whose axis is aligned with the body axis. More flexible is the coordinate system, aligned with the streamlines. If the body surface involves the concave parts, one has to use more complicated coordinate systems; see, for example, the papers by Helliwell *et al* (1980), Tannehill *et al* (1981), Chernii (1982), Chaussee *et al* (1985). It should be stressed that the above papers do not present the comprehensive recommendations since generation of the optimum coordinate system depends to a great extent on the geometric properties of the body and a priori ideas on the flow structure.

Consider, in more detail, the method developed by Chernii (1982,1983). Let x^1, x^2, x^3 be the Cartesian coordinates for which the equation of the body surface is written in the parametric form

$$x^i = x^i(\xi, \eta) \qquad i = 1, 2, 3 \tag{4.29}$$

or in the vector form

$$\boldsymbol{r} = \boldsymbol{r}(\xi, \eta) \tag{4.30}$$

where $\boldsymbol{r}$ is the radius-vector and ξ, η are the independent variables making up the coordinate system on the body surface. At each point K of this surface, see figure 4.2, the moving covariant basis $(\boldsymbol{r}_1, \boldsymbol{r}_2, \boldsymbol{r}_3)$ is introduced, where

$$\boldsymbol{r}_1 = \frac{\partial \boldsymbol{r}}{\partial \xi} \qquad \boldsymbol{r}_2 = \frac{\partial \boldsymbol{r}}{\partial \eta}$$

is the basis of the (ξ, η)-coordinate system; $\boldsymbol{r}_3$ is the unit normal on the body surface

$$\boldsymbol{r}_3 = \frac{\boldsymbol{r}_1 \times \boldsymbol{r}_2}{\sqrt{\gamma}} \qquad \gamma = \gamma_{11}\gamma_{22} - \gamma_{22}^2 \quad \gamma_{\alpha\beta} = \boldsymbol{r}_\alpha \cdot \boldsymbol{r}_\beta \quad \alpha, \beta = 1, 2$$

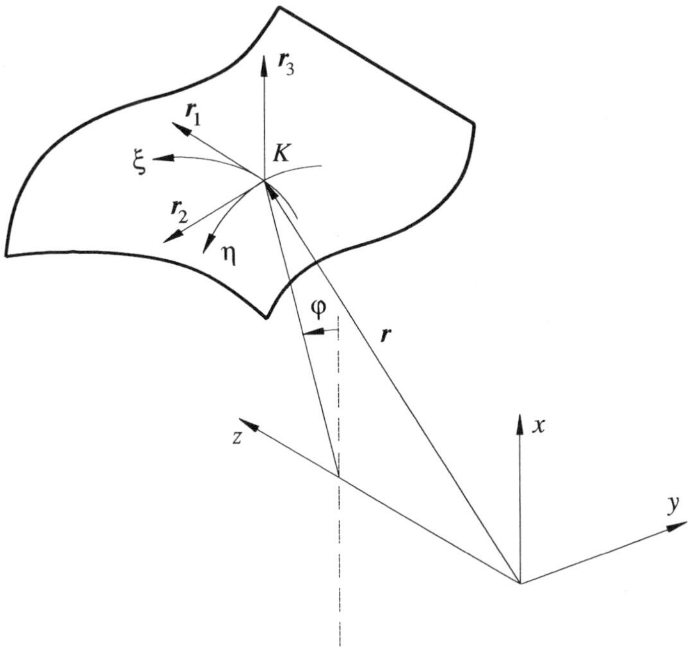

Figure 4.2 On the coordinate system for three-dimensional viscous shock layer calculations by the space-marching methods.

We want to construct the marching cross-section which would be perpendicular at each node to some prescribed vector $\boldsymbol{A}$. The vector field $\boldsymbol{A}$ is determined so that the cross-section would not intersect the previous one and satisfy the above requirements. It can be defined by the formula

$$\boldsymbol{A} = (1 - \lambda)\boldsymbol{N} + \lambda \boldsymbol{V} \tag{4.31}$$

where λ is a continuous function of the marching coordinate, which increases monotonously from zero to unity. Accordingly, with the increasing marching coordinate, the vector field $\boldsymbol{A}$ varies from the vector field $\boldsymbol{N}$, which is orthogonal to the previous marching cross-section, to the gas velocity field $\boldsymbol{V}$ in the current marching cross-section. To determine vector $\boldsymbol{A}$ on the body surface, the velocity vector $\boldsymbol{V}$ is taken from the neighbouring grid points.

Construction of the surfaces on which the solution is searched begins with the construction of their bases on the body surface, which are orthogonal to vector $\boldsymbol{A}$ projections on the planes tangent to the body surface at the points considered. The solution of this problem is given by the quadratures. After simple transformations, one obtains the following formulae for the contravariant components of vector $\boldsymbol{A}_\tau = A_\tau^1 \boldsymbol{r}_1 + A_\tau^2 \boldsymbol{r}_2$ which is tangent to the body surface at point K

$$A_\tau^1 = \frac{T_1\gamma_{22} - T_2\gamma_{12}}{\gamma} \qquad A_\tau^2 = \frac{-T_1\gamma_{12} + T_2\gamma_{11}}{\gamma} \tag{4.32}$$

where the quantities $T_\alpha = \boldsymbol{A} \cdot \boldsymbol{r}_\alpha$ $(\alpha = 1, 2)$ are easily obtained using the presentation of vectors $\boldsymbol{A}$ and $\boldsymbol{r}_\alpha$ on the basis of the Cartesian coordinate system.

The normal to $\boldsymbol{A}_\tau$ in a plane tangent to the body surface is determined by the vector

$$\boldsymbol{B}_\tau = \boldsymbol{r}_3 \times \boldsymbol{A}_\tau = -\frac{T_2}{\sqrt{\gamma}}\boldsymbol{r}_1 + \frac{T_1}{\sqrt{\gamma}}\boldsymbol{r}_2 \tag{4.33}$$

Then, assuming $d\boldsymbol{r} = k\boldsymbol{B}_\tau$, or

$$d\xi \boldsymbol{r}_1 + d\eta \boldsymbol{r}_2 = -\frac{k}{\sqrt{\gamma}} T_2 \boldsymbol{r}_1 + \frac{k}{\sqrt{\gamma}} T_1 \boldsymbol{r}_2$$

one easily obtains the relationship

$$d\xi = -\frac{T_2}{T_1} d\eta \tag{4.34}$$

which gives

$$\xi = \xi_0 - \int_{\eta_0}^{\eta} \frac{T_2}{T_1} d\eta \tag{4.35}$$

At fixed ξ_0 and η_0, formula (4.35) defines the curve on the body surface, which is orthogonal to vector $\boldsymbol{A}_\tau$ and presents the base of the surface on which the solution is sought. The base of the following surface is obtained by evaluation of the integral (4.35) with the same value of η_0 and a new value of ξ_0.

If the body surface is determined by formulae (4.29) or (4.30), the Cartesian coordinate z and the angle φ measured from the plane $y = 0$ (see figure 4.2) can be chosen as independent variables ξ, η. With this choice, relationship (4.35) takes the form

$$z = z_0 - \int_{\varphi_0}^{\varphi} \frac{T_2}{T_1} d\varphi \tag{4.36}$$

The parameter z_0, being the z-coordinate of the intersection point of the plane $\varphi = \varphi_0$ and the base of the surface on which the solution is sought, counts for the marching coordinate and is denoted further by q^1.

Location of the planes φ =const is governed by the coordinate transformation

$$\varphi = \Phi(q^2) \tag{4.37}$$

which takes into account a non-uniformity of the flow field with respect to the circumferential coordinate. Keeping in mind transformation (4.37) and the above notation for the marching coordinate, the equation of the base of the sought computational surface can be rewritten in the following final form

$$z = q^1 - \int_{q_0^2}^{q^2} \frac{T_2}{T_1} \Phi'(q^2) dq^2 \tag{4.38}$$

which provides the grid point distribution along the base of the computational surface. Later on, the surface itself is formed by the third family of coordinate lines. Each point of this surface is located by the radius-vector

$$\boldsymbol{R}(q^1, q^2, s) = \boldsymbol{r}(q^1, q^2) + \int_0^s \boldsymbol{m}(q^1, q^2, s)ds \tag{4.39}$$

where $\boldsymbol{m}$ is the unit tangent vector to coordinate line of the third family, s is the distance along this line from the body surface to the point considered. Normalizing s by the value $\varepsilon(q^1, q^2)$ which corresponds to the bow shock distance from the body surface and using a transformation

$$s/\varepsilon = S(q^3) \tag{4.40}$$

which ensures a proper concentration of the surfaces $s/\varepsilon = const$ in the regions of steep gradients, one can rewrite equation (4.39) as follows

$$\boldsymbol{R}(q^1, q^2, q^3) = \boldsymbol{r}(q^1, q^2) + \varepsilon(q^1, q^2) \int_0^{q^3} S'(q^3)\boldsymbol{m}(q^1, q^2, q^3)dq^3 \tag{4.41}$$

Let us dwell on a choice of the vector field $\boldsymbol{m}$ which determines the tangent directions for the third family coordinate lines q^3. Above, the vector field $\boldsymbol{A}$ orthogonal to the marching cross-section considered has been introduced. To obtain the field $\boldsymbol{m}$, the auxiliary direction $\boldsymbol{E}$ is specified which is tangent to the second family of coordinate lines q^2. Then, one assumes:

$$\boldsymbol{m} = \frac{\boldsymbol{A} \times \boldsymbol{E}}{|\boldsymbol{A} \times \boldsymbol{E}|} \tag{4.42}$$

The above choice of $\boldsymbol{E}$ often suffices well. However, if the cross-section of the body surface has the concave parts as shown in figure 4.3, an intersection of the coordinate lines q^3 may happen at some value of $q^3 = q^3_*$. In that view, the average direction

$$\Lambda \boldsymbol{F} + (1 - \Lambda)\boldsymbol{E}$$

can be used instead of $\boldsymbol{E}$ in the regions where the above intersection is expected. Vector $\boldsymbol{F}$ is perpendicular to the straight line drawn from the grid point considered to the point O located on the body axis, see figure 4.3.

Function

$$\Lambda(q^2, q^3) = \begin{cases} 0 & q^2 \leqslant q^2_1 \\ \left[\dfrac{q^2 - q^2_1}{q^2_* - q^2_1} \dfrac{q^3}{q^3_*}\right]^n & q^2_1 \leqslant q^2 \leqslant q^2_* \\ \left[\dfrac{q^2_2 - q^2}{q^2_2 - q^2_*} \dfrac{q^3}{q^3_*}\right]^n & q^2_* < q^2 \leqslant q^2_2 \\ 0 & q^2_2 < q^2 \end{cases} \tag{4.43}$$

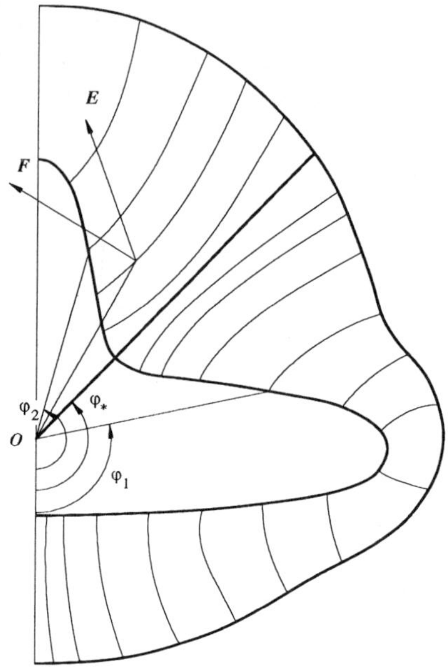

Figure 4.3 On calculation of three-dimensional viscous shock layer flows by the space-marching methods.

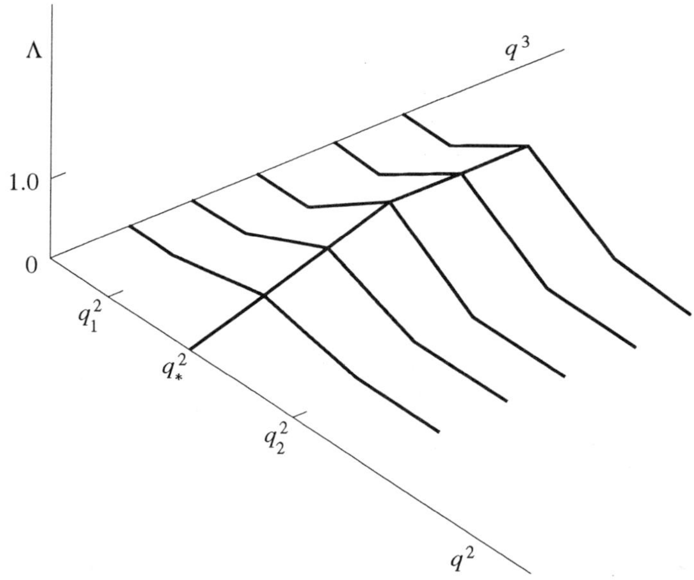

Figure 4.4 On calculation of three-dimensional viscous shock layer flows by the space-marching methods.

shown in figure 4.4 gradually transforms $\boldsymbol{E}$ to $\boldsymbol{F}$ and vise versa within the sector $[\varphi_1, \varphi_2]$ averting the intersection of the third family coordinate lines but leaving them orthogonal to the body surface.

The above method allows one to obtain various coordinate systems using the proper vector fields $\boldsymbol{A}$ and $\boldsymbol{m}$. For instance, assuming $\boldsymbol{A}$ to be aligned with the body axis and $\boldsymbol{m}$ to be the unit vector of the straight line drawn normal to $\boldsymbol{A}$ from a point on the body surface to the body axis, one obtains a cylindrical coordinate system. The vectors $\boldsymbol{m}$ normal to the body surface and to the streamlines provide the body-fitted and streamline-fitted coordinate systems, respectively. The flexibility of the above approach allows its use in calculations of a wide range of three-dimensional flows.

4.3.2 Comments on the parabolized Navier–Stokes equations and the space-marching methods

The parabolized Navier–Stokes equations for three-dimensional flows result from an estimation of the molecular transfer in three directions. This gives a greater variety of the parabolized equations as compared with two-dimensional flows.

In the simplest model, the momentum balance equation in the transversal coordinate does not include any viscous terms and, therefore, does not differ from the inviscid gas equation. Two other projections of the momentum balance equation include only the largest viscous terms, namely, those with the second derivatives of the longitudinal and circumferential velocity components in the transversal coordinate. The energy balance equation includes, besides all inviscid terms, the terms with the second derivative of gas temperature in the transversal coordinate and the squares of the first derivatives of the tangent velocity components in the same direction. The most comprehensive set of parabolized equations differs from the full Navier–Stokes equations only by lack of terms with the second derivatives with respect to the streamwise coordinate. In particular, the above parabolized equations include the derivatives in the circumferential coordinate, which are estimated to be $O(1/Re)$. Finite-volume notation of the parabolized equations can be obtained using the analogous estimates for the components of the molecular momentum and energy fluxes through the sides of a computational cell.

The examples presented in chapter 2 indicate that the parabolized Navier–Stokes equations allow one to calculate not only non-separated flows but the flows with cross-flow separation as well, an accuracy of the results within the separation zones increasing with the use of a more general model.

Using the parabolized equations, one comes to the initial-boundary value problem whose well-posedness within the subsonic regions can be ensured by the methods discussed in subsection 2.2.1. The initial data for space-marching calculations can be obtained in several ways. In the most general case, they can be taken from the calculation of the flow about the body nose with the methods considered in the previous section. The initial data for the calculation of the flows about spherically blunted bodies can be obtained from the calculation of the axisymmetric flow about a sphere. When investigating flows about pointed bodies, the initial data are often calculated using the quasi-conical approach.

In the space-marching methods applied to solve the parabolized equations, the derivatives in the marching coordinate are usually replaced by the first-order accurate difference formulae, which seems especially expedient for the methods in which the initial-boundary value problem in subsonic region is regularized by the truncation error.

In a general case, one has to solve a two-dimensional boundary value problem in each step in the marching coordinate. The boundary conditions in the transversal and circumferential coordinates are the same as those discussed in subsection 4.2.2 for the non-stationary equations. The above boundary value problem can be solved by factored schemes (Li 1981, Tannehill *et al* 1981, Kovenya and Chernii 1983, Chaussee *et al* 1985) or using the iterative methods for solving unsplitted sets of difference equations (Lubard and Helliwell 1974, Helliwell *et al* 1980, Rubin 1981).

As an example, let us consider the space-marching method suggested by Tannehill *et al* (1981). The shock layer flow is calculated using the reduced stationary Navier–Stokes equations in which only viscous terms with the derivatives in the streamwise coordinate are omitted while all terms with the derivatives in two other coordinates are retained. For a general curvilinear coordinate system

$$q^i = q^i(x, y, z) \qquad i = 1, 2, 3 \tag{4.44}$$

the governing equations are written in the following conservation law form

$$\frac{\partial \overline{\boldsymbol{E}}}{\partial q^1} + \frac{\partial \overline{\boldsymbol{F}}}{\partial q^2} + \frac{\partial \overline{\boldsymbol{G}}}{\partial q^3} = 0 \tag{4.45}$$

The flux vector components are defined by the formulae

$$\begin{gathered}
\overline{E}_i = E_i \qquad \overline{F}_i = F_i - \frac{1}{J} q^2_\alpha S_{i\alpha} \qquad \overline{G}_i = G_i - \frac{1}{J} q^3_\alpha S_{i\alpha} \\
(i = 1 - 5; \qquad \alpha = x, y, z) \\
\mathbf{S} = \left\{ \begin{matrix} 0 \quad 0 \quad 0 \\ \mathbf{T} \\ \mathbf{T} \cdot \boldsymbol{u} + \boldsymbol{q} \end{matrix} \right\}
\end{gathered} \tag{4.46}$$

where $\mathbf{T}$ is the viscous stress tensor, $\boldsymbol{q}$ is the heat flux vector, the flux vectors $\boldsymbol{E}, \boldsymbol{F}, \boldsymbol{G}$ look like those in equations (4.12) with $q^i_t \equiv 0$. When writing the equations for the curvilinear coordinate system, the viscous stress tensor and heat flux vector components containing the derivatives with respect to q^1 are omitted. The metric coefficients and Jacobian of the coordinate transformation are defined by formulae (4.13).

The computational domain is bounded by the surface $q^1 = const$ where the initial data are prescribed, the body surface, and the bow shock. The initial-boundary value problem for equations (4.45) is regularized using the method by

Vigneron *et al* (1978) discussed in subsection 2.2.1. In this method, the streamwise pressure gradient terms of the momentum balance equations responsible for an upstream influence in the subsonic region of the shock layer are written in form (2.16) in which the coefficient ω depending on the local Mach number is defined by a relationship similar to (2.9). In the finite-difference presentation, the first term of the right-hand side of (2.16) is approximated in exactly the same way as other unknown functions, while the second term is evaluated via 'sublayer approximation' i.e. assuming the pressure to be constant across the subsonic region of the shock layer.

To implement the above regularization, set (4.45) is rewritten in the form

$$\frac{\partial \overline{\boldsymbol{E}}^*}{\partial q^1} + \frac{\partial \overline{\boldsymbol{P}}}{\partial q^1} + \frac{\partial \overline{\boldsymbol{F}}}{\partial q^2} + \frac{\partial \overline{\boldsymbol{G}}}{\partial q^3} = 0 \tag{4.47}$$

where

$$\overline{\boldsymbol{E}}^* = \overline{\boldsymbol{E}} - \overline{\boldsymbol{P}}$$

$$\overline{\boldsymbol{P}} = \frac{1}{J}\left(q^1_x \boldsymbol{P}_1 + q^1_y \boldsymbol{P}_2 + q^1_z \boldsymbol{P}_3\right) \qquad \boldsymbol{P}_1 = \{0\ (1-\omega)p\ 0\ 0\ 0\}^T$$

$$\boldsymbol{P}_2 = \{0\ 0\ (1-\omega)p\ 0\ 0\}^T \qquad \boldsymbol{P}_3 = \{0\ 0\ 0\ (1-\omega)p\ 0\}^T$$

The computational grid in each step in the marching coordinate consists of N straight lines drawn from the fixed points of the body surface in the directions which are allowed to be non-orthogonal to the body surface. On each line, there are K grid points between the body surface and the bow shock wave, which can be clustered in the regions of steep gradients. In the computational domain, each marching section $q^1 = const$ presents a square $0 \leqslant q^2 \leqslant 1$, $0 \leqslant q^3 \leqslant 1$ in which the computational grid is uniform in both directions.

Equations (4.47) are integrated with the implicit factored finite-difference scheme analogous to the well-known schemes of Beam and Warming (1978), Briley and Mc Donald (1977) for solving the non-stationary Navier–Stokes equations; see also subsection 4.2.1. Below, we consider the two-layer first-order accurate in q^1 scheme (implicit Euler method). It looks as follows:

$$\begin{aligned}
&\left(\mathbf{A}_i + \Delta q^1 \frac{\partial}{\partial q^3}\mathbf{C}_i\right)\mathbf{A}_i^{-1}\left(\mathbf{A}_i + \Delta q^1 \frac{\partial}{\partial q^2}\mathbf{B}_i\right)\Delta \boldsymbol{X} \\
&\quad = -\Delta q^1\left(\frac{\partial \overline{\boldsymbol{F}}}{\partial q^2} + \frac{\partial \overline{\boldsymbol{G}}}{\partial q^3}\right)_i - \Delta \overline{\boldsymbol{P}} \\
&\quad\quad - \Delta q^1/J\left(\boldsymbol{E}^*\frac{\partial q^1_x}{\partial q^1} + \boldsymbol{F}^*\frac{\partial q^1_y}{\partial q^1} + \boldsymbol{G}^*\frac{\partial q^1_z}{\partial q^1}\right)_i
\end{aligned} \tag{4.48}$$

Here $\mathbf{A} = \partial \overline{\boldsymbol{E}}^*/\partial \boldsymbol{X}$, $\mathbf{B} = \partial \overline{\boldsymbol{F}}/\partial \boldsymbol{X}$, and $\mathbf{C} = \partial \overline{\boldsymbol{G}}/\partial \boldsymbol{X}$, are the Jacobi matrices, $\boldsymbol{X} = J^{-1}\{\rho\ \rho u\ \rho v\ \rho w\ \rho E\}^T$ is the conservative variable vector, $\Delta \boldsymbol{X} = \boldsymbol{X}_{i+1} - \boldsymbol{X}_i$, $\boldsymbol{E}^* = \tilde{\boldsymbol{E}} - \boldsymbol{P}_1$, $\boldsymbol{F}^* = \tilde{\boldsymbol{F}} - \boldsymbol{P}_2$, $\boldsymbol{G}^* = \tilde{\boldsymbol{G}} - \boldsymbol{P}_3$. Tildes denote the flux vectors for the Cartesian coordinate system.

The increment $\Delta\boldsymbol{X}$ is determined through two fractional steps

$$\text{(1)} \quad \left(\mathbf{A}_i + \Delta q^1 \frac{\partial}{\partial q^3}\mathbf{C}_i\right)\Delta\boldsymbol{X}^* = \text{R.H.S. of (4.48)} \tag{4.49}$$

$$\text{(2)} \quad \left(\mathbf{A}_i + \Delta q^1 \frac{\partial}{\partial q^2}\mathbf{B}_i\right)\Delta\boldsymbol{X} = \mathbf{A}_i\Delta\boldsymbol{X}^* \tag{4.50}$$

Then, the unknown function vector is evaluated with formula $\boldsymbol{X}_{i+1} = \boldsymbol{X}_i + \Delta\boldsymbol{X}$.

To implement the first fractional step, the increment $\Delta\overline{\boldsymbol{P}}$ is required. This quantity, which differs from zero only in the subsonic region of the shock layer, is evaluated with the first-order accurate difference formulae using the pressure values at the grid points located in the supersonic region of the marching sections $q^1 = q_i^1$ and $q^1 = q_{i-1}^1$. The partial derivatives in the q^2 and q^3-coordinates are replaced by the central second-order accurate difference formulae including the implicit second-order artificial dissipation and the explicit fourth-order artificial dissipation, see subsection 4.2.1. The sets of difference equations for both fractional steps have block tridiagonal matrices of coefficients and are solved by a block Thomas ('progonka') algorithm. The bow shock coordinates and the gas parameters behind the shock front are determined using a procedure elaborated earlier by the authors for inviscid flow calculations. Its description can be found in a paper by Tannehill *et al* (1981) along with formulae for the Jacobi matrix components and the details of grid generation and the finite-difference scheme.

For weakly elliptic problems, for instance hypersonic non-separated flows about smooth bodies, the gas dynamic functions at each step in the streamwise direction can be obtained by marching in the circumferential coordinate. The initial data for this marching procedure are taken from the self-similar solution for a shock layer flow in the symmetry plane. The above calculation procedure has been implemented by Murray and Lewis (1978) and Markov (1989).

The effects of the upstream influence both in the circumferential and longitudinal directions can be taken into account using the global iteration methods which include the multisweep marching calculations. Besides accounting for the above effects, the use of global iterations removes the lower limit of mesh spacing in the marching coordinate inherent to most of the space-marching methods, which allows one to improve the accuracy of the numerical results if needed. Some versions of the global iteration method have been discussed in chapter 2. Their implementation for three-dimensional flows does not involve any new points of principle.

4.3.3 Block marching method

In this method (Afonina *et al* 1983), a three-dimensional viscous shock layer flow is calculated by solving a sequence of the boundary value problems in the overlapping subdomains, each containing N sections. Each boundary value problem is solved by the time-asymptotic technique. If the number of sections calculated simultaneously is equal to their total number in the computational domain, the method presents a conventional time-asymptotic procedure for solving the original problem. If $N = 1$, it presents a space-marching method in which the initial-boundary value problem is regularized using the non-stationary equations at each step in the marching coordinate; see point (f) of subsection 2.2.1.

Numerical implementation of the block marching method is carried out using the block version of a successive relaxation procedure. The difference equations for the grid points located on the lines drawn from the body surface are combined into blocks. The equations of each block are solved by a block Thomas ('progonka') algorithm combined with the Newton iteration procedure. The blocks are calculated in an order of priority corresponding to the main flow direction. Below, the main points of the numerical algorithm are considered.

The problem is solved with the moving curvilinear coordinate system q^i ($i = 1, 2, 3$) in which the q^1, q^2-coordinates present a non-degenerate coordinate system on the body surface and the q^3-coordinate is defined by the formula

$$\boldsymbol{r} = \boldsymbol{r}_w(q^1, q^2) + \varepsilon(t, q^1, q^2) \cdot q^3 \cdot \boldsymbol{a}_3(q^1, q^2) \tag{4.51}$$

where $\boldsymbol{r}$ is the radius-vector of the point considered, $\boldsymbol{a}_3$ is the unit external normal on the body surface, ε is the bow shock distance from the body surface along the $\boldsymbol{a}_3$-direction.

The shock layer flow is calculated using the reduced Navier–Stokes equations which contain all inviscid terms and only the greatest terms responsible for the molecular transport processes. The governing equations are written in the finite-volume notation using the velocity components in the q^i-directions:

$$\frac{\partial}{\partial t}(\boldsymbol{X} v) + \partial_i(S_i \boldsymbol{F}^i) + \boldsymbol{f} v = 0 \tag{4.52}$$

Here $\boldsymbol{X} = \{\rho\ \rho u_3\ \rho u_1\ \rho u_2\ \rho E\}^T$ is the column-vector of unknown functions, v is the volume of the elementary cell bounded by the coordinate surfaces $q^i = const$, S_i is the side area of the cell. The flux vectors $\boldsymbol{F}^i$ and the source term $\boldsymbol{f}$ are defined by the formulae:

$$\boldsymbol{F}^1 = \left\{ \begin{matrix} \rho U \\ \rho u_3 U + p\boldsymbol{a}_3 \cdot \hat{\boldsymbol{S}}_1 \\ \rho u_1 U + p\boldsymbol{a}_1 \cdot \hat{\boldsymbol{S}}_1 \\ \rho u_2 U + p\boldsymbol{a}_2 \cdot \hat{\boldsymbol{S}}_1 \\ (\rho E + p) U \end{matrix} \right\} \qquad \boldsymbol{F}^3 = \left\{ \begin{matrix} \rho V \\ \rho u_3 V + p\boldsymbol{a}_3 \cdot \hat{\boldsymbol{S}}_3 \\ \rho u_1 V + p\boldsymbol{a}_1 \cdot \hat{\boldsymbol{S}}_3 + \tau_1 \boldsymbol{a}_3 \cdot \hat{\boldsymbol{S}}_3 \\ \rho u_2 V + p\boldsymbol{a}_2 \cdot \hat{\boldsymbol{S}}_3 + \tau_2 \boldsymbol{a}_3 \cdot \hat{\boldsymbol{S}}_3 \\ (\rho E + p) V - p\omega + q\boldsymbol{a}_3 \cdot \hat{\boldsymbol{S}}_3 \end{matrix} \right\}$$

$$\boldsymbol{F}^2 = \left\{ \begin{matrix} \rho W \\ \rho u_3 W + p\boldsymbol{a}_3 \cdot \hat{\boldsymbol{S}}_2 \\ \rho u_1 W + p\boldsymbol{a}_1 \cdot \hat{\boldsymbol{S}}_2 \\ \rho u_2 W + p\boldsymbol{a}_2 \cdot \hat{\boldsymbol{S}}_2 \\ (\rho E + p) W \end{matrix} \right\} \qquad \boldsymbol{f} = \left\{ \begin{matrix} 0 \\ -p\dfrac{\partial a_{3\alpha}}{\partial x^\alpha} - \rho \sum\limits_{(\alpha,\beta)} u_\beta u_\alpha \dfrac{\partial a_{3\alpha}}{\partial x^\alpha} \\ -p\dfrac{\partial a_{1\alpha}}{\partial x^\alpha} - \rho \sum\limits_{(\alpha,\beta)} u_\beta u_\alpha \dfrac{\partial a_{1\alpha}}{\partial x^\alpha} \\ -p\dfrac{\partial a_{2\alpha}}{\partial x^\alpha} - \rho \sum\limits_{(\alpha,\beta)} u_\beta u_\alpha \dfrac{\partial a_{2\alpha}}{\partial x^\alpha} \\ 0 \end{matrix} \right\}$$

$$V = \boldsymbol{u} \cdot \hat{\boldsymbol{S}}_3 - \omega \qquad W = \boldsymbol{u} \cdot \hat{\boldsymbol{S}}_2 \qquad U = \boldsymbol{u} \cdot \hat{\boldsymbol{S}}_1$$

$$\omega = q^3 \cdot \boldsymbol{a}_3 \cdot \hat{\boldsymbol{S}}_3 \frac{\partial \varepsilon}{\partial t} \qquad q = -\frac{\lambda}{\varepsilon}\frac{\partial T}{\partial q^3} - \frac{\mu}{2\varepsilon}\frac{\partial}{\partial q^3}\left(u_1^2 + u_2^2\right)$$

$$\tau_1 = -\frac{\mu}{\varepsilon}\frac{\partial u_1}{\partial q^3} \qquad \tau_2 = -\frac{\mu}{\varepsilon}\frac{\partial u_2}{\partial q^3} \qquad u_i = \boldsymbol{u} \cdot \boldsymbol{a}_i$$

In the above formulae, $\hat{\boldsymbol{S}}_i$ is the unit side vector of the elementary volume; $\boldsymbol{a}_i$ is the unit vector of the q^i-direction which, unlike in subsection 4.1.2, is allowed to deviate from the $\boldsymbol{S}_i$-direction; the Greek indices correspond to the Cartesian coordinates x^α; summation is supposed to be made with respect to the repeated indices; all other notations are the same as in previous sections.

The authors investigated the flow about the windward part of the body surface. The computational domain was bounded by the bow shock, body surface and two surfaces $q^1 = const.$ Let the grid point coordinates be (q_i^1, q_j^2, q_k^3) where $0 \leqslant i \leqslant I$, $0 \leqslant j \leqslant J$, $0 \leqslant k \leqslant K$. Averaging and central differencing operators are introduced. For the q^1-coordinate, they are written as follows

$$\begin{aligned} \nu^1_{i,j,k}(\varphi) &= 0.5\left(\varphi_{i+1/2,j,k} + \varphi_{i-1/2,j,k}\right) \\ \delta^1_{i,j,k}(\varphi) &= \varphi_{i+1/2,j,k} - \varphi_{i-1/2,j,k} \end{aligned} \tag{4.53}$$

Similar operators are applied in the q^2 and q^3-coordinates.

In the finite-difference scheme, the first two equations of set (4.52) are written for the elementary volumes $q^1_{i-1/2} \leqslant q^1 \leqslant q^1_{i+1/2}$, $q^2_{j-1/2} \leqslant q^2 \leqslant q^2_{j+1/2}$, $q^3_k \leqslant q^3 \leqslant q^3_{k+1}$ while the rest are written for the volumes $q^1_{i-1/2} \leqslant q^1 \leqslant q^1_{i+1/2}$, $q^2_{j-1/2} \leqslant q^2 \leqslant q^2_{j+1/2}$, $q^3_{k-1/2} \leqslant q^3 \leqslant q^3_{k+1/2}$. Replacing the derivatives with respect to t and q^3 by the difference formulae, one obtains the following set of difference equations:

$$\begin{aligned} & v_D \frac{(X_l)_D^{n+1} - (X_l)_D^n}{\tau} + (X_l)_D^{n+1} h_{k+\xi(l)} \left(\boldsymbol{a}_3 \cdot \boldsymbol{S}_3\right)_D \frac{\varepsilon_{i,j}^{n+1} - \varepsilon_{i,j}^n}{\tau} \\ & \times \sum_{i=1}^{3} \delta_D^i \left(F_l^i S_i\right) - d_l = 0 \qquad (l = 1, \ldots, 5) \end{aligned} \tag{4.54}$$

where

$$d_l = \left\{ \begin{array}{ll} 0, & l = 1,5 \\ \sum\limits_{i=1}^{3} \left\{p_D \delta_D^i (\gamma_l \cdot \boldsymbol{S}_i) + \rho_D \delta_D^i \left[(\boldsymbol{u}_D \cdot \gamma_l)(\boldsymbol{u}_D \cdot \boldsymbol{S}_i)\right]\right\} & l = 2,3,4 \end{array} \right\}$$

$$\gamma_2 = \boldsymbol{a}_3 \qquad \gamma_3 = \boldsymbol{a}_1 \qquad \gamma_4 = \boldsymbol{a}_2$$

$$\xi(l) = \begin{cases} 0.5 & \text{at } l = 1,2 \\ 0 & \text{at } l = 3,4,5 \end{cases} \qquad h_{k+1/2} = \delta^3_{k+1/2}(q^3) \quad h_k = \nu^3_k(h)$$

symbol D denotes three indices $[i, j,\ k + \xi(l)]$.

The components of the flux vectors $\boldsymbol{F}^i$ $(i = 1, 2, 3)$ on the sides of the elementary volumes are determined by the averaging formulae with the weight factors depending on the value and direction of the normal velocity component. These formulae can be found in a paper by Afonina *et al* (1983) along with the formulae for calculating the side vectors and elementary volumes. Set (4.54) is closed by the relationships which present the boundary conditions of the problem. The finite-difference scheme thus obtained has the truncation error $O[\Delta q^1 + \Delta q^2 + (\Delta q^3)^2]$. It is stable provided a condition of the CFL-type is satisfied.

It has been noted above that in the method considered, the set of difference equations is splitted into subsets (blocks) which are solved one after another. Each block involves the difference equations pertaining to the grid points located on the line q^3. The blocks have the following structure which is similar to that of the difference equations of the marching method by Golovachov and Fursenko (1981) described in subsection 2.2.2 (here the indices pertaining to time and q^1, q^2-coordinates are omitted):

$$\begin{aligned} \Phi_{0,l}(z_0) &= 0 \qquad & l &= 1,\ldots,4 & & \\ F_{k,l}(z_{k-1}, z_k, \varepsilon) &= 0 & l &= 1,2 & k &= 1,\ldots,K \\ G_{k,l}(z_{k-1}, z_k, z_{k+1}, \varepsilon) &= 0 & l &= 3,4,5 & k &= 1,\ldots,K-1 \\ \Phi_{K,l}(z_K, \varepsilon) &= 0 & l &= 1,\ldots,5 & & \end{aligned} \tag{4.55}$$

where $z = \{\rho\ u_3\ u_1\ u_2\ T\}^T$. Equations (4.55) are solved using the Newton method. Applying linearization, one obtains, in each iteration, a set of equations with a block tridiagonal matrix of coefficients for the increment of vector z and the bow shock distance from the body surface. This set is solved by the block Thomas ('progonka') algorithm.

4.4 THIN VISCOUS SHOCK LAYER MODEL

This model is an extension of Cheng's two-layer model discussed in chapter 2. It applies to the calculation of hypersonic flows about smooth blunt bodies. The set of thin viscous shock layer equations remains parabolic for three-dimensional flows and is solved by the space-marching methods.

4.4.1 Formulation of the problem

Three-dimensional thin viscous shock layer equations are generally written for the body-fitted curvilinear coordinate system q^i $(i = 1, 2, 3)$ in which q^1, q^2 present a coordinate system on the body surface and the q^3-coordinate is aligned with the normal to this surface. The covariant components of the metric tensor for the above coordinate system are given by the formulae (Shevelev 1986)

$$\begin{aligned} g_{\alpha\beta} &= a_{\alpha\beta} - 2b_{\alpha\beta}q^3 + a^{\gamma\lambda}b_{\gamma\alpha}b_{\lambda\beta}(q^3)^2 \\ g_{3\alpha} &= g_{\alpha 3} = 0 \\ g_{33} &= 1 \end{aligned} \tag{4.56}$$

where $a_{\alpha\beta}$, $b_{\alpha\beta}$ are the components of the symmetric covariant tensors of the first and second quadratic forms of the body surface, respectively. Hereinafter all Greek indices take the values 1,2.

The equations of a stationary three-dimensional thin viscous shock layer flow can be written for the dimensionless quantities as follows (Gershbein 1978, Borodin and Peigin 1989):

$$\begin{aligned}
\frac{\partial}{\partial q^i}\left(\rho u^i \sqrt{\frac{g}{g_{(ii)}}}\right) &= 0 \\
\frac{\partial p}{\partial q^3} &= -\frac{1+\delta}{2}\rho A^3_{\alpha\beta} u^\alpha u^\beta \\
\rho\left(\frac{u^i}{\sqrt{g_{(ii)}}}\frac{\partial u^\alpha}{\partial q^i} + A^\alpha_{\beta\delta} u^\beta u^\delta\right) &= -\frac{2\delta}{1+\delta} g^{\alpha\beta}\sqrt{g_{(\alpha\alpha)}}\frac{\partial p}{\partial q^\beta} \\
&\quad + \frac{\partial}{\partial q^3}\left(\frac{\mu}{K}\frac{\partial u^\alpha}{\partial q^3}\right) \\
\rho\frac{u^i}{\sqrt{g_{(ii)}}}\frac{\partial T}{\partial q^i} &= \frac{2\delta}{1+\delta}\frac{u^\alpha}{\sqrt{g_{(\alpha\alpha)}}}\frac{\partial p}{\partial q^\alpha} + \frac{\mu}{K}\psi_{\alpha\beta}\frac{\partial u^\alpha}{\partial q^3}\frac{\partial u^\beta}{\partial q^3} \\
&\quad + \frac{\partial}{\partial q^3}\left(\frac{\mu}{PrK}\frac{\partial T}{\partial q^3}\right)
\end{aligned} \tag{4.57}$$

Here, u^i are the physical velocity components related to the contravariant components v^i by the formula $u^i = v^i\sqrt{g_{(ii)}}$,

$$K = \delta \cdot Re \qquad \delta = \frac{\gamma-1}{\gamma+1} \qquad Re = \frac{\rho_\infty V_\infty R}{\mu(T^\circ)}$$

$$T^\circ = T_\infty(\gamma-1)M_\infty^2 \qquad g = g_{11}g_{22} - g_{12}^2 \qquad \psi_{\alpha\beta} = \frac{g_{\alpha\beta}}{\sqrt{g_{(\alpha\alpha)}g_{(\beta\beta)}}}$$

$$A^\alpha_{(\alpha\alpha)} = \frac{g_{12}}{g}\left(\frac{\partial\sqrt{g_{(\alpha\alpha)}}}{\partial q^\beta} + \frac{g_{12}}{g_{(\alpha\alpha)}}\frac{\partial\sqrt{g_{(\alpha\alpha)}}}{\partial q^\alpha} - \frac{1}{\sqrt{g_{(\alpha\alpha)}}}\frac{\partial g_{12}}{\partial q^\alpha}\right)$$

$(\alpha \neq \beta)$

$$A^\alpha_{12} = A^\alpha_{21} = \frac{1}{2g}\left[\sqrt{g_{11}g_{22}}\frac{\partial\sqrt{g_{(\alpha\alpha)}}}{\partial q^\beta}\left(1 + \frac{g_{12}^2}{g_{11}g_{22}}\right) - 2g_{12}\frac{\partial\sqrt{g_{(\beta\beta)}}}{\partial q^\alpha}\right]$$

$(\alpha \neq \beta)$

$$A^\beta_{(\alpha\alpha)} = \frac{\sqrt{g_{(\beta\beta)}}}{g}\left(\frac{\partial g_{12}}{\partial q^\alpha} - \sqrt{g_{(\alpha\alpha)}}\frac{\partial\sqrt{g_{(\alpha\alpha)}}}{\partial q^\beta} - \frac{g_{12}}{\sqrt{g_{(\alpha\alpha)}}}\frac{\partial\sqrt{g_{(\alpha\alpha)}}}{\partial q^\alpha}\right)$$

$(\alpha \neq \beta)$

$$A^3_{\alpha\beta} = A^3_{\beta\alpha} = -\frac{1}{2\sqrt{g_{(\alpha\alpha)}g_{(\beta\beta)}}}\frac{\partial g_{\alpha\beta}}{\partial q^3}$$

If the body surface coordinate system (q^1, q^2) is orthogonal, then

$$A^{\alpha}_{(\alpha\alpha)} = A^3_{12} = A^3_{21} = 0 \qquad g = g_{11}g_{22}$$

$$A^{\beta}_{(\alpha\alpha)} = -2A^{\alpha}_{12} = -\frac{1}{\sqrt{g}}\frac{\partial\sqrt{g_{(\alpha\alpha)}}}{\partial q^{\beta}} \quad (\beta \neq \alpha) \qquad A^3_{(\alpha\alpha)} = \frac{b_{(\alpha\alpha)}}{a_{(\alpha\alpha)}}$$

In the above formulae, the summation is supposed to be made with respect to the repeated indices, except for the bracketed ones. The shock layer thickness is supposed to be small as compared with the body nose radius. Therefore, one can assume $g_{\alpha\beta} = a_{\alpha\beta}$

The non-dimensional quantities are defined as follows. The q^3-coordinate, velocity component u^3 and gas density are normalized by δR, δV_∞ and ρ_∞/δ; all other distances, velocity components u^α, pressure, temperature and viscosity are normalized by R, V_∞, $\rho_\infty V_\infty^2(\gamma+1)/\gamma$, T°, $\mu(T^\circ)$, respectively.

Set (4.57) is solved subject to the boundary conditions on the bow shock and the body surface. Taking into account a small thickness of the shock layer, the bow shock boundary conditions (modified Rankine-Hugoniot relationships) are written as follows

$$\begin{aligned}
\rho\left(u^3 - \frac{u^\alpha}{\sqrt{g_{(\alpha\alpha)}}}\frac{\partial q^3_s}{\partial q^\alpha}\right) &= u^3_\infty \\
p &= \frac{1+\delta}{2}\left(u^3_\infty\right)^2 \\
u^3_\infty\left(u^\alpha - u^\alpha_\infty\right) &= \frac{\mu}{K}\frac{\partial u^\alpha}{\partial q^3} \\
\frac{\mu}{PrK}\frac{\partial T}{\partial q^3} &= u^3_\infty\left\{T - \frac{1}{2}\left[\left(u^3_\infty\right)^2 \right.\right. \\
&\left.\left. + \psi_{\alpha\beta}\left(u^\alpha - u^\alpha_\infty\right)\left(u^\beta - u^\beta_\infty\right)\right]\right\}
\end{aligned} \tag{4.58}$$

where $q^3 = q^3_s(q^1, q^2)$ is the equation of the bow shock. The wall boundary conditions are given by the conventional relationships for the velocity components and gas temperature. The velocity slip and temperature jump effects are usually negligible in the range of validity of the thin viscous shock layer model.

4.4.2 Numerical method

Following Borodin and Peigin (1989), let us consider a general case when three-dimensional shock layer flow has no symmetry planes. First, let us dwell on the construction of the coordinate system $\{q^1, q^2\}$ on the body surface. Let $\boldsymbol{l}$ $(\sin\alpha\sin\beta, \sin\alpha\cos\beta, \cos\alpha)$ be the unit vector aligned with the oncoming stream velocity vector $\boldsymbol{V}_\infty$; $F(x^1, x^2, x^3) = 0$ be the equation of the body surface for the Cartesian coordinate system x^α $(\alpha = 1, 2, 3)$, see figure 4.5; the radius-vector $\boldsymbol{x}_0$ locates the stagnation point on the body surface. Here α is the angle of incidence,

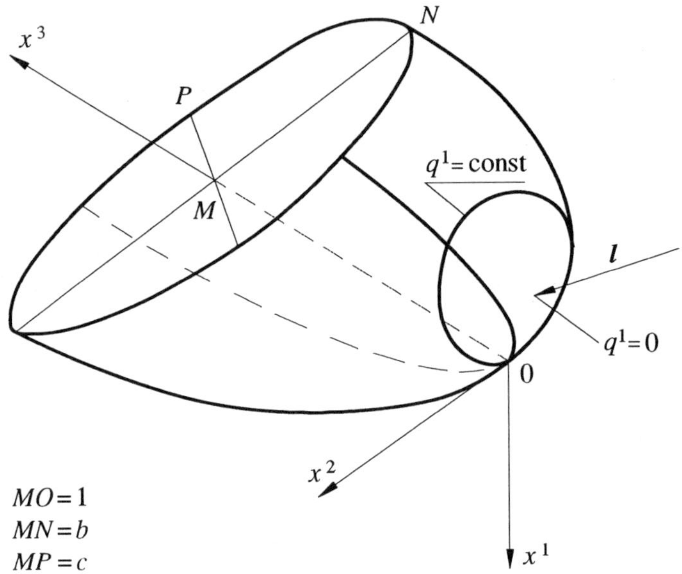

Figure 4.5 Coordinate system for three-dimensional thin viscous shock layer model.

β is the angle of side-slip. Let us introduce another Cartesian coordinate system z^α by the formula

$$\boldsymbol{x} - \boldsymbol{x}_0 = \mathbf{B}(\mathbf{A}\boldsymbol{z}) \tag{4.59}$$

where $\mathbf{B}$ is the diagonal matrix which defines the shrinkage (stretching) transformation, $\mathbf{A}$ is the matrix which defines the rotation of the coordinate axes so that the z^3-axis would be aligned with the inward normal to the body surface. Then, choose the pole C on the z^3-axis and introduce the spherical coordinate system $\{r, q^1, q^2\}$:

$$\begin{aligned} z^1 &= r\sin(mq^1)\cos q^2 \\ z^2 &= r\sin(mq^1)\sin q^2 \\ z^3 - z_C &= -r\cos(mq^1) \end{aligned} \tag{4.60}$$

where $m = m(q^2)$ is the stretching function in the q^1-coordinate. The metric coefficients are calculated by formulae (4.4) taking into account transformations (4.59) and (4.60). The quantities $r(q^1, q^2)$ and $\partial r/\partial q^\alpha$ on the body surface are obtained by substitution of (4.59) and (4.60) into the body surface equation. For the above coordinate system, the q^2-coordinate varies from 0 to 2π, all geometric characteristics of the body surface becoming periodical functions of q^2.

The problem is solved with the use of new independent variables

$$\xi^\alpha = q^\alpha \qquad \zeta = \frac{1}{\Delta}\int_0^{q^3} \rho\sqrt{g}\,dq^3 \qquad \Delta(\xi^1, \xi^2) = \int_0^{q_s^3} \rho\sqrt{g}\,dq^3 \tag{4.61}$$

and functions

$$
\begin{aligned}
\overline{u}^\alpha &= \frac{u^\alpha}{u_*^\alpha} = \frac{\partial \varphi_\alpha}{\partial \zeta} & \overline{T} &= \frac{T}{T_*} \\
p_1 &= \frac{1}{\xi^1}\frac{\partial p}{\partial \xi^1} & p_2 &= \frac{1}{(\xi^1)^2}\frac{\partial p}{\partial \xi^2} \\
\rho u^3\sqrt{g} &= -\frac{\partial}{\partial \xi^\alpha}\left(\Delta \varphi_{(\alpha)}^* \varphi_\alpha\right) - \Delta \varphi_{(\alpha)}^* \frac{\partial \varphi_\alpha}{\partial \zeta}\frac{\partial \zeta}{\partial q^\alpha} \\
\varphi_\alpha^* &= \frac{u_*^\alpha}{\sqrt{g_{(\alpha\alpha)}}} & l &= \frac{\mu \rho g}{\Delta^2 K}
\end{aligned}
\tag{4.62}
$$

Functions $u_*^\alpha(\xi^1,\xi^2)$ and $T_*(\xi^1,\xi^2)$ are introduced to remove the singularity caused by a degeneracy of the coordinate system at $\xi^1 = 0$. Taking into account the orders of magnitude of the metric coefficients, the above functions are assumed to be

$$
u_*^1 = u_\infty^1 \qquad u_*^2 = \xi^1 \qquad T_* = \frac{1}{2}\left(u_\infty^3\right)^2 \tag{4.63}
$$

For the variables of (4.61) and (4.62), the continuity equation is satisfied identically and the other equations of set (4.57) take the form (bars over u and T are omitted):

$$
\begin{aligned}
\frac{\partial}{\partial \zeta}\left(l\frac{\partial u^\alpha}{\partial \zeta}\right) &= Du^\alpha + C_{\beta\delta}^\alpha u^\beta u^\delta + \frac{2\delta}{(1+\delta)\rho} L_\beta^\alpha p_\beta \\
\frac{\partial}{\partial \zeta}\left(\frac{l}{Pr}\frac{\partial T}{\partial \zeta}\right) &= DT + F_\alpha u^\alpha T - \frac{2\delta}{(1+\delta)\rho} u^\beta L_{(\beta)}^3 p_\beta \\
&\quad - lC_{\alpha\beta}^3 \frac{\partial u^\alpha}{\partial \zeta}\frac{\partial u^\beta}{\partial \zeta} \\
\frac{\partial p}{\partial \zeta} &= -\frac{1+\delta}{2} P_{\alpha\beta}^3 u^\alpha u^\beta
\end{aligned}
\tag{4.64}
$$

where

$$
D \equiv \varphi_{(\alpha)}^* u^\alpha \frac{\partial}{\partial \xi^\alpha} - \left(\varphi_{(\alpha)}^* \frac{\partial \varphi_\alpha}{\partial \xi^\alpha} + B_\alpha \varphi_\alpha\right)\frac{\partial}{\partial \zeta}
$$

The equations for functions p_α are derived applying the operators

$$
(\xi^1)^{-1}\frac{\partial}{\partial \xi^1} \quad (\alpha = 1) \qquad \text{and} \qquad (\xi^1)^{-2}\frac{\partial}{\partial \xi^2} \quad (\alpha = 2)
$$

to the last equation of set (4.64). This yields

$$
\frac{\partial p_\alpha}{\partial \zeta} = -\frac{1+\delta}{2}\left(K_{\beta\delta}^\alpha u^\beta u^\delta + M_{\beta\delta}^{(\alpha)} u^\beta \frac{\partial u^\delta}{\partial \xi^\alpha}\right) \tag{4.65}
$$

The boundary conditions on the body surface ($\zeta = 0$) are written as follows

$$
u^\alpha = 0, \quad T = T_w(\xi^1,\xi^2), \quad \Delta\left(B_\alpha \varphi_\alpha + \varphi_{(\alpha)}^* \frac{\partial \varphi_\alpha}{\partial \xi^\alpha}\right) = -G(\xi^1,\xi^2) \tag{4.66}
$$

where G is the mass injection rate. The bow shock boundary conditions at $\zeta = 1$ take the form

$$
\begin{aligned}
l\Phi \frac{\partial u^\alpha}{\partial \zeta} + u^\alpha &= u_s^\alpha \\
p &= \frac{1+\delta}{2}\left(u_\infty^3\right)^2 \\
\frac{l\Phi}{Pr}\frac{\partial T}{\partial \zeta} + T &= 1 + \frac{1}{2} C_{\alpha\beta}^3 \left(u^\alpha - u_s^\alpha\right)\left(u^\beta - u_s^\beta\right) \\
B_\alpha \varphi_\alpha + \varphi_{(\alpha)}^* \frac{\partial \varphi_\alpha}{\partial \xi^\alpha} &= \Phi^{-1}
\end{aligned}
\tag{4.67}
$$

where $\Phi = -\Delta / u_\infty^3 \sqrt{g}$, $u_s^1 = 1$, $u_s^2 = u_\infty^2/\xi^1$. The boundary conditions for the functions p_α are imposed on the bow shock. They are obtained from the second of relationships (4.67) and look as follows

$$
p_1 = (1+\delta)\frac{u_\infty^3}{\xi^1}\frac{\partial u_\infty^3}{\partial \xi^1} \qquad p_2 = (1+\delta)\frac{u_\infty^3}{(\xi^1)^2}\frac{\partial u_\infty^3}{\partial \xi^2} \tag{4.68}
$$

The coefficients of equations (4.64)–(4.67) are defined by the formulae

$$
\begin{aligned}
C_{(\alpha\alpha)}^\alpha &= A_{(\alpha\alpha)}^\alpha u_*^\alpha + \frac{1}{\sqrt{g_{(\alpha\alpha)}}}\frac{\partial u_*^{(\alpha)}}{\partial \xi^\alpha} & B_\alpha &= \frac{1}{\Delta}\frac{\partial(\Delta \varphi_{(\alpha)}^*)}{\partial \xi^\alpha} \\
C_{(\beta\beta)}^\alpha &= A_{(\beta\beta)}^\alpha \frac{(u_*^\beta)^2}{u_*^\alpha} \quad (\alpha \neq \beta) & L_\beta^3 &= \frac{u_*^{(\beta)}(\xi^1)^\beta}{T_* \sqrt{g_{(\beta\beta)}}} \\
C_{\alpha\beta}^3 &= (T_*)^{-1}\psi_{\alpha\beta} u_*^{(\alpha)} u_*^{(\beta)} & L_\beta^\alpha &= \frac{g^{(\alpha\beta)}\sqrt{g_{(\alpha\alpha)}}(\xi^1)^\beta}{u_*^\alpha} \\
P_{\alpha\beta}^3 &= A_{\alpha\beta}^3 \Delta u_*^{(\alpha)} u_*^{(\beta)} & F_\alpha &= \varphi_{(\alpha)}^* \frac{\partial \ln T_*}{\partial \xi^\alpha} \\
K_{\beta\delta}^\alpha &= (\xi^1)^{-\alpha}\frac{\partial P_{\beta\delta}^3}{\partial \xi^{(\alpha)}} & M_{\beta\delta}^\alpha &= \frac{2}{(\xi^1)^\alpha} P_{\beta\delta}^3 \\
C_{12}^\alpha &= C_{21}^\alpha = A_{12}^\alpha u_*^\beta + \frac{1}{2}\varphi_{(\beta)}^* \frac{\partial \ln u_*^\alpha}{\partial \xi^\beta} \quad (\alpha \neq \beta)
\end{aligned}
\tag{4.69}
$$

Calculation of a three-dimensional shock layer flow starts from a self-similar solution on the stagnation streamline. Herein, the non-degenerated curvilinear coordinate system p^i $(i = 1, 2, 3)$ is employed in which the p^3-axis is aligned with the normal to the body surface at the stagnation point and the p^1, p^2-axes are aligned with intersections of the plane tangent to the body surface with the main normal sections of the body. For the above coordinate system, the governing equations and the boundary conditions look like (4.64)–(4.67) with the following non-zero coefficients

$$
\begin{aligned}
B_1 &= C_{11}^1 = R/R_1 & B_2 &= C_{22}^2 = R/R_2 \\
L_1^1 &= L_2^2 = 1 & K_{(\alpha\alpha)}^\alpha &= -2\Delta\left(\frac{R}{R_\alpha}\right)^2
\end{aligned}
$$

where $R_1 \leqslant R_2$ are the main radii of the body surface at the stagnation point. Other coefficients in (4.69) are equal to zero. In the boundary conditions (4.67), $u_s^1 = u_s^2 = 1$; the boundary conditions for the functions p_α take the form

$$p_\alpha = -(1+\delta)R/R_\alpha$$

Having obtained a self-similar solution for the stagnation streamline, one transforms it for the coordinate system (ξ^1, ξ^2, ζ) and uses it as the initial condition for integrating the governing equations along the body surface.

Equations (4.64) and (4.65) subject to boundary conditions (4.66)–(4.68) and the above initial condition are solved using the implicit finite-difference scheme. The derivatives in ξ^1 are replaced by the backward first-order accurate difference formulae. The derivatives in ξ^2 are approximated by the central second-order accurate formulae. The derivatives in ζ are approximated by the fourth-order accurate difference formulae analogous to those of Petukhov (1964).

In each step in the marching coordinate ξ^1, the non-linear set of difference equations is solved by iterations. The calculations required to advance the solution from ξ^1 to $\xi^1 + \Delta\xi^1$ are carried out in the following sequence:

1. Assuming $\Delta(\xi^1 + \Delta\xi^1, \xi^2) = \Delta(\xi^1, \xi^2)$ as an initial guess, the last of the equations (4.66) is solved by the cyclical 'progonka' algorithm, which provides the values of $\varphi_{\alpha w}(\xi^1 + \Delta\xi^1, \xi^2)$. In the case of an impermeable body surface, $\varphi_{\alpha w} \equiv 0$.

2. The following local iterations are carried out on each line ξ^2 of the marching section $\xi^1 + \Delta\xi^1$ to evaluate the functions p, p_α, u^α, T:

(a) the last equation of (4.64) and two equations (4.65) subject to the relevant boundary conditions on the bow shock are employed to calculate, using the Simpson quadrature formulae, the pressure and its derivatives at all grid points of the line considered;

(b) the temperature and velocity components are determined using the 'progonka' algorithm to solve three other equations of set (4.64); after that, one proceeds to point 2(a) again.

The above iterations are carried out until the required accuracy is achieved. The function values at the grid points located on the neighbouring lines are taken from the previous outer loop iteration.

3. Having completed the above calculations along all lines ξ^2, one obtains the values of $\varphi_1(\xi^1 + \Delta\xi^1, \xi^2, 1)$ and $\varphi_2(\xi^1 + \Delta\xi^1, \xi^2, 1)$ by integration of the functions u^α from the body surface to the bow shock.

4. The last of equations (4.67) is solved by the cyclic 'progonka' algorithm in the ξ^2-coordinate to specify the value of $\Delta(\xi^1 + \Delta\xi^1, \xi^2)$.

This terminates the outer loop iteration for the marching section $\xi^1 + \Delta\xi^1$. Later on, one can go back to point 1 or proceed to the next marching step calculations.

4.5 EXAMPLES OF CALCULATIONS

In this section, we consider some examples which demonstrate an application of the above viscous shock layer models and numerical methods to the calculation of three-dimensional flows.

4.5.1 Flow around a blunt body nose

Shock layer flow near the front of a blunted body contains an extensive subsonic region where an upstream disturbance propagation may be important. For this reason, stationary flows of this kind have to be calculated, in a general case, by solving the boundary value problem. Kutler *et al* (1980) and Golovachov *et al* (1985) obtained the time-asymptotic solutions of this problem for the flows around bodies whose shape simulated that taken by the bluntness due to erosion and aerodynamic heating. For the spherical coordinate system, the equation of the body surface was written as follows

$$r_w(\theta,\varphi) = 1 - \delta_{\max}\left[1 - \delta\left|\sin\left(\frac{j\varphi}{2}\right)\right|\right] \times \exp\left[-\frac{1}{2}\left(\frac{\theta - \pi/4}{\beta\pi/4}\right)^2\right]\sin(2\theta) \tag{4.70}$$

Equation (4.70) defines a sphere of the unit radius with j cavities. If j is odd, the body has j symmetry planes $\varphi_n = 2n\pi/j$ $(n = 0, 1, \ldots, j-1)$. The cavity depth amounts to the maximum $\delta_{\max}$ at $\theta = \pi/4$, $\varphi = 2n\pi/j$. On the coordinate surface $\theta = \pi/4$, it varies from $\delta_{\max}$ to $\delta_{\max}(1-\delta)$. The parameter β characterizes the cavity steepness. Thus, the body shape is described by four parameters, namely $\delta_{\max}, \delta, \beta, j$. Equation (4.70) with $j = 0$ corresponds to the body of revolution.

Below, some calculation results of Golovachov *et al* (1985) are discussed which were obtained using the implicit unidirectional scheme. The gas viscosity was calculated with Sutherland's formula (1.8), the specific heat ratio $\gamma = 1.4$, the Prandtl number $Pr = 0.7$, the free stream temperature $T_\infty = 300\ K$, the wall temperature $T_w = 369\ K$. The body shape parameters were chosen to be $\delta_{\max} = 0.2$ and $\delta = 0.75$. Most calculations were carried out on the grid with $K = 12$, $L = 30$, $M = 18$. The grid points were clustered to the body surface by means of transformation (1.43) applied to the q^3-coordinate. The mesh spacing in the q^1- and q^2-coordinates were chosen to be constant. Each time step was fulfilled with three iterations and the stationary solution was achieved after, on average, 300 time steps.

The results are presented in dimensionless form. The skin friction coefficients are defined as the wall shear stress-to-$\rho_\infty V\infty^2/2$ ratio, the wall pressure is normalized by its stagnation point value. The q^1, q^2-coordinates are measured in radians.

A distinct feature of the flows under study arising with increasing cavity steepness is the appearance of a closed subsonic region near the bow shock front. It is clearly seen in figure 4.6 which displays the bow shock wave and the sonic lines in an axisymmetric flow about the body of revolution with an annular hollow, whose shape is described by formula (4.70) with $j = 0$.

The next figures present calculations of flows about the body with three symmetry planes: $q^2 = 0, \pi$; $q^2 = \pi/3, 4\pi/3$ and $q^2 = 2\pi/3, 5\pi/3$. Figure 4.7 depicts the sections of the body, the bow shock and the sonic surface by the conical surface $q^1 = 0.85$ (a) and the meridional plane $q^2 = 0, \pi$ (b) at zero incidence. The

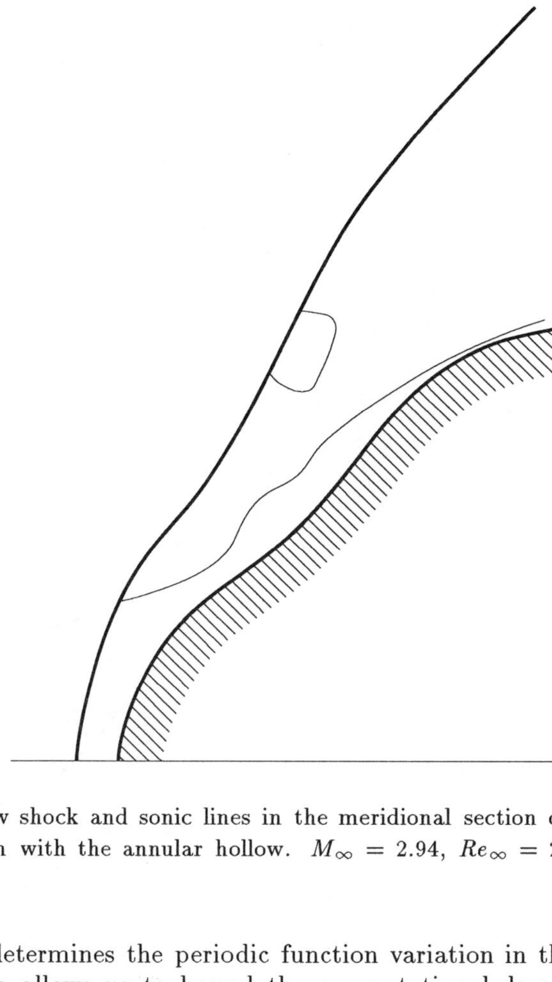

Figure 4.6 Bow shock and sonic lines in the meridional section of a flow about the body of revolution with the annular hollow. $M_\infty = 2.94$, $Re_\infty = 2.8 \times 10^3$, $\beta = 0.6$, $\alpha = 0$.

body shape predetermines the periodic function variation in the circumferential coordinate which allows us to bound the computational domain by the sector $0 \leqslant q^2 \leqslant \pi/3$. Figure 4.8 shows the variation of the skin friction coefficient C_{f2} along the intersections of the body surface and the cones $q^1 = (k - 1/2)\Delta q^1$. The numerals over the curves indicate the k-values, $\Delta q^1 = 0.1309$. The results demonstrate the effect of cavity steepness which is governed by the parameter β. An increase of the cavity steepness (decrease of β) intensifies circumferential flows in the vicinity of the cavity centre.

The next two figures show the skin friction coefficient C_{f1} and the wall pressure distributions in the meridional plane $q^2 = 0, \pi$. Variation of the skin friction coefficient indicates a possibility for longitudinal separation with increasing cavity steepness. The hair line in figure 4.10 depicts calculations of Kutler *et al* (1980) carried out with the implicit factored scheme for the same value of the free stream Mach number and for Reynolds number $Re_\infty = 2.2 \times 10^5$. It is seen that both calculations provide the same pressure distibution on the concave part of the body surface. The discrepancy in the cavity vicinity is apparently due to the different

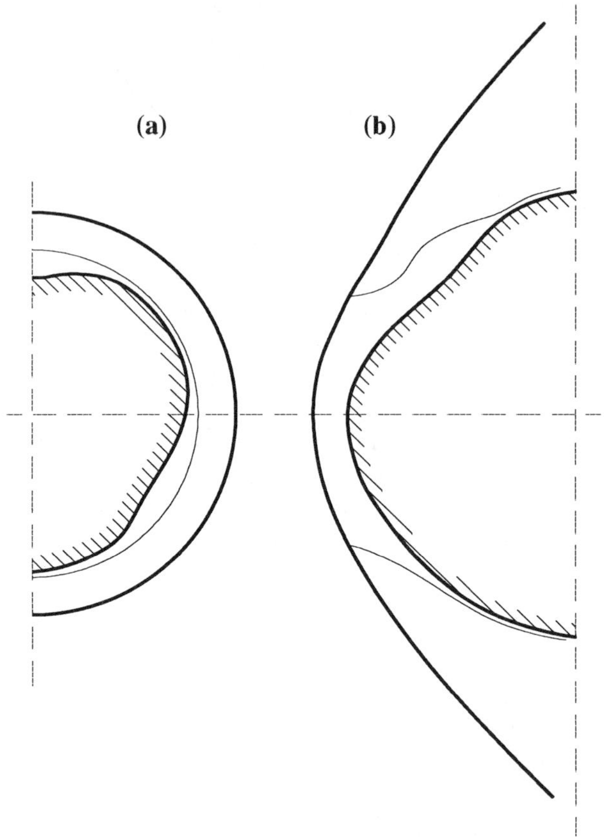

Figure 4.7 Bow shock and sonic lines in a flow about the indented body. $M_\infty = 2.94$, $Re_\infty = 2.8 \times 10^3$, $\beta = 0.5$ $\alpha = 0$.

values of the Reynolds number. The calculations of Kutler *et al* (1980) showed flow separation in the cavities at high Reynolds numbers.

At a hypersonic flight speed, non-separated flow about the nose of a smooth blunt body can be calculated using the thin viscous shock layer model. This is illustrated below by calculations of Borodin and Peigin (1989) which were carried out with the numerical method of subsection 4.4.2. The results pertain to three-dimensional flows of a perfect gas with $\delta = (\gamma - 1)/(\gamma + 1) = 0.1$, $\mu \propto \sqrt{T}$, $Pr = 0.71$.

Figure 4.11 shows the variation of the bow shock distance from the body surface with the circumferential coordinate ξ^2 in the flows about triaxial ellipsoids with the half-axes $b = 0.7$, $c = 0.3$ (full lines) and $b = 2$, $c = 3$ (hair lines); see figure 4.5. The wall temperature $T_w = 0.1T^0$ in the first case and $T_w = 0.15T^0$ in the second one. In both cases, the angles of incidence and side-slip $\alpha = \beta = 45°$. The bow shock distance from the body surface is normalized by its stagnation point value.

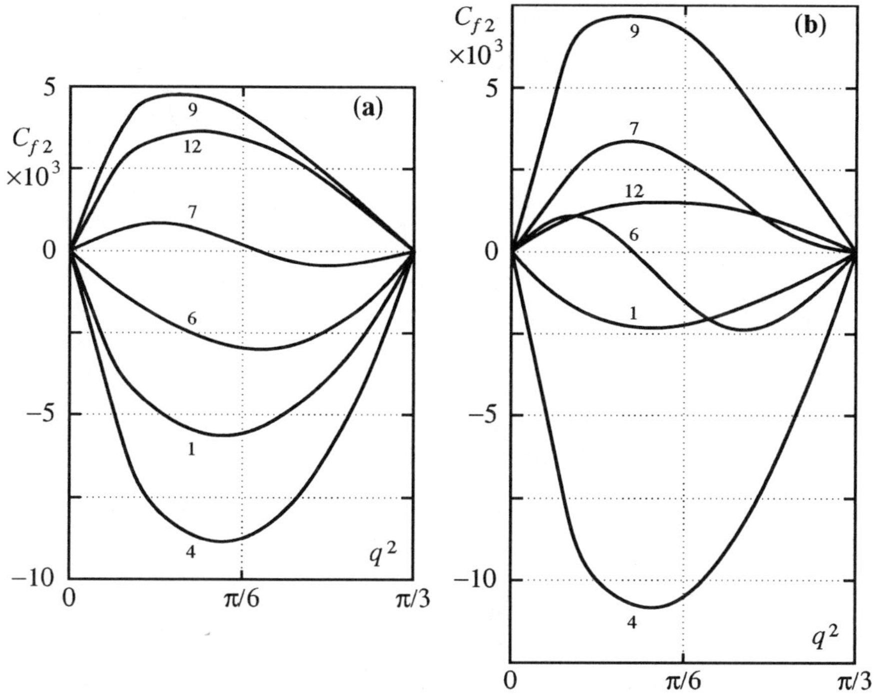

Figure 4.8 Variation of the skin friction coefficient C_{f2} on the indented body. $M_\infty = 2.94$, $Re_\infty = 2.8 \times 10^3$, $\alpha = 0$; (a), $\beta = 1.4$; (b), $\beta = 0.5$.

Curves A through D correspond to the following values of the non-dimensional longitudinal coordinate: $\xi^1 = 0.1, 0.2, 0.3$ and 0.4. One can see a non-monotonous variation of the shock layer thickness with the circumferential coordinate, which essentially depends on the body shape.

The next figures demonstrate the influence of the body shape and the angle of incidence on the wall pressure and heat flux distributions over the body surface. The above quantities are normalized by their stagnation point values.

At $\alpha = \beta = 0$, there are two symmetry planes in which the heat flux and pressure distributions reach local extrema. These distributions are plotted in figure 4.12 where the lines A, C, E correspond to the symmetry plane $\xi^2 = 0$ and the lines B, D, F correspond to the symmetry plane $\xi^2 = \pi/2$. The curves A, B; C, D and E, F pertain to ellipsoids with the half-axes $b = 2$, $c = 3$; $b = 2$, $c = 0.5$, and $b = 0.7$, $c = 0.3$, respectively. It is seen that the kind of extremum in the heat flux distribution depends on the ellipsoid half-axis ratio. At $b = 0.7$, $c = 0.3$, the heat flux decreases monotonously in both symmetry planes, having a maximum at the stagnation point. The contrary heat flux behaviour is seen for $b = 2$, $c = 3$. That is, the heat flux increases in both symmetry planes with a distance from the stagnation point. In an intermediate case $b = 2$, $c = 0.3$, the point $\xi^1 = 0$ is the saddle point in the heat flux distribution. The wall pressure distribution

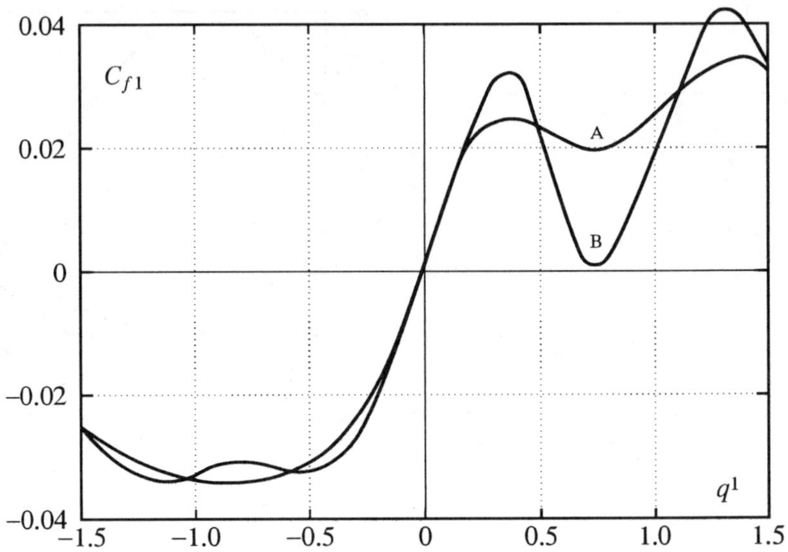

Figure 4.9 Variation of the skin friction coefficient C_{f1} on the indented body in the plane $q^2 = 0, \pi$. $M_\infty = 2.94$, $Re_\infty = 2.8 \times 10^3$, $\alpha = 0$; A, $\beta = 1.4$; B, $\beta = 0.5$.

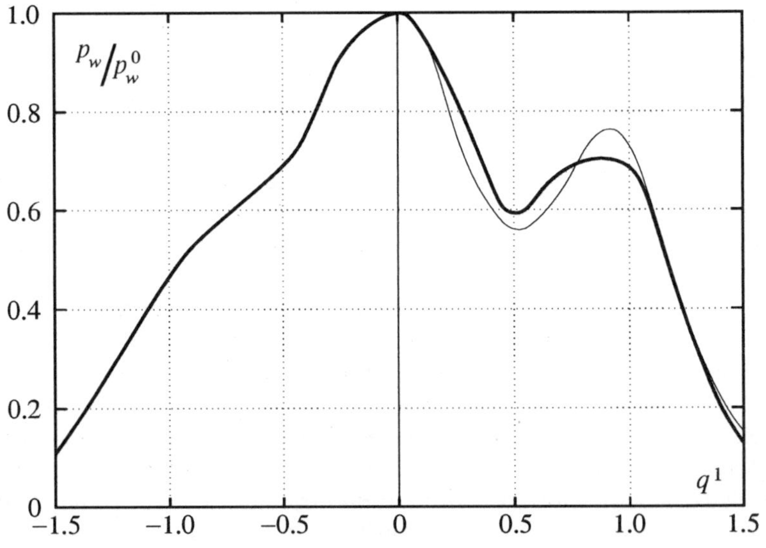

Figure 4.10 Wall pressure variation on the indented body in the plane $q^2 = 0, \pi$. $M_\infty = 2.94$, $\beta = 0.5$, $\alpha = 0$; full line, $Re_\infty = 2.8 \times 10^3$; hair line, $Re_\infty = 2.2 \times 10^5$.

demonstrates a monotonous variation with a maximum at the stagnation point in all the cases considered.

When $\beta = 0$, $\alpha \neq 0$, the flow has only one symmetry plane. The heat flux

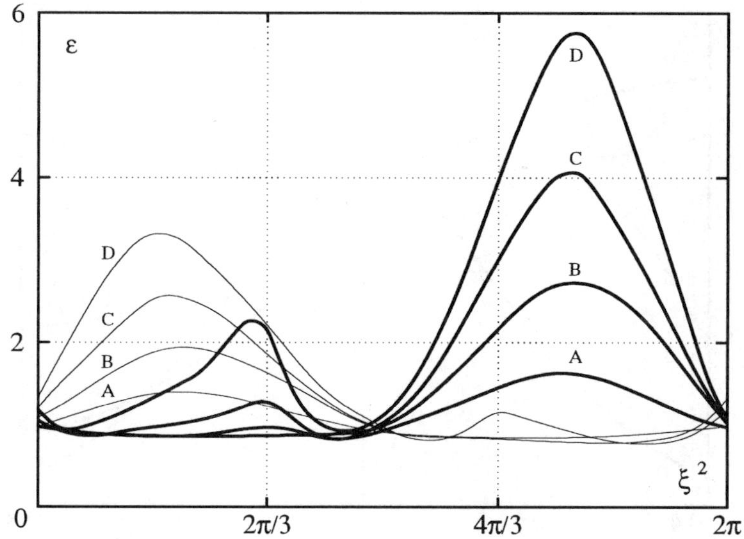

Figure 4.11 Variation of the shock layer thickness in hypersonic flows about triaxial ellipsoids. $Re = 5 \times 10^4$, $\alpha = \beta = 45°$.

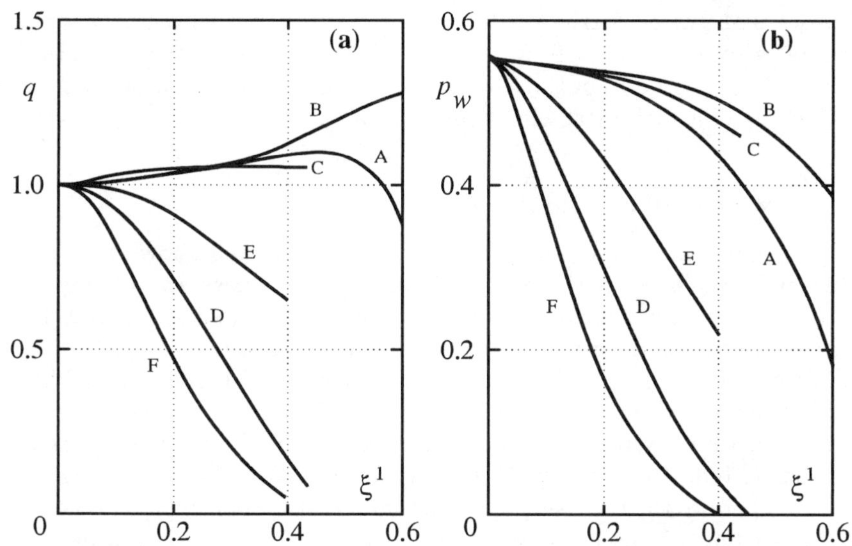

Figure 4.12 Heat flux and wall pressure distributions in the symmetry planes of triaxial ellipsoids. $Re = 5 \times 10^4$, $T_w = 0.15$, $\alpha = \beta = 0$.

distributions in this plane at different angles of attack are shown in figure 4.13 where the positive and negative values of the ξ^1-coordinate correspond to the semi-planes $\xi^2 = 0$ and $\xi^2 = \pi$, respectively. The results pertain to the flow about

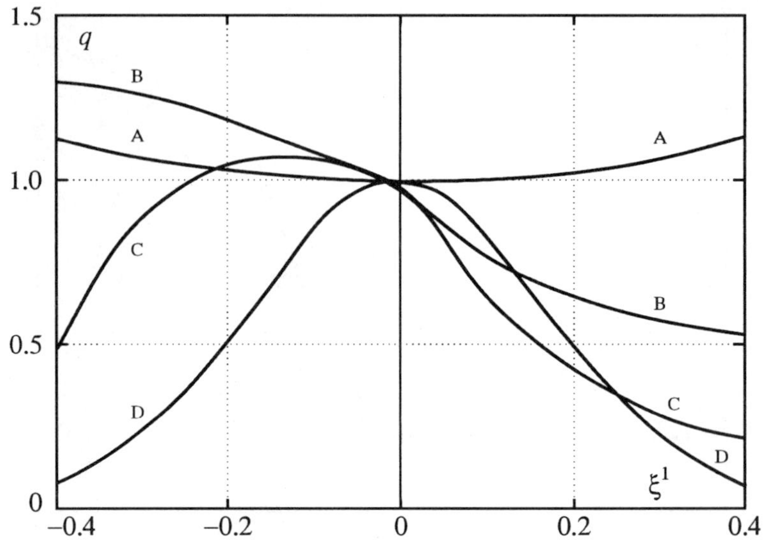

Figure 4.13 Heat flux distributions in the symmetry plane of a triaxial ellipsoid. $Re = 5 \times 10^4$, $T_w = 0.15$, $\beta = 0$.

an ellipsoid with the half-axes $b = 3$, $c = 2$. The curves A through D correspond to the angles of attack $\alpha = 0, 30°, 60°$, and $90°$. It is seen that the stagnation point turns from the point of the heat flux minimum to the point of the maximum with the angle of attack increasing from 0 to 90°.

4.5.2 Flow around an elongated body

Consider, at first, the flows around circular cones. It is well-known that at angles of attack $\alpha \gtrsim 3\theta_c/4$ (θ_c is the cone half-angle), these flows are accompanied by cross-flow separation producing a pair of vortices over the leeward side of the body, which significantly influences the aerodynamic characteristics.

The flows about the circular cones were calculated by Lubard and Helliwell (1974), Lubard and Rakich (1975), Lin and Rubin (1982), Chernii (1983). The above investigations were carried out using the parabolized Navier–Stokes equations which included all terms with the second derivatives and the mixed derivatives with respect to the transversal and circumferential coordinates. A body-fitted coordinate system was used. At moderate incidence, the cross-flow separation appears at some distance from the cone apex. Accordingly, the initial data were obtained using the models and methods developed for the calculation of non-separated flows.

In calculations by Lubard and Helliwell (1974), Lubard and Rakich (1975), Lin and Rubin (1982) the initial-boundary value problem was regularized in the subsonic region by an explicit backward difference approximation of the streamwise pressure gradient term. The linearized set of difference equations in each marching

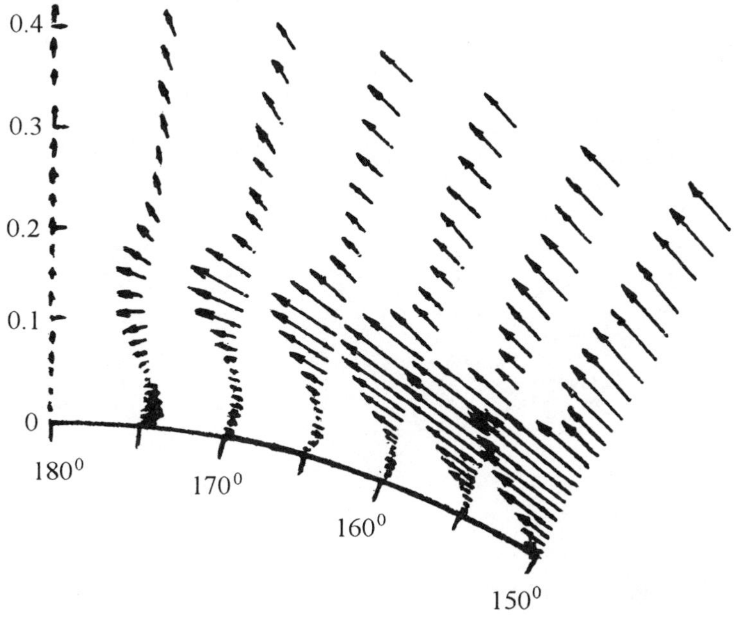

Figure 4.14 Cross-flow velocity map near the leeward side of a pointed circular cone. $M_\infty = 8$, $Re_\infty = 1.13 \times 10^6/ft$, $T_w = 0.41T^*$, $\theta_c = 10°$ $\alpha = 12°$.

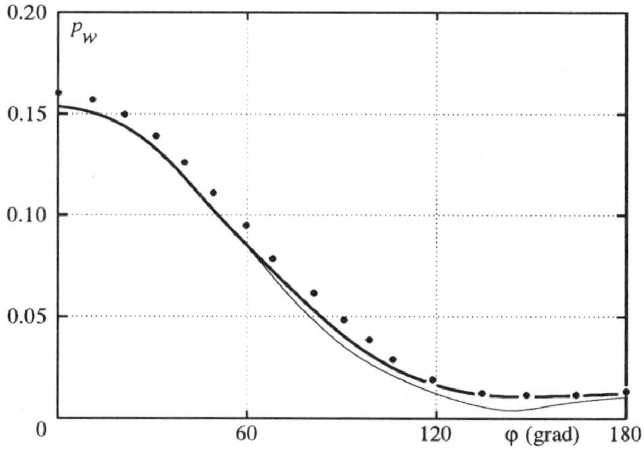

Figure 4.15 Wall pressure distribution over a pointed circular cone. $M_\infty = 8$, $Re_\infty = 1.13 \times 10^6/ft$, $T_w = 0.41T^*$, $\theta_c = 10°$, $\alpha = 12°$. Full and hair lines, calculations of Lubard and Helliwell (1974) for $x = 0.34\ ft$ and $x = 2\ ft$, respectively; circles, experimental data of Tracy (1963) for $x = 0.33\ ft$.

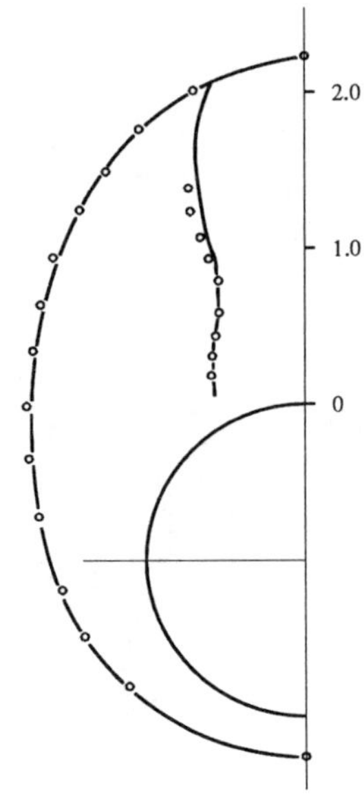

Figure 4.16 Bow and inner shocks in the flow about a pointed circular cone. $M_\infty = 8$, $Re_\infty = 1.13 \times 10^6/ft$, $T_w = 0.41T^*$, $\theta_c = 10°$, $\alpha = 24°$. Full lines, calculations of Lin and Rubin (1982); circles, measurements by Tracy (1963).

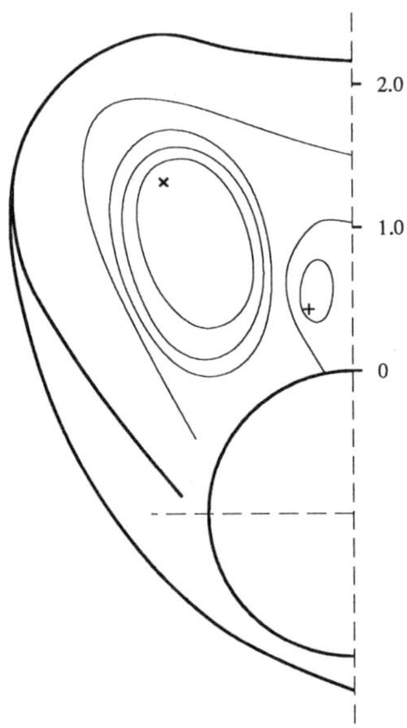

Figure 4.17 Flow structure in a cross section of the shock layer on the pointed cone. $M_\infty = 8$, $Re_\infty = 1.13 \times 10^6/ft$, $T_w = 0.41T^*$, $\theta_c = 10°$, $\alpha = 45°$.

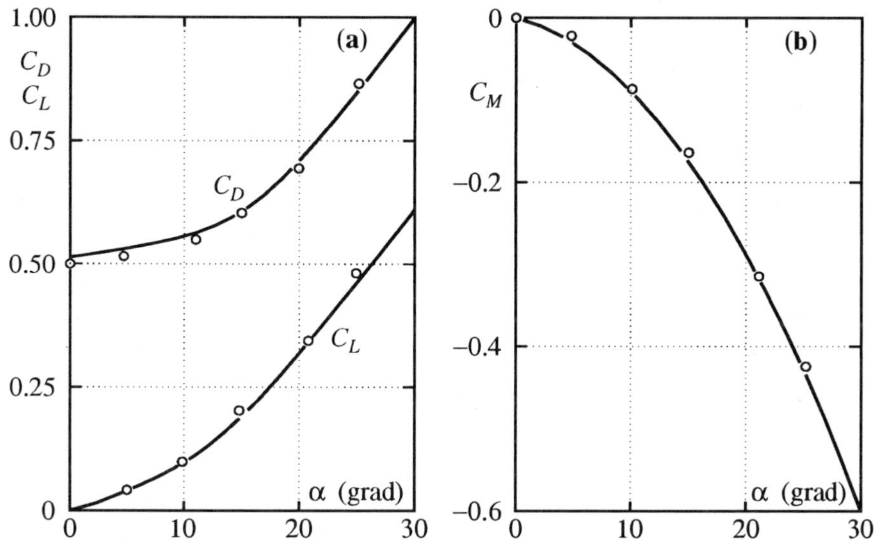

Figure 4.18 Dependence of aerodynamic characteristics of a blunted cone on the angle of attack. $M_\infty = 13$, $Re_\infty = 4100$, $T_w = 0.25T^*$, $\theta_c = 5°$. The lines, calculations by Chernii (1983); circles, experiments of Krasilschikov and Nosov (1971) and Kiselyov (1982).

step was solved using the line relaxation and incomplete factorization (strong implicit procedure by Stone (1968)) methods. The method of Chernii (1983) presents a straightforward generalization of the two-dimensional space-marching procedure described in subsection 2.2.2 with an additional operator splitting in the q^2- and q^3-coordinates.

Figures 4.14 and 4.15 show calculation results pertaining to the flow about a pointed cone with a half-angle $\theta_c = 10°$ at an angle of attack $\alpha = 12°$. Here the gas specific heat ratio $\gamma = 1.4$, the Prandtl number $Pr = 0.75$, the viscosity was evaluated from Sutherland's formula (1.8). Figure 4.14 depicts a velocity map in the marching section $x = 2ft$, where x is the distance from the cone apex along the generatrix. Figure 4.15 presents a comparison of the calculation results with the experimental data. Shown are the wall pressure distributions in the sections $x = 0.34ft$ (full line) and $x = 2ft$ (hair line). The dark circles stand for the measurements at $x = 0.33ft$. The maximum discrepancy appears on the windward part of the cone. To the authors' mind, it is due to an excess of about 0.5° of the model half-angle over the value $\theta_c = 10°$. A weak dependence of the wall pressure distribution on the distance from the cone apex evidences for the conical behaviour of the flow under study.

Lin and Rubin (1982) investigated supersonic flows about the circular cones at high angles of attack. In this case, flow deceleration in the leeward part of the shock layer gives rise to the formation of inner shocks. They are clearly seen in figure 4.16 where the calculation results are compared with the experimental data

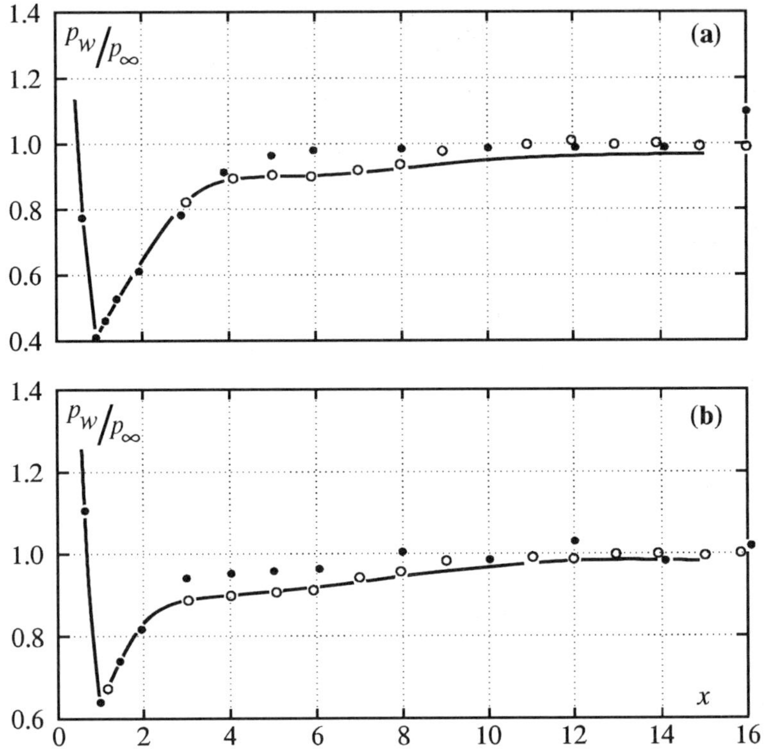

Figure 4.19 Wall pressure distributions in the symmetry plane of a turbulent flow about a blunted cylinder. $M_\infty = 1.4$, $Re_\infty = 2 \times 10^5$, $\alpha = 5°$; (a), $\varphi = 0$; (b), $\varphi = 180°$. Full lines and bright circles, calculations of Schiff and Steger (1980) by the time-asymptotic and space-marching methods, respectively; dark circles, experimental data of Hsieh (1976).

of Tracy (1963). The picture is nearly independent of the distance from the cone apex due to conical properties of the flow under study.

At angles of attack $\alpha \gtrsim 3\theta_c$, the calculations showed the secondary vortices near the leeward part of the flow symmetry plane and a local minimum in the variation of the shock layer thickness with the circumferential coordinate. These features are clearly seen in figure 4.17 where the calculated cross-flow structure is presented. Here the bow and inner shocks are depicted along with the streamlines of the main and secondary recirculating flows. The crosses stand for the centres of the vortices. The calculations provide such flow structure at $x \lesssim 60$ cm. When further advancing the solution in the marching coordinate, the bow shock wave acquires the usual shape with the maximum distance from the body surface in the leeward part of the flow symmetry plane. The above change of the bow shock shape with a distance from the cone apex corresponds to a transition from the anomalous to the regular location of the bow shock (Buchmanova *et al* 1973).

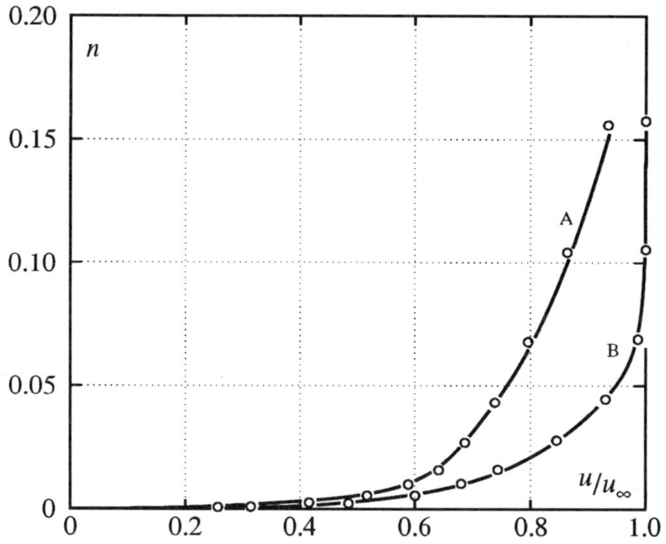

Figure 4.20 Profiles of the longitudinal velocity component in the symmetry plane of a turbulent flow about the blunted cylinder (Schiff and Steger 1980). $M_\infty = 1.4$, $Re_\infty = 2 \times 10^5$, $\alpha = 5°$, $x = 6.98$; A, $\varphi = 0$; B, $\varphi = 180°$; lines, time-asymptotic solution; circles, space-marching solution.

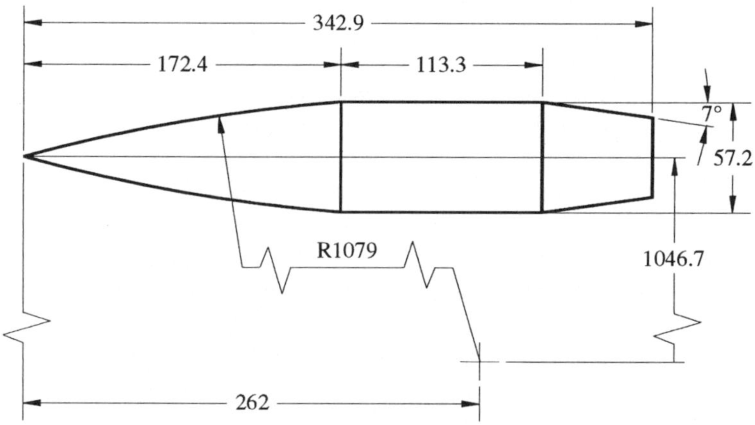

Figure 4.21 Body of revolution with an ogival nose and taper tail. All distances are in mm.

It is also interesting to note that recent investigations revealed an asymmetric and non-stationary periodic vortical flows over the leeward side of slender conical bodies at high angles of attack; see for example the paper by Kandil *et al* (1990).

Lubard and Rakich (1975) and Chernii (1983) investigated three-dimensional flows about spherically blunted cones. Figure 4.18, reproduced from the paper by

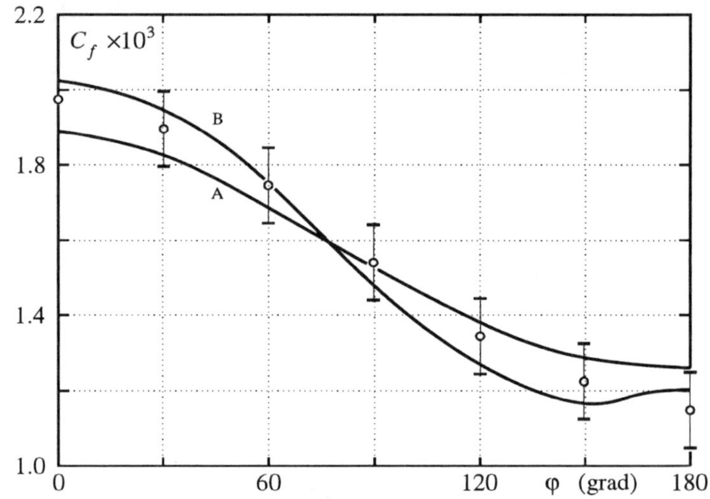

Figure 4.22 Calculated and measured variation of the total skin friction coefficient over the body shown in figure 4.21. $M_\infty = 3$, $Re_\infty = 7.3 \times 10^6$, $x = 254$ mm; A, $\alpha = 4.2°$; B, $\alpha = 6.3°$.

Chernii (1983), shows the dependence of aerodynamic characteristics of a spherically blunted cone with the half-angle $\theta_c = 5°$ and length $L = 6.98R$ on the angle of attack. Here, the gas specific heat ratio $\gamma = 1.4$, the viscosity $\mu \propto T^{0.75}$, the Prandtl number $Pr = 0.72$. The Reynolds number is based on the body nose radius R. The figure shows the variation of the drag, lift force, and pitching moment coefficients. Calculation results are compared with the experimental data of Krasilschikov and Nosov (1971) and Kiselyov (1982). It is seen that calculations overestimate, by a few per cent, the drag coefficient C_D at small angles of attack and underestimate the lift force coefficient C_L at large angles of attack as compared with the measurements. The discrepancy is explained by neglecting the base pressure in the calculation of the aerodynamic coefficients.

The next two figures present the results of Schiff and Steger (1980) for the turbulent flow about a spherically blunted cylinder. The shock layer flow was calculated with the reduced Reynolds equations, similar to (4.12), which were closed with the algebraic turbulence model by Baldwin and Lomax (1978).

Figures 4.19 and 4.20 show the wall pressure distributions and the longitudinal velocity component profiles in the windward ($\varphi = 0$) and leeward ($\varphi = 180°$) parts of the flow symmetry plane. Here the x-coordinate is measured from the stagnation point of the cylinder along its axis, the n-coordinate is normal to the cylinder surface. Both coordinates are normalized by the body nose radius which is also used to evaluate the Reynolds number. The cylinder surface is supposed to be adiabatic. The calculation results are shown by the full lines and bright circles. The lines depict the solution of the boundary value problem with the time-asymptotic method of subsection 4.2.1. The bright circles depict the results obtained with the

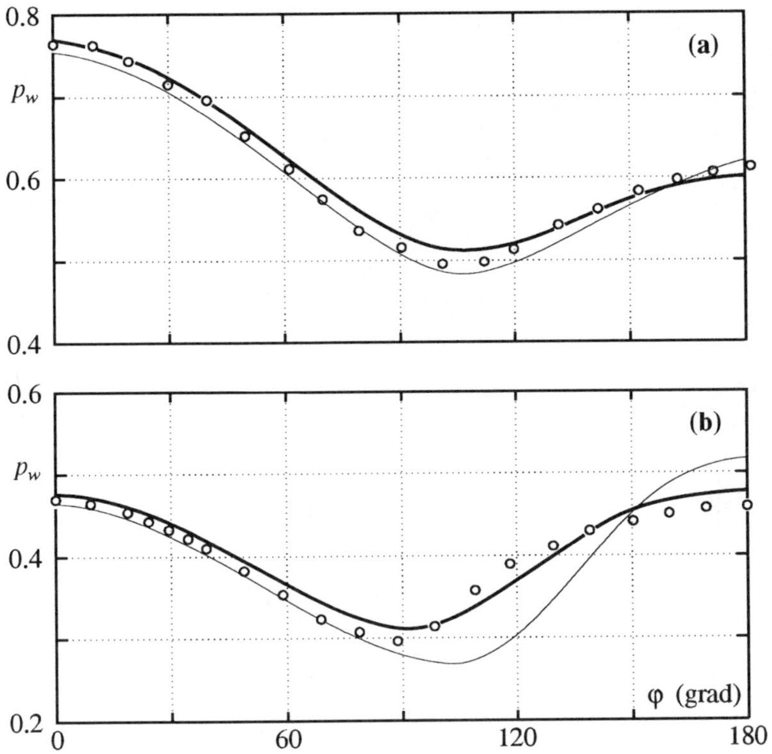

Figure 4.23 Wall pressure distributions over the body shown in figure 4.21. $M_\infty = 3$, $Re_\infty = 7.3 \times 10^6$, $\alpha = 6.3°$; A, $x = 279$ mm, B, $x = 330$ mm. Full lines, viscous shock layer calculation; hair lines, inviscid flow calculation; circles, wind tunnel experiments.

space-marching method discussed in subsection 2.2.2b with reference to a plane flow. Space-marching calculations were started from $x = 3.07$ with the initial data borrowed from the time-asymptotic solution. The dark circles stand for the experimental data of Hsieh (1976). The figures demonstrate a good agreement between the results obtained by both numerical methods. At $8 \lesssim x \lesssim 14$, the wall pressure distributions calculated by the space-marching method are in a better agreement with the measurements, which can be explained by the use of a finer computational mesh in the space-marching method at a large distance from the stagnation point.

To illustrate the calculation of shock layer flow about a more complicated shape, we consider below the results of Sturek and Schiff (1982) who investigated turbulent flow about a body of revolution with an ogival nose and taper tail; see figure 4.21. The viscous shock layer calculations are compared with the results obtained from the inviscid gas model and the wind tunnel measurements. The turbulent flow regime was initiated in the experiments by the installation of a

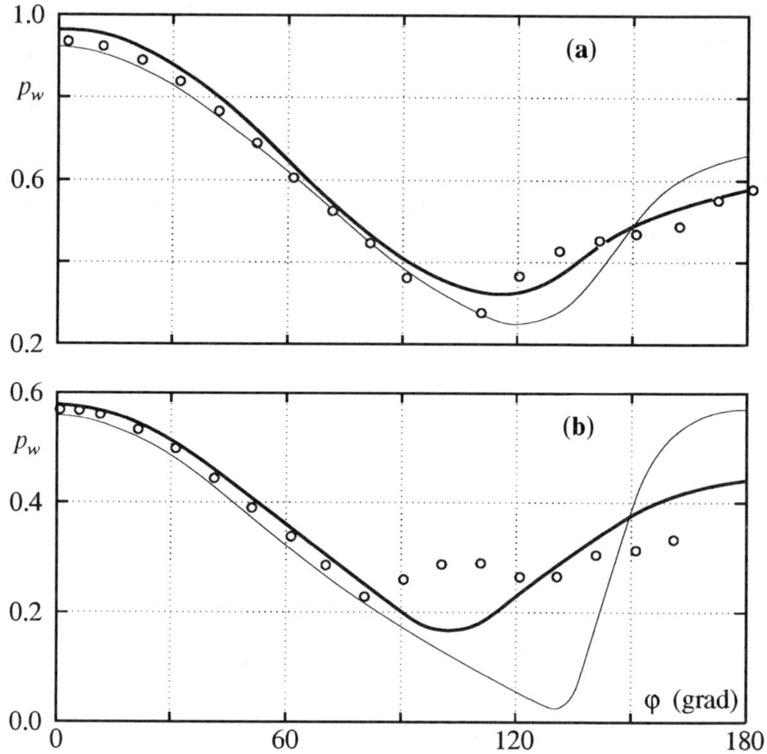

Figure 4.24 Wall pressure distributions over the body shown in figure 4.21. $M_\infty = 3$, $Re_\infty = 7.3 \times 10^6$, $\alpha = 10.4°$. The notations are the same as in the previous figure.

turbulizer on the nose of the model. The shock layer flow was calculated using the space-marching method by Schiff and Steger (1980). The initial data for the marching calculations were obtained assuming the flow near the nose tip of the body to be conical. The Reynolds number is evaluated with the model length. The wall pressure and shear stress are normalized by $\rho_\infty c_\infty^2$, where c_∞ is the speed of sound in the free stream.

Figure 4.22 shows the variation of the skin friction coefficient over the body surface at two angles of attack. The circles represent measurements at $\alpha = 4.2°$. The calculations fit well the experimental data within the measurement uncertainty. The shift of a minimum of the total skin friction coefficient from the symmetry plane at $\alpha = 6.3°$ indicates the appearance of a cross-flow separation.

The next two figures show the wall pressure distributions at $x = 279$ mm (a) and $x = 330$ mm (b) at two angles of attack. The first cross section is located within the cylindrical part of the body, the second one within the tail. In figure 4.23, one can see a good agreement of the viscous shock layer calculations with the wind tunnel measurements both on the cylindrical and tail parts of the body at

$\alpha = 6.3°$. With an increasing angle of incidence α, the discrepancy of the results appears on the leeward side of the body, which is especially large on the tail. To the authors' mind, the above discrepancy may be caused by the use of reduced equations (omission of the viscous terms with the derivatives in the circumferential coordinate) and by the inadequacy of the turbulent transport description in the separated flow region. The results of inviscid flow calculations shown by the hair lines demonstrate an error on the leeward side of the body, which rapidly increases with the intensification of the separated flow. At $\alpha = 10.4°$, the inviscid flow calculations predict inner shocks which are corroborated by neither experiments nor viscous shock layer calculations.

4.5.3 Flow around a spinning body

Shock layer flows around spinning bodies were calculated mainly with a view to investigate the Magnus effect, i.e. the appearance of a side force due to violation of the flow symmetry with regard to the plane containing the body axis and the oncoming stream velocity vector. Though the Magnus force is usually small, it can noticeably influence the flight trajectory and violate the stability. The results which are discussed below have been obtained using the parabolized Navier–Stokes equations and the space-marching numerical methods.

Figure 4.25 presents the calculations of a turbulent flow about the pointed body of revolution shown in figure 4.21 (Sturek and Schiff 1982). Variation of the side force coefficient with a distance from the nose tip is depicted. The hair line shows the results of the inviscid gas and boundary layer calculations. The circle stands for the measured value of C_y. It is seen that the use of inviscid gas and boundary layer models overestimates noticeably the side force on the taper tail of the body. The calculations revealed a dependence of the Magnus effect on wall temperature, demonstrating a decrease of the side force with body surface cooling.

Kim and Lewis (1983) investigated laminar flows about a spherically blunted cone with mass injection on its spinning conical part. The spherical part of the body was supposed to be non-spinning and impermeable. The initial data for the space-marching calculation were prescribed on the surface containing the sphere-cone conjugation line with the use of available data on the axisymmetric flow about a sphere. The streamwise component of the pressure gradient was assumed to be equal to zero. The space-marching procedure is similar, in the main, to the method by Lubard and Helliwell (1974) and Helliwell *et al* (1980), differing only in the determination of the wall pressure and the location of the bow shock.

The calculations presented in figures 4.26 and 4.27 pertain to the flow about a cone with a half-angle $\theta_c = 7°$ and length $L = 25$ cm. The mass injection rate $G = \rho_w V_w / \rho_\infty V_\infty = 0.00125$, the angular velocity of the body rotation $\Omega = 8000$ rpm.

Figure 4.26 demonstrates the spin and mass injection effects on the side force. The distance from the stagnation point along the body surface s is normalized by the body nose radius. The results shown by the curves A through C were calculated taking into account mass injection, body spin, and both processes, respectively. One can see that the mass injection results in a significant increase of the side

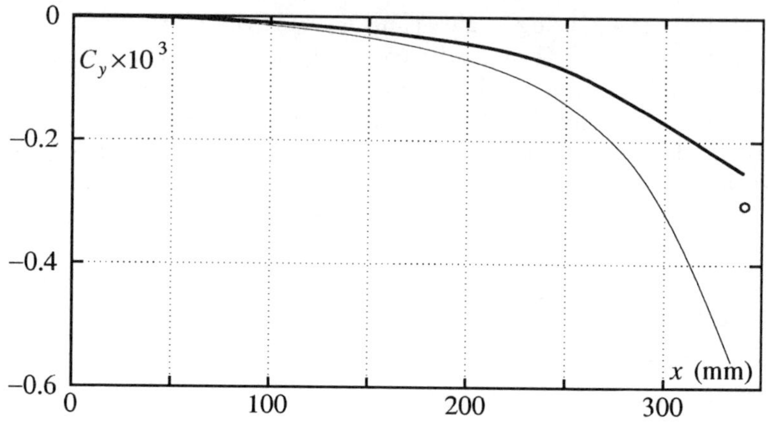

Figure 4.25 Variation of the side force coefficient with a distance from the tip of the spinning body shown in figure 4.21. $M_\infty = 3$, $Re_\infty = 7.3 \times 10^6$, $\alpha = 2°$, $\Omega = 20000$ rpm. Full line, viscous shock layer calculation; hair line, inviscid gas and boundary layer calculations; circle, measurement.

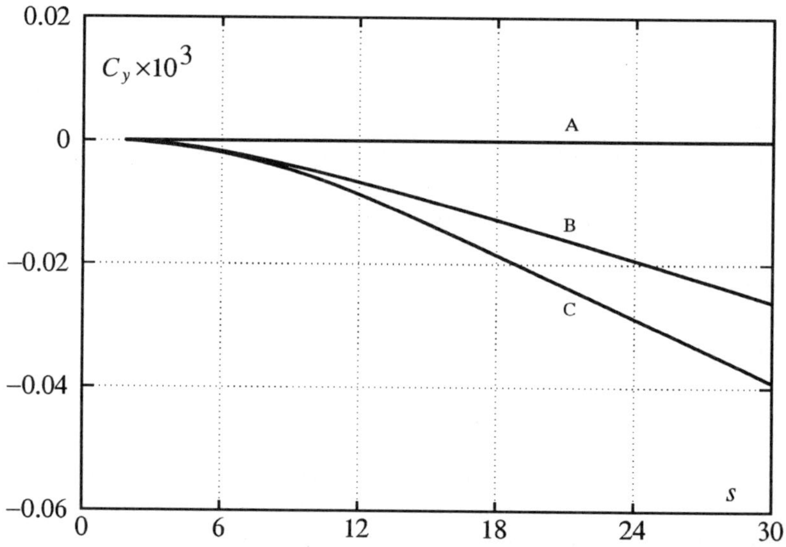

Figure 4.26 Variation of the side force coefficient of a spherically blunted cone. $M_\infty = 18$, $Re_\infty = 1.66 \times 10^5$ /m, $T_w = 0.127T^*$, $\theta_c = 7°$, $\alpha = 15°$; A, with mass injection; B, with spin; C, with mass injection and spin.

force on the spinning cone, which indicates a non-linear interaction of the above processes.

Calculations show that like the flows about sharp cones, the main contribution

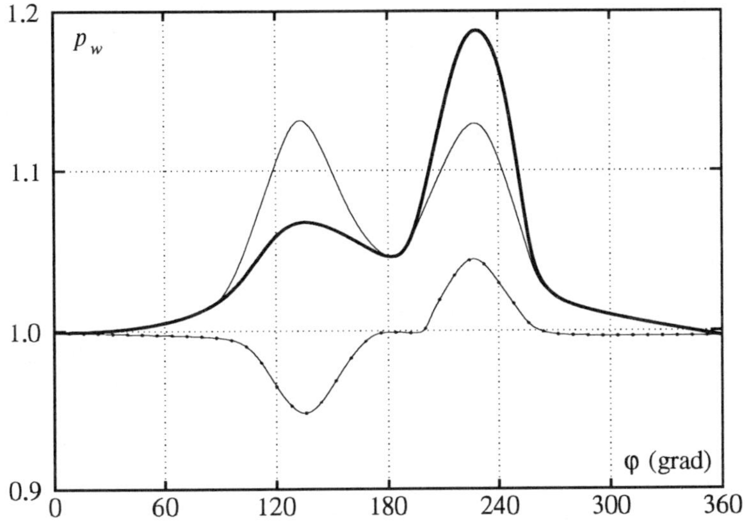

Figure 4.27 Wall pressure distribution on a spherically blunted cone at $s = 30.06$ with mass injection (hair line), spin (hair-dotted line), mass injection and spin (full line). The flight conditions are the same as in the previous figure.

to the Magnus force is provided by an asymmetry of the wall pressure distribution. Figure 4.27 demonstrates the mass injection and spin effects on the wall pressure distribution. The pressure is normalized by its value on a non-spinning impermeable cone. The results taking into account only mass injection, only spin, and both processes are plotted by the hair, hair-dotted and full lines, respectively. The influence of mass injection and spin on the heat flux to the body surface was found to be small under the flight conditions considered. Other calculation results on three-dimensional shock layer flows about the spinning bodies can be found in the papers by Lin and Rubin (1974), Agarwal and Rakich (1978), Agarwal and Rakich (1982).

4.5.4 Flow around a delta wing

The calculations which are discussed below were carried out for supersonic flows around plane delta wings with a spherical nose and cylindrical leading edge. The over and side views of the wing are presented in figure 4.28. Numerical solutions were obtained with the parabolized Navier–Stokes equations and the space-marching schemes of subsections 4.3.2 and 4.3.3.

Tannehill *et al* (1981) calculated the flow about a wing with a sweep angle $\beta = 70°$ and a radius of the edge bluntness $R = 9.525$ mm. The viscosity was evaluated by Sutherland's formula (1.8), the gas specific heat ratio $\gamma = 1.4$, the Prandtl number $Pr = 0.72$. Calculations were carried out for the conditions of the experiments by Bertram and Everhart (1963). Figure 4.29 shows the calculated

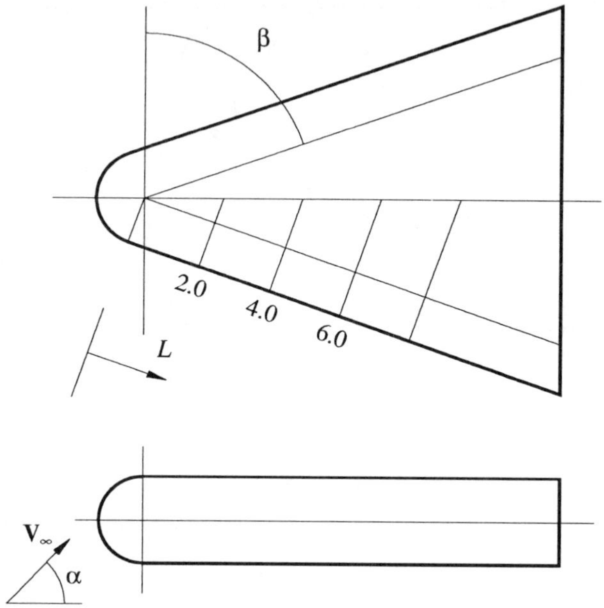

Figure 4.28 Plane delta wing with blunted leading edge.

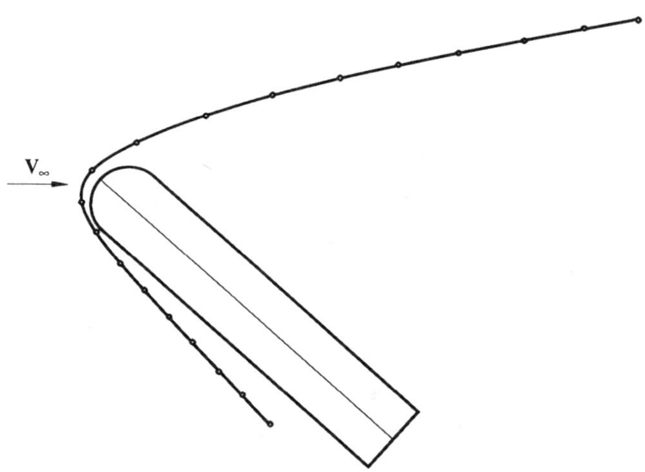

Figure 4.29 Bow shock in the symmetry plane of the flow about a blunt delta wing. $M_\infty = 9.6$, $Re_\infty = 0.4 \times 10^5$, $\alpha = 41.5°$. Line, calculations by Tannehill *et al* (1981); circles, measurements by Bertram and Everhart (1963).

and measured bow shocks in the flow symmetry plane at an angle of incidence $\alpha = 41.5°$. Figures 4.30 and 4.31 depict the wall pressure and heat flux distributions

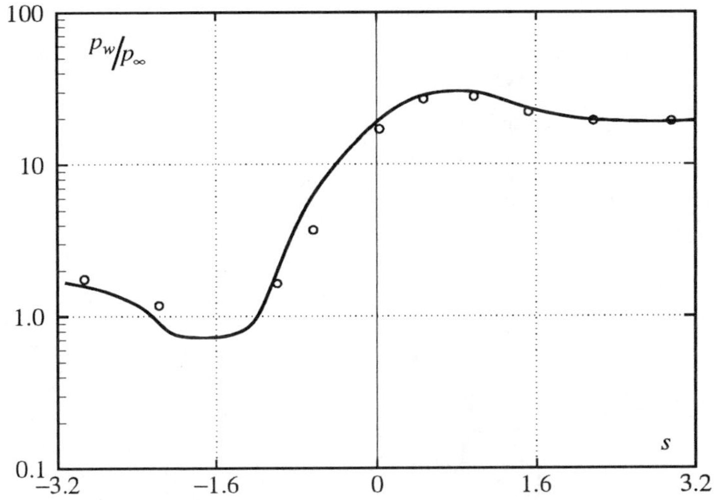

Figure 4.30 Wall pressure distribution on a heat-insulated blunt delta wing at $L = 4$. $M_\infty = 9.6$, $Re_\infty = 0.4 \times 10^5$, $\alpha = 20°$. Line, calculations by Tannehill *et al* (1981); circles, measurements by Bertram and Everhart (1963).

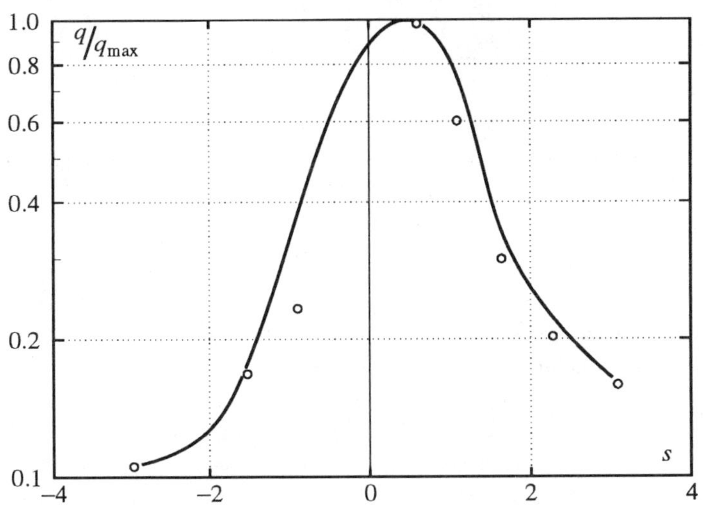

Figure 4.31 Heat flux distribution on a blunt delta wing at $L = 4$. $M_\infty = 9.6$, $Re_\infty = 0.4 \times 10^5$, $T_w = 0.343T^*$, $\alpha = 5°$. Line, calculations by Tannehill *et al* (1981); circles, measurements by Bertram and Everhart (1963).

along the wing intersection with the plane $L = 4$; see figure 4.28. Here s is the distance measured along the wing contour from the symmetry plane, $s < 0$

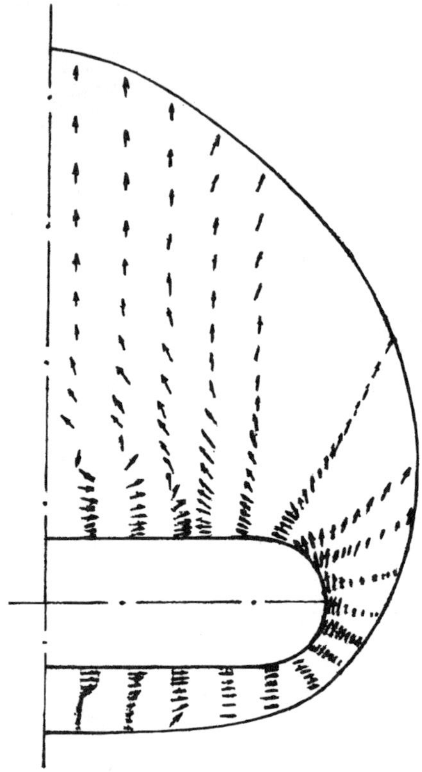

Figure 4.32 Cross-flow velocity map in the shock layer on a heat-insulated blunt delta wing at $L = 10.05$. $M_\infty = 9.6$, $Re_\infty = 0.4 \times 10^5$, $\alpha = 20°$.

corresponding to the leeward side and $s > 0$ to the windward side of the wing. All distances are normalized by the bluntness radius. The agreement of calculation and experimental data appears to be quite good.

Figure 4.32 shows the calculated cross-flow velocity map at $L = 10.05$ and the angle of incidence $\alpha = 20°$ in the case of an adiabatic wing surface. One can see an extensive region of recirculating flow over the leeward side of the wing. A detailed analysis of the calculation results reveals also a weak secondary vortex near the wing surface.

Afonina *et al* (1983) calculated the flow past the windward side of a plane blunted delta wing with a sweep angle $\beta = 75°$ at angles of attack $\alpha = 40°$ and $60°$. Using the method of subsection 4.3.3, the time-asymptotic solution for the flow about spherical bluntness was first obtained and then the gas dynamic functions were calculated within the overlapping domains which advanced along the wing. At $\alpha = 40°$, it was found to be admissible to use the domains containing only three sections, namely, the calculated one and two neighbours. To ensure stability of

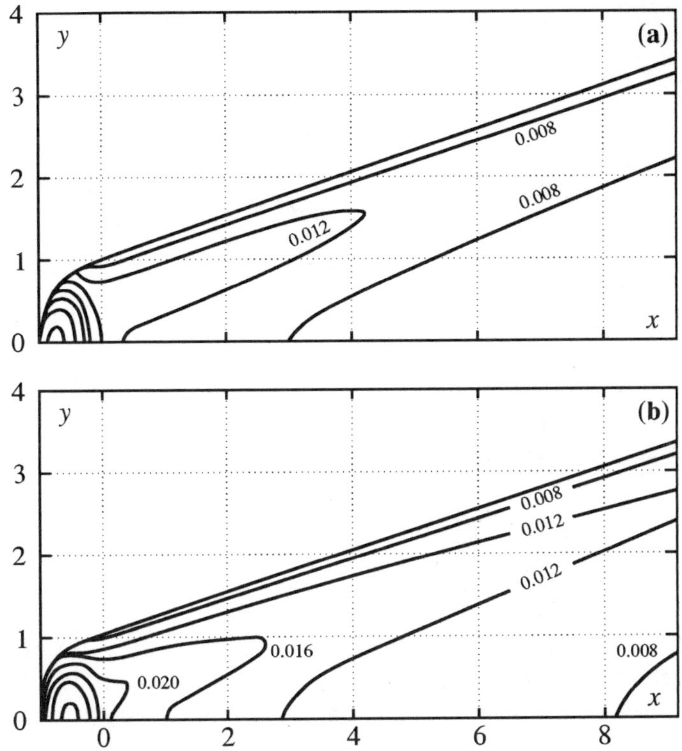

Figure 4.33 Heat flux isolines on the windward side of a blunt delta wing. $M_\infty = 6$, $Re_\infty = 10^4$, $T_w = 0.2T^*$; (a), $\alpha = 40°$; (b), $\alpha = 60°$.

the marching procedure at $\alpha = 60°$, one had to include no less than four sections in a local computational domain.

Figure 4.33 depicts the isolines of the non-dimensional heat flux on the windward side of the wing surface. Here $\gamma = 1.4$, $Pr = 0.71$, $\mu \propto T^{0.7}$. It is seen that at both angles of attack, the heat flux maximum is achieved on the spherical bluntness in the vicinity of the stagnation point. The heat flux maximum over the wing span is achieved at the wing edge. The heat flux monotonously decreases with the longitudinal coordinate x varying more rapidly in the symmetry plane as compared with the leading edge.

4.5.5 Flow around a combined body

As an example of such computations, let us consider the results of Chernii (1983) which have been obtained with the use of parabolized Navier–Stokes equations and the implicit factored space-marching scheme based on the operator splitting in the spatial coordinates and the physical processes; see subsection 2.2.2(c). The governing equations differed from the full Navier–Stokes equations only by the

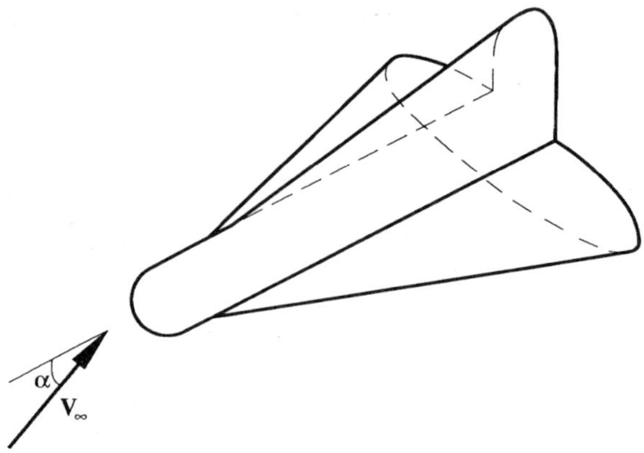

Figure 4.34 Combined body 'blunted cylinder — bielliptic cone'.

lack of viscous terms with derivatives in the marching coordinate. The initial-boundary value problem was regularized in subsonic regions with the method of subsection 2.2.1(b). The initial data were borrowed from available solutions for the flow about a sphere.

The results discussed below pertain to the flow about a spherically blunted cylinder transforming itself into a bielliptic cone; see figure 4.34. The Cartesian coordinates of the lateral part of the body surface are given by the formulae

$$x = r(z,\varphi)\cos\varphi \qquad y = -r(z,\varphi)\sin\varphi \qquad z = z \tag{4.71}$$

where

$$\begin{aligned} r(z,\varphi) &= \left(a^{-2}\cos^2\varphi + b^{-2}\sin^2\varphi\right)^{-1/2} \\ a(z,\varphi) &= \begin{cases} 1, & 0 \leqslant \varphi \leqslant 135^\circ \\ F(z), & 135 < \varphi \leqslant 180^\circ \end{cases} \\ b(z,\varphi) &= \begin{cases} F(z), & 0 \leqslant \varphi \leqslant 135^\circ \\ 1, & 135 < \varphi \leqslant 180^\circ \end{cases} \\ F(z) &= 1 + \int_0^z f(z)dz \\ f(z) &= \begin{cases} \tan(0.35z), & 0 \leqslant z < 1 \\ \tan(0.35), & z \geqslant 1 \end{cases} \end{aligned} \tag{4.72}$$

z, r, φ are the cylindrical coordinates, the z-coordinate aligning with the body axis.

The coordinate system and computational grid were generated by the method described in subsection 4.3.1 using the function $\lambda(q^1)$ within the segment [1,10] and the identity $\varphi = q^2$ as coordinate transformation (4.37). Intersection of the coordinate lines q^3 belonging to the same marching section was prevented by the function $\Lambda(q^2, q^3)$ determined by formulae (4.43) with the following parameters: $q_1^2 = 100°$, $q_*^2 = 135°$, $q_2^2 = 180°$, $q_*^3 = 0.6$, $n = 0.5$.

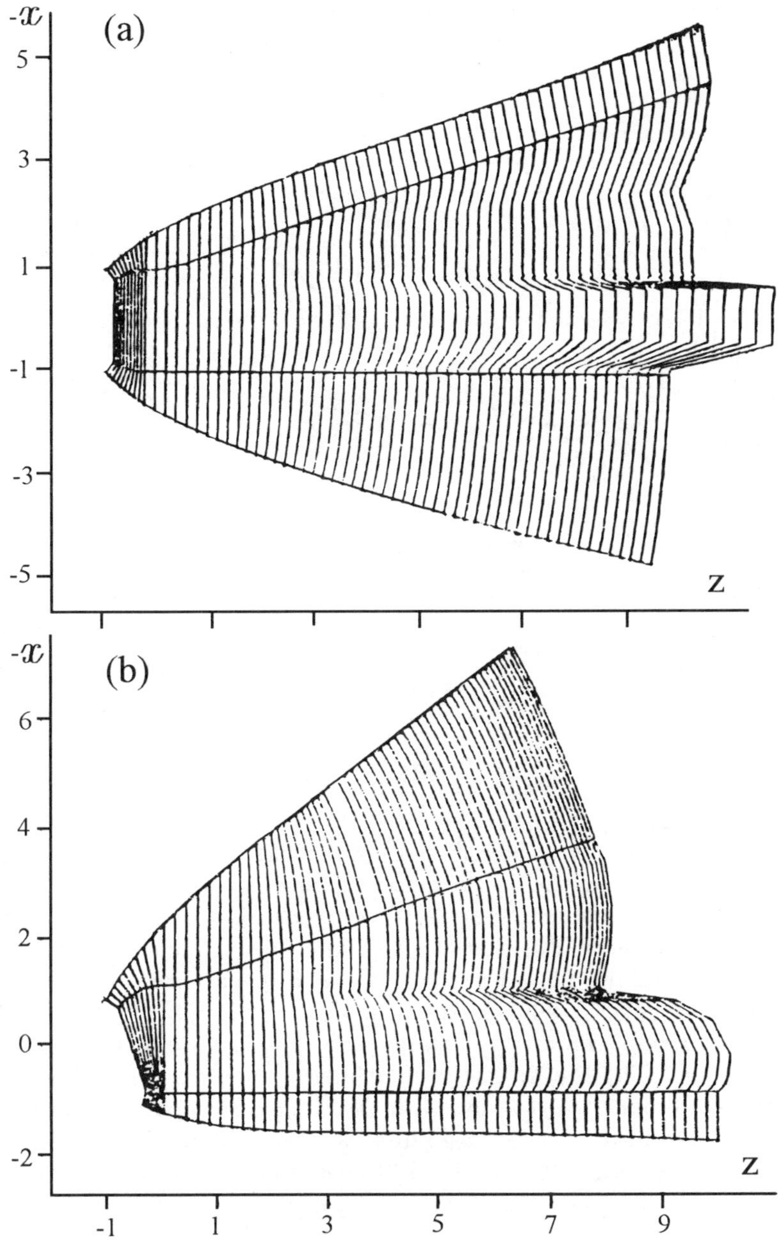

Figure 4.35 Flow about a combined body. $M_\infty = 10$, $Re_\infty = 10^5$, $T_w = 0.5T^*$; (a), $\alpha = 0$; (b), $\alpha = 20°$.

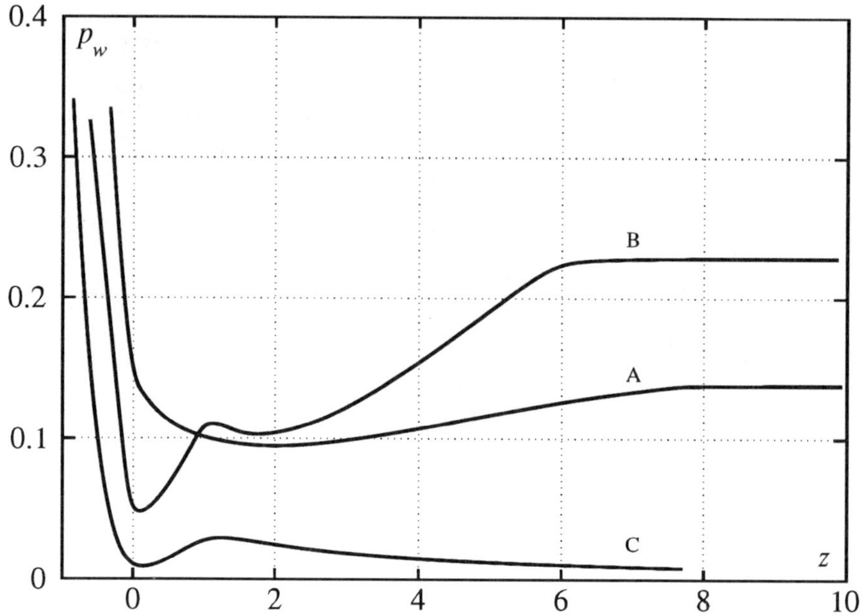

Figure 4.36 Wall pressure distribution over a combined body. $M_\infty = 10$, $Re_\infty = 10^5$, $T_w = 0.5T^*$, $\alpha = 20°$; A, $\varphi = 0$; B, $\varphi = 90°$; C, $\varphi = 180°$.

The calculations were carried out for a perfect gas with $\gamma = 1.4$, $Pr = 0.72$, $\mu \propto T^{0.75}$.

Figure 4.35 displays the body shape, bow shocks, marching sections and projections of their bases on the flow symmetry plane at two angles of attack. Being orthogonal to the streamlines, the bases of the marching sections provide valuable information on the flow near the body surface. Distortion of these lines indicates a variation of the flow direction. At $\alpha = 0$, one can see the gas spread from the wing edge ($\varphi = 90°$) towards the bottom ($0 < \varphi < 90°$) and 'canopy' ($135° < \varphi < 180°$). From the 'canopy', the gas is flowing towards the wing and downstream. The breaks of the marching section bases near the middle of the 'canopy' and along its junction with the wing correspond to two convergent streamlines. The third of such streamlines is located at the bottom in the plane $\varphi = 0$. With increasing angle of attack, the streamlines on the 'canopy' become divergent. At $\alpha = 20°$, one can see only two convergent streamlines, namely $\varphi = 0$ and $\varphi = 135°$.

Figure 4.36 shows the wall pressure variation in the meridional planes $\varphi = 0, 90°$, and $180°$ at $\alpha = 20°$. It is seen that the pressure maximum is achieved on the wing edge. With increasing α, the 'canopy' effect diminishes and, at $\alpha = 20°$, it is relatively small. A more detailed discussion of the numerical solutions and some other examples of calculation of the flows about combined aerodynamic shapes can be found in the monograph by Kovenya *et al* (1990).

4.6 SELF-SIMILAR SOLUTIONS

The section which follows deals with self-similar solutions which can be obtained for both the full and reduced Navier–Stokes equations. The data provided by self-similar solutions of the three-dimensional problems are of a considerable interest since numerical simulation of such flows requires, in a general case, a large amount of computations.

4.6.1 Flow in the vicinity of a three-dimensional stagnation point

Calculation of this flow at zero incidence and side-slip is one of the simplest three-dimensional viscous shock layer problems. Keeping in mind the data discussed in subsection 2.4.3, one can suppose the self-similar solutions to be quite accurate for three-dimensional stagnation streamline flows at hypersonic flight speeds.

When searching for self-similar solutions, the governing equations are written for the coordinate system q^i $(i = 1, 2, 3)$ conforming to the body surface which is supposed to have two symmetry planes. The q^1, q^2-coordinates are chosen to be the Cartesian coordinates x^1, x^2 of the body surface points; see figure 4.5. The q^3-coordinate is aligned with the normal to the body surface. The Cartesian coordinates x^α are related to the q^i-coordinates by the formulae

$$
\begin{aligned}
x^\alpha &= q^\alpha + q^3 f'_\alpha \Big/ \sqrt{a} \qquad (\alpha = 1, 2) \\
x^3 &= f(q^1, q^2) - q^3 \Big/ \sqrt{a}
\end{aligned}
\tag{4.73}
$$

where

$$
a = 1 + (f'_1)^2 + (f'_2)^2 \qquad f'_\alpha = \frac{\partial f}{\partial q^\alpha}
$$

$x^3 = f(q^1, q^2) = f(x^1, x^2)$ is the equation of the body surface.

The metric coefficients in the vicinity of the stagnation streamline $(q^1, q^2 \to 0)$ are calculated with formulae (4.4) assuming $f'_\alpha \approx \kappa_\alpha q^\alpha$, where κ_α $(\alpha = 1, 2)$ are the main curvatures of the body surface at the stagnation point.

A self-similar solution is obtained in the same way as described in subsection 2.5.1 for an axisymmetric flow, using the following series expansions

$$
\begin{aligned}
u &= u_1(q^3)q^1/R_1 + \ldots & w &= w_1(q^3)q^2/R_2 + \ldots \\
v &= v_0(q^3) + \ldots & T &= T_0(q^3) + \ldots \\
\mu &= \mu_0(q^3) + \ldots & \rho &= \rho_0(q^3) + \ldots \\
p &= p_0(q^3) + p_2^1(q^3)\left(\frac{q^1}{R_1}\right)^2 + p_2^2(q^3)\left(\frac{q^2}{R_2}\right)^2 + \ldots
\end{aligned}
\tag{4.74}
$$

Here u, w, v are the velocity vector projections on the q^1, q^2, q^3-axes, respectively; $R_\alpha = 1/\kappa_\alpha$ are the main radii of the body surface at the stagnation point; other notations are the same as in the previous sections.

Substituting series expansions (4.74) into the Navier–Stokes equations (4.1)–(4.3) and applying the truncation procedure of Kao (1964), one obtains, in the case of a stationary flow, the set of ordinary differential equations for the expansion coefficients. For hypersonic flows ($M_\infty \gg 1$), this set can be written as follows (Peigin and Rusakov 1985):

$$\begin{gathered}
\rho\sqrt{g}\left(\frac{u}{1+z}+\frac{\kappa w}{1+\kappa z}\right)+(\rho\sqrt{g}v)'=0\\
\rho v v'+p'-\frac{4}{3Re}(\mu v')'=0\\
\rho\left[\frac{u}{1+z}(u+v)+vu'\right]+\frac{2}{1+z}p_2^1-\frac{1}{\sqrt{g}Re}(\sqrt{g}\mu u')'=0\\
\rho\left[\frac{\kappa w}{1+\kappa z}(w+v)+vw'\right]+\frac{2\kappa}{1+\kappa z}p_2^2-\frac{1}{Re}(\mu w')'=0\\
\rho v T'-vp'-\frac{1}{Re}\left(\frac{\mu}{Pr}T'\right)'-\frac{4}{3}\frac{\mu}{Re}(v')^2=0\\
(p_2^1)'+\frac{1}{2}p'-\frac{\rho v}{2}v'-\frac{\rho u(u+v)}{1+z}=0\\
(p_2^2)'+\frac{1}{2}p'-\frac{\rho v}{2}v'-\frac{\kappa}{1+\kappa z}\rho w(w+v)=0
\end{gathered}\tag{4.75}$$

Here the q^3-coordinate is denoted by z, $\kappa = R_1/R_2$, $\sqrt{g} = (1+z)(1+\kappa z)$, the subscripts are retained only for the coefficients p_2^α ($\alpha = 1, 2$). Set (4.75) is written for the non-dimensional variables. All distances are normalized by R_1, the velocity components, density, pressure and temperature are normalized by V_∞, ρ_∞, $\rho_\infty V_\infty^2$, and $T^\circ = (\gamma-1)M_\infty^2 T_\infty$, respectively.

If the outer boundary of the computational domain is located in a free stream, the proper boundary conditions are

$$\begin{gathered}
u = w = -v = 1 \qquad \rho = 1\\
T = \frac{1}{(\gamma-1)M_\infty^2} \qquad p_2^\alpha = 0 \qquad (\alpha = 1, 2)
\end{gathered}\tag{4.76}$$

The boundary conditions on the body surface look as follows

$$u = w = 0 \qquad \rho v = (\rho v)_w = G \qquad T = T_w \tag{4.77}$$

With the above formulation of the problem, Peigin and Rusakov (1985) and Gershbein and Scherbak (1987) calculated the flows on the stagnation streamline ahead of a three-dimensional blunt body.

With the new independent variable

$$\zeta = \int_0^z \rho\sqrt{g}\, dz$$

and the stream functions φ_1 and φ_2

$$\varphi_1' = \frac{u}{1+z} \qquad \varphi_2' = \frac{w}{1+\kappa z} \qquad \varphi = \varphi_1 + \kappa\varphi_2 = -\frac{v}{z'}$$

the set of governing equations takes the form

$$\begin{aligned}
\left[\frac{\mu}{Re}\left(\varphi_1' + \frac{1+z}{z'}\varphi_1''\right)\right]' &= \frac{2z'}{1+z}p_2^1 - \frac{\varphi}{1+\kappa z}\varphi_1'' \\
&\quad + \frac{\varphi_1'}{\sqrt{g}}\left[(1+z)\varphi_1' - 2\varphi z'\right] \\
\left[\frac{\mu}{Re}\left(\kappa\varphi_2' + \frac{1+\kappa z}{z'}\varphi_2''\right)\right]' &= \frac{2\kappa z'}{1+\kappa z}p_2^2 - \frac{\varphi}{1+z}\varphi_2'' \\
&\quad + \frac{\kappa\varphi_2'}{\sqrt{g}}\left[(1+\kappa z)\varphi_2' - 2\varphi z'\right] \\
\left[\frac{4\mu}{3Re}\left(\varphi' + \varphi\frac{z''}{z'}\right)\right]' &= -\left[\varphi^2 - \frac{\gamma-1}{\gamma}\frac{T}{(z')^2}\right]\frac{z''}{\sqrt{g}} - \frac{\varphi\varphi'}{\sqrt{g}}z' \\
&\quad - \frac{\gamma-1}{\gamma}\frac{1}{\sqrt{g}}\left[\frac{1}{z'}T' - T\frac{1+\kappa+2\kappa z}{\sqrt{g}}\right] \\
\left(\frac{\mu}{RePrz'}T'\right)' &= -\frac{\varphi}{\gamma\sqrt{g}}T' - \frac{\gamma-1}{\gamma}\frac{\varphi z'}{\sqrt{g}}\left[\frac{z''}{(z')^2} + \frac{1+\kappa+2\kappa z}{\sqrt{g}}\right]T \\
&\quad - \frac{4\mu}{3Re\, z'}\left(\varphi' z' + \varphi z''\right)^2 \\
(p_2^1)' &= \frac{1}{\sqrt{g}}\left[-\frac{\gamma-1}{2\gamma}\left(\frac{T'}{z'} - T\left(\frac{z''}{(z')^2} + \frac{1+\kappa+2\kappa z}{\sqrt{g}}\right)\right)\right. \\
&\quad \left. + \varphi_1'\left((1+z)\varphi_1' - \varphi z'\right) + \frac{\varphi}{2}\left(\varphi' z' + \varphi z''\right)\right] \\
(p_2^2)' &= \frac{1}{\sqrt{g}}\left[-\frac{\gamma-1}{2\gamma}\left(\frac{T'}{z'} - T\left(\frac{z''}{(z')^2} + \frac{1+\kappa+2\kappa z}{\sqrt{g}}\right)\right)\right. \\
&\quad \left. + \kappa\varphi_2'\left((1+\kappa z)\varphi_2' - \varphi z'\right) + \frac{\varphi}{2}\left(\varphi' z' + \varphi z''\right)\right]
\end{aligned} \tag{4.78}$$

The boundary conditions are rewritten as follows

$$\zeta = 0: \qquad \varphi_1' = \varphi_2' = 0, \qquad \varphi = -(\rho v)_w = -G$$
$$z = 0 \qquad T = T_w \tag{4.79}$$

$$\begin{aligned}
\zeta = \zeta_\Gamma: \qquad (1+z)\varphi_1' &= 1 \qquad \sqrt{g}\, z' = 1 \\
(1+\kappa z)\varphi_2' &= 1 \qquad \varphi z' = 1 \\
p_2^1 = p_2^2 &= 0 \qquad T = 1/((\gamma-1)M_\infty^2)
\end{aligned} \tag{4.80}$$

Here $\zeta = \zeta_\Gamma$ is the coordinate of the outer boundary of the computational domain located in the free stream.

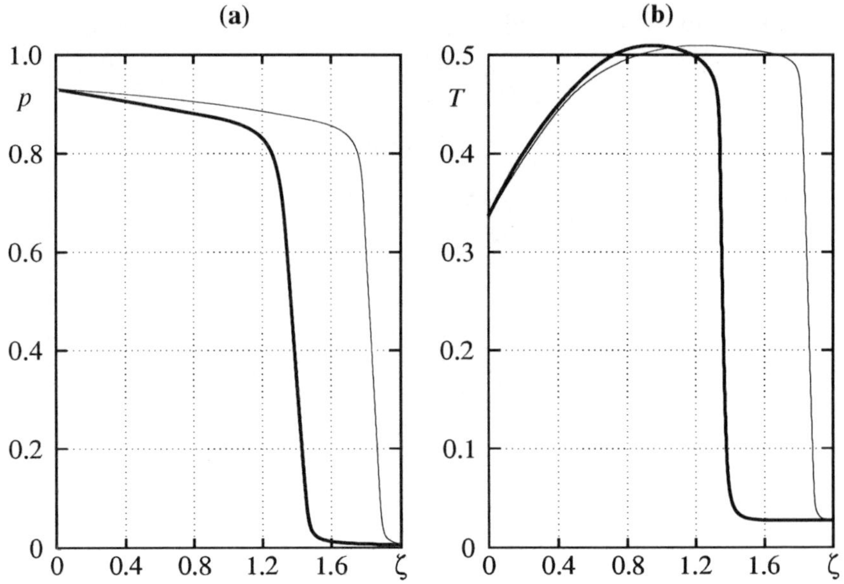

Figure 4.37 Pressure and temperature profiles on the stagnation streamline in front of a three-dimensional blunt body. $M_\infty = 10$, $Re_\infty = 500$. Full lines, $\kappa = 0.5$; hair lines, $\kappa = 0.1$.

Like in case of an axisymmetric flow, the point corresponding to the undisturbed free stream and the points with $v = 0$ are the singular node points of set (4.78).

Numerical solution of the problem was obtained by the method similar to that described in subsection 3.4.1 using the fourth-order accurate finite-difference scheme (Petukhov 1964) and the Simpson quadrature formula.

Below, we consider the results which pertain to the flows of a perfect gas with $\gamma = 1.4$, $\mu \propto \sqrt{T}$, $Pr = 0.7$. Figure 4.37 demonstrates the body shape effect on the profiles of the gas dynamic functions on the stagnation streamline. Figure 4.38 shows a dependence of the stagnation point heat flux on the main radii ratio $\kappa = R_1/R_2$. Here, the heat flux is normalized by its value at the stagnation point of a body of revolution. The dependence is close to a linear one. It is weakly influenced by the Reynolds number and wall temperature, changing only by a few per cent with the Reynolds number decreasing down to $Re_\infty = 100$ and the wall temperature increasing up to $T_w = 0.35T^\circ$.

Investigating hypersonic stagnation line flows at moderate and high Reynolds numbers, it seems expedient to apply the two-layer thin viscous shock layer model. For the above coordinate system with a new independent variable of Dorodnitsin's type and the stream functions, the thin viscous shock layer equations for the stagnation line flow in front of a three-dimensional blunt body are written as follows (Bruikina and Rusakov 1988)

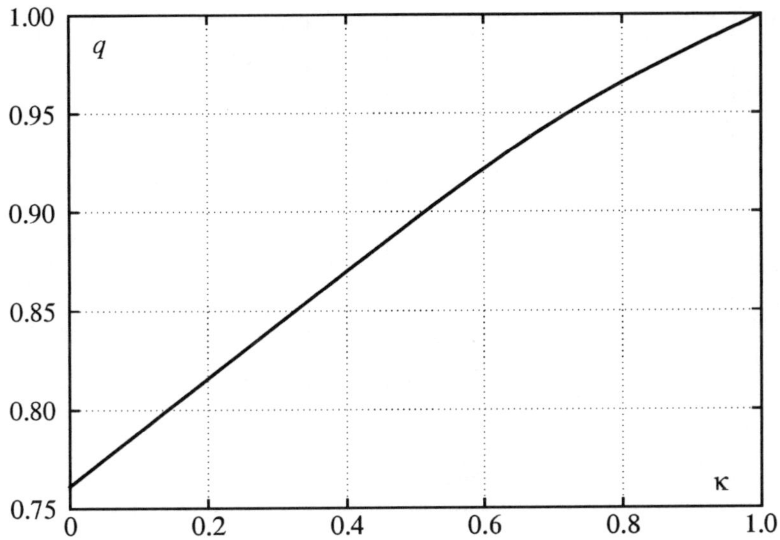

Figure 4.38 Dependence of the stagnation point heat flux on the ratio of the body main radii. $M_\infty = 10$, $Re_\infty = 500$, $T_w = 0.15T^\circ$.

$$\begin{gathered}
\left(\frac{\mu\rho}{Re\Delta^2}\varphi_\alpha''\right)' + \varphi\varphi_\alpha'' = 2d_\alpha\frac{p_2^\alpha}{\rho} + d_\alpha(\varphi_\alpha')^2 \\
\left(\frac{\mu\rho}{RePr\Delta^2}T'\right)' + \varphi T' = 0 \\
(p_2^\alpha)' = d_\alpha\Delta\,(\varphi_\alpha')^2 \\
p = 1
\end{gathered} \qquad (4.81)$$

Here

$$\varphi_1' = u \qquad \varphi_2' = w \qquad \varphi = \varphi_1 + \kappa\varphi_2 = -\rho v$$

$$d_1 = 1 \qquad d_2 = \kappa$$

$$Re = R_1\rho_\infty V_\infty/\mu(T^*) \qquad T^* = V_\infty^2/2c_p \qquad p_2^\alpha = \frac{1}{2(d_\alpha)^2}\frac{\partial^2 p}{\partial(\xi^\alpha)^2}\bigg|_{\xi^1=\xi^2=0}$$

$$\xi^\alpha = \frac{q^\alpha}{R_1} \qquad \zeta = \frac{1}{R_1\Delta}\int_0^{q^3}\rho dq^3 \qquad \Delta = \frac{1}{R_1}\int_0^{q_s^3}\rho dq^3$$

$$\alpha = 1, 2$$

the prime denotes differentiation in ζ, q_s^3 is the bow shock coordinate, and summation with respect to the repeated indices is not supposed to be made.

Equations (4.81) are written for non-dimensional variables. The density, pressure, temperature, viscosity and velocity components in the q^1, q^2, q^3-directions are normalized by ρ_∞, $\rho_\infty V_\infty^2$, T^*, $\mu(T^*)$, $V_\infty q^1$, $V_\infty\kappa q^2$ and V_∞, respectively. The smaller of the two body main radii R_1 is chosen as the characteristic length.

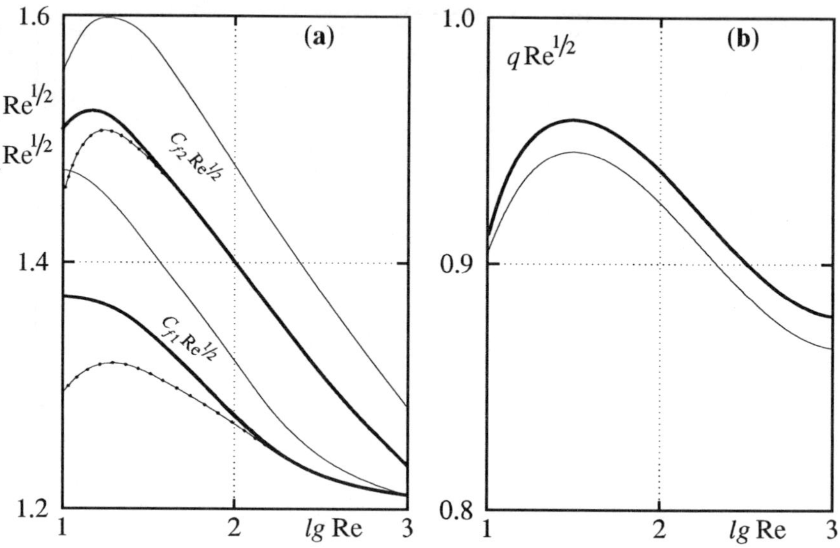

Figure 4.39 Dependence of the skin friction coefficients and heat flux at the stagnation point of a three-dimensional blunt body on the Reynolds number. $M_\infty = 10$, $T_w = 0.1T^*$, $\kappa = 0.1$. Hair lines, thin viscous shock layer model (Gershbein and Yunitskii 1979); full and hair-dotted lines, bow shock capturing calculations by Gershbein and Scherbak (1987) with no-slip and slip wall boundary conditions, respectively.

The boundary conditions are given by the following relationships:
body surface ($\zeta = 0$)

$$\varphi'_\alpha = 0 \qquad \varphi = -(\rho v)_w = -G \qquad T = T_w \tag{4.82}$$

bow shock wave ($\zeta = 1$)

$$\begin{aligned} &\frac{\mu\rho}{Re\Delta}\varphi''_\alpha + \varphi'_\alpha = 1 \qquad \varphi = 1/\Delta \\ &\frac{\mu\rho}{RePr\Delta}T' + T = 1 \qquad p_2^\alpha = -1 \end{aligned} \tag{4.83}$$

Within the framework of the above model, the hypersonic stagnation line flows in front of the three-dimensional bodies were investigated by Gershbein and Yunitskii (1979), Markov (1980), Gershbein *et al* (1981), Bruikina and Rusakov (1988), and Ankudinov (1989).

Figure 4.39 presents the calculation results for the skin friction coefficients and heat flux. It was assumed in these calculations that $\gamma = 1.25$, $\mu \propto \sqrt{T}$, $Pr = 0.71$. The results of Gershbein and Yunitskii (1979) are plotted by the hair curves. The full lines depict the solution of set (4.75) subject to boundary conditions (4.76) and (4.77) as obtained by Gershbein and Scherbak (1987). The hair-dotted lines depict their results calculated with the velocity slip and temperature jump wall boundary conditions.

On the basis of numerical solutions, Gershbein and Scherbak (1987) and Bruikina and Rusakov (1988) suggested to determine the stagnation point heat flux $q(\kappa)$, taking it to be equal to the stagnation point heat flux of the body of revolution, whose curvature is equal to half the sum of the main curvatures of the three-dimensional body. Since the body curvature is involved in the heat flux calculation only through the Reynolds number, one obtains:

$$q(\kappa, Re) = q(1, Re_0) \quad \text{where} \quad Re_0 = \frac{2Re}{1+\kappa} \tag{4.84}$$

The function $q(\lg Re)$ for an axisymmetric flow being available (see figure 3.24), one can obtain the stagnation point heat flux of a three-dimensional body from (4.84) shifting the abscissa axis by the quantity $\lg(2/(1+\kappa))$. The viscous shock layer calculations carried out for $0 \leqslant \kappa \leqslant 1$ and a wide range of Reynolds numbers have proved the accuracy of formula (4.84) within some per cent. It is worth noting as well that the above formula agrees with the relationship

$$q(\kappa)/q(1) = \sqrt{\frac{1+\kappa}{2}} \tag{4.85}$$

which follows from the boundary layer theory (Bruikina *et al* 1980).

4.6.2 Flow in the vicinity of the symmetry plane

Self-similar solution for the flow symmetry plane can be obtained using the power or trigonometric series expansions of the unknown functions in the circumferential coordinate. Such solutions are expected to be most accurate at hypersonic flight speed. So, let us consider the problem with reference to the thin viscous shock layer model.

Let the coordinate system and the body surface equation be given by formulae (4.73). Without loss of generality, one can suppose the flow symmetry plane containing the free stream velocity vector to be the plane $x^2 = 0$ and the origin of the curvilinear coordinate system q^i $(i = 1, 2, 3)$ to be placed at the stagnation point on the body surface. The solution in the vicinity of the symmetry plane $q^2 = 0$ is searched for in the form

$$\begin{aligned} \Phi(q^1, q^2, q^3) &= \Phi_0(q^1, q^3) + \Phi_2(q^1, q^3)(q^2)^2 + \ldots \\ u^2(q^1, q^2, q^3) &= u^2_1(q^1, q^3)q^2 + \ldots \end{aligned} \tag{4.86}$$

where Φ is any of the functions u^1, u^3, p, ρ, T.

Substitution of series expansions (4.86) into the three-dimensional thin viscous shock layer equations (4.57) and use of the truncation procedure of Kao (1964) result in closed sets of equations for the series coefficients; see Gershbein and Yunitskii (1984). For Dorodnitsin's variables, the set of equations including two coefficients for the pressure and only the first coefficients for other gas dynamic functions can be written as follows (Bruikina and Rusakov 1989):

$$
\begin{aligned}
&\beta_3 u^1 \frac{\partial u^1}{\partial \xi} - S\frac{\partial u^1}{\partial \zeta} + \beta_1 (u^1)^2 = -\frac{p_1}{\rho} + \frac{\partial}{\partial \zeta}\left(\frac{\mu\rho}{Re\Delta^2}\frac{\partial u^1}{\partial \zeta}\right) \\
&\beta_3 u^1 \frac{\partial u^2}{\partial \xi} - S\frac{\partial u^2}{\partial \zeta} + \beta_2 (u^2)^2 + \beta_4 (u^1)^2 + \beta_8 u^1 u^2 \\
&= -\frac{g}{\rho}(p_2 - \beta_0 p_1) + \frac{\partial}{\partial \zeta}\left(\frac{\mu\rho}{Re\Delta^2}\frac{\partial u^2}{\partial \zeta}\right) \\
&\beta_3 u^1 \frac{\partial H}{\partial \xi} - S\frac{\partial H}{\partial \zeta} = \frac{\partial}{\partial \zeta}\left[\frac{\mu\rho}{RePr\Delta^2}\frac{\partial}{\partial \zeta}\left(H - (1-Pr)\beta_0 (u^1)^2\right)\right] \qquad (4.87) \\
&\sqrt{g}\frac{\partial p}{\partial \zeta} = \beta_4 \Delta (u^1)^2 \\
&\sqrt{g}\frac{\partial p_1}{\partial \zeta} = g\beta_1\beta_3 u^1\left(u^1\frac{d\Delta}{d\xi} + 2\Delta\frac{\partial u^1}{\partial \xi}\right) + \beta_5 \Delta (u^1)^2 \\
&\sqrt{g}\frac{\partial p_2}{\partial \zeta} = \Delta\left[\beta_6 (u^1)^2 + 4\beta_7 u^1 u^2 + 2\beta_2^2 (u^2)^2\right]
\end{aligned}
$$

where

$$
\xi = q^1 \qquad \zeta = \frac{1}{\Delta}\int_0^{q^3} \rho \, dq^3
$$

$$
\Delta = \int_0^{q_s^3} \rho \, dq^3 \qquad Re = \frac{R\rho_\infty V_\infty}{\mu(T^*)}
$$

$$
u^1 = \frac{\partial \varphi_1}{\partial \zeta} \qquad u^2 = \frac{\partial \varphi_2}{\partial \zeta}
$$

$$
g = 1 + f_1'^2 \qquad S = \beta_1\varphi_1 + \beta_2\varphi_2 + \beta_3\left(\frac{\partial \varphi_1}{\partial \xi} + \frac{\varphi_1}{\Delta}\frac{d\Delta}{d\xi}\right)
$$

$$
p_1 = \frac{1}{g\beta_3}\frac{\partial p}{\partial q^1}\bigg|_{q^2=0} \qquad p_2 = \frac{1}{g\beta_2}\frac{\partial^2 p}{(\partial q^2)^2}\bigg|_{q^2=0}
$$

$$
\beta_0 = \frac{f_1'^2}{g} \qquad \beta_1 = \frac{f_{11}''}{g^2}
$$

$$
\beta_2 = \frac{f_{22}''}{g} \qquad \beta_3 = \frac{f_1'}{g}
$$

$$
\beta_4 = f_1'^2 \beta_1 \qquad \beta_5 = \frac{f_{111}'''}{g^2} f_1' + \beta_1^2 g(6 - 4g)
$$

$$
\beta_6 = \frac{1}{\beta_2}\left[\frac{f_{1122}^{IV}}{g^2}\beta_0 + \beta_1\left(\beta_7(6-4g) + 6\beta_2^2(1-g)\right)\right]
$$

$$
\beta_7 = \frac{f_{122}'''}{g^2} f_1' \qquad \beta_8 = \beta_3\frac{f_{122}'''}{f_{22}''} - 2\beta_4
$$

The density, pressure, total enthalpy per unit mass, viscosity and velocity vector components in the q^1, q^2-directions are normalized by ρ_∞, $\rho_\infty V_\infty^2$, $H_\infty \approx V_\infty^2/2$, $\mu(T^*)$, $V_\infty u_\infty^\alpha$, respectively. The distances are normalized by the body radius at the stagnation point in the symmetry plane. Subscripts 1 and 2 at the derivatives of the body shape function f indicate the coordinate (q^1 or q^2) in which differentiation is carried out.

Set (4.87) is parabolic. It can be solved by the space-marching methods considered in chapter 3. The initial data in the ξ-coordinate are taken from the solution for the stagnation streamline. The boundary conditions are obtained by substitution of series expansions (4.86) into the wall boundary conditions and the modified Rankine–Hugoniot relationships. This yields the following relationships for the retained terms of the series:

$$\begin{aligned} \zeta = 0: \quad & u^\alpha = 0 \quad (\alpha = 1, 2) \\ & \beta_3 \varphi_1 \frac{d\Delta}{d\xi} + \Delta\left(\beta_1\varphi_1 + \beta_2\varphi_2 + \beta_3 \frac{\partial \varphi_1}{\partial \xi}\right) = 0 \\ & H = H_w \end{aligned} \tag{4.88}$$

$$\begin{aligned} \zeta = 1: \quad & u^\alpha = 1 - \frac{\mu\rho\sqrt{g}}{Re\Delta} \frac{\partial u^\alpha}{\partial \zeta} \quad (\alpha = 1, 2) \\ & \beta_3 \varphi_1 \frac{d\Delta}{d\xi} + \Delta\left(\beta_1\varphi_1 + \beta_2\varphi_2 + \beta_3 \frac{\partial \varphi_1}{\partial \xi}\right) = \frac{1}{\sqrt{g}} \\ & H = 1 - \frac{\mu\rho\sqrt{g}}{Re\,\mathrm{Pr}\,\Delta} \frac{\partial}{\partial \zeta}\left[H - (1 - Pr)\beta_0 (u^1)^2\right] \\ & p = \frac{1}{g} \quad p_1 = -2\beta_1 \quad p_2 = -\frac{2}{g}\left(\beta_2 + \frac{\beta_7}{\beta_2}\right) \end{aligned} \tag{4.89}$$

With such a formulation of the problem, three-dimensional viscous shock layer flows in the symmetry plane were calculated by Ankudinov (1983), Gershbein and Yunitskii (1984), and Bruikina and Rusakov (1989). Figure 4.40 shows the calculation results pertaining to hypersonic flows of a perfect gas with $\mu \propto \sqrt{T}$, $Pr = 0.71$ about elliptic paraboloids. For the Cartesian coordinate system shown in figure 4.5, the body surface is defined by the equation

$$x^3 = \frac{1}{2}\left[(x^1)^2 + \kappa (x^2)^2\right] \tag{4.90}$$

The figure depicts the temperature and tangent velocity component profiles in the symmetry plane $x^1 = 0$ in the flow past a paraboloid with $\kappa = 0.4$ at zero incidence. The curves A correspond to the stagnation line $q^2 = 0$, the curves B correspond to $q^2 = 3$. The temperature maximum at $q^2 = 3$ appears because the near-wall region is filled with the high-temperature gas flowing from the stagnation region of the shock layer.

The next figures present calculations of the flows about spherically blunted circular cones (Kumar and Graves 1977). These authors used full viscous shock layer equations similar to those suggested by Davis (1970) for an axisymmetric flow.

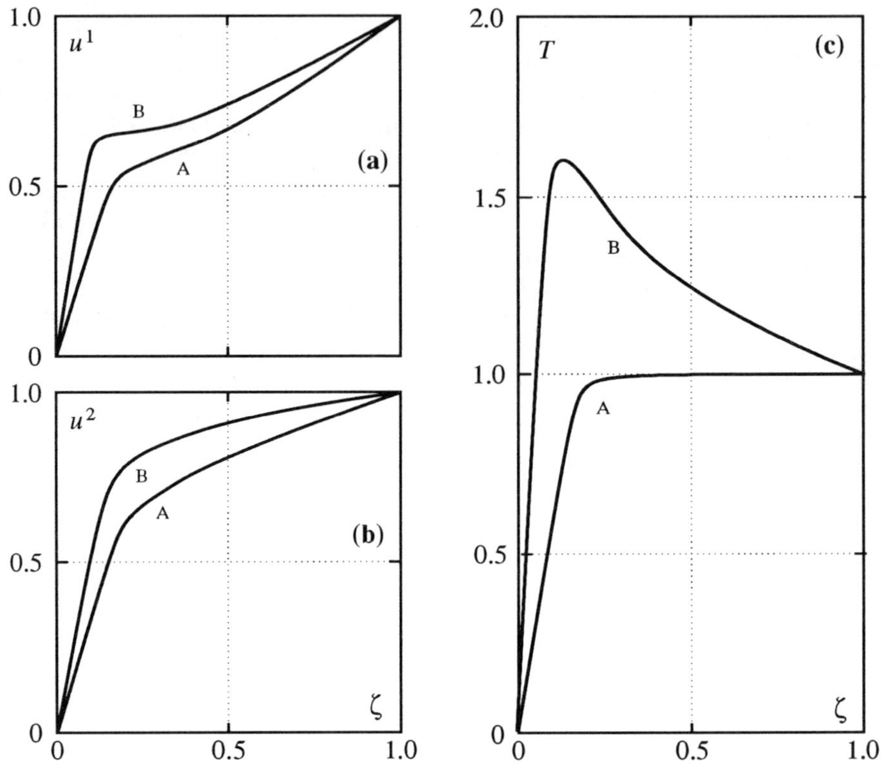

Figure 4.40 Profiles of gas dynamic functions in the shock layer symmetry plane in hypersonic flow past an elliptic paraboloid. $\gamma = 1.25$, $Re = 5 \times 10^3$, $T_w = 0.1T^*$, $\alpha = 0$; A, $q^2 = 0$; B, $q^2 = 3$.

The governing equations were written for the body-conforming orthogonal coordinate system (s, n, φ) where s was the distance along the body contour measured from the point of its intersection with the body axis, n was the distance along the normal to the body surface, and angle φ was measured from the leeward part of the flow symmetry plane. The equations for the flow in the symmetry plane were derived using the symmetry relationships and differentiation of the circumferential momentum balance equation in the φ-coordinate, which is, in essence, equivalent to the use of series expansions (4.86). In contrast to the previous shock layer model, the set of governing equations included only five differential equations. It was written for the non-dimensional variables as follows

$$\frac{\partial \boldsymbol{X}}{\partial t} + \frac{\partial \boldsymbol{F}}{\partial s} + \frac{\partial \boldsymbol{G}}{\partial n} + \boldsymbol{Q} = 0 \tag{4.91}$$

where

$$\boldsymbol{X} = H_s \begin{Bmatrix} \rho \\ \rho u \\ \rho v \\ \rho w \\ \rho H - p \end{Bmatrix} \qquad \boldsymbol{F} = \begin{Bmatrix} \rho u \\ \rho u^2 + p \\ \rho uv \\ \rho uw \\ \rho uH \end{Bmatrix}$$

$$\boldsymbol{G} = H_s \begin{Bmatrix} \rho v \\ \rho uv - \tau_1 \\ \rho v^2 + p \\ \rho vw - \tau_2 \\ \rho vH - \dfrac{\mu}{RePr}\dfrac{\partial h}{\partial n} - u\tau_1 \end{Bmatrix}$$

$$\boldsymbol{Q} = \frac{H_s \sin\theta}{H_\varphi}\boldsymbol{F} + \frac{\cos\theta}{H_\varphi}\boldsymbol{G}$$

$$+ \frac{1}{H_\varphi} \begin{Bmatrix} H_s \rho w \\ H_s \rho uw + H_\varphi \kappa \rho uv - H_\varphi \kappa \tau_1 - pH_s \sin\theta \\ H_s \rho vw - H_\varphi \kappa \rho u^2 - p(H_s \cos\theta + H_\varphi \kappa) \\ H_s \left(\frac{\partial^2 p}{\partial \varphi^2} + 2\rho w^2\right) + H_s \rho w(u \sin\theta + v\cos\theta) - H_s \tau_2 \cos\theta \\ H_s \rho H w \end{Bmatrix}$$

$$\tau_1 = \frac{\mu}{Re}\left(\frac{\partial u}{\partial n} - \frac{\kappa u}{H_s}\right) \qquad \tau_2 = \frac{\mu}{Re}\left(\frac{\partial w}{\partial n} - \frac{w\cos\theta}{H_\varphi}\right)$$

In the above formulae, u and v are the velocity components in the s- and n-directions; w is the derivative of the circumferential velocity component with respect to φ; $H_s = 1+\kappa n$, $H_\varphi = r+n\cos\theta$ are the Lamé coefficients; other notations are the same as in the previous sections.

To close set (4.91), one has to know the quantity $\partial^2 p/\partial\varphi^2$ (coefficient p_2 in the series expansions (4.86)). It was determined with the following approximation for the pressure variation in the circumferential coordinate

$$p(s,n,\varphi) = \frac{p(s,n,0)+p(s,n,\pi)}{2} + \frac{p(s,n,0)-p(s,n,\pi)}{2}\cos\varphi \tag{4.92}$$

Differentiation of this formula gives

$$\begin{aligned} \frac{\partial^2 p}{\partial\varphi^2}(s,n,0) &= -\frac{p(s,n,0)-p(s,n,\pi)}{2} \\ \frac{\partial^2 p}{\partial\varphi^2}(s,n,\pi) &= \frac{p(s,n,0)-p(s,n,\pi)}{2} \end{aligned} \tag{4.93}$$

The computational domain was bounded by the bow shock wave, the body surface and the two lines $s = const$ located in the windward and leeward parts of the flow symmetry plane far enough from the stagnation line. The stationary solution was obtained using Mac Cormack's (1969) finite-difference scheme.

Figures 4.41 through 4.43 present a comparison of the calculated and measured variation of the shock layer thickness, wall pressure, and heat transfer coefficients.

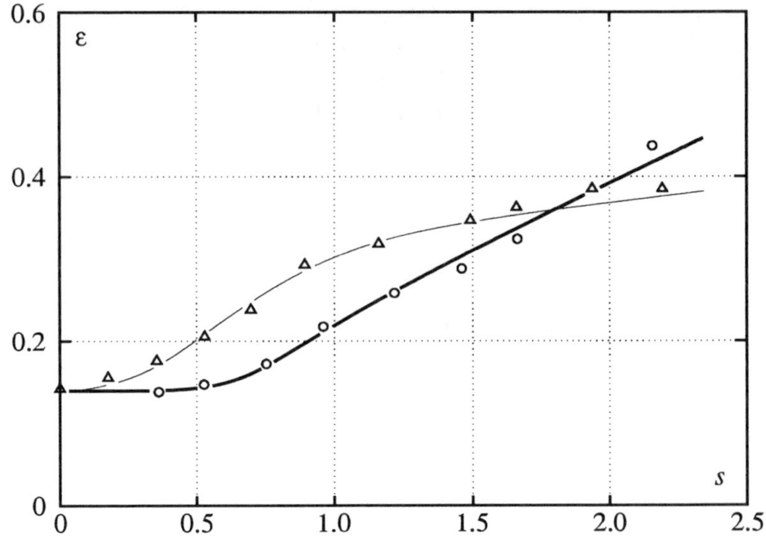

Figure 4.41 Variation of the bow shock distance from a spherically blunted cone in the flow symmetry plane. Comparison of calculations (lines) and measurements (circles, triangles). $M_\infty = 5.9$, $\gamma = 1.4$, $Re_\infty = 3.95 \times 10^4$, $T_w = 0.7T^*$, $R = 0.0127$ m, $\theta_c = 45°$, $\alpha = 10°$.

The shock layer thickness and the distance from the stagnation point are normalized by the body nose radius, and the pressure and heat transfer coefficients are normalized by their stagnation point values. In figures 4.41 and 4.42 the calculation results for the windward and leeward parts of the flow symmetry plane are plotted by full and hair lines. The proper experimental data are shown by the circles and triangles. The agreement between the calculated and experimental results appears to be quite good. It is worth noting that the bow shock distance from the windward part of the cone becomes larger than that from the leeward one at a sufficiently large distance from the stagnation point.

Eaton and Kaestner (1973) and Eaton and Larson (1974) investigated shock layer flows in the symmetry plane using set (4.91) with the normal momentum balance equation simplified in accordance with the thin shock layer model, i.e. supposing the pressure variation across the shock layer to be caused only by a centrifugal force. The function $p_2(s, n)$, as well as the bow shock boundary condition for the velocity component $w(s, n)$, were determined from available data on the inviscid flows. Calculations were carried out with the space-marching implicit finite-difference scheme of Crank–Nicolson's type. The initial profiles of the unknown functions were generated using the wall boundary conditions and available data on the inviscid gas flows.

Under the flight conditions considered, the shock layer flow transits from the laminar to the turbulent regime. The viscosity within the laminar region was evaluated by Sutherland's formula (1.8), the Prandtl number $Pr = 0.7$. The eddy

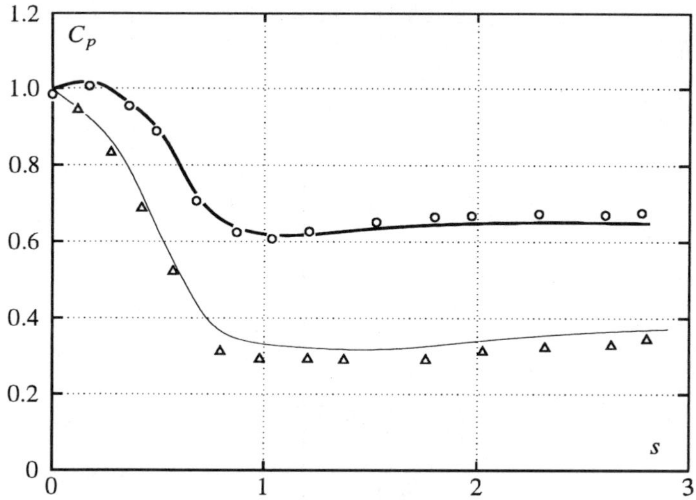

Figure 4.42 Variation of the wall pressure coefficient in the symmetry plane of a spherically blunted cone. Comparison of calculations (lines) and measurements (circles, triangles). $M_\infty = 5.9$, $\gamma = 1.4$, $Re_\infty = 3.95 \times 10^4$, $T_w = 0.7T^*$, $R = 0.0127$ m, $\theta_c = 40°$, $\alpha = 8°$.

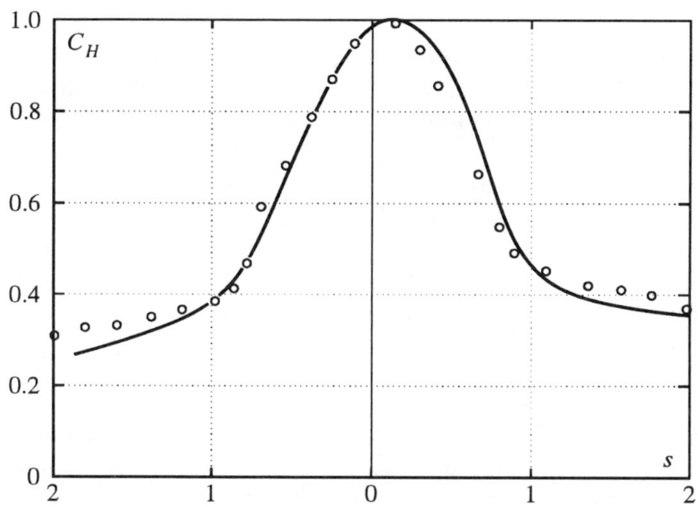

Figure 4.43 Variation of the heat transfer coefficient in the symmetry plane of a spherically blunted cone. Comparison of calculations (line) and measurements (circles). $M_\infty = 10.33$, $\gamma = 1.4$, $Re_\infty = 113050$, $T_w = 0.32T^*$, $R = 0.03175$ m, $\theta_c = 45°$, $\alpha = 5°$.

viscosity was evaluated from the algebraic mixing length model (Mayne 1973), the turbulent Prandtl number $Pr_t = 0.9$. The finite length of the transition region

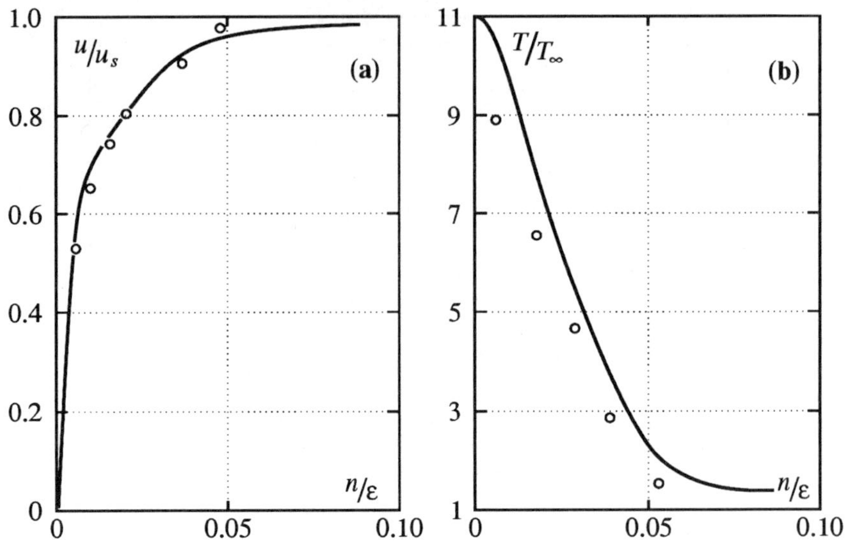

Figure 4.44 Longitudinal velocity and temperature profiles in the windward part of the symmetry plane of a pointed circular cone. $M_\infty = 8$, $Re_\infty = 1.1 \times 10^7/\text{m}$, $T_w = 570\ K$, $\theta_c = 5°$, $\alpha = 5°$. Lines, calculations of Eaton and Larson (1974); circles, measurements of Martellucci *et al* (1973).

was taken into account by the intermittency factor calculated with the formulae of Dhawan and Narasimha (1958). Figure 4.44 presents the calculation results for shock layer flow in the windward part of the symmetry plane of a pointed circular cone. Shown are the profiles of the longitudinal velocity component and gas temperature near the body surface in the turbulent flow region at a distance $l = 73$ cm from the cone apex. Here, the abscissa represents the distance from the cone surface normalized by the shock layer thickness, u_s is the longitudinal velocity component value behind the bow shock. The calculations are compared with the measurements by Martellucci *et al* (1973).

4.6.3 Local conical approximation

This approach is applied mainly in the investigations of supersonic flows about sharp conical bodies. Within its framework, the gas parameters are assumed to be invariable along the straight lines drawn from the cone apex. Accordingly, the derivatives in these directions are taken to be zero. As a result, the calculation of a stationary three-dimensional flow reduces to the solution of a two-dimensional problem. Considerations on the validity of this approach discussed in subsection 2.5.3 hold true for three-dimensional flows as well. Strictly speaking, the conical similarity inheres only in the inviscid gas flows. The viscous gas dynamics equations written on a conical assumption include the characteristic length involved in the Reynolds number. Therefore, the viscous gas flows are considered

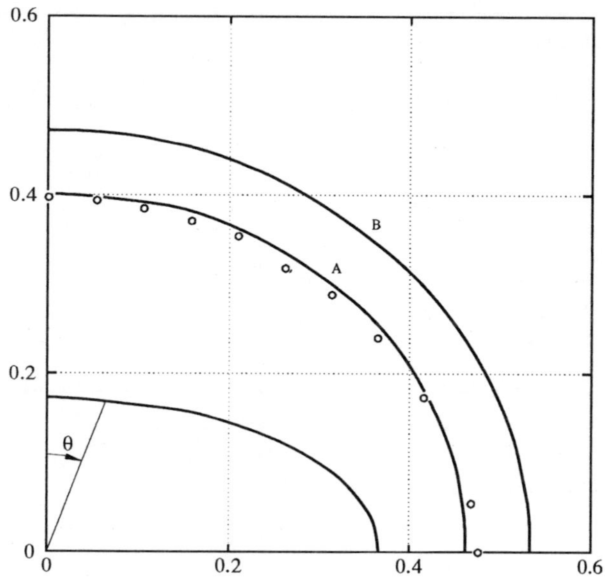

Figure 4.45 Bow shock in supersonic flow past a pointed elliptical cone. $M_\infty = 3.5$, $\gamma = 1.4$, $\alpha = 0$; $A, Re_\infty = 21000$; $B, Re_\infty = 700$. Lines, local conical solution; circles, wind tunnel measurements.

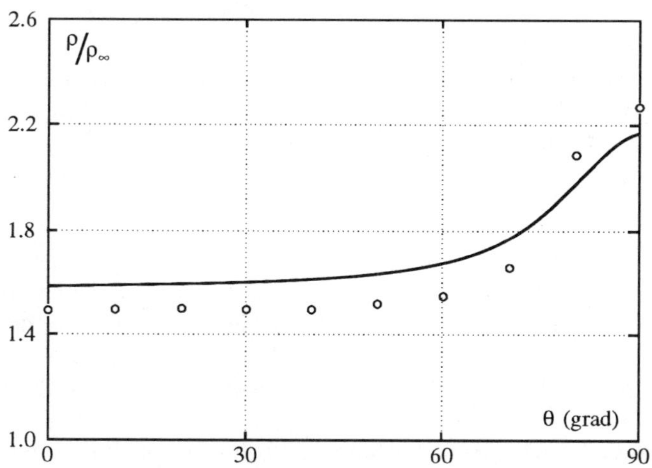

Figure 4.46 Density distribution over a pointed elliptical cone. $M_\infty = 3.5$, $\gamma = 1.4$, $Re_\infty = 21000$, $T_w = 0.38T^*$, $\alpha = 0$. Line, local conical solution; circles, wind tunnel measurements.

as the local conical ones being characterized in each cross section by a Reynolds number based on the distance from the body apex. The local conical approximation applies also to the calculation of flows past pointed bodies whose shape

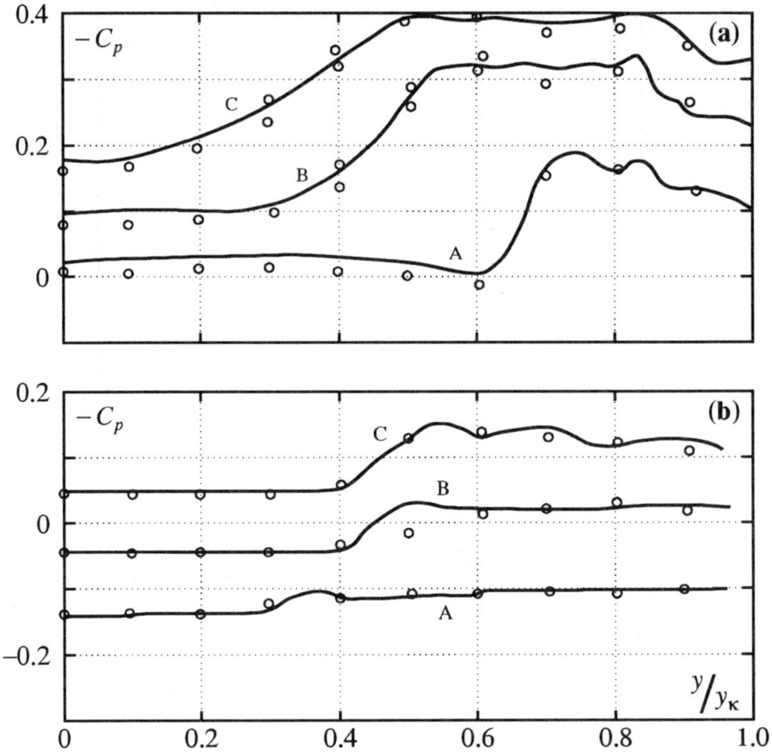

Figure 4.47 Pressure coefficient distribution on the leeward side of thin delta wings. (a) $M_\infty = 1.7$, $\beta = 75°$; A, $\alpha = 4°$; B, $\alpha = 12°$; C, $\alpha = 20°$; (b) $M_\infty = 2.8$, $\alpha = 8°$; A, $\beta = 60°$; B, $\beta = 67.5°$; C, $\beta = 75°$. Lines, local conical solution (Thomas and Newsome 1989); circles, experiment (Miller and Wood 1985).

differs from the conical one. This being the case, the dependence on the distance from the body apex is determined not only by the Reynolds number but the local geometric properties of the body as well. Below, some examples are presented which illustrate the accuracy of the local conical solutions for three-dimensional flows.

Figures 4.45 and 4.46 present local conical solutions of the Navier–Stokes equations for supersonic flow past a pointed elliptical cone with an ellipse axis ratio 2.1 and a half-angle in the plane with the greater axis of 20° at zero incidence. The calculations were carried out using the implicit unidirectional scheme of Golovachov and Zemlyakov (1994) combining the polynomial and exponential difference approximation of the governing equations. Calculation results are compared with wind tunnel measurements by Mozhayev (1967). Figure 4.45 displays the body contour and the bow shock at two Reynolds numbers. One can see that the bow shock location calculated at $Re = 21000$ agrees well with the experimental data.

The figure displays the rarefaction effect on the shock layer thickness as well. Figure 4.46 presents density variation over the cone. The calculations agree, on the whole, with the experimental data. A discrepancy results from the difference in the wall temperature which was not kept to be constant in the wind tunnel experiments.

Figure 4.47 depicts the results pertaining to the flows about thin delta wings (Thomas and Newsome 1989). The figure demonstrates the pressure coefficient variation on the leeward side of the wings at various sweep angles β and angles of attack α. Here y_k is the Cartesian coordinate of the wing edge. The local Reynolds numbers Re_∞ for the wings with sweep angles $\beta = 75°, 67.5°, 60°$ are equal to 3.56×10^6, 2.25×10^6 and 1.56×10^6, respectively. Numerical solutions were obtained using the time-asymptotic technique. The circles show the experimental data of Miller and Wood (1985).

A local conical approach is often applied to obtain the initial data for space-marching calculations using the parabolized Navier–Stokes equations . This being the case, it is convenient to obtain the local conical solution with the same marching algorithm which is applied later to solve the parabolized viscous shock layer equations. Computations are carried out in the following way. Starting from rather arbitrary data in the marching section $q^1 = q^1_0$, one calculates the solution in the section $q^1_0+\Delta q^1$. After proper scaling of the transversal coordinate, this solution is attributed to the section $q^1 = q^1_0$ and is used again to find the solution at $q^1_0+\Delta q^1$. The computations are repeated until the local conical solution is obtained which does not change when marching from q^1_0 to $q^1_0 + \Delta q^1$. The only distinction of the above procedure from the conventional space-marching algorithm consists in the need to recalculate the shock layer thickness in the previous marching section. The recalculation can be carried out with the formula

$$(\varepsilon^n)^{\nu+1} = (\varepsilon^{n+1})^{\nu+1} x^n / x^{n+1} \tag{4.94}$$

where x is the distance from the body apex, n is the marching section number, ν is the iteration number. The above iteration procedure for obtaining the local conical solution is equivalent to the use of a time-asymptotic technique with a time increment equal to the mesh spacing in the marching coordinate.

The accuracy of the local conical approximation for parabolized Navier–Stokes equations was investigated by Schiff and Sturek (1980) with reference to the turbulent flow around a sharp circular cone. The algebraic model of turbulence (Baldwin and Lomax 1978) was used. The local conical solutions were obtained for two sections located at the distances $x_1 = 15.2$ mm and $x_2 = 61$ mm from the cone apex. Later on, the first solution was exploited as the initial data for the parabolized shock layer equations which were integrated in x within the segment $x_1 \leqslant x \leqslant x_2$ by the space-marching method of Schiff and Steger (1980). Figures 4.48 and 4.49 show a comparison of the calculation results with the local conical solution at $x = x_2$. The wall pressure distribution and profiles of the longitudinal velocity component in the windward and leeward parts of the shock layer are presented. One can see an excellent agreement between the calculations using the space-marching method and the local conical solution.

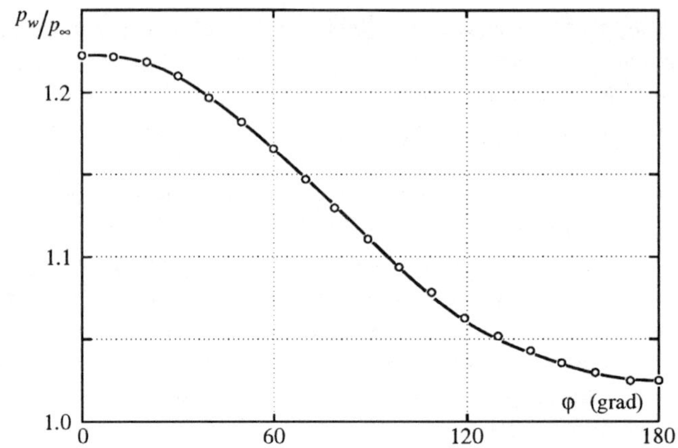

Figure 4.48 Wall pressure distribution on a sharp circular cone. $M_\infty = 3$, $Re_\infty = 2.53 \times 10^7$/m, $\theta_c = 10°$, $\alpha = 2°$. Line, space-marching method; circles, local conical solution.

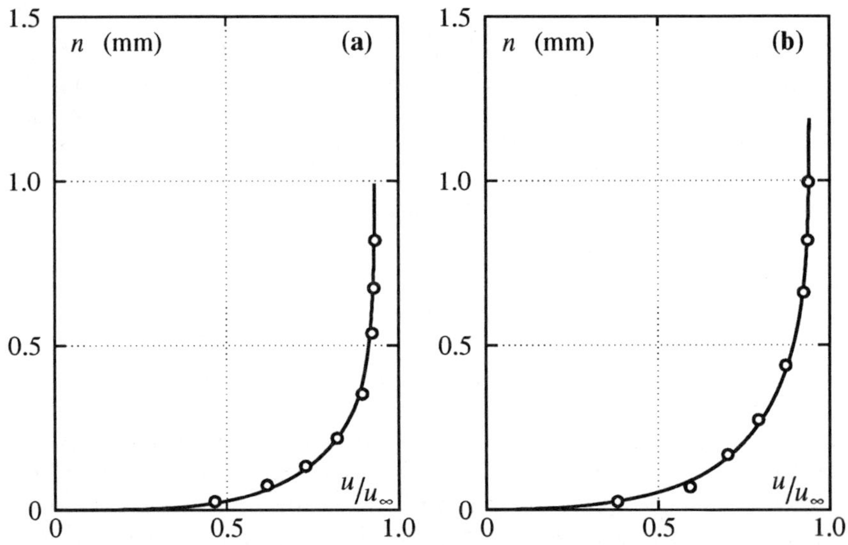

Figure 4.49 Profiles of the longitudinal velocity component in the flow past a sharp circular cone. $M_\infty = 3$, $Re_\infty = 2.53 \times 10^7$ /m, $\theta_c = 10°$, $\alpha = 2°$; (a), $\varphi = 0$; (b), $\varphi = 180$. Lines, space-marching method; circles, local conical solution.

Comparison with available experimental data and numerical solutions of the three-dimensional boundary value problem (Thomas and Newsome 1989) evidences that the local conical approximation, as applied to the flows past slender bodies at small and moderate angles of attack, ensures an acceptable accuracy for the flow field and its evolution with a change of the body geometry and flight conditions. The local conical approach ensures also quite accurate prediction for the aerodynamic characteristics.

4.7 DEGENERATE THREE-DIMENSIONAL FLOWS

In this section we consider stationary three-dimensional shock layer flows depending on only two spatial coordinates. Such flows occur near the plane arrow-shaped wings of an infinite span and spinning bodies of revolution at zero incidence.

4.7.1 Flow around a blunted arrow-shaped wing of an infinite span

Consider supersonic flow about an infinite cylinder supposing the oncoming stream velocity vector $\boldsymbol{V}_\infty$ to be arranged at an angle of side-slip β to the plane which is normal to the body generatrix. For the case of hypersonic flight speed, the problem was investigated by Bruikina and Gershbein (1979), Markov (1980,1988), Gershbein and Scherbak (1984) and Gershbein *et al* (1984a,b).

Let us consider, at first, the self-similar solution for the flow symmetry plane at zero incidence (Gershbein and Scherbak 1984). By analogy with an axisymmetric flow (see subsection 2.5.1), variation of gas dynamic functions in the vicinity of the flow symmetry plane is described with trigonometric series expansions:

$$
\begin{aligned}
u &= V_\infty \cos\beta \, [u_1(n)\cos\theta(s) + \ldots] \\
v &= V_\infty \cos\beta \, [v_0(n)\sin\theta(s) + \ldots] \\
w &= V_\infty \sin\beta \, [w_0(n) + \ldots] \\
\rho &= \rho_\infty \, [\rho_0(n) + \ldots] \\
h &= V_\infty^2 \cos^2\beta \, [h_0(n) + \ldots] \\
p &= \rho_\infty V_\infty^2 \cos^2\beta \left[p_0(n) + p_2(n)\cos^2\theta(s) + \ldots\right] \\
\mu &= \mu_\infty \, [\mu_0(n) + \ldots]
\end{aligned}
\tag{4.95}
$$

Here u, v, w are the velocity components for the orthogonal body-conforming coordinate system (s, n, z) in which the z-axis is aligned with the body generatrix, the s, n-coordinates are aligned with the body cross section contour and its outer normal, respectively; θ is the angle between the projection of the free stream velocity vector $\boldsymbol{V}_\infty$ on the plane normal to the wing generatrix and the tangent to the wing contour. Substitution of series expansions (4.95) into the Navier–Stokes equations and use of the truncation procedure (Kao 1964) yield a set of ordinary differential equations for the coefficients of series (4.95). Then, Dorodnitsin's variables and the stream function are introduced. With zero angle of side-slip, the result does not differ from set (3.16) with $\nu = 0$. At a non-zero side-slip and $M_\infty \gg 1$, the set of equations for the flow in the symmetry plane is written for non-dimensional variables as follows:

$$\left(\frac{\mu\rho}{Re_\infty} r u' + f u\right)' = u^2 + u(u+v) + 2\frac{p_2}{\rho}$$
$$\left(\frac{\mu\rho}{Re_\infty} r w' + f w\right)' = f w$$
$$\left(\frac{4}{3}\frac{\mu\rho}{Re_\infty} r v' + f v - r p\right)' - u v + p/\rho = 0$$
$$\left\{\frac{\mu\rho}{Re_\infty}\frac{r}{Pr}\left[h' + Pr\left(\frac{4}{3}\left(\frac{v^2}{2}\right)' + \left(\frac{w^2}{2}\right)' \tan^2\varphi\right)\right]\right.$$
$$\left. + f\left(h + \frac{v^2}{2} + \frac{w^2}{2}\tan^2\varphi\right)\right\}' = u\left(h + \frac{v^2}{2} + \frac{w^2}{2}\tan^2\varphi\right) \tag{4.96}$$
$$p_2' = -p' + \frac{u(u+v)}{r} \qquad u = f'$$

Here $r = 1 + n$; the prime denotes differentiation in Dorodnitsin's coordinate $\eta = \int_0^n \rho dn$; the subscripts indicating the expansion term numbers are omitted, except for the second coefficient of the pressure expansion. All distances are normalized by the radius R of the wing contour at the stagnation point. The Mach number M_∞ and the Reynolds number Re_∞ are evaluated here using the free stream velocity component $V_\infty \cos\beta$ which is normal to the wing generatrix.

Set (4.96) is solved subject to the boundary conditions on the body surface ($\eta = 0$)

$$u = 0 \qquad -f = (\rho v)_w = G \qquad w = 0 \qquad h = h_w \tag{4.97}$$

and in the free stream ($\eta = \eta_\Gamma$)

$$u = 1 \qquad f = -r \qquad w = 1$$
$$h = \frac{1}{(\gamma - 1)M_\infty^2} \qquad p = \frac{1}{\gamma M_\infty^2} \qquad p_2 = 0 \tag{4.98}$$

The numerical method is quite similar to that described in subsection 2.5.1. The calculations were carried out for a perfect gas with the specific heat ratio $\gamma = 1.4$, the viscosity $\mu \propto \sqrt{T}$, the Prandtl number $Pr = 0.75$. The body surface was supposed to be impermeable ($G = 0$).

The side-slip effect is demonstrated in figure 4.50 which presents the temperature profiles at various angles of side-slip. The temperature is normalized by $(\gamma - 1)M_\infty^2 T_\infty$. It is worth noting the appearance of a distinct temperature maximum in the near-wall region at large angles of side-slip, which is caused by the dissipation of the kinetic energy of the flow along the wing generatrix.

Gershbein *et al* (1984a) and Markov (1988) investigated flows in the vicinity of the wing leading edge at zero incidence using the two-layer thin viscous shock layer model. Their calculations agree well with the above results obtained by the bow shock capturing method. Markov (1988) investigated also the non-stationary effects caused by the variation of the mass injection rate and wall temperature.

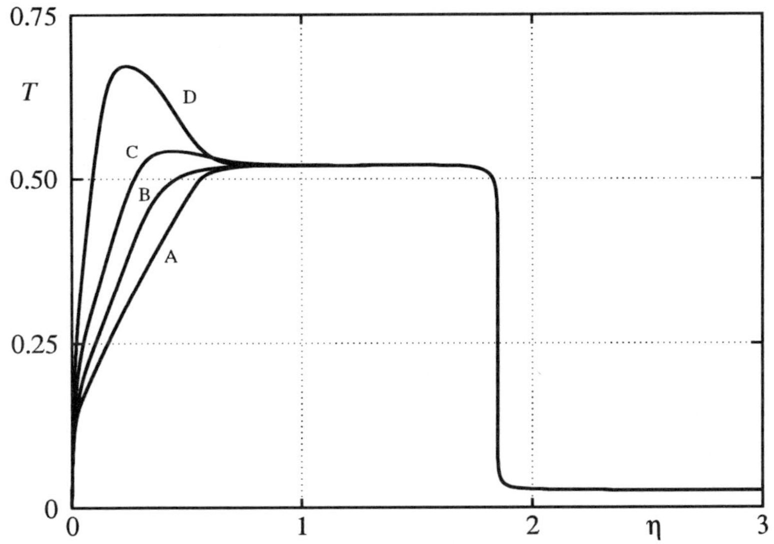

Figure 4.50 Temperature profiles in the symmetry plane of hypersonic flow about an infinite cylinder at zero incidence and various angles of side-slip. $M_\infty = 10$, $Re_\infty = 10^3$, $T_w = 0.3T^*$; A, $\beta = 0$; B, $\beta = 40°$; C, $\beta = 50°$; D, $\beta = 60°$.

Hypersonic flows around the leading edge of the arrow-shaped wings at angles of attack were calculated with the use of the thin and full viscous shock layer equations by Bruikina and Gershbein (1979), Gershbein *et al* (1984b), and Vershinin *et al* (1991). Golovachov and Leontieva (1993) obtained a numerical solution of the problem using the full Navier–Stokes equations and taking into account the velocity slip and temperature jump on the wing surface at small Reynolds numbers. Their results are presented in figures 4.51 through 4.54.

Figures 4.51 and 4.52 present a comparison with the results of Gershbein *et al* (1984a) obtained from the thin viscous shock layer model. The curves A correspond to the flow about an elliptic cylinder with the ellipse axis ratio equal to 2; the curves B correspond to the flow about a parabolic cylinder. For comparison, the solutions of the full Navier–Stokes equations shown by the full curves were obtained subject to the no-slip wall boundary conditions. The Mach number M_∞ and the stagnation temperature T^* are based on the free stream velocity component $V_\infty \cos\beta$ which is normal to the body generatrix. The gas viscosity $\mu \propto \sqrt{T}$, the Prandtl number $Pr = 0.71$, the Reynolds number $Re = \rho_\infty V_\infty \cos\beta R/\mu(T^*)$.

Figure 4.51 depicts the temperature profiles on the stagnation streamline at zero incidence. Here η is Dorodnitsin's coordinate normalized by its value at the bow shock. Figure 4.52 shows the heat flux distributions over the wing surface at an angle of attack $\alpha = 45°$ and zero side-slip. The heat flux is normalized by its stagnation point value at zero incidence. The distance s from the stagnation point is normalized by the body nose radius. The comparisons indicate that the thin viscous shock layer model gives a large error even at a hypersonic flight speed,

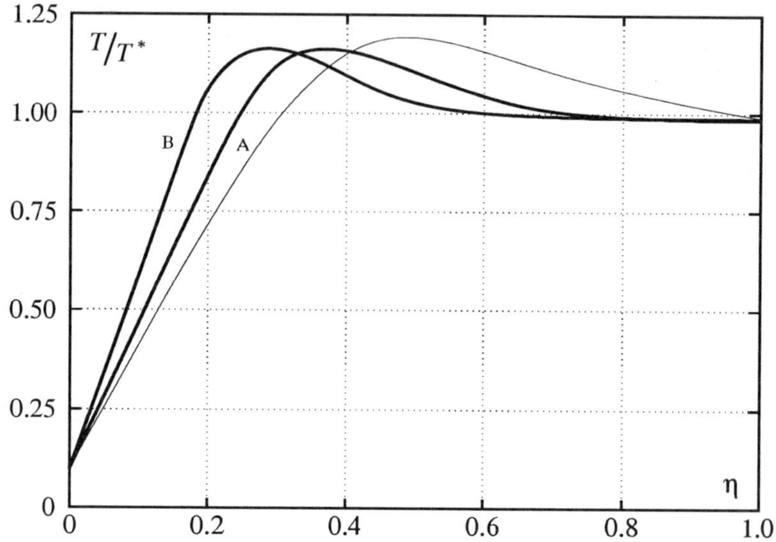

Figure 4.51 Temperature profiles on the stagnation streamline in hypersonic flows about elliptic (A) and parabolic (B) cylinders. $M_\infty = 12.5$, $\gamma = 1.25$, $Re = 50$, $T_w = 0.1T^*$, $\alpha = 0$, $\beta = 60°$. Full lines, full Navier–Stokes equations; hair line, thin viscous shock layer model.

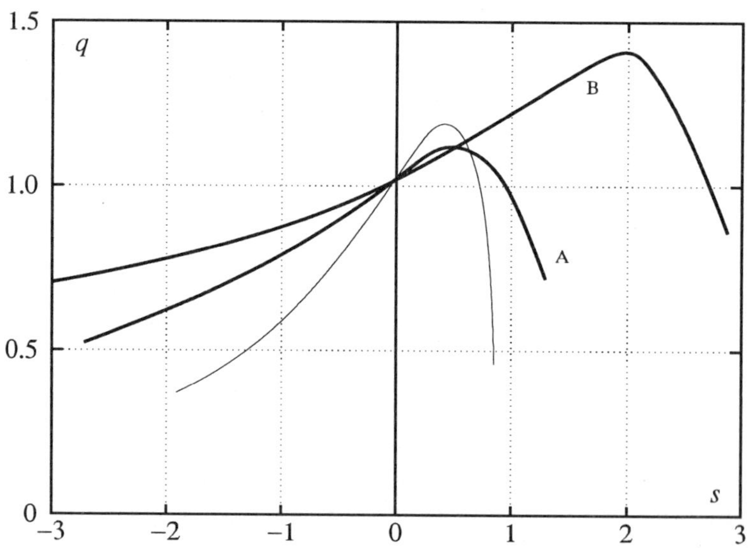

Figure 4.52 Heat flux distributions over elliptic (A) and parabolic (B) cylinders. $M_\infty = 12.5$, $\gamma = 1.25$, $Re = 50$, $T_w = 0.1T^*$, $\alpha = 45°$, $\beta = 0$. Full lines, full Navier–Stokes equations; hair line, thin viscous shock layer model.

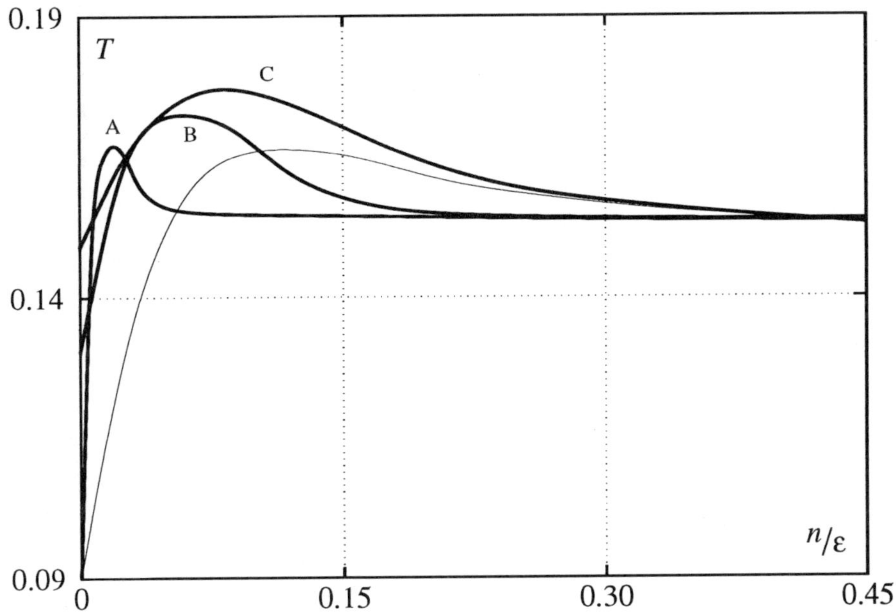

Figure 4.53 Temperature profiles on the stagnation streamline ahead of a parabolic cylinder. $M_\infty = 4$, $\gamma = 1.4$, $T_w = 0.238T^*$, $\alpha = 0$, $\beta = 60°$; A, $Re_\infty = 2000$; B, $Re_\infty = 100$; C, $Re_\infty = 30$; hair curve, no-slip wall boundary conditions.

which is, probably, caused by a substantial increase of the shock layer thickness as compared with axisymmetric flows.

The use of full Navier–Stokes equations allows an extension of the flight condition range towards smaller values of the Mach and Reynolds numbers. The next two figures demonstrate calculations of the flow about a parabolic cylinder at Mach number $M_\infty = 4$. Here, unlike the previous figures, the Mach and Reynolds numbers and the stagnation temperature are defined with the modulus of the free stream velocity vector V_∞. The viscosity was evaluated by Sutherland's formula (1.8), the Prandtl number $Pr = 0.71$, the gas specific heat ratio $\gamma = 1.4$.

The calculations showed that, like in the case of a hypersonic flight speed, the side-slip effect on the shock layer flow field is most prominent for the gas density, temperature, and tangent velocity component. Figure 4.53 presents the temperature profiles on the stagnation streamline at three Reynolds numbers. Here, the distance from the body surface is normalized by the shock layer thickness. The temperature is normalized by $V_\infty^2/2R^*$ where R^* is the specific gas constant. To illustrate the velocity slip and temperature jump effects, the hair curve depicts the results obtained with no-slip wall boundary conditions. It is seen that the side-slip effect on the temperature profile manifests itself in the near-wall region of the shock layer. With decreasing Reynolds number, this region expands and the maximum temperature increases.

Figure 4.54 shows the heat flux distributions over the wing surface at zero

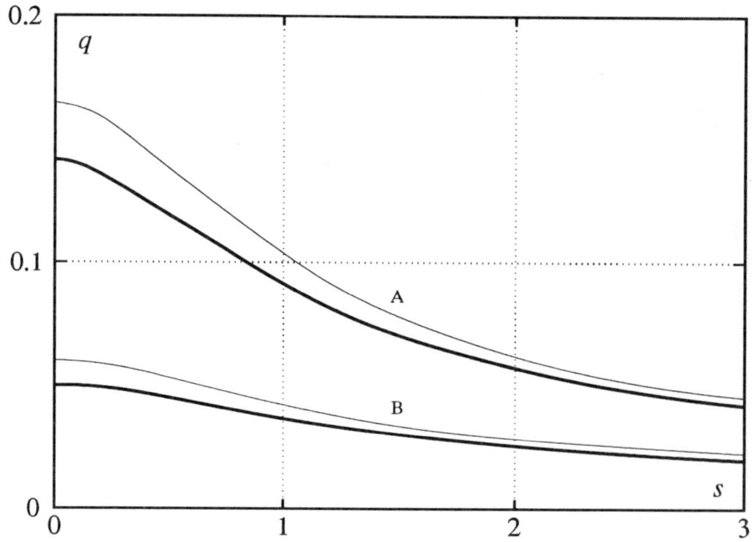

Figure 4.54 Heat flux distribution over a parabolic cylinder. $M_\infty = 4$, $\gamma = 1.4$, $Re_\infty = 30$, $T_w = 0.238T^*$, $\alpha = 0$; A, $\beta = 0$; B, $\beta = 60°$; hair lines, no-slip wall boundary conditions.

incidence. The heat flux and the s-coordinate are normalized by $\rho_\infty V_\infty^3$ and the body nose radius R. One can see a substantial reduction of the heat flux in the flow with side-slip.

4.7.2 Spin effects in shock layer flow

Consider supersonic flow around a blunted body of revolution spinning round its axis with the angular velocity Ω. The oncoming stream, whose translation velocity V_∞ is aligned with the body axis, revolves round this axis with the angular velocity Ω_1. The oncoming stream parameters are given by the formulae

$$\rho_\infty = \text{const} \qquad p_\infty = p_\infty^0 + \frac{\rho_\infty w_\infty^2}{2} \tag{4.99}$$

where $w_\infty = \Omega_1 r$, r is the distance from the flow axis, p_∞^0 is the gas pressure on the axis. Hypersonic flows of this kind were investigated by Markov (1984, 1987), Krasilov *et al* (1986a,b), and Gershbein and Peigin (1986) using the thin viscous shock layer model. The governing equations written for the immovable body-conforming curvilinear orthogonal coordinate system (s, n, φ) are the following:

$$\begin{aligned}
&\frac{\partial}{\partial s}(\rho r u) + \frac{\partial}{\partial n}(\rho r v) = 0 \\
&\rho\left(u\frac{\partial u}{\partial s} + v\frac{\partial u}{\partial n} - \frac{\sin\theta}{r}w^2\right) = -\delta\frac{\partial p}{\partial s} + \frac{\partial}{\partial n}\left(\frac{\mu}{K}\frac{\partial u}{\partial n}\right) \\
&\rho\left(u\frac{\partial w}{\partial s} + v\frac{\partial w}{\partial n} + \frac{\sin\theta}{r}uw\right) = \frac{\partial}{\partial n}\left(\frac{\mu}{K}\frac{\partial w}{\partial n}\right) \\
&\rho\left(u\frac{\partial T}{\partial s} + v\frac{\partial T}{\partial n}\right) = 2\delta u\frac{\partial p}{\partial s} + 2\frac{\mu}{K}\left[\left(\frac{\partial u}{\partial n}\right)^2 + \left(\frac{\partial w}{\partial n}\right)^2\right] \\
&\quad + \frac{\partial}{\partial n}\left(\frac{\mu}{PrK}\frac{\partial T}{\partial n}\right) \\
&\frac{\partial p}{\partial n} = \rho\left(\kappa u^2 + \frac{\cos\theta}{r}w^2\right)
\end{aligned} \tag{4.100}$$

Here $K = \delta Re$, $\delta = (\gamma - 1)/(2\gamma)$, $Re = \rho_\infty V_\infty R/\mu(T^*)$, $T^* = V_\infty^2/(2c_p)$; R is the body nose radius; $V_\infty u$, $\delta V_\infty v$, $V_\infty w$ are the velocity vector components in the s, n, φ-coordinates, respectively; $\rho_\infty V_\infty^2 p$, $\delta^{-1}\rho_\infty \rho$, T^*T are the pressure, density and temperature; κ is the body surface curvature, r is the distance from the body axis to its surface, θ is the slope of the tangent to the body surface to the body axis. The n-coordinate is normalized by δR, all other distances are normalized by R. The angular velocities Ω and Ω_1 are normalized by V_∞/R.

The wall boundary conditions read

$$u = 0 \qquad w = \Omega r(s) \qquad T = T_w(s) \qquad \rho v = G(s) \tag{4.101}$$

The boundary conditions on the bow shock wave are given by the modified Rankine–Hugoniot relationships. Taking into account a hypersonic speed of the oncoming stream and using (4.99), they are written as follows

$$\begin{aligned}
&\rho\left(v - u\frac{\partial n_s}{\partial s}\right) = v_\infty \qquad p = v_\infty^2 + \frac{w_\infty^2}{2} \\
&v_\infty(u - u_\infty) = \frac{\mu}{K}\frac{\partial u}{\partial n} \qquad v_\infty(w - w_\infty) = \frac{\mu}{K}\frac{\partial w}{\partial n} \\
&v_\infty\left(H - H_\infty - v_\infty^2\right) = \frac{\mu}{PrK}\frac{\partial T}{\partial n} + \frac{2\mu}{K}\left(u\frac{\partial u}{\partial n} + w\frac{\partial w}{\partial n}\right)
\end{aligned} \tag{4.102}$$

where $n_s(s)$ is the bow shock coordinate, $u_\infty = \cos\theta$, $v_\infty = -\sin\theta$, and $H = T + u^2 + w^2$.

It is convenient to solve the problem using new independent variables of Dorodnitsin's type

$$\xi = s \qquad \zeta = \frac{1}{\Delta}\int_0^n \rho dn \qquad \Delta = \int_0^{n_s} \rho dn$$

and new functions f, φ, ψ

$$\frac{u}{u_\infty} = \frac{\partial f}{\partial \zeta} \qquad \frac{w}{w_0} = \frac{\partial \varphi}{\partial \zeta} \qquad \rho r v = -\frac{\partial \psi}{\partial s} \qquad \psi = r\Delta u_\infty f$$

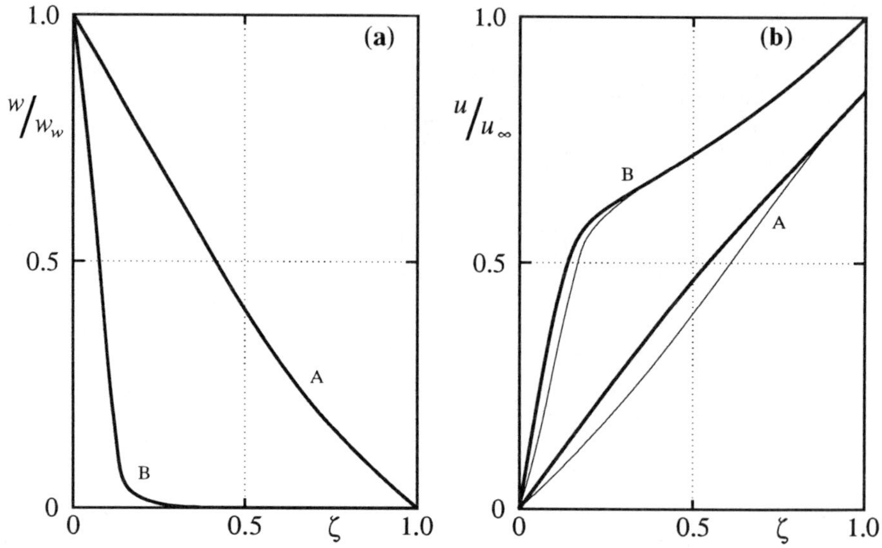

Figure 4.55 Profiles of the velocity components on the stagnation streamline in hypersonic flow about a spinning blunt body. $T_w = 0.1T^*$, $G = 0$; A, $Re = 50$; B, $Re = 5000$. Full lines, $\Omega = 1$; hair lines, $\Omega = 0$.

where

$$w_0 = \begin{cases} w_\infty & \text{at } \Omega_1 \neq 0 \\ w_w & \text{at } \Omega \neq 0 \end{cases}$$

Numerical solutions of the problem were obtained with the implicit space-marching finite-difference schemes starting from the stagnation line $\xi = 0$. The solution on this line was obtained from the set of ordinary differential equations resulting from an evaluation of the limit $\xi \to 0$ in coefficients of the basic equations. As for the pressure gradient term $p_2 = 1/u_\infty(\partial p/\partial \xi)$, Krasilov *et al* (1986a,b) and Gershbein and Peigin (1986) determined it from the solution of the first-order ordinary differential equation which resulted from differentiation in ξ of the normal momentum balance equation (the last one of set (4.100)). The equation for p_2 was integrated from the bow shock to the body surface using the Simpson quadrature formula. The same method was applied to integrate the equation for the gas pressure. The boundary condition for p_2 was obtained by differentiating in ξ the proper relationship of (4.102).

Consider, at first, the results of Krasilov *et al* (1986a,b) for the case of a non-swirling oncoming stream ($\Omega_1 = 0$). The authors assumed $\delta = 0.1$, $\mu \propto \sqrt{T}$, $Pr = 0.71$. Figure 4.55 depicts the profiles of the velocity components on the stagnation streamline. The hair curves show the results for a non-spinning body. It is seen that at high Reynolds numbers the spin effect is localized near the body surface. At $Re = 50$ the molecular transport is important throughout the shock layer and, for this reason, the body spin causes gas rotation in the whole region bounded by the bow shock.

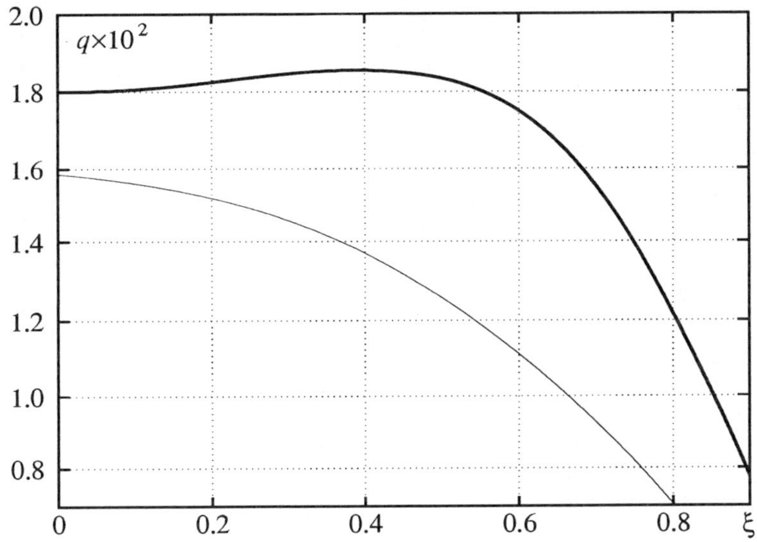

Figure 4.56 Heat flux distribution over the spinning (full line) and non-spinning (hair line) sphere in a hypersonic stream. $Re = 5000$ $T_w = 0.1T^*$, $G = 0$.

Calculations show a small spin effect on the temperature profile on the stagnation streamline. However, in the downstream region of the shock layer, the temperature is substantially higher as compared with the flow about an immovable body, which is caused by the dissipation of the kinetic energy of the flow in the circumferential direction.

The next remark concerns the spin effect on the drag coefficient. With increasing angular velocity Ω, the wall pressure on the lateral part of the body falls, while the shear stress increases, so that variation of the total drag coefficient turns out to be relatively small. For instance, the drag coefficient of the front of a paraboloid of length 2.4 R at $Re = 500$, $\Omega = 1.0$ is less by 5 per cent as compared with the drag coefficient of the immovable body.

Figure 4.56 demonstrates the spin effect on the heat flux distribution over the sphere surface. One can see a significant increase of the heat flux to the spinning body and displacement of its maximum away from the stagnation point. The bow shock distance from the body surface was found to be noticeably influenced by the body spin at Reynolds numbers $Re \lesssim 10^3$. Because of increase of the tangent velocity, the spin reduces the shock layer thickness in the vicinity of the stagnation point. At the same time, the bow shock distance from the spinning body increases with distance from the stagnation point more rapidly because of the pressure reduction and temperature increase in the downstream part of the shock layer as compared with the case of a non-spinning body.

Krasilov *et al* (1986a) investigated also the combined spin and mass injection effect on the shock layer flow. As an example, figure 4.57 presents the heat flux distributions over a paraboloid whose nose ($\xi \leqslant 1$) is spinning while the rest

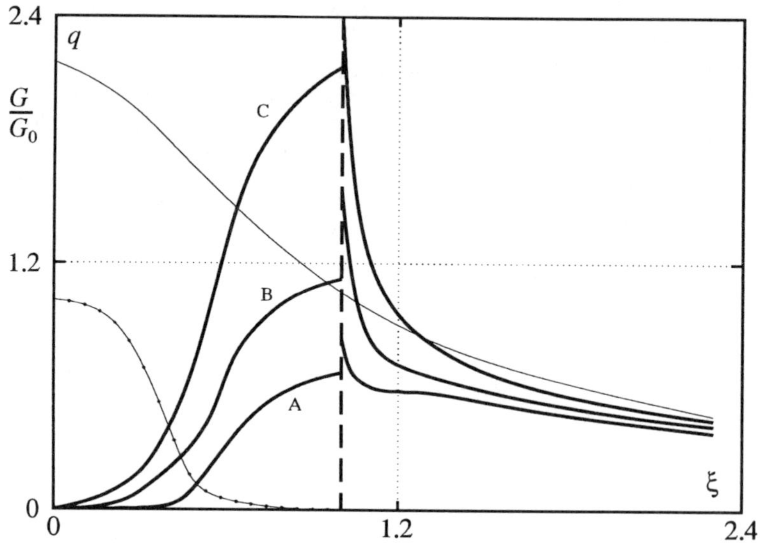

Figure 4.57 Heat flux (full and hair curves) and mass injection rate (hair-dotted curve) distributions over a paraboloid with a spinning nose. $Re = 5000$, $T_w = 0.1T^*$; A, $\Omega = 0.3$; B, $\Omega = 0.6$; C, $\Omega = 1.0$; hair line, $G = \Omega = 0$.

of the body is immovable. Mass injection is supposed to be made through the nose part of the body, the injected gas being identical with that of the oncoming stream. The mass injection rate is determined by a formula similar to (3.20) as a function of the distance z measured from the stagnation point along the paraboloid axis. Calculation results shown by the full lines pertain to $G_0 = 0.2$, $z_* = 0.14$ and several values of the angular velocity of the body nose. The hair curve depicts calculations for an impermeable and non-spinning body. The heat flux values are normalized by $\rho_\infty V_\infty^3/2\sqrt{Re}$. The hair-dotted curve shows the distribution of the mass injection rate G/G_0. One can see in figure 4.57 that under the flight conditions considered, the heat flux maximum is achieved on the edge of the spinning part of the body at $\xi = 1$. It is worth noting, however, that in the vicinity of this edge, which is a line of discontinuity of the wall boundary condition for the circumferential velocity component w, the thin viscous shock layer model may be inaccurate.

Effects of the oncoming stream swirling are illustrated below by the numerical solutions of Gershbein and Peigin (1986), which pertain to hypersonic flows about impermeable non-spinning bodies at $\delta = (\gamma - 1)/(\gamma + 1) = 0.1$.

Figure 4.58 shows the variation of the stagnation point heat flux and skin friction coefficient

$$q = \frac{2\sqrt{Re}}{\rho_\infty V_\infty^3}\lambda\frac{\partial T}{\partial n} \qquad C_{f1} = \frac{2\sqrt{Re}}{\rho_\infty V_\infty^2}\mu\frac{\partial u}{\partial n}$$

with Reynolds number. The results indicate an increase of the shear stress and heat flux maxima due to swirling of the oncoming stream. It is worth noting also

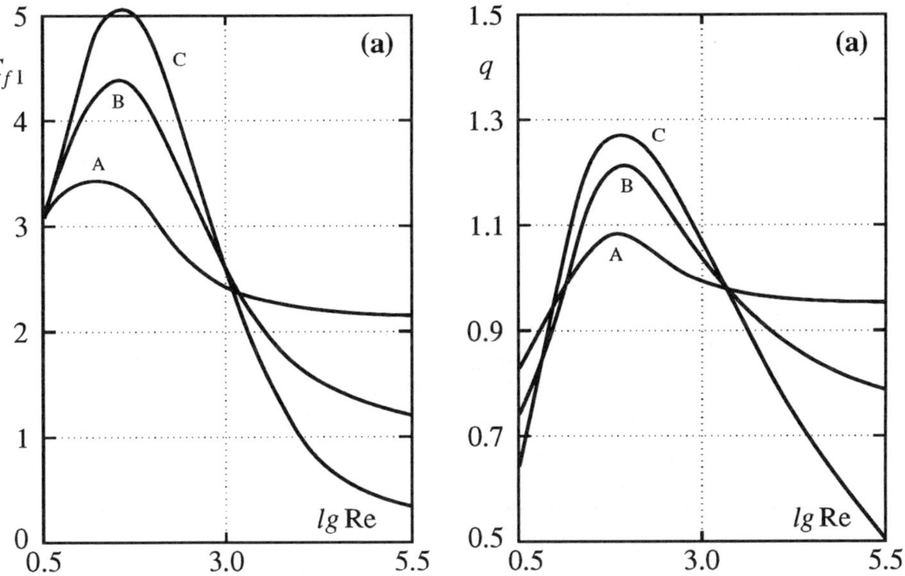

Figure 4.58 Dependence of the stagnation point skin friction coefficient and heat flux on Reynolds number in the hypersonic swirling flow about a blunt body. $T_w = 0.1T^*$; A, $\Omega_1 = 0$; B, $\Omega_1 = 1.5$; C, $\Omega_1 = 2.1$.

that the swirling effect changes sign with increasing Reynolds number.

Figure 4.59 presents the heat flux distributions over the sphere surface. Here, the heat flux is normalized by its stagnation point value. The full, hair and hair-dotted curves depict calculations at various Reynolds numbers. The curves A and B depict calculations at various angular velocities Ω_1. It is seen that, like the body spin, the swirling of the oncoming stream can result in a qualitative change of the heat flux behaviour and displacement of the heat flux maximum downstream from the stagnation point. This effect manifests itself to the greatest extent at small Reynolds numbers; however, it was revealed in boundary layer calculations as well (Peigin 1985).

An interesting feature of the shock layer flows under study caused by the centrifugal force is the existence of a threshold value of the angular swirling velocity Ω_1^* at which a drastic change of the shock layer flow happens. At $\Omega_1 < \Omega_1^*$ an increase of the angular swirling velocity results in a monotonous reduction of the shock layer thickness in the vicinity of the stagnation point, producing a distinct maximum in the profile of the longitudinal velocity near the bow shock. Exceeding the critical value Ω_1^* gives rise to an abrupt increase of the shock layer thickness, substantial change of the wall pressure distribution, additional extrema in the longitudinal velocity profiles, and recirculating flow in the shock layer. This happens because the total gas pressure turns to be insufficient to overcome the pressure increase induced by the centrifugal force downstream from the stagnation region of the shock layer. The above effect was reported by Smith (1973). In figure 4.60

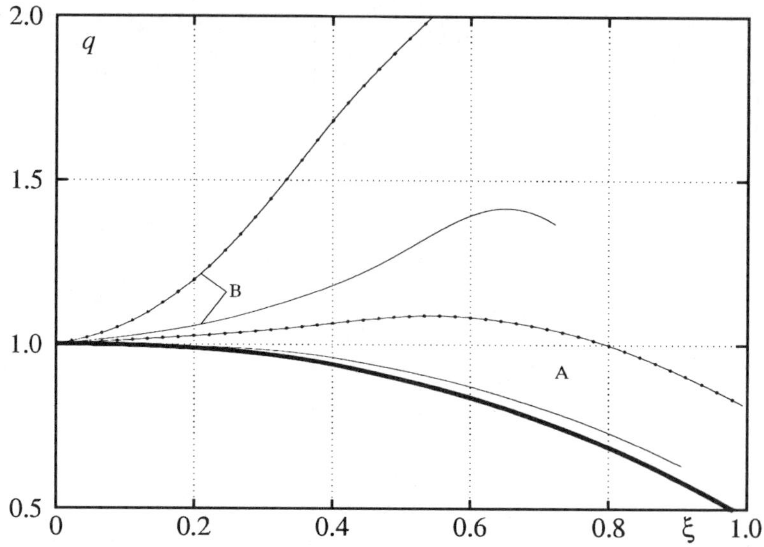

Figure 4.59 Heat flux distributions over the sphere surface in a hypersonic swirling flow. $T_w = 0.1T^*$; A, $\Omega_1 = 1$; B, $\Omega_1 = 2$. Full line, $Re = 10^4$; hair lines, $Re = 500$; hair-dotted lines, $Re = 25$.

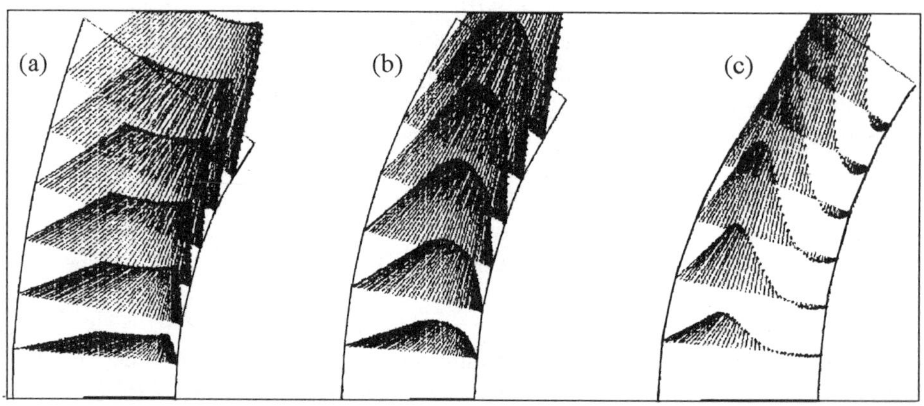

Figure 4.60 Longitudinal velocity map in the shock layer ahead of a sphere in a swirling supersonic stream. $M_\infty = 2$, $\gamma = 1.4$ $Re = 10^3$; (a), $\Omega_1 = 0.0$; (b), $\Omega_1 = 1.0$; (c), $\Omega_1 = 1.95$.

it is illustrated by the numerical solutions of the full Navier–Stokes equations for supersonic flow around a heat-insulated sphere (Golovachov *et al* 1994).

4.8 APPROXIMATE METHODS FOR PREDICTION OF THREE-DIMENSIONAL FLOW CHARACTERISTICS

Together with the numerical simulation of shock layer flows, it seems expedient to use approximate methods for the estimation of aerodynamic and heat transfer characteristics, which saves a large amount of computer resources and allows one to speed up the research.

4.8.1 Series expansions with respect to the angle of attack

Consider this method following the paper by Karyakin (1981) who investigated three-dimensional supersonic flows about blunted bodies of revolution. Formulation of the problem is the same as in subsection 4.2.2, except for the boundary conditions on the bow shock wave where the conventional shock relationships are used. The shock layer flow is described by the reduced Navier–Stokes equations (4.19) written for the moving curvilinear coordinate system q^i $(i = 1, 2, 3)$ related to the spherical and Cartesian coordinate systems by formulae (4.17) and (4.18). Solution of the three-dimensional problem is sought for in terms of the power series expansions of unknown functions with respect to the angle of attack α

$$X = \sum_{i=0}^{\infty} X_i \alpha^i \tag{4.103}$$

The coefficients X_i are given by the Fourier series in the circumferential coordinate q^2. In this case, it is convinient to use the velocity vector components V_i written for the cylindrical coordinate system

$$V_1 = u_1 \quad V_2 = u_2 \cos q^2 + u_3 \sin q^2 \quad V_3 = -u_2 \sin q^2 + u_3 \cos q^2 \tag{4.104}$$

Here u_α $(\alpha = 1, 2, 3)$ are the velocity components for the Cartesian coordinate system x^α. Taking into account the flow symmetry relative to the plane $q^2 = 0, \pi$, one can write the following expansions

$$X^{(1)} = \sum_{i=0}^{\infty} \sum_{k=0}^{\infty} X_{ik}^{(1)} \alpha^i \cos kq^2 \quad X^{(2)} = \sum_{i=0}^{\infty} \sum_{k=1}^{\infty} X_{ik}^{(2)} \alpha^i \sin kq^2 \tag{4.105}$$

where $X^{(1)} = \{V_1, V_2, T, \rho, r_s\}$, $X^{(2)} = V_3$. When using expansions (4.105), one ought to exclude the coefficients which are surely equal to zero. Substitution of the expansions into the governing equations and the boundary conditions shows that $X_{ik}^{(1)} = X_{ik}^{(2)} = 0$, if $k > i$, the functions $X_{ik}^{(r)}$ $(r = 1, 2)$ for different k being obtained independently of each other and of the functions with larger indices i.

Retaining the terms up to α^2, one can write (4.105) in the form

$$\begin{aligned} X^{(1)} &= X_{00}^{(1)} + \alpha X_{11}^{(1)} \cos q^2 + \alpha^2 \left(X_{20}^{(1)} + X_{22}^{(1)} \cos 2q^2 \right) + \ldots \\ X^{(2)} &= \alpha X_{11}^{(2)} \sin q^2 + \alpha^2 X_{22}^{(2)} \sin 2q^2 + \ldots \end{aligned} \tag{4.106}$$

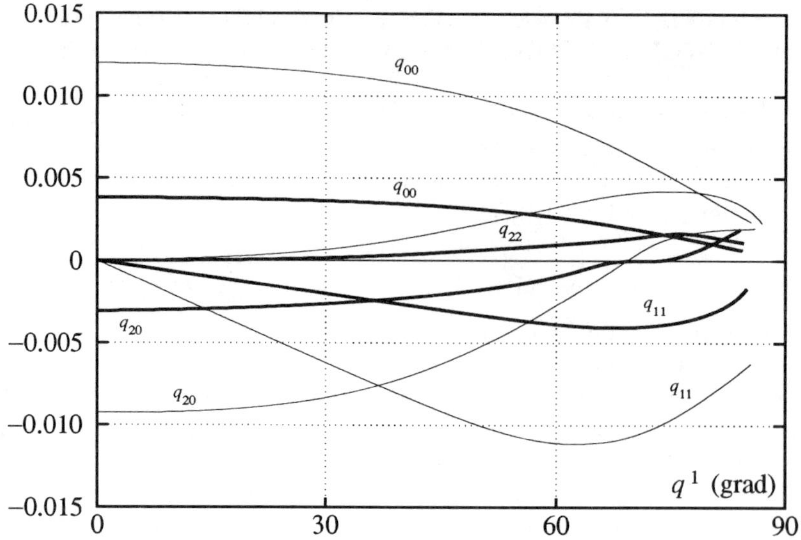

Figure 4.61 Variation of the series coefficients of the heat flux expansion in supersonic flow about an ellipsoid. $M_\infty = 4$, $T_w = 0.8T^*$, $\alpha = 10°$. Full lines, $Re_\infty = 3333$; hair lines, $Re_\infty = 333$.

Using the cylindrical coordinates to represent the tensor derivatives and metric coefficients, one can derive the power series expansions of the governing equations and the boundary conditions with respect to α by operating formally with series (4.106) and equating the coefficients at $\alpha^i \cos kq^2$, $\alpha^i \sin kq^2$ in the left- and the right-hand sides of the equations. Retaining only the first terms of the above expansions, one obtains the axisymmetric flow problem. For the higher order expansion coefficients, linear equations involving two spatial coordinates are obtained, which allows a successive determination of the functions $X_{11}(q^1, q^3)$, $X_{20}(q^1, q^3) \ldots$. The series expansions are performed by the computer simultaneously with a difference approximation of the equations using the procedures which operate with series (4.105). The equations resulting from the series expansions are solved with a two-dimensional version of the numerical algorithm described in subsection 4.2.2.

The above method was applied to investigate supersonic flows around ellipsoids of revolution and spherically blunted cones. Consider some calculation results which pertain to the flow about an ellipsoid with the axis ratio equal to 1.5 at an angle of attack $\alpha = 10°$ relative to the smaller axis. The calculations were carried out for a perfect gas with the specific heat ratio $\gamma = 1.4$, the viscosity $\mu \propto \sqrt{T}$, the Prandtl number $Pr = 0.75$.

Figure 4.61 presents the distribution of the heat flux expansion coefficients over the body surface at two Reynolds numbers. The graphs demonstrate a diminution of the relative contributions of the series expansion terms with increasing power index of α. The accuracy of the above approximate method was checked by comparison with the solution of the three-dimensional problem. Maximum discrepancies in the heat flux and skin friction coefficient values were found to be less than 7 per cent for the approximate solution including the terms linear in α and less than 3.5 per cent for that with the terms quadratic in α. The discrepancies in the gas dynamic function values in the shock layer do not exceed 3.5 per cent even for the approximate solution including only the linear terms. The CPU-time required to obtain the approximate solution quadratic in α was found to be five times less as compared with the solution of the three-dimensional problem.

4.8.2 Approximate solutions of thin viscous shock layer equations

In this subsection we follow the papers by Gershbein (1978), Gershbein *et al* (1981), Bruikina and Rusakov (1988, 1989), and Tirskii *et al* (1989) which pertain to hypersonic shock layer flows in the vicinity of the three-dimensional stagnation point, the symmetry plane, and about second-order surfaces at zero incidence. The body surface is given for the Cartesian coordinate system by the equation $z = f(x, y)$. The z-coordinate is aligned with the longitudinal body axis. The coordinate origin is located at the stagnation point; the x, y-axes are arranged in the planes of the body main curvatures at this point. The thin viscous shock layer equations are solved using the curvilinear coordinate system q^i $(i = 1, 2, 3)$ in which q^3 is the normal distance from the body surface; q^1, q^2 are chosen to be the Cartesian coordinates of the body surface points. A set of governing equations, along with the bow shock and wall boundary conditions, are written for Dorodnitsin's variables; see subsection 4.4.2.

An approximate analytic solution is the first iteration of the integral iterative method described in detail for a two-dimensional flow in subsection 2.4.2. In accordance with the basic idea of the above method, the equations for the tangent velocity components u^α $(\alpha = 1, 2)$ and the energy balance equation are integrated twice in ζ subject to the proper boundary conditions, first from ζ to 1 and then from 0 to ζ. The normal momentum balance equation and the equations for the tangent components of the pressure gradient p_α are integrated from ζ to 1 subject to the bow shock boundary conditions. The integral-differential equations thus obtained are solved using the iteration algorithm in which the unknown functions in each iteration are expressed through the integrals of the previous iteration data. The analytical solution in the first iteration is obtained using the linear velocity and temperature profiles satisfying the bow shock and wall boundary conditions.

In accordance with the determination of the initial data, the above analytical solution is expected to be accurate only at sufficiently small Reynolds numbers. The figures which follow present a comparison of the analytical solutions of the three-dimensional thin viscous shock layer equations with numerical results. The numerical solutions were obtained by methods based on Petukhov's (1964) finite-difference scheme.

Figure 4.62 depicts the variation of the heat transfer coefficient C_H (Stanton number) and the skin friction coefficient C_f^1 at the three-dimensional stagnation point with Reynolds number $Re = \rho_\infty V_\infty R/\mu(T^*)$. Here R is the smaller of

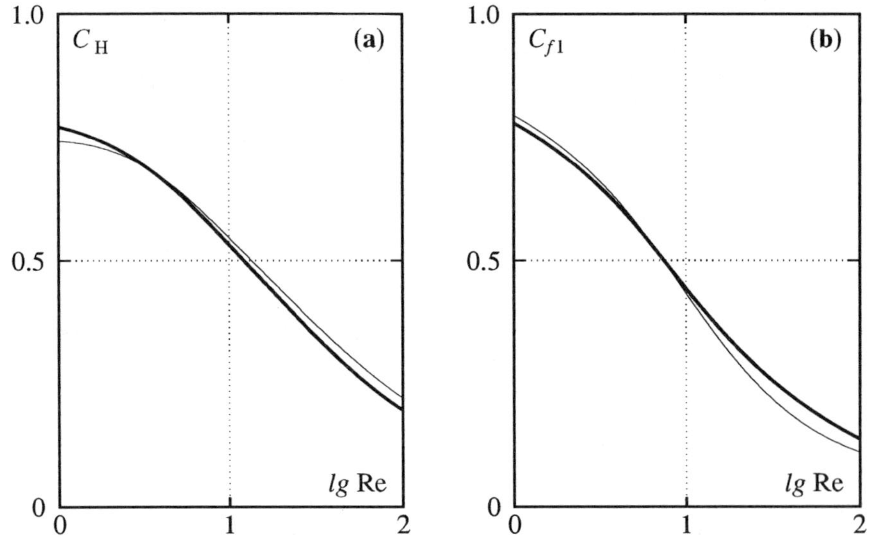

Figure 4.62 Comparison of numerical and analytical solutions of the thin viscous shock layer equations for hypersonic three-dimensional stagnation line flow. $\gamma = 1.25$, $T_w = 0.1T^*$. Full lines, numerical solution; hair lines, analytical solution.

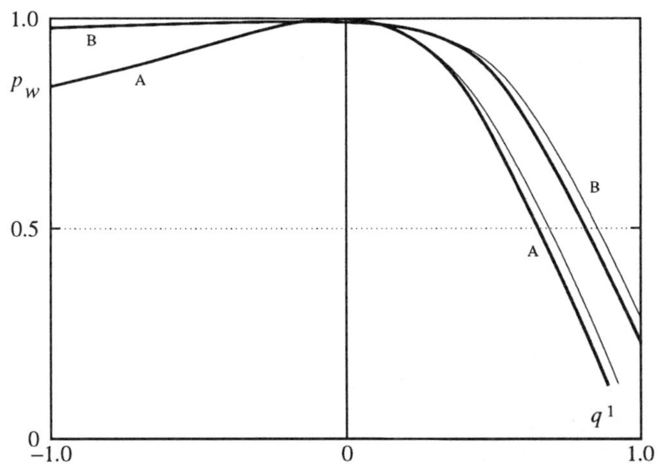

Figure 4.63 Comparison of numerical and analytical solutions of the thin viscous shock layer equations for a hypersonic flow in the symmetry plane of blunt bodies of revolution. $\gamma = 1.4$, $Re = 100$, $T_w = 0.1T^*$, $\alpha = 30°$. Full lines, numerical solution; hair lines, analytical solution.

the main body radii, the main curvature ratio $\kappa = 0.1$, and the shear stress is normalized by $\rho_\infty V_\infty^2$. The results pertain to a perfect gas flow with $\mu \propto \sqrt{T}$, $Pr = 0.71$.

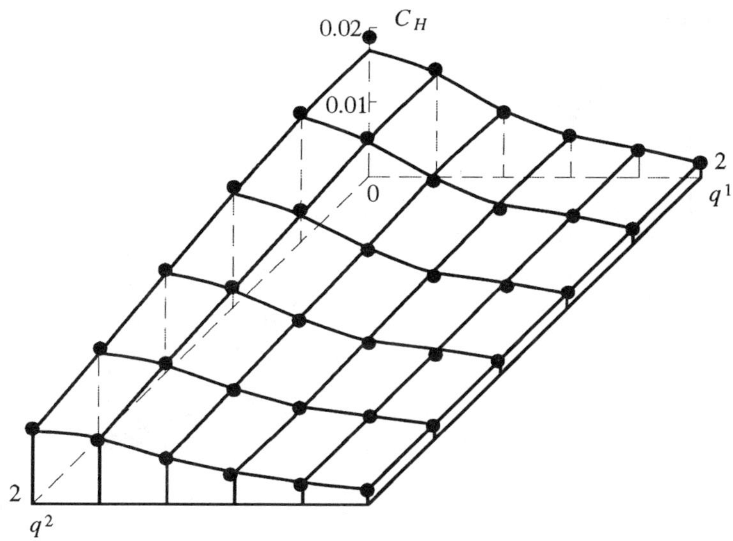

Figure 4.64 Comparison of numerical and analytical results for the heat flux distribution over an elliptic paraboloid in a hypersonic flow. $Re = 100$, $T_w = 0.1T^*$, $\alpha = 0$. Full lines, numerical solution; circles, analytical solution.

The results presented in figure 4.63 pertain to a hypersonic flow in the symmetry plane. The lines A and B show the wall pressure variations on the elliptic paraboloid $z = 0.5(x^2 + 0.8y^2)$ and the hyperboloid $z = (1 + x^2 + y^2)^{1/2}$-1 at the angle of attack $\alpha = 30°$. Here $q^1 = 0$ is the stagnation point coordinate, R is the smaller of body main radii. The results of numerical and analytical solutions are plotted by full and hair lines, respectively.

Figure 4.64 shows the Stanton number distribution over an elliptic paraboloid with the stagnation point main curvature ratio $\kappa = 0.4$ at zero incidence. The flight conditions and the gas properties are the same as in the previous figure. One can see an excellent agreement of numerical and analytical results.

On the basis of the analytical and numerical solutions of the three-dimensional thin viscous shock layer equations, Bruikina and Rusakov (1989) and Tirskii *et al* (1989) suggested approximate formulae for the evaluation of the heat flux distribution over smooth blunt bodies

$$\frac{q}{q^\circ} = \sqrt{\frac{H \cos^3 \alpha}{H_0 \lambda}} \qquad \lambda = 1 + \frac{4}{15} \frac{f''_{11} f'^2_1 + 2f''_{12} f'_1 f'_2 + f''_{22} f'^2_2}{g^{3/2} H} \tag{4.107}$$

Here α is the angle between the oncoming stream velocity vector and the normal to the body surface; H is the local average body surface curvature equal to half the sum of the main curvatures; the function $f(q^1, q^2)$ defines the body shape; the parameter λ takes into account a deviation of the wall pressure distribution from the Newtonian one; the subscript $\circ$ denotes the quantities at the stagnation point.

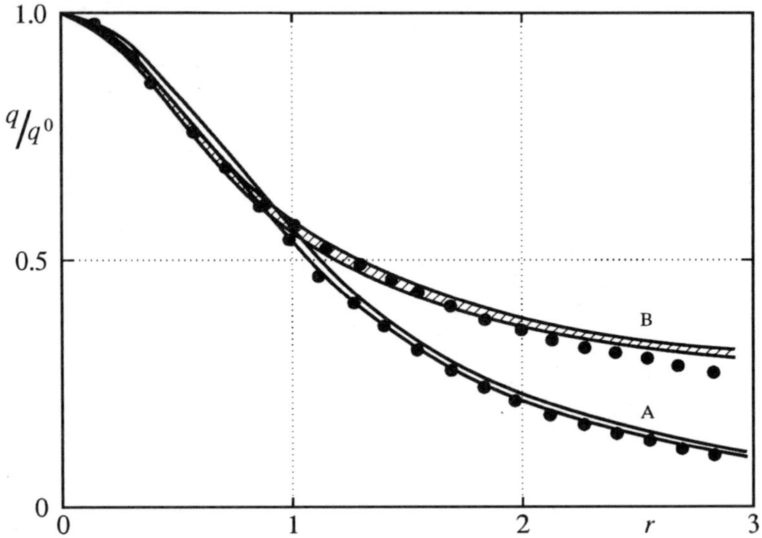

Figure 4.65 Heat flux distributions over smooth three-dimensional bodies in hypersonic flow at zero incidence. Strips, numerical solutions of the thin viscous shock layer equations; circles, formulae (4.107).

For the body-conforming curvilinear coordinate system (4.73), which is usually exploited in the thin viscous shock layer calculations, one has

$$\cos\alpha = \frac{1}{\sqrt{g}} \qquad H = \frac{1}{2g^{3/2}}\left[f''_{11}\left(1+f'^2_2\right) + f''_{22}\left(1+f'^2_1\right) - 2f''_{12}f'_1f'_2\right] \tag{4.108}$$

where

$$g = g_{11}g_{22} - g^2_{12} \qquad g_{11} = 1 + f'^2_1$$

$$g_{12} = f'_1 f'_2 \qquad g_{22} = 1 + f'^2_2 \qquad f'_\alpha = \frac{\partial f}{\partial q^\alpha}$$

Formulae (4.107) were derived for sufficiently high Reynolds number flows ($Re \gtrsim 10^2$) in which the normalized heat flux values are only slightly influenced by the Reynolds number, gas properties and wall temperature. The accuracy of the above formulae is illustrated in figure 4.65 which presents the heat flux distributions along the meridional section $\varphi = 45°$ on an elliptic paraboloid with the stagnation point main curvature ratio $\kappa = 0.4$ and on a hyperboloid with $\kappa = 0.5$ and an asymptotic half-angle in the symmetry plane equal to 40°. The normalized heat flux values obtained from calculations of the hypersonic flows about these bodies at $10^2 \leqslant Re \leqslant 10^4$, $1.15 \leqslant \gamma \leqslant 1.667$, $0.01 \leqslant T_w/T^* \leqslant 0.25$ are within the strips A and B, respectively. The circles show the relative heat flux evaluated by formulae (4.107). The abscissa r represents the distance from the body surface to its longitudinal axis normalized by the smaller of the body main radii at the stagnation point.

4.8.3 Axisymmetric analogy

Within the framework of this approach, the calculation of three-dimensional flows reduces to the solution of the axisymmetric problem. The simplest model of this kind is the equivalent cone approximation applied to estimate the flow characteristics on the windward side of pointed and blunted cones at small angles of attack. In this approximation, the gas parameters in the windward part of the flow symmetry plane are assumed to be the same as in an axisymmetric flow around an equivalent cone, whose half-angle exceeds that of a real cone by the angle of attack. The effects of the gas run from the symmetry plane are ignored. Bureyev and Zinchenko (1991) investigated the accuracy of the equivalent cone approximation by comparison of the results with self-similar solutions of the viscous shock layer equations for the flow in the windward part of the symmetry plane of a spherically blunted cone with half-angle $\theta_c = 5°$. The calculations showed that the equivalent cone approximation and the self-similar solution gave nearly the same values of the shock layer thickness, wall pressure and heat flux to the body surface at the angle of attack $\alpha = 10°$. However, the discrepancy becomes large with α increasing. At $\alpha = 20°$, the equivalent cone approximation resulted in a large error in the prediction of the flow characteristics.

The approximate method of Tirskii *et al* (1989) and Bruikina and Rusakov (1990) is based on solutions of the thin viscous shock layer equations for hypersonic flows about three-dimensional smooth blunt bodies at zero incidence. The results of numerical and approximate analytical solutions indicate that the Stanton number on the lateral part of the body surface is determined, in the main, by two parameters, i.e. the angle α between the oncoming stream velocity vector and the normal to the body surface and the combination $Re/(H \cos\alpha)$ where H is the local average body surface curvature. The quantities $\cos\alpha$ and H are found from (4.108).

For any line drawn on the body surface through the stagnation point, the equivalent axisymmetric body is sought with the same variation of the angle $\tilde{\alpha}$ between the normal to its generatrix and the oncoming stream direction as a variation of the α-angle along the chosen line on the three-dimensional body considered. Provided the values of the second of the above parameters coincide, i.e. if

$$\tilde{Re}/\tilde{H} \cos\tilde{\alpha} = Re/H \cos\alpha, \tag{4.109}$$

the heat flux along the chosen line on the three-dimensional body is assumed to be equal to that on the equivalent axisymmetric body. Since $\tilde{\alpha} = \alpha$, it follows from (4.109) that in calculating the heat flux on an axisymmetric body, one has to use the quantity

$$\tilde{Re} = Re\tilde{H}/H \tag{4.110}$$

as the Reynolds number. Here $\tilde{H}$ is the local average curvature of the equivalent axisymmetric body.

It is convenient to choose, as reference lines on the surface of a three-dimensional body, its sections with the meridional planes and to use the cylindrical coordinates

(z, r, φ) for the construction of an effective axisymmetric body. Let the surface of the equivalent axisymmetric body be defined for the cylindrical coordinates by the equation $\tilde{r} = \tilde{r}(z)$. Then

$$\cos\tilde{\alpha} = \frac{\tilde{r}'_z}{\sqrt{1+\tilde{r}'^2_z}} \qquad \tilde{H} = \frac{\tilde{r}\tilde{r}''_{zz} - \left(1+\tilde{r}'^2_z\right)}{2\tilde{r}\left(1+\tilde{r}'^2_z\right)^{3/2}} \tag{4.111}$$

To generate the generatrix of the equivalent axisymmetric body corresponding to a fixed meridional section of the three-dimensional body, first the derivative $\tilde{r}'_z$ is obtained from the equation

$$\frac{\tilde{r}'_z}{\sqrt{1+\tilde{r}'^2_z}} = \frac{1}{\sqrt{g}} \tag{4.112}$$

Integration and differentiation of the above quantity give the function $\tilde{r}(z)$ and the second derivative $\tilde{r}''_{zz}$ which are necessary to evaluate the curvature $\tilde{H}$.

Thus, the heat flux (Stanton number) distribution along the meridional section $q^1 = q^1(q^2)$ of the three-dimensional body is determined from the relationship

$$C_H(Re) = \tilde{C}_H(\tilde{Re}) \tag{4.113}$$

where $\tilde{C}_H$ is the Stanton number obtained from the axisymmetric flow calculations in which the constant Reynolds number Re is replaced by the variable Reynolds number $\tilde{Re} = Re\tilde{H}/H$ depending on the local geometric characteristics of the three-dimensional body.

At high Reynolds numbers, the heat flux on the body surface is inversely proportional to $\sqrt{Re}$. In this case, the axisymmetric analogy takes the form

$$C_H = \sqrt{\frac{H}{\tilde{H}}}\tilde{C}_H \tag{4.114}$$

Here the Stanton number $\tilde{C}_H$ is defined at the same constant Reynolds number Re as the Stanton number C_H.

The above axisymmetric analogy was applied by Tirskii *et al* (1989) to calculate the heat flux on elliptic paraboloids, hyperboloids and triaxial ellipsoids. The results were compared with numerical solutions of the three-dimensional thin viscous shock layer equations. A comparison proved that the error of the approximate formula (4.113) did not exceed 15 per cent. The axisymmetric analogy was found to be most accurate in the flows about elliptic paraboloids. This being the case, the accuracy of the approximate formula is within some per cent and nearly independent of such flow parameters as Reynolds number, gas specific heat ratio, and wall temperature.

Figure 4.66 shows the Stanton number distributions along several meridional sections $\varphi = const$ of an elliptic paraboloid with the body main curvature ratio at the stagnation point $\kappa = 0.25$. Here the gas specific heat ratio $\gamma = 1.4$, the

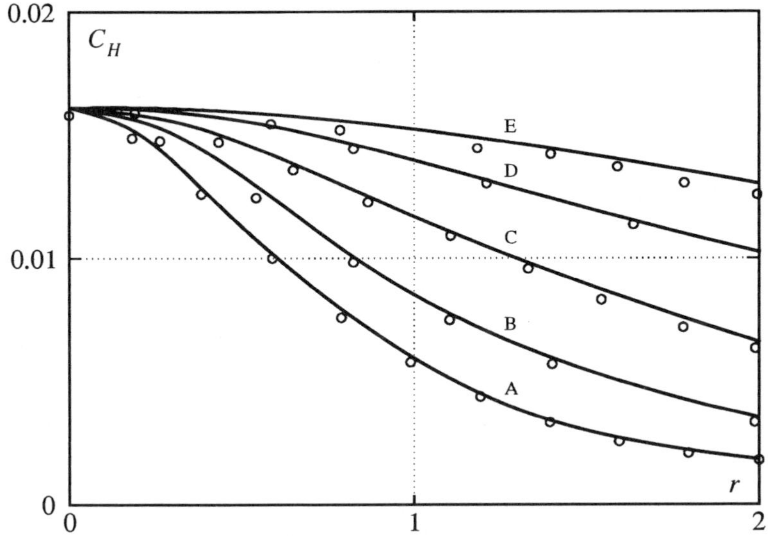

Figure 4.66 Distribution of the heat transfer parameter on an elliptic paraboloid in a hypersonic flow. $Re = 10^4$, $T_w = 0.1T^*$, $\alpha = 0$; A–E, $\varphi = 0, 45°, 63.4°, 76°$ and $90°$. Lines, numerical solution of thin viscous shock layer equations; circles, axisymmetric analogy.

viscosity $\mu \propto \sqrt{T}$, the Prandtl number $Pr = 0.71$. The abscissa r is the distance from the body surface to its axis which coincides with the z-axis of the cylindrical coordinate system (z, r, φ). The distance r is normalized by the smaller of the body main radii at the stagnation point.

Thus, the above approximate method allows a rather accurate estimation of the heat flux values on three-dimensional bodies using the codes elaborated for axisymmetric flow calculations. In conclusion, it is worth noting that in contrast to the axisymmetric analogy for the boundary layer flows (Shevelev 1986), the above approach is not needed in preliminary calculations of the inviscid flow and applies both at high and low Reynolds numbers.

5

Physical and chemical effects

At high supersonic flight speed, gas heating within the bow wave gives rise to such physical and chemical processes as excitation of the internal energy of the molecules, dissociation, chemical reactions, ionization, and radiation. Intensive aerodynamic heating of the vehicle often results in destruction of the heat protection material. The chapter which follows deals with numerical simulation of the viscous shock layer flows coupled with the above physical and chemical processes which considerably influence the shock layer flow field and heat transfer. The examples of calculations are presented pertaining to spacecraft flight in the Earth and other planet atmospheres.

5.1 MATHEMATICAL DESCRIPTION OF HIGH-TEMPERATURE FLOWS

Basic equations of multicomponent chemically reacting and radiating gas mixtures have been discussed in detail in a number of monographs; see for example Hirschfelder *et al* (1954), Shih I Pai (1966), Zeldovich and Rizer (1966), Lapin and Strelets (1989), Pilugin and Tirskii (1989). So, we shall restrict ourselves to reference data required for solving the viscous shock layer problems.

5.1.1 Governing equations

Let us consider, at first, the governing equations for non-radiating shock layer flows which happen at a moderate hypersonic flight speed. Treating the shock layer gas as a reacting multicomponent mixture of perfect gases, ignoring ionization and assuming equilibrium excitation of the internal energy of the molecules, one can use the set of equations which includes the mass, momentum and energy balance equations for the mixture, the mass balance (continuity) equations for the species and the equation of state. The mass, momentum and energy balance equations can be written as follows

$$\frac{\partial \rho}{\partial t} + \nabla \cdot (\rho \boldsymbol{V}) = 0 \qquad (5.1)$$

$$\frac{\partial}{\partial t}(\rho c_i) + \nabla \cdot (\rho \boldsymbol{V} c_i + \boldsymbol{J}_i) = \dot{w}_i \ (i = 1, ..., N) \tag{5.2}$$

$$\frac{\partial}{\partial t}(\rho \boldsymbol{V}) + \nabla \cdot (\rho \boldsymbol{V}\boldsymbol{V} - \mathbf{T}) + \nabla p = \rho \sum_{i=1}^{N} c_i \boldsymbol{F}_i \tag{5.3}$$

$$\frac{\partial}{\partial t}(\rho E) + \nabla \cdot [(\rho E + p)\boldsymbol{V} - \mathbf{T} \cdot \boldsymbol{V} + \boldsymbol{q}] = \rho \sum_{i=1}^{N} c_i \boldsymbol{F}_i \cdot (\boldsymbol{V} + \boldsymbol{V}_i) \tag{5.4}$$

The equation of state reads

$$p = \frac{\rho R T}{m} \tag{5.5}$$

Here $\mathbf{T}$ is the viscous stress tensor, $\boldsymbol{F}_i$ is the external mass force, $E = e + V^2/2$ is the total specific energy of the mixture, e is the specific internal energy, $c_i = \rho_i/\rho$ is the mass fraction of the i-th species, $\dot{w}_i$ is the mass production rate of the i-th species due to chemical reactions, $\boldsymbol{V}$ is the mass-averaged mixture velocity, $\boldsymbol{V}_i$ is the diffusion velocity of the i-th species, $\boldsymbol{J}_i = \rho_i \boldsymbol{V}_i$ is the mass diffusion flux of the i-th species, N is the total number of species, and R is the universal gas constant. The molar mass of the mixture m is evaluated from the species molar masses m_i by the formula

$$m = \left(\sum_{i=1}^{N} \frac{c_i}{m_i} \right)^{-1} \tag{5.6}$$

The viscous stress tensor is determined by the Newton law

$$\mathbf{T} = 2\mu \mathbf{S} - \left(\frac{2}{3} \mu \mathrm{div} \boldsymbol{V} \right) \mathbf{I} \tag{5.7}$$

where $\mathbf{S}$ is the tensor of rate strains, $\mathbf{I}$ is the unit tensor, μ is the mixture viscosity. The heat flux vector for a multicomponent mixture of chemically reacting gases is written as follows

$$\boldsymbol{q} = -\lambda \nabla T + \sum_{i=1}^{N} h_i \boldsymbol{J}_i \tag{5.8}$$

where λ is the mixture heat conductivity, h_i is the specific enthalpy of the i-th species including the production energy. Formula (5.8) omits the diffusion thermo-effect whose contribution to the energy transfer is usually negligible for the flows under study.

The mass production rates of the species are determined by the law of acting masses

$$\dot{w}_i = m_i \sum_{(s)} (\nu''_{is} - \nu'_{is}) \left[k_{fs} \prod_{j=1}^{N} \left(\frac{\rho}{m_j} c_j \right)^{\nu'_{js}} - k_{rs} \prod_{j=1}^{N} \left(\frac{\rho}{m_j} c_j \right)^{\nu''_{js}} \right] \tag{5.9}$$

where ν'_i, ν''_i are the stoichiometric coefficients and k_f, k_r are the rate constants of the forward and reverse processes. Summation is supposed to be made over all

s reactions with the i-th species. The rate constants of the forward and reverse reactions are related to each other through the equilibrium constant

$$K_p = \frac{k_f}{k_r}(RT/m)^{\sum_{j=1}^{N}(\nu_j''-\nu_j')}$$

which can be evaluated by the methods of statistic thermodynamics. The results of such calculations are presented, for instance, in the monograph by Gurvich *et al* (1962).

If the characteristic times of all processes considered are by far greater than the flow characteristic time (frozen flow), one can suppose $\dot{w}_i = 0$ ($i = 1, \ldots, N$). This being the case, all the continuity equations become homogeneous. In a general case, it appears to be convenient to replace some of the equations (5.2) by the homogeneous mass balance equations for the chemical elements. The mass fraction c_k^* and mass diffusion flux $\boldsymbol{J}_k^*$ of the k-th element are found as follows

$$c_k^* = \sum_{i=1}^{N} a_{ki}\frac{m_k}{m_i}c_i \qquad \boldsymbol{J}_k^* = \sum_{i=1}^{N} a_{ki}\frac{m_k}{m_i}\boldsymbol{J}_i \tag{5.10}$$

where a_{ki} is the number of the k-th element atoms in the i-th species molecule.

Note that continuity equations (5.2) can be written also for particles of the same sort differing in the internal energy excitation and, for this reason, considered as the independent species. When investigating ionized gas flows, one has to keep in mind that the electron temperature can differ from the heavy particle translational temperature because of a great mass difference. In this case, set (5.1)–(5.5) is complemented by a balance equation for the electron energy. Besides, one has to take into account the electric field caused by the charge separation. When ionization is important, the Debye length is usually small as compared with the flow characteristic length and, therefore, one can exclude the electric field vector from the governing equations exploiting the mixture quasi-neutrality assumption. As a result, the governing equations take the same form as for a mixture of neutral species. When the quasi-neutrality assumption becomes invalid, the set of governing equations has to be complemented with Poisson equation for the electric field; see for instance Chung *et al* (1975), Pilugin and Tirskii (1989).

Taking into account the diffusion and chemistry effects gives the additional similarity criteria, i.e. the Schmidt and Damkohler numbers. The Schmidt number $Sc_i = \mu_i/\rho D_i$ (D_i is the reference diffusivity value) presents a diffusion analogue of the Prandtl number, characterizing a correlation between the molecular momentum and mass transfer of the i-th species. The Damkohler number $Da_i = (t_f/t_{ch})_i$ is the ratio of the characteristic time of hydrodynamic mass transfer of the i-th spesies $(t_f)_i$ to that of its production due to chemical reactions $(t_{ch})_i$. The limits $Da_i = 0$ and $Da_i = \infty$ correspond to the frozen and equilibrium chemistry, respectively. Sometimes, the Damkohler number is introduced which presents the ratio of the energy release in chemical reactions to the reference gas enthalpy.

Physical and chemical processes occurring on the body surface are taken into account with proper wall boundary conditions which present the balance of mass and energy fluxes. When calculating ionized and rarefied flows, the wall boundary conditions take into account violation of the mixture quasi-neutrality and continuity within the near-wall Knudsen layer. A conventional way to formulate the wall boundary conditions for such problems is an approximate consideration of the transport processes within the near-wall region on the basis of the kinetic theory and matching the results with those obtained using the continuous medium approach for the outer flow.

If the computational domain is bounded by a bow shock wave, the proper boundary conditions are given by Rankine–Hugoniot relationships. In a non-equilibrium flow, the characteristic length of physical and chemical processes is usually large as compared with the bow shock thickness which is equal, within an order of magnitude, to several mean free paths of the molecules. For this reason, the modified Rankine–Hugoniot relationships can be written assuming the physical and chemical processes to be frozen within the bow shock wave. In the bow shock capturing methods, the outer boundary conditions are given by the free stream parameters.

5.1.2 Diffusion fluxes and molecular transport coefficients

The mass diffusion flux vectors and transport coefficients of multicomponent mixtures can be obtained using the molecular kinetic theory. To evaluate the mass diffusion flux vectors, Stephan–Maxwell relationships are usually applied, which can be written in the following form (Pilugin and Tirskii 1989):

$$\sum_{j=1}^{N} \frac{m^2}{\rho m_i m_j D_{ij}(1) f_{ij}(\xi)} (c_i \boldsymbol{J}_j - c_j \boldsymbol{J}_i) = \nabla\left(\frac{m}{m_i} c_i\right)$$
$$+ \left(\frac{m}{m_i} - 1\right)\frac{c_i}{p}\nabla p - \frac{\rho c_i}{p}\left(\boldsymbol{F}_i - \sum_{j=1}^{N} c_j \boldsymbol{F}_j\right) + k_{Ti}(\xi)\frac{\nabla T}{T} \tag{5.11}$$
$$i = 1, 2, \ldots, N-1,$$
$$\sum_{i=1}^{N} \boldsymbol{J}_i = 0$$

Here $D_{ij}(\xi)$ are the binary diffusion coefficients, $k_{Ti}(\xi)$ are the thermo-diffusion ratios which can be expressed through the binary diffusion coefficients, thermo-diffusion coefficients, concentrations and molar masses of the species; ξ is the number of retained terms of the Sonine polynomial series expansion for the perturbed partition function. The coefficients $k_{Ti}(\xi)$ and the factors $f_{ij}(\xi)$ are found from the sets of linear algebraic equations of order $N(\xi - 1)$ and are given by their determinant ratios.

For a mixture of neutral species, it proves to be quite sufficient to retain only the first terms of the Sonine expansions providing non-zero values of the transport coefficients. The minimum number of retained terms is equal to one for the diffusion coefficients and to two for the thermo-diffusion coefficients. This being the case, relationships (5.11) are equivalent to those presented by Hirschfelder *et al* (1954). The ionized mixtures contain charged particles with greatly different masses and obeying Coulomb interaction law. In this case, the Sonine polynomial series converge by far slower and the transport coefficients have to be calculated taking into account the higher-order terms (Devoto 1966, 1976). Such calculations were carried out, for instance, by Sokolova and Tirskii (1987).

The kinetic theory of gas mixtures gives the following formula for the binary diffusion coefficients D_{ij} (1):

$$D_{ij}(1) = 0.843 \times 10^{-4} \sqrt{\frac{T^3(m_i + m_j)}{2m_i m_j}} \Big/ p\overline{\Omega}_{ij}^{(1,1)}(T) \frac{\mathrm{cm}^2}{\mathrm{s}} \tag{5.12}$$

where m_i is the molar mass in kg/mol, p is the mixture pressure in kPa, T is the temperature in Kelvins, $\overline{\Omega}_{ij}^{(1,1)}$ are the reduced particle collision integrals in nm^2. The collision integrals depend on the particle interaction law which is described with one or another model of the interaction potential.

In the shock layer flows, baro-diffusion and thermo-diffusion are usually small as compared with mass transfer due to species concentration gradients. Ignoring these processes and mass forces and calculating the binary diffusion coefficients with only the first term of Sonine series expansion, the Stephan–Maxwell relationships can be written as follows

$$\sum_{j=1}^{N} \frac{m^2}{\rho m_i m_j D_{ij}(1)} (c_i \boldsymbol{J}_j - c_j \boldsymbol{J}_i) = \nabla \left(\frac{m}{m_i} c_i \right)$$

$$i = 1, 2, \ldots, N-1, \tag{5.13}$$

$$\sum_{i=1}^{N} \boldsymbol{J}_i = 0$$

It is seen from (5.11) or (5.13) that one has to inverse the matrix of order $(N-1)$ to obtain the explicit formulae for the diffusion flux vectors. It requires a large amount of computations if the number of species is not small. Hence, approximate methods are often used to describe the diffusion in multicomponent mixtures with a large number of species.

An analysis of the diffusion matrix shows that its non-diagonal elements are proportional to the difference between the molar masses and diffusion coefficients of the species. In the simplest approximate methods, the non-diagonal elements of the diffusion matrix are ignored. In this case, the diffusion mass flux of the i-th species is written in Fick law form

$$\boldsymbol{J}_i = -\rho D_i \nabla c_i \tag{5.14}$$

where the effective diffusion coefficient D_i is supposed to be the same for all species (binary model) or is evaluated with the formula (Wilke 1950a)

$$D_i = \left(\frac{1}{m} - \frac{c_i}{m_i}\right) \Bigg/ \sum_{\substack{j=1 \\ j \neq i}}^{N} \frac{c_j}{m_j D_{ij}} \tag{5.15}$$

A drawback of the above approach is that it does not ensure the identity $\sum_{(j)} \boldsymbol{J}_j = 0$, which may violates the mass balance for the species and mixture on the whole.

The binary model can be generalized in the following way. All species are divided into several groups. The molar masses and binary diffusion coefficients of the species attributed to the same group are assumed to be the same and determined by averaging over all species of the group. This simplifies the structure and accordingly the inversion of the diffusion matrix allowing at the same time to take into account its non-diagonal elements.

The diffusion flux of each species can be expressed from (5.13) through the concentration gradient of this species and the diffusion fluxes of all other species. Therefore, the multicomponent diffusion can be evaluated using the iteration procedure in which the diffusion fluxes of all other species are calculated with the previous iteration data. This procedure, which proves to be especially effective for the flows of gas mixtures with close diffusion properties, can be combined with the iterations necessitated by a non-linearity of the governing equations.

In accordance with the kinetic theory, the viscosity and heat conductivity of a multicomponent mixture is given by the high-order determinant ratio. In practice, the approximate formulae are usually applied. The wide-spread formulae were suggested by Wilke (1950b) and Mason and Saxena (1958)

$$\mu = \sum_{i=1}^{N} \mu_i \Bigg/ \sum_{k=1}^{N} G_{ik} \frac{c_k}{c_i} \frac{m_i}{m_k} \tag{5.16}$$

$$\lambda = \sum_{i=1}^{N} \lambda_i \Bigg/ \left(1.065 \sum_{k=1}^{N} G_{ik} \frac{c_k}{c_i} \frac{m_i}{m_k} - 0.065\right) \tag{5.17}$$

where

$$G_{ik} = \left[1 + \sqrt{\frac{\mu_i}{\mu_k}\sqrt{\frac{m_k}{m_i}}}\right]^2 \Bigg/ \sqrt{8\left(1 + \frac{m_i}{m_k}\right)}$$

For the viscosity and heat conductivity of the species, the kinetic theory gives

$$\mu_i = 0.844 \times 10^{-6} \sqrt{T m_i} \Big/ \overline{\Omega}_i^{(2,2)}(T) \quad \mathrm{Pa \cdot s} \tag{5.18}$$

$$\lambda_i = \frac{15}{4} R_i^* \mu_i \left(0.115 + 0.354 \frac{c_{pi}}{R_i^*}\right) \tag{5.19}$$

In formula (5.18), the molar mass, temperature and reduced collision integral $\overline{\Omega}_i^{(2,2)}$ are taken in the same units as in (5.12). In formula (5.19), R_i^* is the specific gas constant of the i-th component, and the second term in the brackets presents the Eucken correction for an equilibrium excitation of the particle internal energy.

The reduced collision integrals are the products

$$\overline{\Omega}_{ij}^{(1,1)} = \sigma_{ij}^2 \Omega_{ij}^{(1,1)} \left(T_{ij}^*\right) \qquad \overline{\Omega}_{i}^{(2,2)} = \sigma_{i}^2 \Omega_{i}^{(2,2)} \left(T_{i}^*\right) \tag{5.20}$$

in which σ_i, σ_{ij} are the effective collision diameters; $T_i^* = kT/\varepsilon_i$, $T_{ij}^* = kT/\varepsilon_{ij}$ are the characteristic temperatures; ε_i/k, ε_{ij}/k are the parameters of the particle interaction potential; $\Omega_i^{(2,2)}$ and $\Omega_{ij}^{(1,1)}$ are the collision integrals for the momentum and mass transfer characterizing the deviation from the rigid sphere model for which $\Omega_i^{(2,2)} = \Omega_{ij}^{(1,1)} \equiv 1$; k is the Boltzmann constant. The values of σ_i, ε_i for several kinds of molecules were presented by Hirschfelder *et al* (1954) and Lapin and Strelets (1989). The extended data on more than two hundred kinds of molecules were gathered by Svehla (1961). The parameters σ_{ij}, ε_{ij} can be calculated with the combinatorial formulae

$$\sigma_{ij} = \frac{1}{2}\left(\sigma_i + \sigma_j\right) \qquad \varepsilon_{ij} = \sqrt{\varepsilon_i \varepsilon_j} \tag{5.21}$$

The collision integrals $\Omega_i^{(2,2)}$ and $\Omega_{ij}^{(1,1)}$ for the widely-used Lennard–Jones potential can be evaluated with the following approximations (Anfimov 1962)

$$\Omega_{ij}^{(1,1)} = 1.074 \left(T_{ij}^*\right)^{-0.1604} \qquad \Omega_i^{(2,2)} = 1.157 \left(T_i^*\right)^{-0.1472} \tag{5.22}$$

In numerical investigations, it is more convenient to use the reduced collision integrals $\overline{\Omega}_i^{(2,2)}$ and $\overline{\Omega}_{ij}^{(1,1)}$. Their values for high-temperature gas mixtures corresponding to the composition of planetary atmospheres were presented by Yun and Mason (1962) and Kalinin *et al* (1971).

5.1.3 Equilibrium flow

If the characteristic times of physical and chemical processes are small as compared with the flow characteristic time, i.e. all Damkohler numbers $Da_i \gg 1$, one can assume the shock layer gas to be in local thermodynamic equilibrium. This substantially simplifies the calculation of multicomponent reacting flows. This being the case, $(N - L)$ continuity equations are replaced by the chemical equilibrium relationships

$$\prod_{j=1}^{N} \left(\frac{c_j m}{m_j}\right)^{\nu_j'' - \nu_j'} = \frac{K_{pi}(T)}{p^\alpha} \tag{5.23}$$

where $\alpha = \sum\limits_{j=1}^{N} \left(\nu_j'' - \nu_j'\right)$, L is the number of basic species (chemical elements), $K_{pi}(T)$ are equilibrium constants of the basic chemical reactions, $i = 1, \ldots, N - L$.

With relationships (5.23), the concentrations of all species are determined by the mixture temperature, pressure and the chemical element concentrations which are obtained, as before, from the homogeneous continuity equations. This allows one to exclude the concentration gradients ∇c_i and diffusion fluxes $\boldsymbol{J}_i$ of $(N - L)$ species from the Stephan–Maxwell relationships (5.11) and heat flux formula (5.8), expressing the above quantities through the concentration gradients ∇c_j^* and diffusion fluxes $\boldsymbol{J}_j^*$ of the elements $(j = 1, \ldots, L)$ and the mixture temperature and pressure gradients. As a result, the number of independent Stephan–Maxwell relationships, which determine now only the element diffusion fluxes, reduces to $L - 1$ and the heat flux vector can be written in the form (Pilugin and Tirskii 1989):

$$\boldsymbol{q} = -\lambda_t \nabla T + \sum_{j=1}^{L} b_j^* \boldsymbol{J}_j^* \tag{5.24}$$

where $\lambda_t = \lambda + \lambda_{ch}$ is the total heat conductivity taking into account the heat transfer due to the mixture heat conductivity proper and the diffusion of the reacting species. The coefficients λ, λ_{ch}, b_j^* are expressed by rather cumbersome formulae which can be found in the monograph by Pilugin and Tirskii (1989).

Further simplification of the mathematical description of equilibrium flows is achieved with a binary diffusion model which provides quite accurate results for a number of practical applications. Within this model, the element concentrations in the flows without mass sources are the known constant quantities and the diffusion fluxes of all elements are equal to zero. So the chemical composition of an equilibrium multicomponent mixture, its thermodynamic properties and transport coefficients turn to be functions of only two variables, for instance temperature and pressure. For a number of gas mixtures of interest, there are analytical approximations of the equilibrium composition and thermodynamic functions; see, for example, Hansen (1959), Mikhailov (1961), Kraiko (1964), Sinchenko (1968). The calculations of the viscosity and total heat conductivity for some equilibrium gas mixtures were carried out by Hansen (1959), Yos (1963), Zdunkevich and Leonas (1972), Sokolova (1973). Within the framework of the above approach, the set of governing equations for a flow with equilibrium physical and chemical processes is written in the same form as for a non-reacting gas flow. The distinction lies in the use of more complicated formulae (or tables) for the internal energy, enthalpy, viscosity, and heat conductivity which become the functions of not only temperature but pressure as well. Instead of temperature and pressure, one can choose another pair of independent thermodynamic variables.

5.1.4 Radiation transfer

Investigation of radiating shock layer flows is generally carried out using the concept of geometric optics in which electromagnetic wave propagation is considered as a propagation of rays with fixed energy and direction $\boldsymbol{\Omega}$, each contributing to the total radiation. In the above approach, the radiation is characterized by specific intensity $I_\nu(\boldsymbol{r}, \boldsymbol{\Omega}, t)$. The quantity $I_\nu d\nu d\boldsymbol{\Omega}$ presents the energy transfer per unit time in the spectral interval $d\nu$ within the solid angle $d\boldsymbol{\Omega}$ by the vector $\boldsymbol{\Omega}$ through the unit area normal to $\boldsymbol{\Omega}$ at the point with radius-vector $\boldsymbol{r}$. The radiation intensity being known, one can evaluate the radiative heat flux, pressure, and energy of the radiation field. The estimates presented, for instance, by Zeldovich and Rizer (1966), indicate that in aerodynamic problems the pressure and energy of the radiation field are negligible as compared with the pressure and energy of a gas. However, radiative and gas dynamic heat fluxes can be of the same order of magnitude due to a large value of the light speed. For this reason, radiation can substantially influence the energy balance of the shock layer and body surface heating.

It follows from the above that an account of gas radiation in the flow analysis reduces to an evaluation of the radiative heat flux which is expressed through the specific intensity by the formula

$$\boldsymbol{q}_r = \int_0^\infty \boldsymbol{q}_{r\nu} d\nu = \int_0^\infty \int_{(\boldsymbol{\Omega})} I_\nu \boldsymbol{\Omega} d\boldsymbol{\Omega} d\nu \tag{5.25}$$

The specific intensity I_ν is obtained from the radiation transport equation

$$\frac{1}{c}\frac{\partial I_\nu}{\partial t} + \boldsymbol{\Omega} \cdot \nabla I_\nu = \kappa_\nu (S_\nu - I_\nu) + \beta_\nu \int_{(\boldsymbol{\Omega}')} I_\nu(\boldsymbol{\Omega}') \gamma_\nu(\boldsymbol{\Omega}, \boldsymbol{\Omega}') d\boldsymbol{\Omega}' \tag{5.26}$$

Here c is the light speed, κ_ν is the volumetric absorption coefficient including the induced radiation, S_ν is the source function, β_ν is the scattering coefficient, γ_ν is the scattering indicatrice. Since the light speed is by far greater than the gas velocity, one can consider radiation transfer as a quasi-stationary process and neglect the first term of the left-hand side of (5.26). The scattering is important if the shock layer contains the particles whose dimensions are comparable with the radiation wavelength. Such particles can appear in the shock layer because of destruction of the heat protection materials. In most shock layer flows, the scattering processes can be neglected and one can use the radiation transport equation (5.26) without the second term of its right-hand side.

The source function S_ν is determined by the population of the energy states of the particles and by the probabilities of the radiation transitions. In local thermodynamic equilibrium, it is the equilibrium radiation intensity B_ν evaluated at the local temperature

$$S_\nu = B_\nu = \frac{2h\nu^3}{c^2}\left(e^{h\nu/kT} - 1\right)^{-1} \tag{5.27}$$

where h and k are the Planck and Boltzmann constants, respectively.

To obtain the absorption coefficient, a detailed analysis of the emission and absorption processes in the atomic systems is required. The methods for calculating the absorption coefficients of high-temperature gases are discussed, for instance, in the monographs by Zeldovich and Rizer (1966), Avilova *et al* (1970), Kamenschikov *et al* (1971), Pilugin and Tirskii (1989). The spectral absorption coefficients of equilibrium mixtures with a constant chemical element composition can be calculated as functions of temperature and pressure. The results of such calculations were presented by Avilova *et al* (1970) and Kamenschikov *et al* (1971).

Equation (5.26) is written for the intensity of radiation with frequency ν, which spreads in the direction $\boldsymbol{\Omega}$, while the energy balance equation of a radiating gas involves a divergence of the total radiative flux integrated over the spectrum and the solid angle. Because of the complexity of the spectral and spatial distributions of the shock layer radiation of gas mixtures of interest, the integration is often carried out using approximate methods. Thus, for example, to take into account the wave length dependence of the radiation field, the whole spectral range is divided into N intervals, each being characterized by an average absorption coefficient independent of the wave length. Spatial distribution of the radiation field can be described within the local one-dimensional tangent slab model which seems well appropriate to the problems under study since the function variation occurs mainly across the shock layer. In this model, the radiating volume is replaced by an infinite one-dimensional slab whose thickness and gas parameter distributions are the same as those of the shock layer on the computational ray considered. An applicability of this model was substantiated by Cheng and Vincenti (1967) and Lebedev and Fomin (1969) by a comparison of the results calculated with one- and two-dimensional radiation transport equations. An accuracy of the local one-dimensional slab approximation is also discussed in the monograph by Pilugin and Tirskii (1989).

In the above model, the solution of the radiation transport equation subject to zero radiation intensity of the slab boundaries yields the following formulae for the radiative heat flux and its divergence (Shih I Pai 1966)

$$\boldsymbol{q}_r = 2\pi \int_0^\infty \int_0^{\tau_{\nu s}} S_\nu E_2\left(|\tau_\nu - t_\nu|\right) dt_\nu d\nu \tag{5.28}$$

$$\mathrm{div} \cdot \boldsymbol{q}_r = 2\pi \int_0^\infty \kappa_\nu \left[2S_\nu - \int_0^{\tau_{\nu s}} S_\nu E_1\left(|\tau_\nu - t_\nu|\right) dt_\nu \right] d\nu \tag{5.29}$$

Here $\tau_\nu = \int_0^n \kappa_\nu dn$ is the optical coordinate for the frequency ν, n is the physical coordinate measured across the shock layer, $\tau_{\nu s}$ is the optical shock layer thickness, E_n is the integro-exponential function

$$E_n(t) = \int_1^\infty w^{-n} \exp(-wt) dw \tag{5.30}$$

The first term in the square brackets of (5.29) represents the emission and the second one the absorption of radiation. The above formula can be easily modified to take into account non-zero boundary conditions for the radiation intensity. Thus, the wall boundary condition can be formulated taking into account reflection of the shock layer radiation and the body surface emission. However, the reflection effect on the shock layer flow field is usually small since the typical heat protection materials are nearly black in the spectral range where the absorption of radiation within the shock layer is significant (Chang and Sutton 1969). The wall radiation is small as compared with the shock layer radiation because of a relatively low wall temperature. Beside that, the maximum wall radiation falls on the wave lengths for which the shock layer is nearly transparent. The oncoming stream radiation is usually negligible. The above considerations substantiate the use of formulae (5.28) and (5.29).

The set of governing equations for radiating gas flows involves two additional similarity criteria.

The first criterion, which is often referred to as the Bouger number, represents an optical thickness of the volume under study. It characterizes the gas-radiation interaction and is defined as a product of the reference flow length L and the reference absorption coefficient value κ_0 (or the ratio of L to the reference radiation free path $l_0 = 1/\kappa_0$)

$$\tau = \kappa_0 L = \frac{L}{l_0} \tag{5.31}$$

If $\tau \ll 1$ for all frequencies, the volume is optically thin. This being the case, the absorption is negligible and the radiation field is isotropic. Using the local thermodynamic equilibrium assumption, the divergence of the integral radiative heat flux is determined by the formula which follows directly from the radiation transport equation

$$\text{div}\boldsymbol{q}_r = 4\kappa_P \sigma T^4 \tag{5.32}$$

where σ is the Stephan–Boltzmann constant and $\kappa_P = \int_0^\infty \kappa_\nu B_\nu d\nu / \int_0^\infty B_\nu d\nu$ is the Planck mean absorption coefficient.

In the opposite limit $\tau \gg 1$ for all frequencies, the volume is optically thick and, because of multiple emission and absorption processes, the radiation energy transfer is similar to heat conductivity. If the local thermodynamic equilibrium assumption is valid, the integral radiative heat flux vector is given by the formula

$$\boldsymbol{q}_r = -\frac{1}{3\kappa_R} \cdot \text{grad} B \tag{5.33}$$

where $B = \sigma T^4/\pi$ is the total equilibrium radiation intensity and

$$\kappa_R = \int\limits_0^\infty \frac{dB_\nu}{dT} d\nu \Big/ \int\limits_0^\infty \frac{1}{\kappa_\nu} \frac{dB_\nu}{dT} d\nu$$

is the Rosseland mean absorption coefficient.

It is worth noting that in the above limits the use of the Planck and Rosseland mean absorption coefficients ensures an exact accounting of radiation dependence on the wave length.

The second additional similarity criterion in the radiation gas dynamics equations characterizes the relative importance of the radiative and gas dynamic energy transfers. This quantity, which is usually denoted by the symbol Γ, is defined as follows

$$\Gamma = \frac{q_{r_0}}{\rho_0 U_0 H_0} \tag{5.34}$$

where q_{r_0} is the reference value of the radiative heat flux; ρ_0, U_0, H_0 are the reference values of the gas density, velocity, and total enthalpy. In accordance with the radiation transport equation, one can assume

$$q_{r_0} = 4\pi\kappa_0 B_0 L = 4\pi B_0 \tau$$

where B_0 is the reference value of the total equilibrium radiation intensity. With this definition of q_{r_0}, the quantity inverse to Γ is sometimes referred to as the Boltzmann number.

If $\Gamma \ll 1$, the radiation effect on the flow field is small and the radiative energy transport can be evaluated using the gas dynamic functions calculated without the radiation. When $\Gamma \gtrsim 1$, the radiation transfer significantly influences the shock flow field.

5.2 NON-EQUILIBRIUM CHEMICALLY REACTING FLOW

An influence of various physical and chemical processes on the shock layer flow field and heat transfer depends on the flight condition range. At first, let us consider the flows about blunt bodies at flight speed $4 \lesssim V_\infty \lesssim 7$ km/s and ambient density $10^{-5} \lesssim \rho_\infty \lesssim 10^{-2}$ kg/m^3, which correspond to the spacecraft flight in the Earth's atmosphere at altitudes $60 \lesssim H \lesssim 100$ km. For $CO_2 + N_2$ mixtures, the same flight conditions correspond to the space vehicle entry into the Martian atmosphere.

Under the above flight conditions, the thermodynamic state of a shock layer gas is determined by collisional processes. The characteristic times of these processes depend on the gas density i.e. on the Reynolds number whose variation along the flight trajectory is governed mainly by the ambient density variation. For the flight conditions considered, the Reynolds number Re_s varies within the range $10 \lesssim Re_s \lesssim 10^4$. With decreasing Reynolds number, both molecular transport and non-equilibrium thermodynamic and chemical effects become important throughout the shock layer. The dependence of the rates of physical and chemical processes on the shock layer parameters allows one to take into account non-equilibrium effects only for some of these processes, assuming the others to be in equilibrium or frozen.

5.2.1 Comment on the computation of non-equilibrium flows

Within the flight condition range considered, the thermodynamic state of the gas mixture in the shock layer can change from the frozen to the equilibrium one. Approaching equilibrium, the calculation of the shock layer flows is hampered by stiffness of the species continuity equations which is caused by increasing reaction rates. The stability of calculation procedures is often ensured using the elements of the Newton method.

To elucidate the problem, let us consider the set of equations

$$\rho v \boldsymbol{C}' - (\rho \mathbf{D} \boldsymbol{C}')' = \dot{\boldsymbol{W}} \tag{5.35}$$

which describes the variation of the species concentrations in a one-dimensional non-equilibrium flow with fixed distributions of the gas dynamic functions. Here ρ is the mixture density, $\boldsymbol{C}$ is the column-vector of the species mass fractions, $\mathbf{D}$ is the diffusion matrix, $\dot{\boldsymbol{W}}$ is the column-vector of the species mass production rates.

Let the numerical solution of set (5.35) be obtained using the central three-point finite-difference scheme. The difference approximation yields the following non-linear set of algebraic equations with a block tridiagonal matrix of coefficients

$$\mathbf{P}_j \boldsymbol{C}_{j+1} + \mathbf{R}_j \boldsymbol{C}_j + \mathbf{Q}_j \boldsymbol{C}_{j-1} = \dot{\boldsymbol{W}}_j \tag{5.36}$$

where j is the number of the grid point. Keeping in mind the above peculiarity of the species continuity equations, the iteration procedure for solving set (5.36) takes into account the dependence of vector $\dot{\boldsymbol{W}}_j$ on the unknown function vector $\boldsymbol{C}_j$ in the current iteration. With that end in view, the Taylor series expansion of the vector $\dot{\boldsymbol{W}}_j$ is exploited

$$\dot{\boldsymbol{W}}_j^n = \dot{\boldsymbol{W}}_j^{n-1} + \left(\frac{\partial \dot{\boldsymbol{W}}}{\partial \boldsymbol{C}}\right)_j^{n-1} (\boldsymbol{C}_j^n - \boldsymbol{C}_j^{n-1}) + \dots \tag{5.37}$$

where n is the iteration number. As a result, one obtains the following set of linear equations, which determines the vector $\boldsymbol{C}$ at the interior mesh points in the n-th iteration

$$\begin{aligned} \mathbf{P}_j \boldsymbol{C}_{j+1}^n + \left[\mathbf{R} - \left(\frac{\partial \dot{\boldsymbol{W}}}{\partial \boldsymbol{C}}\right)^{n-1}\right]_j \boldsymbol{C}_j^n + \mathbf{Q}_j \boldsymbol{C}_{j-1}^n = \dot{\boldsymbol{W}}_j^{n-1} \\ - \left(\frac{\partial \dot{\boldsymbol{W}}}{\partial \boldsymbol{C}}\right)_j^{n-1} \boldsymbol{C}_j^{n-1} \end{aligned} \tag{5.38}$$

With iteration procedure (5.38), the species concentrations can be found without restriction on the reaction rates. When all reactions are frozen, the species concentrations vary due to convection and diffusion. This being the case, (5.38) represents an unconditionally stable implicit finite-difference scheme for solving the relevant homogeneous equation. In the opposite limit of equilibrium chemistry, (5.38) corresponds to the use of the Newton method for the calculation of the mixture composition at each grid point from the non-linear set of chemical equilibrium relationships

$$\dot{\boldsymbol{W}}_j(\boldsymbol{C}) = 0 \tag{5.39}$$

In a general case, the solution of set (5.38) complemented with the boundary conditions can be obtained in each iteration by the block 'progonka' algorithm.

Consider now implementation of the wall boundary conditions. For an N-component mixture they can be written as follows

$$\mathbf{F}\boldsymbol{C}' + \boldsymbol{G}(\boldsymbol{C}) = 0 \tag{5.40}$$

where $\mathbf{F}$ is the $(N-1) \times (N-1)$ matrix, $\boldsymbol{G}$ is the column-vector with $(N-1)$ elements which depend in a non-linear way on the species concentrations on the body surface.

The difference approximation of the derivative $\boldsymbol{C}'$ on the body surface can be carried out using the series expansion

$$(\rho\mathbf{D}\boldsymbol{C}')_{1/2} = (\rho\mathbf{D}\boldsymbol{C}')_0 + \frac{h}{2}(\rho\mathbf{D}\boldsymbol{C}')'_0 + \mathrm{O}(h^2) \tag{5.41}$$

and equation (5.35). Assuming, for simplicity, $v_0 = 0$ and $\dot{\boldsymbol{W}}_0 = 0$, one obtains from (5.41)

$$\boldsymbol{C}'_0 \approx (\rho\mathbf{D})_0^{-1}(\rho\mathbf{D}\boldsymbol{C}')_{1/2} \approx (\rho\mathbf{D})_0^{-1}(\rho\boldsymbol{D})_{1/2}\frac{\boldsymbol{C}_1 - \boldsymbol{C}_0}{h} + \mathrm{O}(h^2) \tag{5.42}$$

where h is the mesh spacing, the indices 0, 1/2, 1 denote the function values on the body surface and at distances $h/2$ and h from the body surface. It is worth noting that formula (5.42) involving the function values at only two grid points ensures the second-order accuracy for the derivative $\boldsymbol{C}'$ on the body surface.

The vector $\boldsymbol{G}$ is linearized using the series expansion of type (5.37). As a result, the difference approximation of the wall boundary condition (5.40) is written with the use of (5.42) as follows

$$\tilde{\mathbf{F}}_1^{n-1}\boldsymbol{C}_1^n + \left[\tilde{\mathbf{F}}_0 + \frac{\partial \boldsymbol{G}}{\partial \boldsymbol{C}_0}\right]^{n-1}\boldsymbol{C}_0^n + \left[\boldsymbol{G} - \frac{\partial \boldsymbol{G}}{\partial \boldsymbol{C}_0}\,\boldsymbol{C}_0\right]^{n-1} = 0 \tag{5.43}$$

Equation (5.43) yields the first 'progonka' coefficients.

The above assumptions $v_0 = \dot{\boldsymbol{W}}_0 = 0$, correspond to an impermeable and non-ablating body with either a fully catalytic or a non-catalytic surface. A similar second-order accurate difference approximation for the derivative $\boldsymbol{C}'_0$ can be derived in a more general case as well.

The outer boundary conditions are formulated supposing the chemical reactions to be frozen within the bow shock wave. Their numerical implementation does not differ from that for a non-reacting gas flow.

In a general case, the above calculation procedure is fulfilled using the block 'progonka' algorithm which requires a large amount of computations for a multicomponent mixture with a large number of species. In the calculation of such flows, a partially implicit approximation of the source terms of the species continuity equations can be expedient, in which the $(\dot{\boldsymbol{W}}_j)_i$ dependence only on the

i-th species concentration is taken into account implicitly in the i-th continuity equation. As a result, the matrix $(\partial \dot{\boldsymbol{W}}/\partial \boldsymbol{C})_j^{n-1}$ becomes diagonal, which allows to solve set (5.38) by successive scalar 'progonka' procedures, provided that a proper description of the multicomponent diffusion and implementation of the boundary conditions are applied.

Another way to solve the stiff set of the relaxation-diffusion species continuity equations is to use operator splitting in the physical processes, see, for example, the paper by Gorinov and Magomedov (1973). In this approach, chemical reactions are taken into account through a separate fractional step operating with a set of ordinary differential equations, which is solved using special methods, such as the method by Gear (1971).

5.2.2 Self-similar solution for the stagnation streamline

Self-similar problems for non-equilibrium shock layer flows are formulated in the same way as for a non-reacting gas with constant heat capacity; see sections 2.5 and 4.6. Below, the self-similar solutions for axisymmetric stagnation line flows are considered. Such calculations have been carried out by many authors, see, for instance, the papers of Chung (1961), Cheng (1963), Chung *et al* (1968), Ladnova (1969), Voronkin (1971), Arkhipov and Polenov (1971), Kumar and Jain (1975), Zalogin (1974). In view of hypersonic flight speed, most calculations were performed within the thin viscous shock layer model. The results have elucidated the kinetics of non-equilibrium processes in the gas mixtures corresponding to the planet atmospheres and the effects of vibrational relaxation, mass injection, wall catalycity, and the Knudsen layer.

As an example, let us dwell on the calculation of a hypersonic flow of a mixture containing by volume 95% CO_2 + 5% N_2 (Golovachov 1978). In these calculations, the chemical kinetics model of Mc Kenzie and Arnold (1967) complemented by reactions with oxygen molecules was employed. The shock layer gas was considered as a 9-component mixture in which the following reactions were going on

$$
\begin{array}{llll}
(1) & \mathrm{CO_2 + M \rightleftarrows CO + O + M} & (7) & \mathrm{N_2 + O \rightleftarrows NO + N} \\
(2) & \mathrm{N_2 + M \rightleftarrows N + N + M} & (8) & \mathrm{CO + N \rightleftarrows CN + O} \\
(3) & \mathrm{O_2 + M \rightleftarrows O + O + M} & (9) & \mathrm{N_2 + C \rightleftarrows CN + N} \\
(4) & \mathrm{NO + M \rightleftarrows N + O + M} & (10) & \mathrm{CO + NO \rightleftarrows CO_2 + N} \\
(5) & \mathrm{CO + M \rightleftarrows C + O + M} & (11) & \mathrm{CN + O \rightleftarrows C + NO} \\
(6) & \mathrm{CN + M \rightleftarrows C + N + M} & (12) & \mathrm{NO + O \rightleftarrows O_2 + N} \\
 & & (13) & \mathrm{CO_2 + O \rightleftarrows CO + O_2}
\end{array}
$$

where M denotes any particle. The reaction rate constants are presented in table 5.1. For the 5th reaction with M being any of the atoms C, O, N or the molecule CO, the recombination rate constant is given. In the formula for the 6th reaction, $[CO_2]_\infty$ is the molar carbon dioxide concentration in the free stream.

The rotational and vibrational states of all the molecules are supposed to be in equilibrium with the gas translational temperature. The equilibrium constants for

Table 5.1

Reaction number	M	Forward reaction rate constant ($cm^3/mol{\cdot}s$)
1	CO_2	$k' = 6k''$
	the rest	$k'' = 1.2 \times 10^{11} T^{1/2} \exp(-36850/T)$
2	C, O, N	$k' = 4.3 \times 10^{22} T^{-3/2} \exp(-113200/T)$
	N_2	$k'' = 4.8 \times 10^{17} T^{-1/2} \exp(-113200/T)$
	CO_2, O_2, CO, NO, CN	$k''' = 1.9 \times 10^{17} T^{-1/2} \exp(-113200/T)$
3	CO_2	$k' = 0.77 \times 10^{17} T^{-1/2} \exp(-59400/T)$
	CO, N_2	$k'' = 0.36 \times 10^{17} T^{-1/2} \exp(-59400/T)$
	O_2, NO, CN, N	$k''' = 1.1 \times 10^{25} T^{-5/2} \exp(-59400/T)$
	C, O	$k'''' = 0.92 \times 10^{20} T^{-1} \exp(-59400/T)$
4	NO, C, O, N	$k' = 20k''$
	the rest	$k'' = 4 \times 10^{20} T^{-3/2} \exp(-75510/T)$
5	CO, C, O, N	$k_r = 2.5 \times 10^{28} T^{-7/2}$, $cm^6/(mol^2 \cdot s)$
	the rest	$k = 3.5 \times 10^{12} T^{1/2} \exp(-92800/T)$
6	CN, C, O, N	$k' = 3.6 \times 10^{19} T^{-1} \exp(-66000/T)$
	the rest	$k'' = 32 \times [CO_2]_{\infty}^{5/2} k'$
7		$k = 7 \times 10^{13} \exp(-38000/T)$
8		$k = 4.3 \times 10^{12} T^{1/2} \exp(-35060/T)$
9		$k = 1.5 \times 10^{11} T^{1/2} \exp(-26000/T)$
10		$k = 10^3 T^2 \exp(-21000/T)$
11		$k = 2 \times 10^{10} T \exp(-28000/T)$
12		$k = 0.32 \times 10^{10} T \exp(-19700/T)$
13		$k = 0.61 \times 10^{12} \exp(-17600/T)$

the basic reactions K_{pi} (atm) and the molar enthalpies of the species h_i (J/mol) are calculated with the following analytical approximations (Mikhailov 1961)

$$
\begin{aligned}
K_1 &= T^{-1/2} \exp\left(22.44 - \frac{65201}{T}\right) \\
K_2 &= 10^{-15} T^2 \exp\left(32.092 - \frac{107180}{T}\right) \\
K_3 &= \exp\left(16.283 - \frac{62006}{T}\right) \\
K_4 &= 10^{-10} T \exp\left(28.598 - \frac{74435}{T}\right) \qquad (5.44)
\end{aligned}
$$

$$K_5 = T^{1/2} \exp\left(12.856 - \frac{131374}{T}\right)$$

$$K_6 = T^{3/2} \exp\left(0.523 - \frac{90000}{T}\right)$$

$$\begin{aligned}
h_{CO_2} &= 35.88T + 26.67\sqrt{(T-273)^2 + 6.22 \times 10^5} \\
&\quad + 0.703\sqrt{(T-273)^2 + 3.84 \times 10^4} \\
&\quad - 4.187\sqrt{(T-773)^2 + 3.84 \times 10^4} - 21735 \\
h_{N_2} &= 35.58T + 4.48\sqrt{(T-800)^2 + 4.47 \times 10^6} \\
&\quad + 1.813\sqrt{(T-1000)^2 + 6 \times 10^4} \\
&\quad + 3.559\sqrt{(T-9000)^2 + 10^6} - 44152 \\
h_{O_2} &= 30.14T + 13.99\sqrt{(T-500)^2 + 3.355 \times 10^6} \\
&\quad - \sqrt{(T-1400)^2 + 10^4} - 24971 \\
h_{CO} &= 47.31T + 9.63\sqrt{(T-10^4)^2 + 2.51 \times 10^6} + 170830 \\
h_{NO} &= 38.94T + 75785 \\
h_{CN} &= 42.71T + 781713 \\
h_C &= 45.64T + 21.98\sqrt{(T-12000)^2 + 6.4 \times 10^5} + 828607 \\
h_O &= 22.19T + 0.837\sqrt{(T-6000)^2 + 7.9 \times 10^5} + 241171 \\
h_N &= 26.8T + 4.187\sqrt{(T-7000)^2 + 3 \times 10^6} + 433354
\end{aligned} \tag{5.45}$$

The mixture viscosity and heat conductivity are calculated using the approximate formulae (5.16)-(5.19). The mass diffusion fluxes are evaluated dividing all species into three groups: (1) CO_2; (2) N_2, O_2, CO, NO, CN; (3) C, O, N and assuming the diffusion properties of the species attributed to the same group to be identical; see subsection 5.1.2. In this case, the Stephan–Maxwell relationships (5.13) involve only three molar masses and six binary diffusion coefficients, which substantially simplifies the inversion of the diffusion matrix and explicit presentation of the diffusion fluxes through the species concentration gradients. The collision integral data are borrowed from the papers by Yun and Mason (1962) and Kalinin *et al* (1971). The average values are presented in table 5.2 in nm^2.

The thin viscous shock layer equations read

$$\begin{aligned}
\frac{\partial}{\partial s}(\rho u s) + \frac{\partial}{\partial n}(\rho v s) &= 0 \\
\rho u \frac{\partial c_i}{\partial s} + \rho v \frac{\partial c_i}{\partial n} + \frac{\partial J_i}{\partial n} &= \dot{w}_i \\
\rho u \frac{\partial u}{\partial s} + \rho v \frac{\partial u}{\partial n} + \frac{dp}{ds} &= \frac{\partial}{\partial n}\left(\mu \frac{\partial u}{\partial n}\right) \\
\rho u \frac{\partial h}{\partial s} + \rho v \frac{\partial h}{\partial n} &= \frac{\partial}{\partial n}\left(\lambda \frac{\partial T}{\partial n} - \sum_{i=1}^{N} h_i J_i\right)
\end{aligned} \tag{5.46}$$

Table 5.2

T (K)	$\overline{\Omega}^{(1,1)}_{1,1}$	$\overline{\Omega}^{(1,1)}_{1,2}$	$\overline{\Omega}^{(1,1)}_{1,3}$	$\overline{\Omega}^{(1,1)}_{2,2}$	$\overline{\Omega}^{(1,1)}_{2,3}$	$\overline{\Omega}^{(1,1)}_{3,3}$	$\overline{\Omega}^{(2,2)}_{1}$	$\overline{\Omega}^{(2,2)}_{2}$	$\overline{\Omega}^{(2,2)}_{3}$
1000	0.1340	0.1060	0.0778	0.0860	0.0755	0.0615	0.1518	0.1044	0.0705
2000	0.1195	0.0950	0.0692	0.0808	0.0653	0.0527	0.1366	0.0941	0.0580
3000	0.1118	0.0882	0.0633	0.0755	0.0578	0.0470	0.1277	0.0880	0.0522
4000	0.1062	0.0833	0.0588	0.0690	0.0508	0.0425	0.1216	0.0810	0.0479
5000	0.1019	0.0794	0.0555	0.0634	0.0453	0.0392	0.1171	0.0748	0.0448
6000	0.0985	0.0763	0.0530	0.0605	0.0425	0.0368	0.1133	0.0714	0.0425
7000	0.0958	0.0740	0.0510	0.0584	0.0404	0.0349	0.1101	0.0690	0.0405
8000	0.0935	0.0721	0.0495	0.0566	0.0388	0.0334	0.1077	0.0669	0.0391
9000	0.0914	0.0704	0.0482	0.0549	0.0375	0.0323	0.1056	0.0653	0.0378
10000	0.0895	0.0687	0.0470	0.0533	0.0365	0.0312	0.1037	0.0638	0.0367
11000	0.0879	0.0673	0.0461	0.0520	0.0355	0.0302	0.1019	0.0623	0.0357
12000	0.0865	0.0660	0.0455	0.0508	0.0345	0.0293	0.1003	0.0610	0.0346
13000	0.0855	0.0650	0.0450	0.0497	0.0336	0.0285	0.0989	0.0599	0.0336
14000	0.0846	0.0641	0.0446	0.0487	0.0329	0.0278	0.0978	0.0590	0.0328
15000	0.0839	0.0634	0.0443	0.0478	0.0324	0.0272	0.0968	0.0582	0.0321

For the stagnation line flow, these equations, rewritten for Dorodnitsin's variables with the use of the Stephan–Maxwell relationships (5.13), take the form

$$\frac{1}{\eta_s^2}(l\varphi')' + f\varphi' - \frac{\varphi^2}{2} + \frac{\rho_\infty}{\rho}\left(1 - \frac{\rho_\infty}{\rho_s}\right) = 0 \tag{5.47}$$

$$\frac{1}{\eta_s^2}(\mathbf{A}\boldsymbol{X}')' + \frac{1}{\eta_s}\mathbf{B}\boldsymbol{X}' + \boldsymbol{D} = 0 \tag{5.48}$$

where

$$\eta = \frac{s u_s}{2\sqrt{\xi}}\int_0^n \rho dn \qquad \xi = \int_0^s \rho_w \mu_w u_s s^2 ds \qquad \zeta = \eta/\eta_s \qquad u_s = \frac{V_\infty s}{R}$$

$$l = \frac{\rho\mu}{\rho_w \mu_w} \qquad \varphi = \frac{u}{u_s} \qquad f = \int_0^\zeta \varphi d\zeta$$

The prime denotes differentiation in ζ, the subscript s denotes the gas parameters just behind the bow shock wave, R is the body nose radius. Matrix equation (5.48) contains the energy balance and species continuity equations. The column-vectors $\boldsymbol{X}$ and $\boldsymbol{D}$ are

$$\boldsymbol{X} = \{T\ c_{CO_2}\ c_{N_2}\ c_{O_2}\ c_{CO}\ c_{NO}\ c_{CN}\ c_C^*\ c_O^*\}^T$$

$$\boldsymbol{D} = \{0\ \dot{w}_{CO_2}\ \dot{w}_{N_2}\ \dot{w}_{O_2}\ \dot{w}_{CO}\ \dot{w}_{NO}\ \dot{w}_{CN}\ 0\ 0\}^T$$

where c_C^*, c_O^* are the mass fractions of the chemical elements. Matrices **A** and **B** are not written out since they are given by evident but rather cumbersome formulae.

The wall boundary conditions for the gas velocity and temperature read

$$\varphi(0) = 0 \qquad T(0) = T_w \tag{5.49}$$

The body surface is supposed to be impermeable and fully catalytic. Therefore, the species concentrations on the body surface are determined by the chemical equilibrium relationships assuming the diffusion fluxes of the chemical elements to be equal to zero. These boundary conditions look like (5.40).

The bow shock boundary conditions are given by the modified Rankine–Hugoniot relationships taking the chemical reactions to be frozen within the bow shock wave. This yields

$$\begin{aligned} &l\varphi' + \eta_s^2 f(1)(\varphi - 1) = 0 \qquad \eta_s = \frac{\rho_\infty}{f(1)}\sqrt{\frac{V_\infty R}{2\rho_w \mu_w}} \\ &\mathbf{A}\boldsymbol{X}' + \eta_s^2 f(1)\mathbf{H}\boldsymbol{X} = \left\{ \begin{matrix} 1 \\ \boldsymbol{C}_\infty \end{matrix} \right\} \end{aligned} \tag{5.50}$$

where $\boldsymbol{C}_\infty$ is the column-vector of the molecular species and elements C, O concentrations in the free stream.

Equations (5.47) and (5.48), subject to boundary conditions (5.49), (5.40) and (5.50), are solved by the finite-difference method similar to that described in subsection 2.5.2 for a non-reacting gas flow. In each iteration, the momentum balance equation is solved first by the scalar 'progonka' algorithm and then the energy balance and species continuity equations are solved together using the block 'progonka' algorithm. The source terms of the continuity equations and the non-linear wall boundary conditions are treated following the methods of subsection 5.2.1. In the bow shock boundary conditions (5.50), the derivatives are approximated by the one-sided difference formulae including the function values at the grid points $j = M$, $M-1$ and $M-2$, where M corresponds to the bow shock. Using the 'progonka' relationships, one can easily derive the following formulae for the function values behind the bow shock wave

$$\varphi_M = \frac{2\Delta\zeta\eta_s^2 f(1) - l_M\left[q_M(p_{M-1} - 4) + q_{M-1}\right]}{2\Delta\zeta\eta_s^2 f(1) + l_M\left[p_M(p_{M-1} - 4) + 3\right]} \tag{5.51}$$

$$\begin{aligned} \boldsymbol{X}_M = &\left\{\mathbf{A}\left[3\mathbf{I} + (\mathbf{P}_{M-1} - 4\mathbf{I})\mathbf{P}_M\right] + 2\Delta\zeta\eta_s^2 f(1)\mathbf{H}\right\}^{-1} \\ &\times \left\{2\Delta\zeta\eta_s^2 f(1)\left\{ \begin{matrix} 1 \\ \boldsymbol{C}_\infty \end{matrix} \right\} - \mathbf{A}\left[\mathbf{Q}_{M-1} + (\mathbf{P}_{M-1} - 4\mathbf{I})\mathbf{Q}_M\right]\right\} \end{aligned} \tag{5.52}$$

where p, q and $\mathbf{P}, \mathbf{Q}$ are the scalar and block 'progonka' coefficients for equations (5.47) and (5.48), respectively; $\mathbf{I}$ is the identity matrix. The calculations were carried out with non-uniform grids generated with coordinate transformation (1.43). In most calculations, the number of grid points was equal to 50. The number of

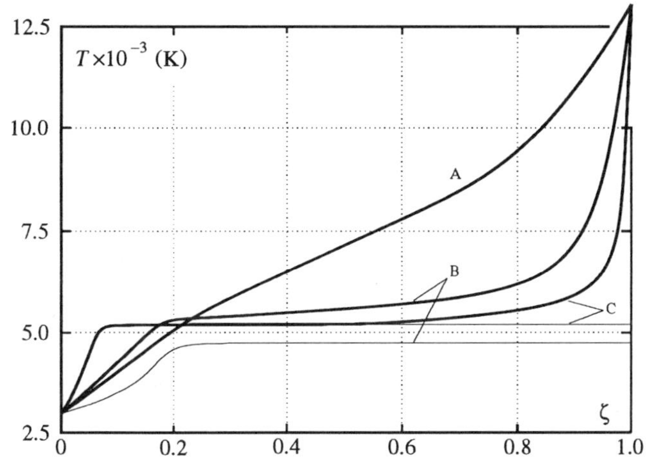

Figure 5.1 Temperature profiles on the stagnation streamline. $95\%CO_2 + 5\%N_2$, $V_\infty = 6$ km/s, $R = 1$ m, $T_w = 2900$K; A, $p = 1$ kPa; B, $p = 10$ kPa; C, $p = 100$ kPa. Dashed curves, local equilibrium assumption.

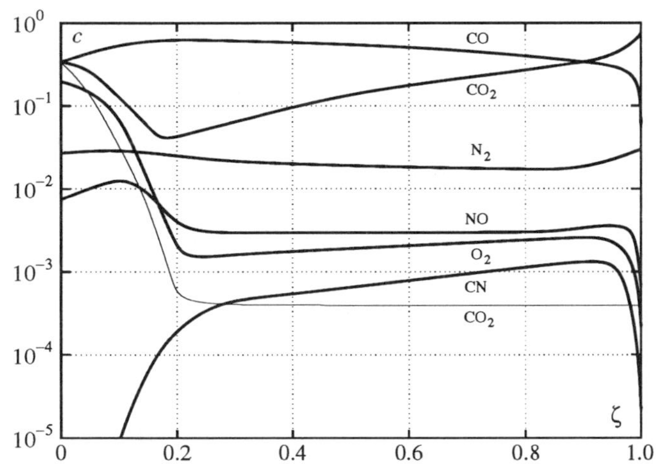

Figure 5.2 Species concentration profiles on the stagnation streamline. $95\%CO_2 + 5\%N_2$, $V_\infty = 6$ km/s, $p = 10$ kPa, $R = 1$ m, $T_w = 2900$K. Hair curve, local equilibrium assumption.

iterations did not exceed ten in the whole range of flight conditions considered. Below, some of the calculation results are discussed.

Figures 5.1 and 5.2 show the variation of the temperature profile with the shock layer pressure and the species concentration profiles at $p = 10$ kPa. The hair lines depict the profiles calculated on the local thermodynamic equilibrium assumption. The range of flight conditions covers the flow regimes from a fully merged shock

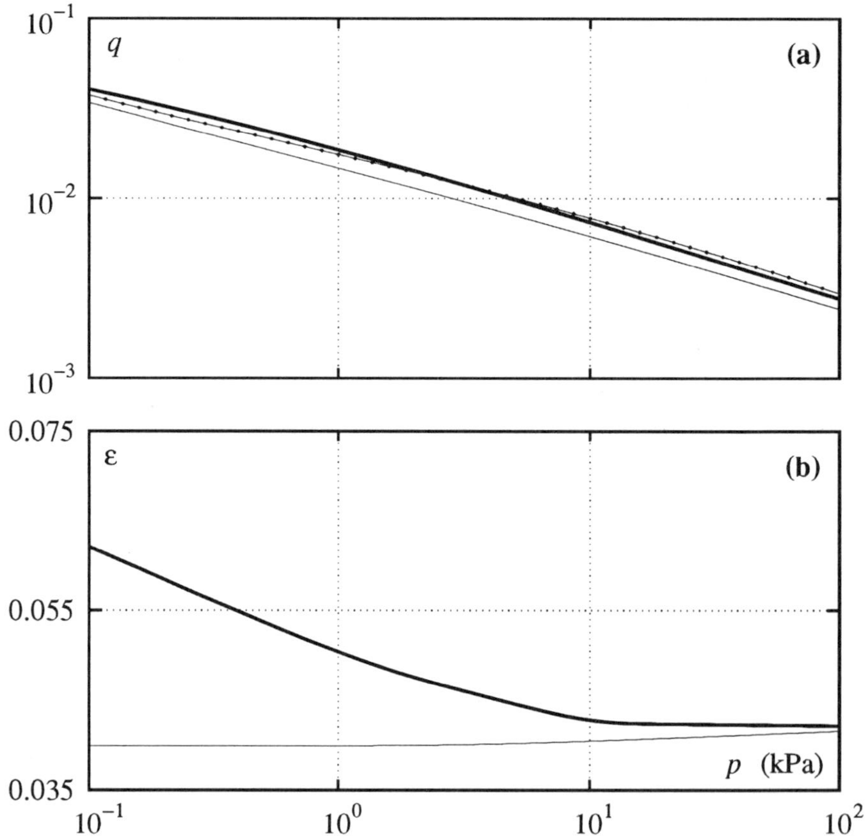

Figure 5.3 Dependence of the stagnation point heat flux and the shock layer thickness on the shock layer pressure. $95\%CO_2 + 5\%N_2$, $V_\infty = 6$ km/s, $R = 1$ m, $T_w = 2900$K. Hair curves, local equilibrium assumption; hair-dotted curve, $Sc_i = 0.5$.

layer to a flow with a thin boundary layer. One can see that chemical relaxation results in a significant increase of the gas temperature. At $p = 10^2$ kPa, the gas mixture is in equilibrium in most of the shock layer. At $p = 10$ kPa, the gas temperature and species concentrations substantially differ from the equilibrium values throughout the shock layer. At $p = 1$ kPa, the shock layer flow becomes nearly frozen. In this case, the mixture composition varies across the shock layer mainly because of the wall catalycity.

Figure 5.3 shows the variation of the shock layer thickness and the stagnation point heat flux with the shock layer pressure. The hair curves depict calculations made on the local thermodynamic equilibrium assumption. One can see a significant increase of the shock layer thickness due to the non-equilibrium chemistry at low pressure. The heat flux values on the fully catalytic wall are close to the equilibrium ones. The hair-dotted curve shows the heat flux dependence calculat-

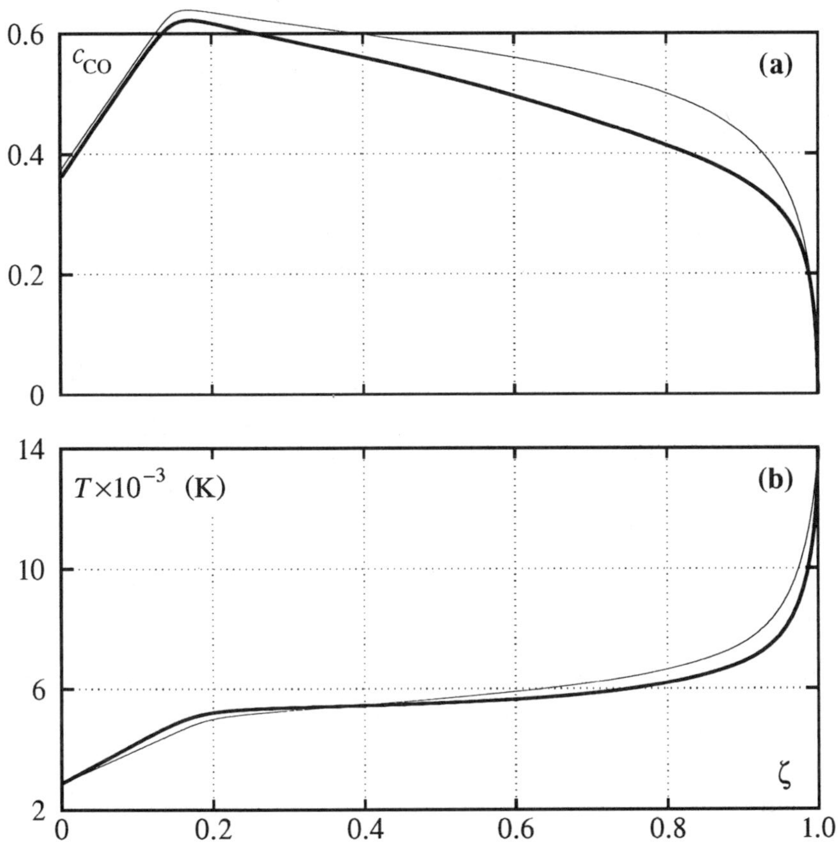

Figure 5.4 The effects of free stream nitrogen concentration of the shock layer parameters on the stagnation stream line. $V_\infty = 6$ km/s, $p = 10$ kPa, $R = 1$ m, $T_w = 2900$K. Full lines, 95%CO_2 + 5%N_2; hair lines, 100%CO_2.

ed using the description of the multicomponent diffusion by Fick law (5.14) with constant effective Schmidt numbers $Sc_i = 0.5$ of all species. One can see that this simplification results in an error which does not exceed 10%. The element concentrations coincide with their free stream values within some per cent differing only within the immediate vicinities of the body surface and the bow shock wave.

Figure 5.4 illustrates the effect of the free stream nitrogen concentration on the shock layer parameters. The temperature and carbon monoxide concentration profiles are compared for pure CO_2 and 95%CO_2 + 5%N_2 flows. The temperature in 100%CO_2 flow is somewhat higher near the bow shock and slightly lower near the wall, which is explained by the larger free stream molar mass and by a higher chemical activity of the pure CO_2 flow, respectively. The difference in CO concentrations is as much as 15%. The nitrogen concentration effects on the shock layer thickness and heat flux to the body surface were found to be small.

5.2.3 Comparison of non-equilibrium shock layer models

The above results were obtained within the framework of the thin viscous shock layer model. Its applicability to the calculation of non-equilibrium flows was investigated by Kumar and Jain (1975) who compared the results with those obtained using the full Navier–Stokes equations. The comparison proved an acceptable accuracy of the thin viscous shock layer model at Reynolds numbers $Re \gtrsim 50$. For non-equilibrium flows with smaller Reynolds numbers, the full Navier–Stokes equations provide a higher gas temperature within the shock layer, which increases the chemical reaction rates.

The solution results provided by various models of a non-equilibrium blunt body flow at small Reynolds numbers were compared by Gupta and Simmonds (1986) and Scherbak (1990). The calculation results were compared also with available data of a direct statistic simulation by means of the Monte–Carlo technique (Moss and Bird 1984). The results of Scherbak (1990), which are presented below, pertain to an axisymmetric air flow about the hyperboloid, which simulates a flow in the windward part of the Space Shuttle symmetry plane. Non-equilibrium air is considered as a 5-component mixture (N_2, O_2, NO, N, O) in which the dissociation, recombination and atom exchange processes occur. The internal energy of all particles is supposed to be in equilibrium. The author obtained numerical solutions of the problem using the two-layer thin viscous shock layer model and the reduced Navier–Stokes equations. The latter included viscous terms of the order $O(1/\sqrt{Re})$ and were integrated in the domain including the bow shock wave. The boundary conditions on the body surface were formulated following Scott (1975), taking into account the catalytic atom recombination, velocity slip, and temperature jump effects:

$$
\begin{aligned}
u &= \frac{2-\theta}{\theta}\sqrt{\frac{\pi}{2RT}}\frac{\mu}{\rho}\left(\sum_{i=1}^{N}\frac{c_i}{\sqrt{m_i}}\right)^{-1}\frac{\partial u}{\partial n} \qquad v = 0 \\
\frac{T}{T_w} &= \left(1+\frac{2-\theta}{\theta}\frac{1}{E}\sum_{i=1}^{N}\frac{J_i}{m_i}\right)\left(1-\frac{2-\theta}{\theta}\frac{\lambda}{2RET}\frac{\partial T}{\partial n}\right)^{-1} \\
E &= \rho\sqrt{\frac{2RT}{\pi}}\sum_{i=1}^{N}\frac{c_i}{m_i^{3/2}} \\
J_i &= -\frac{2}{2-\gamma_i}k_{wi}\sqrt{\frac{T}{T_w}}\,\rho c_i \\
k_{wi} &= \gamma_i(T_w)\sqrt{\frac{RT_w}{2\pi m_i}} \qquad i = 1,\ldots,\mathrm{N}-L; \\
J_i^* &= 0 \qquad i = 1,\ldots,L-1
\end{aligned}
\tag{5.53}
$$

Here R is the universal gas constant; θ, γ_i are the accommodation and catalytic recombination coefficients; $\mathrm{N} = 5$ is the number of the species; $L = 2$ is the number of the chemical elements.

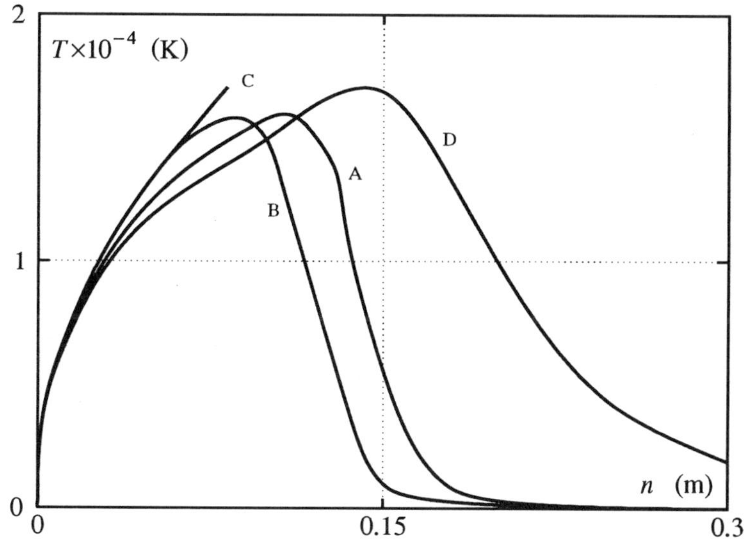

Figure 5.5 Temperature profiles on the stagnation streamline. Non-equilibrium air, $V_\infty = 7.5$ km/s, $R = 1.296$ m, $H = 92.35$ km, $T_w = 1043$K. A, B, two-dimensional and self-similar solutions of reduced Navier–Stokes equations, respectively; C, thin viscous shock layer model; D, Monte–Carlo simulation.

Figures 5.5 and 5.6 show the temperature and species concentration profiles on the stagnation streamline ahead of a hyperboloid with nose radius $R = 1.296$ m and asymptotic half-angle $\theta_c = 40.65°$. Here, the body surface is taken to be non-catalytic with the accommodation coefficient $\theta = 1$. It is seen that under the flight conditions considered, the thin viscous shock layer model does not work in a considerable part of the perturbed flow field. Nevertheless, the function profiles agree well with solutions of the reduced Navier–Stokes equations near the body surface. It is seen also that, like in the case of a non-reacting flow, the self-similar solution underestimates the extent of the perturbed zone. Besides that, it underestimates the dissociation degree. The continuous medium models give a substantially smaller thickness of the perturbed zone and a smaller maximum of the dissociation degree as compared with the solution of the Boltzmann equation, which is partly explained by ignoring the three-particle collisions in the Monte–Carlo calculations of Moss and Bird (1984).

For the gliding flight trajectory considered, the stagnation point heat flux values obtained from the reduced Navier–Stokes and Boltzmann equations differ by no more than a few per cent at the altitudes $H = 90 \div 110$ km. The thin viscous shock layer model overestimates the stagnation point heat flux at altitudes $H \lesssim 100$ km and underestimates it at $H \gtrsim 100$ km. The discrepancy amounts to 15%. The velocity slip and temperature jump noticeably reduce the heat flux to the body surface at $H > 90$ km. These effects increase the gas temperature in the shock layer, dissociation degree and the extent of the perturbed flow region.

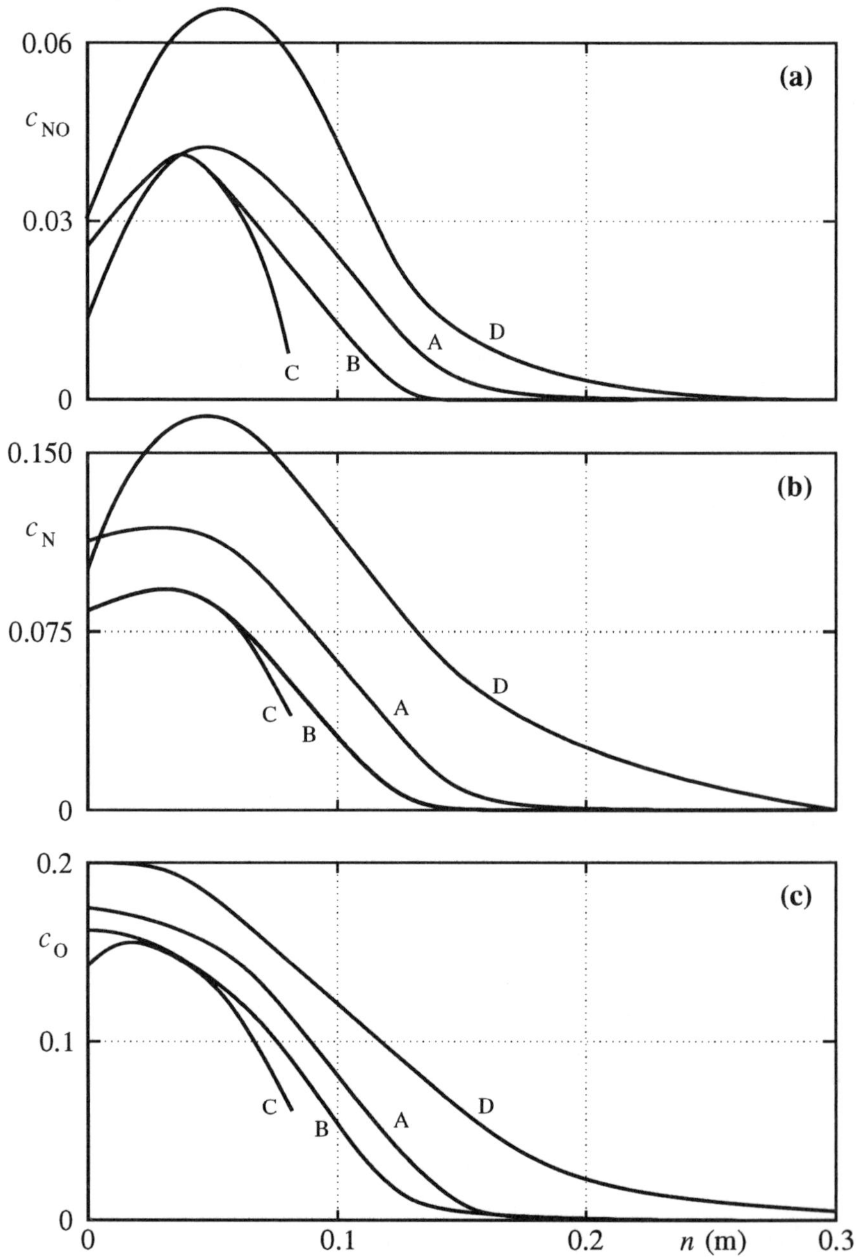

Figure 5.6 Species concentration profiles on the stagnation streamline. Non-equilibrium air, $V_\infty = 7.5$ km/s, $R = 1.296$ m, $H = 92.35$ km, $T_w = 1043$K. A, B, two-dimensional and self-similar solutions of reduced Navier–Stokes equations, respectively; C, thin viscous shock layer model; D, Monte–Carlo simulation.

5.2.4 Two-dimensional flows

To illustrate the solution of a two-dimensional problem, we shall consider non-equilibrium carbon dioxide and air flows about bodies of revolution at zero incidence (Golovachov 1979a, 1981a; Golovachov and Leontieva 1982). The calculations were carried out using the full viscous shock layer equations and the implicit unidirectional finite-difference method of subsection 1.2.2 with the modifications of subsection 5.2.1.

When investigating carbon dioxide flows, the following reactions are taken into account

$$\mathrm{CO_2 + M \rightleftarrows CO + O + M} \qquad \mathrm{O_2 + M \rightleftarrows O + O + M}$$
$$\mathrm{CO + M \rightleftarrows C + O + M} \qquad \mathrm{CO_2 + O \rightleftarrows CO + O_2}$$

The dissociation rate constant of CO_2 molecules is evaluated with the formulae recommended by Afonina and Gromov (1978)

$$\begin{aligned} k' &= 0.2 \times 10^{27.84} T^{-3.21} \exp\left(-\frac{63400}{T}\right) \frac{\mathrm{cm}^3}{\mathrm{mol}\cdot\mathrm{s}} \qquad && \mathrm{M} = \mathrm{O_2, CO} \\ k'' &= 5k' && \mathrm{M} = \mathrm{CO_2, O, C} \end{aligned} \tag{5.54}$$

The rate constants of other chemical reactions are the same as those of table 5.1. The equilibrium constants are calculated with formulae (5.44). Besides the non-equilibrium chemistry, the vibrational relaxation of CO_2 molecules is taken into account. Following Taylor and Bitterman (1969), it is supposed that the bending vibrational mode of CO_2 molecules is excited first due to the particle collisions and then its energy is instantaneously redistributed between all vibrational modes. Using the harmonic oscillator model, the vibrational relaxation time is determined by the formula

$$\frac{1}{\tau} = f P_{10}^{(2)} \left[1 - \exp\left(-\frac{T_*^{(2)}}{T}\right)\right] \tag{5.55}$$

where f is the particle collision frequency, P_{10} is the probability of a collision quenching the first vibrational state, T_* is the characteristic vibrational temperature, and the superscript (2) indicates the second (bending) vibrational mode. The collision frequency can be found with the formula (Hirschfelder *et al* 1954)

$$f = p\sigma^2 \sqrt{\frac{8\pi}{mkT}} \tag{5.56}$$

where σ is the collision diameter, m is the reduced mass of the collision partners, k is the Boltzmann constant. Assuming $\sigma = 3.75$ Å and using an analytic approximation of the measured $P_{10}^{(2)}$ values, one can obtain the following final formula for the vibrational relaxation time of CO_2 molecules

$$\frac{1}{\tau} = 10^9 \frac{p}{T} \exp\left(-107.3 T^{-1/3} + 5.69\right) \frac{1}{\mathrm{s}} \tag{5.57}$$

Here p is the pressure in kPa, T is the temperature in Kelvins. The vibrational states of other molecules are assumed to be in equilibrium with the translational temperature. The same is accepted for the rotational states of all molecules.

The diffusion is described by the Fick law with constant effective Schmidt numbers $\mathrm{Sc}_i = 0.5$. In this approach, the chemical element concentrations in the shock layer remain equal to their free stream values. As has been noted in the previous subsection, a more accurate description of the multicomponent diffusion gives nearly the same results. The mixture viscosity and heat conductivity are obtained from formulae (5.16) and (5.17). The molar enthalpies of the species are determined by formulae (5.45).

The flow is calculated using the full viscous shock layer equations which include terms of the order $O(1/\sqrt{Re})$; see subsection 2.2.1. The mixture continuity and momentum balance equations do not differ from the proper equations (1.1)–(1.3). The balance equations for the mixture energy and the vibrational energy of CO_2-molecules, as well as the species continuity equations, include only those components of the molecular energy and mass fluxes which are normal to the body surface. For the body-conforming coordinate system (s, n), they are written as follows

$$\begin{aligned}&\frac{\partial h}{\partial t} - \frac{1}{\rho}\frac{\partial p}{\partial t} + v\left(\frac{\partial h}{\partial n} - \frac{1}{\rho}\frac{\partial p}{\partial n}\right) + \frac{u}{1+\kappa n}\left(\frac{\partial h}{\partial s} - \frac{1}{\rho}\frac{\partial p}{\partial s}\right)\\ &\quad + \frac{1}{\rho}\left[\frac{\partial q_n}{\partial n} + g_2 q_n - \mu\left(\frac{\partial u}{\partial n} - 2\frac{\kappa u}{1+\kappa n}\right)\frac{\partial u}{\partial n}\right] = 0\end{aligned} \tag{5.58}$$

$$\begin{aligned}&\frac{\partial e_v}{\partial t} + v\frac{\partial e_v}{\partial n} + \frac{u}{1+\kappa n}\frac{\partial e_v}{\partial s} - \frac{1}{\rho}\left[\frac{\partial}{\partial n}\left(\frac{\mu}{\mathrm{Sc}_v}\frac{\partial e_v}{\partial n}\right)\right.\\ &\quad \left. + \frac{\mu}{\mathrm{Sc}_v}\frac{\partial e_v}{\partial n}\left(g_2 + \frac{2}{c_{CO_2}}\frac{\partial c_{CO_2}}{\partial n}\right)\right] + \frac{e_v - e_v^{\mathrm{eq}}}{\tau} = 0\end{aligned} \tag{5.59}$$

$$\begin{aligned}&\frac{\partial c_i}{\partial t} + v\frac{\partial c_i}{\partial n} + \frac{u}{1+\kappa n}\frac{\partial c_i}{\partial s} - \frac{1}{\rho}\left[\frac{\partial}{\partial n}\left(\frac{\mu}{\mathrm{Sc}_i}\frac{\partial c_i}{\partial n}\right) + g_2\frac{\mu}{\mathrm{Sc}_i}\frac{\partial c_i}{\partial n}\right]\\ &\quad - \frac{\dot{w}_i}{\rho} = 0 \qquad (i = CO_2, CO, O_2)\end{aligned} \tag{5.60}$$

Here q_n is the total heat flux vector component which is normal to the body surface, e_v is the specific vibrational energy of CO_2 molecules; other notations do not differ from the previously accepted.

Taking into account non-equilibrium excitation of the vibrational states of CO_2 molecules, the total heat flux vector is defined by the formula

$$\boldsymbol{q} = -\lambda\nabla T - \sum_{i=1}^{\mathrm{N}} h_i \frac{\mu}{\mathrm{Sc}_i}\nabla c_i - c_{CO_2}\frac{\mu}{\mathrm{Sc}_v}\nabla e_v \tag{5.61}$$

where the species enthalpies h_i are contributed only by the equilibrium modes of the internal energy. If the vibrational states of all molecules are in equilibrium, the total heat flux vector is given by formula (5.61) without the last term.

Keeping in mind a two-fold degeneracy of the bending mode and applying the harmonic oscillator model, one can write out the following formula for the specific equilibrium vibrational energy of CO_2 molecules

$$
\begin{aligned}
e_v^{\mathrm{eq}} = R^* \Bigg[& \frac{T_*^{(1)}}{\exp\left(T_*^{(1)}/T\right) - 1} + \frac{2T_*^{(2)}}{\exp\left(T_*^{(2)}/T\right) - 1} \\
& + \frac{T_*^{(3)}}{\exp\left(T_*^{(3)}/T\right) - 1} \Bigg]
\end{aligned}
\tag{5.62}
$$

where R^* is the specific gas constant. The characteristic vibrational temperatures are: $T_*^{(1)} = 1920\mathrm{K}$, $T_*^{(2)} = 959\mathrm{K}$, $T_*^{(3)} = 3380\mathrm{K}$. The effective Schmidt number for the diffusion transport of CO_2 vibrational energy is assumed to be the same as in the species continuity equations, i.e. $\mathrm{Sc}_v = \mathrm{Sc}_i = 0.5$. The continuity equations are written for the molecular species. The atom concentrations are calculated through the chemical element concentrations.

The bow shock boundary conditions for the mixture continuity, momentum, and energy balance equations are given by relationships (1.9) in which the heat flux vector is defined by formula (5.61). In the boundary conditions for equations (5.59) and (5.60), the vibrational excitation and chemical processes are supposed to be frozen within the bow shock wave. These boundary conditions read

$$
\frac{\mu}{\mathrm{Sc}_v}\left(\frac{de_v}{d\nu} + \frac{e_v}{c_{CO_2}}\frac{dc_{CO_2}}{d\nu}\right) = \rho_\infty \left(V_{\infty\nu} - D_\nu\right)\left(e_v - \frac{e_{v\infty}}{c_{CO_2}}\right) \tag{5.63}
$$

$$
\frac{\mu}{\mathrm{Sc}_i}\frac{dc_i}{d\nu} = \rho_\infty \left(V_{\infty\nu} - D_\nu\right)\left(c_i - c_{i\infty}\right) \tag{5.64}
$$

$$
(i = CO_2,\ CO,\ O_2)
$$

Here V_ν, D_ν are the gas and bow shock velocity components normal to the bow shock front, $d/d\nu$ is the derivative in this direction; see subsection 1.1.2.

The boundary conditions on the body surface are formulated as follows. Both velocity components are taken to be zero, the body surface is supposed to be fully catalytic with a fixed constant temperature T_w. The vibrational energy of CO_2 molecules is taken to be equal to its equilibrium value $e_v^{\mathrm{eq}}(T_w)$. The conditions on the flow axis and downstream boundary of the computational domain are given by the symmetry relationships and approximate formulae (1.12), respectively.

Most calculations were carried out using the grid with 12 lines $s = const$, each containing 31 grid points. Their distribution along the lines $s = const$ was governed by coordinate transformation (1.44). Integration in time was carried out with the constant time increment $\Delta t = 0.1R/V_\infty$. The stationary flows were calculated without iterations, except for several first time steps. Later on, each time step was calculated using only the previous time level data. When starting from linear function profiles, the stationary solution was achieved, at an average, after 200 - 300 time steps.

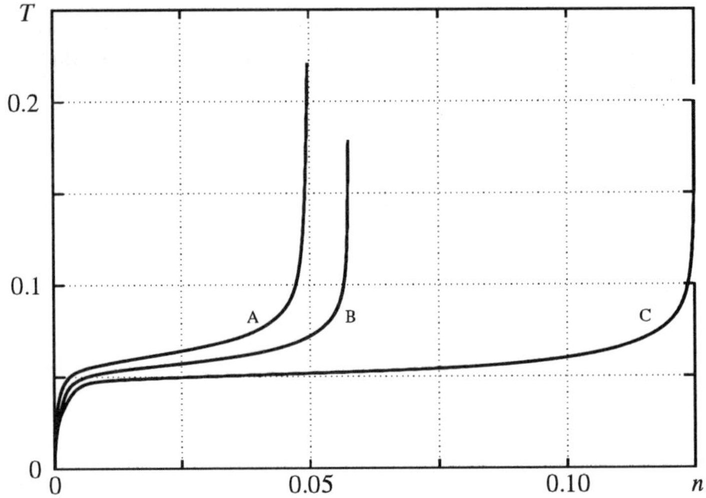

Figure 5.7 Translational temperature profiles in non-equilibrium carbon dioxide flow past a spherically blunted cone. $V_\infty = 5.5$ km/s, $\rho_\infty = 3 \times 10^{-7}$ g/cm^3, $R = 1$ m, $\theta_c = 60°$, $T_w = 1100$K; A, $s = 0.0$; B, $s = 0.52$; C, $s = 1.92$.

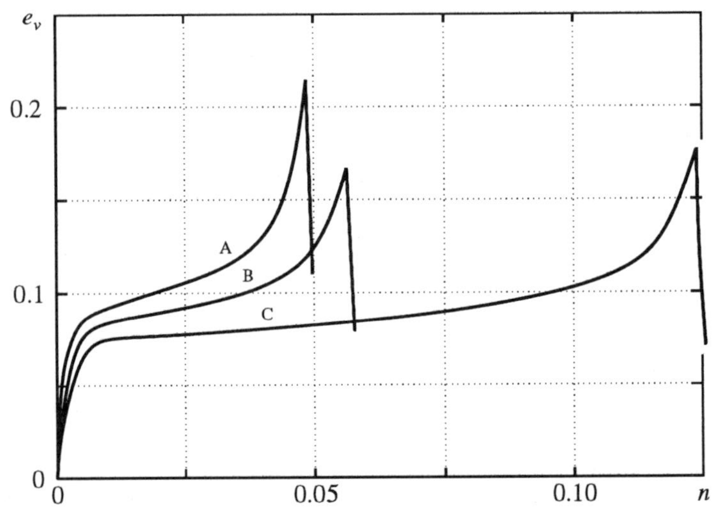

Figure 5.8 Profiles of the specific vibrational energy of CO_2 in non-equilibrium carbon dioxide flow past a spherically blunted cone. $V_\infty = 5.5$ km/s, $\rho_\infty = 3 \times 10^{-7}$ g/cm^3, $R = 1$ m, $\theta_c = 60°$, $T_w = 1100$K; A, $s = 0.0$; B, $s = 0.52$; C, $s = 1.92$.

Let us dwell on the calculation results for flows about spherically blunted cones with large half-angles, which corresponds to the shape of planet entry probes. Below, the temperature and specific vibrational energy are normalized by $V_\infty^2/2R_\infty^*$ and V_∞^2, where R_∞^* is the specific gas constant of the free stream.

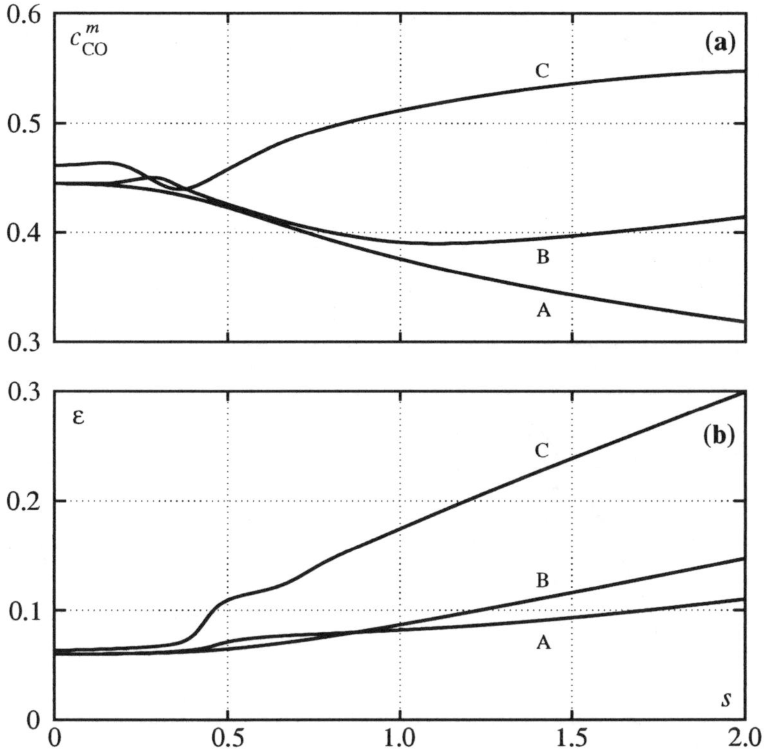

Figure 5.9 Variation of carbon monoxide concentration and the shock layer thickness in non-equilibrium carbon dioxide flows past spherically blunted cones. $V_\infty = 6$ km/s, $\rho_\infty = 2 \times 10^{-7}$ g/cm^3, $R = 0.2$ m, $T_w = 2500$K; A, $\theta_c = 50°$; B, $\theta_c = 60°$; C, $\theta_c = 70°$.

Figures 5.7 and 5.8 show the temperature and CO_2 vibrational energy profiles on the lines $s = const$ for one of the computational runs. At the flight conditions considered, the shock layer flow is nearly in equilibrium. It is characterized by steep gradients within the near-wall boundary layer and the relaxation zone adjacent to the bow shock front. It is also worth to note a non-monotonous variation of the maximum temperature and CO_2 vibrational energy along the shock layer, which is caused by the effect of a discontinuity of the body generatrix curvature on the bow shock shape.

Figures 5.9 and 5.10 present calculations of carbon dioxide flows past cones with various half-angles. Here, c^m_{CO} is the maximum carbon monoxide concentration on a line $s = const$. One can see that the flow past a cone with half-angle $\theta_c = 70°$ is characterized by a substantial increase of the shock layer thickness near the lateral part of the cone and an essentially non-monotonous variation of carbon monoxide concentration along the shock layer. The stagnation point heat flux is less than that for the smaller half-angle cones. The heat flux dependence on the cone half-angle on the lateral part of the body surface is non-monotonous. The skin friction

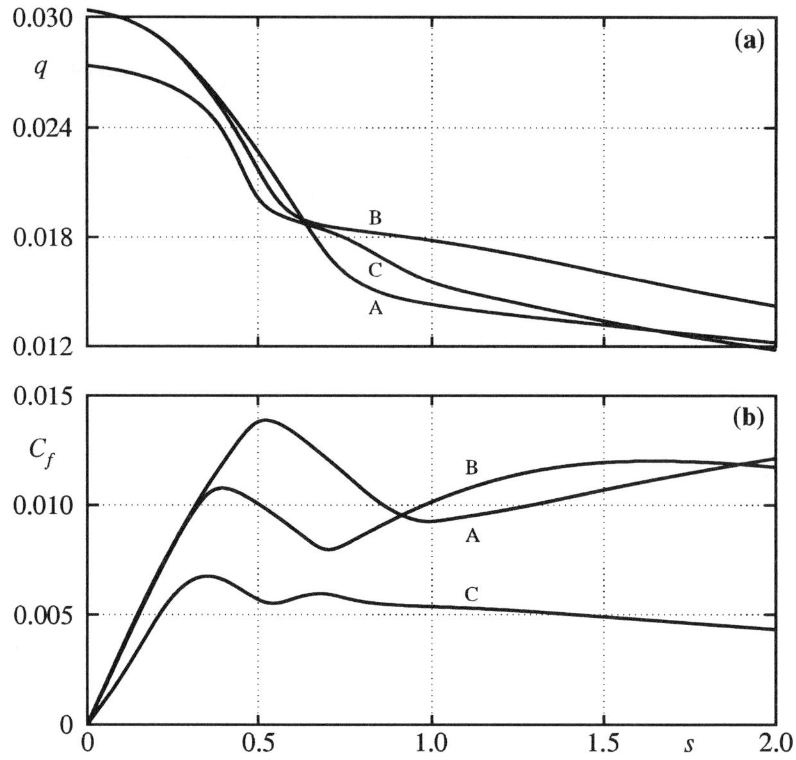

Figure 5.10 Heat flux and skin friction coefficient variation over spherically blunted cones in the carbon dioxide stream. $V_\infty = 6$ km/s, $\rho_\infty = 2 \times 10^{-7}$ g/cm^3, $R = 0.2$ m, $T_w = 2500$K; A, $\theta_c = 50°$; B, $\theta_c = 60°$; C, $\theta_c = 70°$.

coefficient distributions are also non-monotonous and greately dependent on the cone half-angle.

Figure 5.11 shows the effect of vibrational relaxation of CO_2 molecules on the shock layer parameters on the stagnation streamline. One can see that under the flight conditions considered, the vibrational relaxation zone occupies a rather small part of the shock layer and its influence is localized in the vicinity of the bow shock wave.

Figure 5.12 presents a comparison with the solution of the full Navier–Stokes equations calculated with the bow shock capturing method (Afonina and Gromov 1978). Here, the function profiles on the line $s = 1.57$ are plotted. Good agreement of the results indicates an applicability of the reduced Navier–Stokes equations and the bow shock fitting.

Let us consider now calculation results pertaining to non-equilibrium air flows. The calculations are carried out taking into account the following reactions

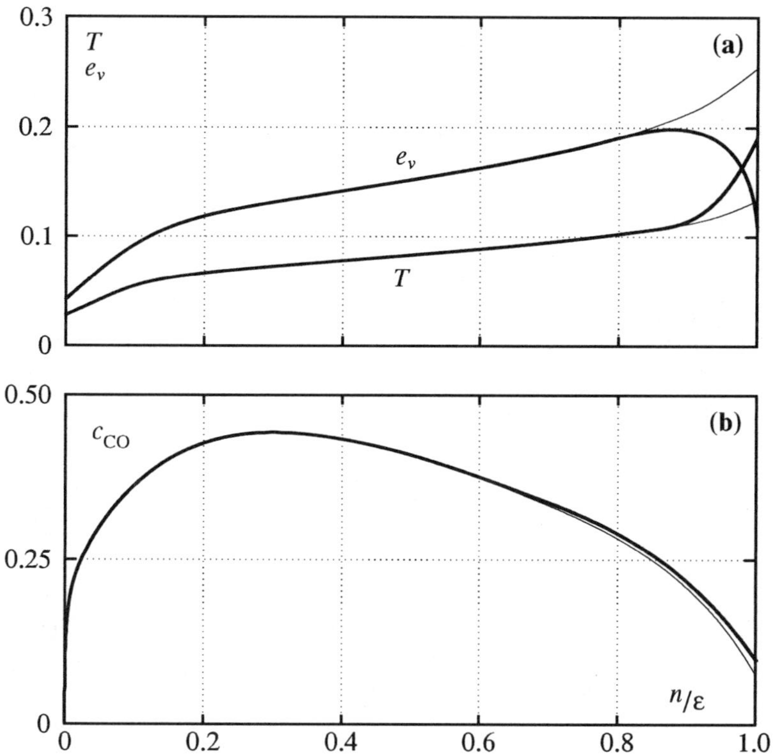

Figure 5.11 Profiles of translational temperature, specific vibrational energy of CO_2 and CO concentration on the stagnation streamline calculated with non-equilibrium (full lines) and equilibrium (hair lines) vibrational excitation of CO_2 molecules. $V_\infty = 6$ km/s, $\rho_\infty = 2 \times 10^{-7}$ g/cm^3, $R = 0.2$ m, $\theta_c = 60°$, $T_w = 2500$K.

$$N_2 + M \rightleftarrows N + N + M \qquad N_2 + O \rightleftarrows NO + N$$
$$O_2 + M \rightleftarrows O + O + M \qquad NO + O \rightleftarrows O_2 + N$$
$$NO + M \rightleftarrows N + O + M$$

The dissociation rate constants are evaluated with the formulae presented by Dellinger (1971), which slightly differ from those of table 5.1. The rotational and vibrational states of all molecules are assumed to be in equilibrium with the translational temperature.

Formulation of the problem differs from the above only in the wall boundary conditions for the species continuity and energy balance equations. Besides the limits of a fully catalytic and non-catalytic wall with fixed temperature, the calculations are carried out taking into account a finite rate of heterogeneous recombination on the body surface whose temperature was determined from the balance of heat fluxes. Supposing the element composition of the 5-component mixture under study to be constant, one has to solve only three species continuity equations. If they are written for O, N and NO concentrations, the wall boundary

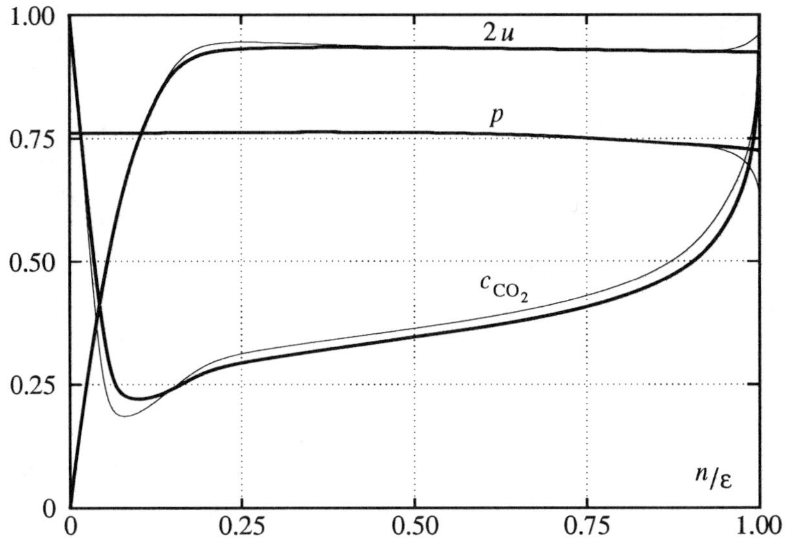

Figure 5.12 Function profiles on the line $s = 1.57$ in non-equilibrium carbon dioxide flow past a spherically blunted cone. $V_\infty = 5.5$ km/s, $\rho_\infty = 3.83 \times 10^{-7}$ g/cm^3, $R = 1$ m, $\theta_c = 60°$, $T_w = 1000$K. Full lines, reduced Navier–Stokes equations (Golovachov 1979a); hair lines, full Navier–Stokes equations (Afonina and Gromov 1978).

conditions take the form

$$J_i = \dot{r}_{wi} \qquad i = \mathrm{O}, \mathrm{N}, \mathrm{NO} \tag{5.65}$$

$$-\lambda \frac{\partial T}{\partial n} + \sum_{(i)} h_i J_i + \varepsilon \sigma T^4 = 0 \tag{5.66}$$

where $\dot{r}_{wi}$ is the mass production rate of the i-th species due to the heterogeneous reactions, ε is the wall emissivity, σ is the Stephan–Boltzmann constant. The third term of the left-hand side of (5.66) represents the wall radiation.

At relatively low wall temperatures, the heterogeneous processes can be considered as first-order reactions proceeding in one way. This yields

$$\dot{r}_{wi} = -\rho k_{wi} c_i \tag{5.67}$$

where k_{wi} are the effective wall catalycity coefficients which are evaluated using the models of gas-surface interaction or the experimental data. The limits $k_{wi} \to \infty$ and $k_{wi} = 0$ ($i = \mathrm{O}$, N, NO) correspond to a fully catalytic ($c_\mathrm{O} = c_\mathrm{N} = c_\mathrm{NO} = 0$) and non-catalytic wall. Besides the above limits, the following values of k_{wi} are used

$$k_{w\mathrm{O}} = k_{w\mathrm{N}} = 1.2\mathrm{m/s} \qquad\qquad k_{w\mathrm{NO}} = 0 \tag{5.68}$$

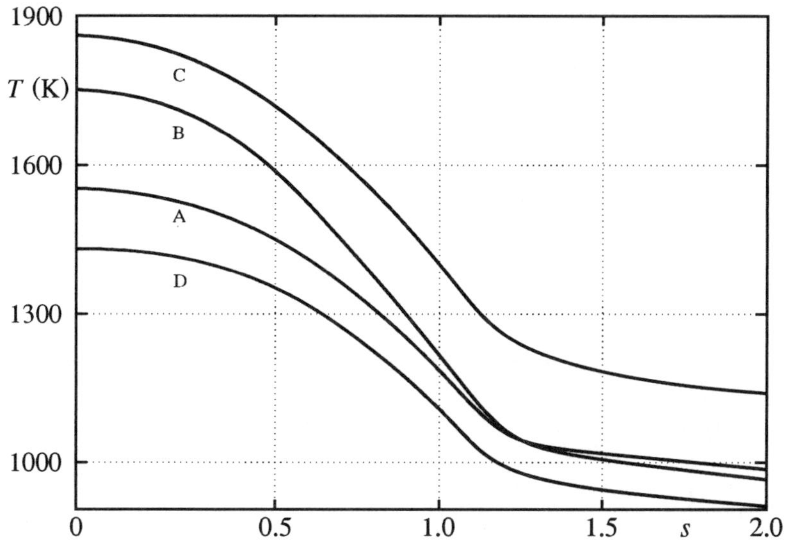

Figure 5.13 Wall temperature of a spherically blunted cone. Non-equilibrium air, $V_\infty = 6.77$ km/s, $H = 71$ km, $R = 1$ m, $\theta_c = 20°$; A, wall catalycity (5.68); B, wall catalycity (5.69); C, fully catalytic wall; D, non-catalytic wall.

$$k_{wi} = \sqrt{\frac{R^* T}{2\pi}} a_i \exp\left(-E_i/T\right)$$

$$i = \mathrm{O}, \mathrm{N} \qquad k_{w\mathrm{NO}} = 0 \tag{5.69}$$

$$a_\mathrm{O} = 16 \qquad a_\mathrm{N} = 0.071$$

$$E_\mathrm{O} = 10271\mathrm{K} \qquad E_\mathrm{N} = 2219\mathrm{K}$$

The effective wall catalycity coefficients of (5.68) were obtained from the measurements of the atom recombination rate on quartz heat protection materials (Gordeyev *et al* 1985). Formulae (5.69) were suggested by Scott (1980, 1981).

Figure 5.13 demonstrates the wall catalycity effect on the wall temperature determined with boundary condition (5.66). The results pertain to the flow about a spherically blunted cone with wall emissivity $\varepsilon = 0.9$.

Figures 5.14 through 5.16 compare the results calculated for the limits of a fully catalytic and non-catalytic wall with a fixed temperature. The calculations carried out on the local thermodynamic equilibrium assumption are also presented in these figures. One can see that the shock layer flow under study is essentially a non-equilibrium one. The wall catalycity influences the function profiles only near the body surface. The wall pressure and shock layer thickness do not depend on the wall catalycity. The heat flux to the non-catalytic wall is, approximately, half its value on the fully catalytic surface. The calculations carried out for a wide range of body nose radii and free stream densities proved the validity of binary scaling for the flight conditions considered.

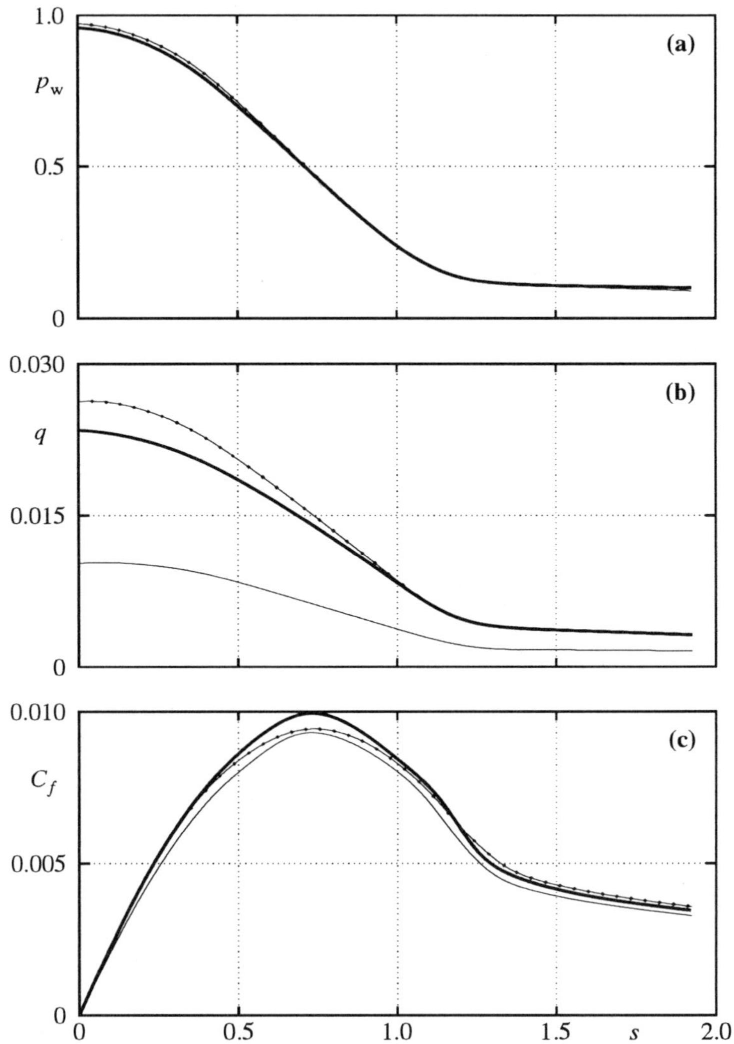

Figure 5.14 Wall pressure, heat flux and skin friction coefficient distributions over a spherically blunted cone. Non-equilibrium air, $V_\infty = 6.77$ km/s, $H = 71$ km, $R = 1$ m, $\theta_c = 20^\circ$, $T_w = 861$K. Full lines A, fully catalytic wall, hair lines B, non-catalytic wall; hair-dotted lines C, local equilibrium assumption.

Figures 5.17 and 5.18 show calculation results for the flows about spherically blunted cones with various half-angles. The body surface is assumed to be fully catalytic with temperature $T_w = 1000$K. The graphs show the variation of the shock layer thickness, heat flux and maximum NO concentration. One can see a coincidence of the numerical results for all cones in the vicinity of the stagnation point. The shock layer thickness at the lateral part of the cone varies with the

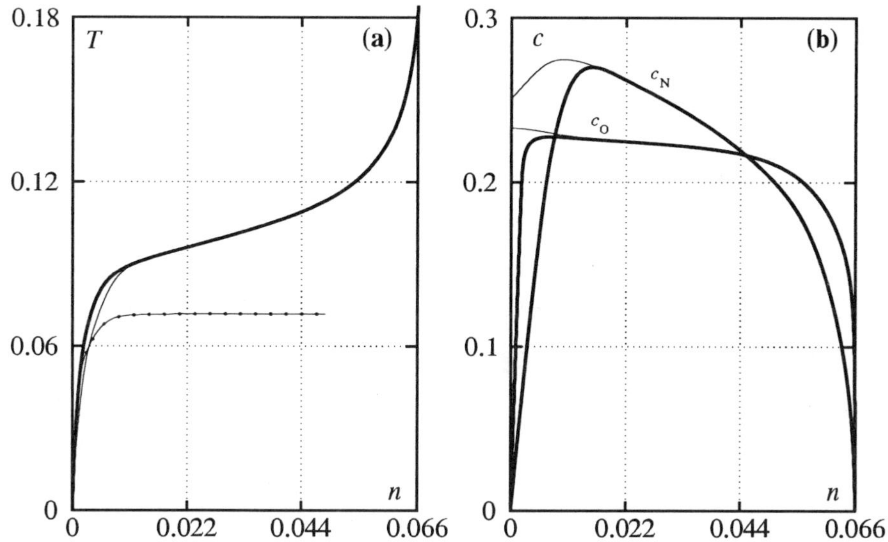

Figure 5.15 Temperature and species concentration profiles on the stagnation streamline ahead of a spherically blunted cone. Flight conditions and notations are the same as in figure 5.14.

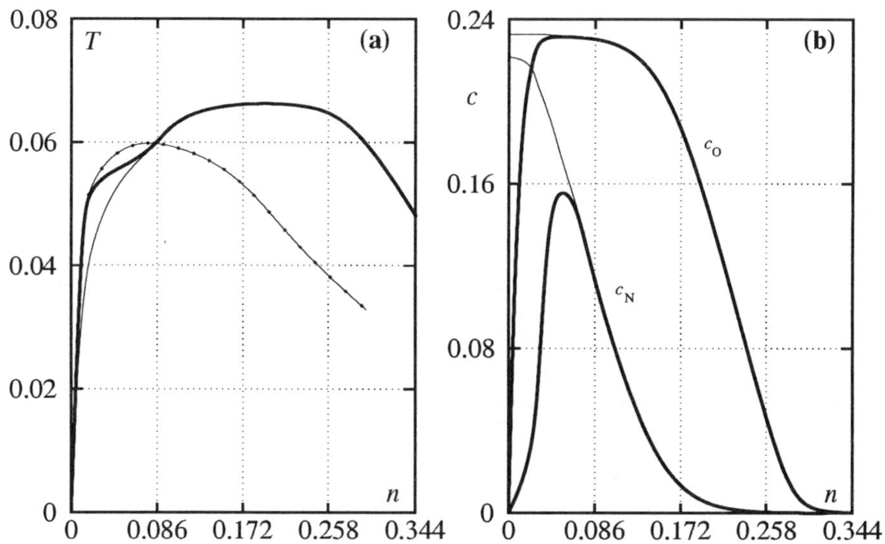

Figure 5.16 Temperature and species concentration profiles on the line $s = 1.92$. Flight conditions and notations are the same as in figure 5.14.

cone half-angle in a non-monotonous way. Variation of the heat flux and NO concentration are also strongly dependent on the cone half-angle.

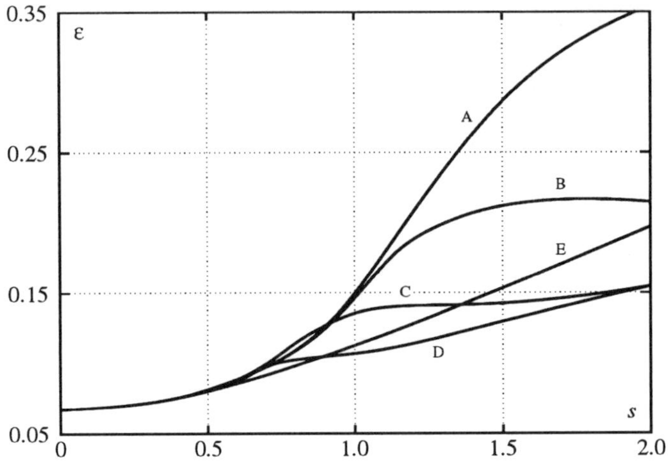

Figure 5.17 Variation of the shock layer thickness on spherically blunted cones with various half-angles. Non-equilibrium air, $V_\infty = 7$ km/s, $\rho_\infty = 2.1 \times 10^{-7}$ g/cm^3, $R = 0.2$ m, $T_w = 1000$K; A–E, $\theta_c = 20°, 30°, 40°, 50°, 60°$.

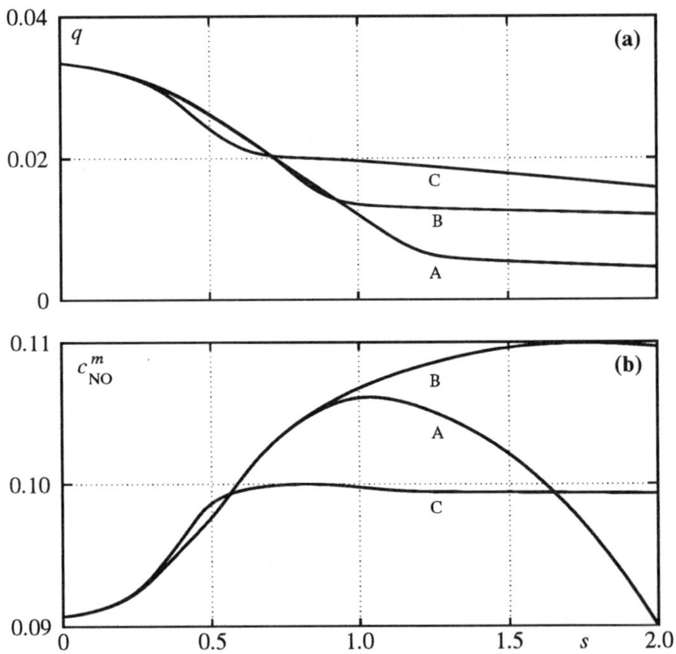

Figure 5.18 Variation of the heat flux and NO concentration in non-equilibrium air flows about spherically blunted cones with various half-angles. $V_\infty = 7$ km/s, $\rho_\infty = 2.1 \times 10^{-7}$ g/cm^3, $R = 0.2$ m, $T_w = 1000$K; A, $\theta_c = 20°$; B, $\theta_c = 40°$; C, $\theta_c = 60°$.

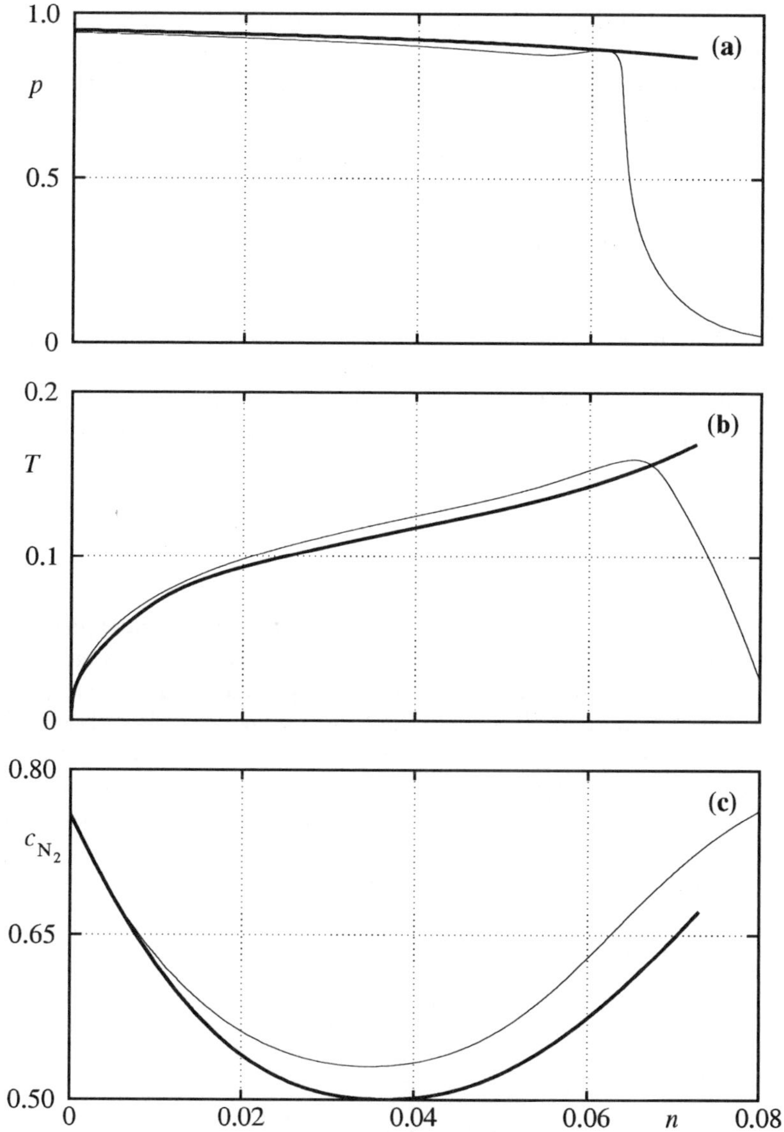

Figure 5.19 Function profiles on the stagnation streamline. Non-equilibrium air, $V_\infty = 7.93$ km/s, $\rho_\infty = 2.2 \times 10^{-7}$ g/cm^3, $R = 0.0305$ m, $T_w = 1500$K, fully catalytic wall. Full lines, full viscous shock layer equations; hair lines, self-similar solution of full Navier–Stokes equations.

Figure 5.19 compares the calculation results for the stagnation streamline with the self-similar solution of the full Navier–Stokes equations in a domain including a bow shock wave (Dellinger 1971). The discrepancies between these two solutions are accounted for, mainly, by the error related to the self-similar approach which underestimates the shock layer thickness. Nevertheless, the comparison indicates the applicability of the bow shock fitting and the frozen chemistry assumption within the bow shock wave.

In conclusion, it should be noted that axisymmetric supersonic non-equilibrium flows about bodies of revolution have been investigated by many authors using various viscous shock layer models and numerical methods. For instance, Davis (1970b), Dellinger (1970), and Miner and Lewis (1975) integrated full viscous shock layer equations by global iteration methods. The full Navier–Stokes equations were solved by Li (1973), Scott (1975), Petrova (1976), Afonina and Gromov (1978). The thin viscous shock layer model was applied by Kang and Dunn (1972), Scherbak (1989), Provotorov and Ryabov (1981). Non-equilibrium flows about the lateral part of elongated bodies of revolution were calculated with parabolized Navier–Stokes equations and space-marching methods (Voronkin 1979).

5.2.5 Three-dimensional flows

Let us begin the consideration of three-dimensional non-equilibrium viscous shock layer flows with the particular cases of self-similar and degenerate problems. The solutions of this kind can be obtained for flows in the vicinity of a stagnation point and a symmetry plane and for flows about the wings of an infinite span (Miner and Lewis 1976, Gershbein *et al* 1984b, 1985a,b, Scherbak 1987, Kazakov and Peigin 1988, Bruikina *et al* 1989, Tirskii and Scherbak 1990). As an example of such problems, a hypersonic non-equilibrium air flow in the vicinity of the stagnation point of a three-dimensional body is considered below, following the papers by Scherbak (1987) and Tirskii and Scherbak (1990).

The problem is solved using the reduced Navier–Stokes equations written for the curvilinear body-fitted coordinate system q^i $(i = 1, 2, 3)$ and non-dimensional variables

$$\frac{\partial}{\partial q^i}\left(\rho u^i \sqrt{\frac{g}{g_{(ii)}}}\right) = 0 \tag{5.70}$$

$$\begin{aligned} \rho D u^\alpha + \rho A^\alpha_{\beta\gamma} u^\beta u^\gamma = & -\sqrt{g_{(\alpha\alpha)}} g^{\alpha\beta} \frac{\partial p}{\partial q^\beta} \\ & + \sqrt{\frac{g_{(\alpha\alpha)}}{g}} \frac{\partial}{\partial q^3} \left(\frac{\mu\sqrt{g}}{Re_\infty} \frac{\partial u^\alpha / \sqrt{g_{(\alpha\alpha)}}}{\partial q^3} \right) \end{aligned} \tag{5.71}$$

$$\rho D u^3 + \rho A^3_{\beta\gamma} u^\beta u^\gamma = -\frac{\partial p}{\partial q^3} + \frac{4}{3} \frac{1}{\sqrt{g}} \frac{\partial}{\partial q^3} \left(\frac{\mu\sqrt{g}}{Re_\infty} \frac{\partial u^3}{\partial q^3} \right) \tag{5.72}$$

$$\rho c_p DT = Dp + \frac{1}{\sqrt{g}}\frac{\partial}{\partial q^3}\left(\frac{\mu c_p\sqrt{g}}{Re_\infty Pr}\frac{\partial T}{\partial q^3}\right)$$
$$+\frac{\mu}{Re_\infty}g_{\alpha\beta}\frac{\partial u^\alpha/\sqrt{g_{(\alpha\alpha)}}}{\partial q^3}\frac{\partial u^\beta/\sqrt{g_{(\beta\beta)}}}{\partial q^3}+\frac{4}{3}\frac{\mu}{Re_\infty}\left(\frac{\partial u^3}{\partial q^3}\right)^2 \tag{5.73}$$
$$-\sum_{i=1}^{N}h_i\dot{w}_i-\left(\sum_{i=1}^{N}c_{pi}J_i\right)\frac{\partial T}{\partial q^3}$$

$$\rho Dc_j + \frac{1}{\sqrt{g}}\frac{\partial\sqrt{g}J_j}{\partial q^3} = \dot{w}_j \qquad j=1,\dots,N-L \tag{5.74}$$

$$\rho Dc_k^* + \frac{1}{\sqrt{g}}\frac{\partial\sqrt{g}J_k^*}{\partial q^3} = 0 \qquad k=1,\dots,L-1 \tag{5.75}$$

where

$$h=\sum_{i=1}^{N}c_ih_i \qquad c_p=\sum_{i=1}^{N}c_ic_{pi} \qquad Re_\infty=\frac{\rho_\infty V_\infty R}{\mu_\infty}$$
$$D=\frac{u^i}{\sqrt{g_{(ii)}}}\frac{\partial}{\partial q^i} \qquad g=\det\|g_{ij}\|$$
$$A^i_{jk}=\frac{1}{2\sqrt{g_{(jj)}g_{(kk)}}}\left[\sqrt{g_{(ii)}}g^{il}\left(\frac{\partial g_{kl}}{\partial q^j}+\frac{\partial g_{jl}}{\partial q^k}-\frac{\partial g_{jk}}{\partial q^l}\right)\right.$$
$$\left.-\delta^i_j\frac{\partial\sqrt{g_{(jj)}}}{\partial q^k}-\delta^i_k\frac{\partial\sqrt{g_{(kk)}}}{\partial q^j}\right]$$

In the above equations, J_j is the mass diffusion flux component which is normal to the body surface, δ^i_j is the Kronecker symbol, the asterisk denotes the concentrations and diffusion fluxes of the chemical elements. The covariant components $g_{\alpha\beta}$ of the metric tensor can be expressed through the components $a_{\alpha\beta}$, $b_{\alpha\beta}$ of the tensors which determine the first and second quadratic forms of the body surface by formulae (4.56). The Latin indices take the values 1,2,3; the Greek ones take the values 1,2. The summation is supposed to be made with respect to the pairs of the same indices, except for the bracketed pairs. The temperature, specific heat capacity, specific enthalpy, and viscosity are normalized by $T_0 = V_\infty^2/c_{p\infty}$, $c_{p\infty}$, $c_{p\infty}T_0$, and μ_∞, respectively. Other variables are normalized by the conventional reference values.

Equations (5.71)–(5.73) differ from the full Navier–Stokes equations by lack of some terms which are expected to be small at hypersonic flight speed. The energy balance equation (5.73) is written on the assumption of equilibrium excitation of the internal energy of the particles. The diffusion fluxes are obtained from the Stephan–Maxwell relationships including baro-diffusion which may be important in the vicinity of a bow shock wave at small Reynolds numbers (Zalogin 1976).

The outer boundary of the computational domain is located in the free stream. The body surface is supposed to be impermeable. The wall boundary conditions take into account the catalytic recombination and velocity slip. They are given by formulae (5.53) in which the boundary condition for the gas temperature is replaced by the condition of heat flux balance on the body surface. Being written for the non-dimensional variables, it takes the form

$$\frac{\mu c_p}{PrRe_\infty}\frac{\partial T}{\partial q^3} - \sum_{i=1}^{N} h_i J_i + \frac{\mu}{Re_\infty}\frac{u^\alpha}{\sqrt{a_{(\alpha\alpha)}}}a_{\alpha\beta}\frac{\partial u^\beta/\sqrt{a_{(\beta\beta)}}}{\partial q^3} - \frac{\varepsilon\sigma T_0^4}{\rho_\infty V_\infty^3}T_w^4 = 0 \tag{5.76}$$

where ε is the wall emissivity and σ is the Stephan–Boltzmann constant.

The body surface in the vicinity of the stagnation point is defined by equation (4.90) of the elliptic paraboloid with the main radius ratio κ. The solution is searched for in the form

$$\begin{aligned} u^\alpha &= d_{(\alpha)} q^{(\alpha)} u_0^\alpha(q^3) + \dots \\ u^3 &= \left[1 - \frac{(q^1)^2}{2} - \frac{(\kappa q^2)^2}{2}\right] u_0^3(q^3) + \dots \\ p &= p_0(q^3) + p_2^1(q^3)\cdot(q^1)^2 + \kappa p_2^2(q^3)\cdot(q^2)^2 + \dots \\ \Phi &= \Phi_0(q^3) + \dots \end{aligned} \tag{5.77}$$

where Φ is any of the functions ρ, T, μ, c_j, c_k^* ($j = 1, \dots, N$; $k = 1, \dots, L$); $d_1 = 1$, $d_2 = \kappa$. Substitution of (5.77) into equations (5.70)–(5.75) and truncation of the series yield a set of ordinary differential equations for the leading terms of expansions (5.77) and the equations for the pressure gradient components p_2^α. The wall boundary conditions are obtained by substitution of (5.77) into relationships (5.53) and (5.76). On the outer boundary of the computational domain, $p_2^\alpha = 0$, other functions are given by the free stream parameters. The above equations and boundary conditions are derived as described in subsection 4.6.1.

Numerical solutions were obtained for the spacecraft glide trajectory in the Earth's atmosphere assuming $T_\infty = 250$K and the following density variation

$$\rho_\infty = 1.225 \times 10^{-3} \exp(-0.142H) \text{g/cm}^3 \tag{5.78}$$

where the altitude H is taken in km (Masek *et al* 1973). Variation of the flight speed with the altitude is shown in figure 5.20.

Figures 5.21 through 5.24 show calculation results for the flows about a body with the smaller of two main radii at the stagnation point $R_1 = 0.5$ m, and the wall emissivity $\varepsilon = 0.85$. Numerical solutions were obtained for the fully catalytic wall, wall catalycity coefficients (5.69), and the non-catalytic wall. Non-equilibrium air was considered as a mixture of five components: N_2, O_2, NO, N, O.

Figure 5.21 demonstrates a variation of the gas temperature on the body surface with the main radius ratio κ and the effects of wall catalycity, velocity slip

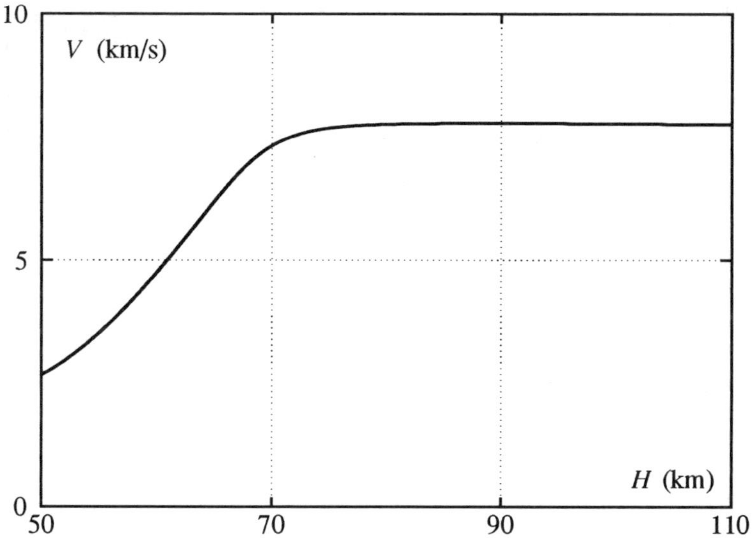

Figure 5.20 Flight speed-altitude dependence for the spacecraft glide trajectory in the Earth's atmosphere.

and temperature jump. The temperature is normalized by its value on an axisymmetric body ($\kappa = 1$). One sees a decrease of the gas temperature with κ diminishing. The wall catalycity effect on the gas temperature is small. Velocity slip and temperature jump effects manifest themselves in a more rapid decrease of the gas temperature, which is caused by a decrease of the temperature jump with κ diminishing. At $H \lesssim 90$ km, the velocity slip and temperature jump effects were found to be negligible.

Figure 5.22 shows the temperature profiles on the stagnation streamline at several points of the flight trajectory. The q^3-coordinate is normalized by the radius R_1. The Reynolds numbers Re_∞ corresponding to the altitudes $H = 90, 70, 60, 50$ km are the following: 845, 1.33×10^4, 3.6×10^4, and 8.65×10^4. In these calculations, the wall catalycity was prescribed by formulae (5.69). The figure demonstrates a transition from the merged flow regime to the flow involving an inviscid region, a thin boundary layer, and a thin bow shock wave. Because of the high temperature, the chemical reactions are significant even in the merged flow regime. With the altitude decreasing, the dissociation degree increases due to an increase of the free stream density. This results in a steep temperature gradient behind the bow shock wave. At $H \lesssim 70$ km, an inviscid flow and a near-wall boundary layer appear. With further vehicle gliding, the dissociation degree decreases due to a decrease of the flight speed. At the altitude $H = 50$ km, the flow structure becomes consistent with a non-reacting gas model.

Tirskii and Scherbak (1990) complemented the above analysis by taking into account the vibrational relaxation. Since the characteristic vibrational temperatures of N_2, O_2 and NO molecules are relatively close, the average vibrational

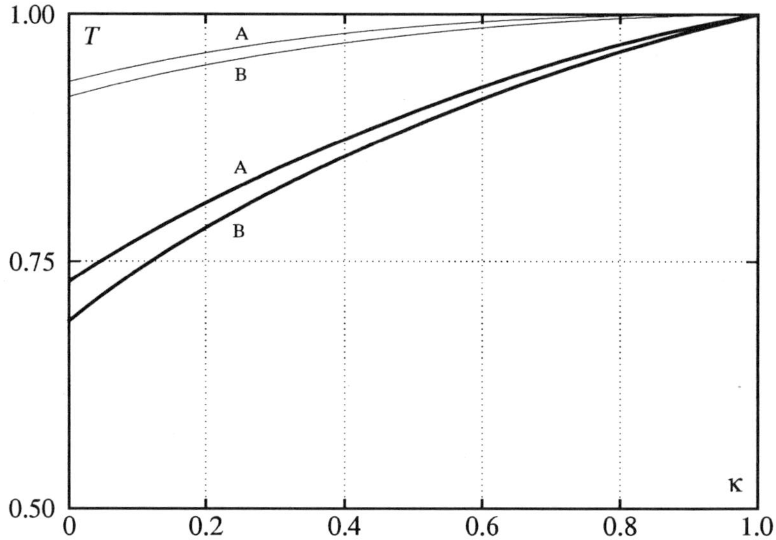

Figure 5.21 Dependence of gas temperature at the three-dimensional stagnation point on the body main radius ratio. Flight trajectory of figure 5.20, $H = 100$ km; A, fully catalytic wall; B, non-catalytic wall. Hair curves, the velocity slip and temperature jump ignored.

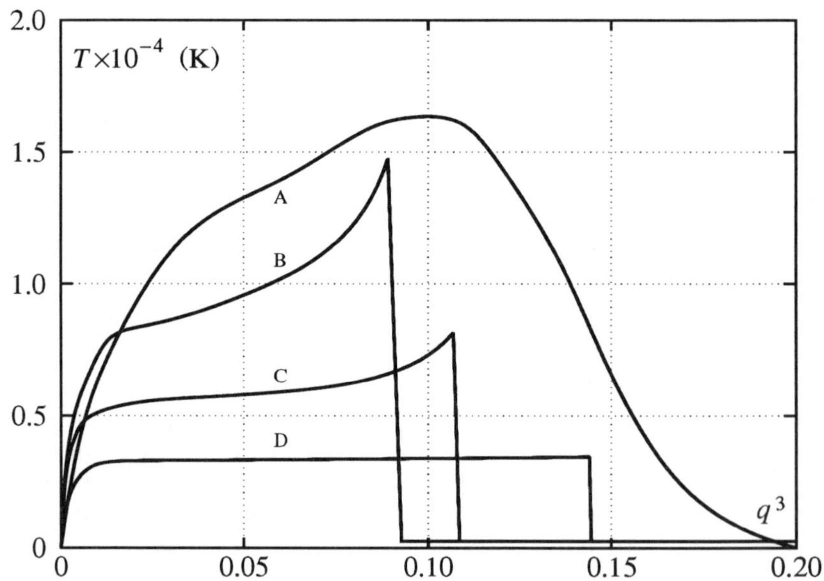

Figure 5.22 Temperature profiles on the stagnation streamline. Flight trajectory of figure 5.20, $\kappa = 0.4$; A, $H = 90$ km; B, $H = 70$ km; C, $H = 60$ km; D, $H = 50$ km.

temperature T_v is used which is taken to be the same for all molecules. In this approach, an inclusion of vibrational relaxation effects in the flow analysis reduces to a modification of the mixture energy balance equation (5.73) and a supplement of set (5.70)–(5.75) with an equation for the vibrational temperature T_v, which results from a summation of the vibrational energy balance equations of the molecular species. In the formulation of the problem, the effect of vibrational relaxation on the dissociation rate and the dissociation effect on the vibrational energy were taken into account with the technique elaborated by Hammerling *et al* (1959), Treanor and Marrone (1962), Marrone and Treanor (1963). The parameter characterizing the decrease of a probability of dissociation from the low vibrational states was taken to be $T_D/3$, where T_D is the dissociation temperature. The wall boundary condition for the vibrational temperature was given by either the equilibrium relationship

$$T_v = T_w \tag{5.79}$$

or the assumption on the body surface non-catalycity with respect to the vibrational degrees of freedom:

$$\frac{\partial T_v}{\partial q^3} = 0 \tag{5.80}$$

The calculations show that for the flight trajectory and body dimensions considered, the vibrational temperature noticeably differs from the translational one throughout the perturbed flow region at the altitudes $H \gtrsim 65$ km. Figures 5.23 depict the translational and vibrational temperature profiles on the stagnation streamline at two trajectory points. In these calculations, the body surface was assumed to be fully catalytic with respect to both the chemical reactions and the vibrational excitation. The calculation results are compared with those obtained on the assumption of vibrational equilibrium ($T_v = T$). One can see that vibrational relaxation may result in a significant increase of the translational temperature and the extent of the perturbed flow region ahead of the body.

Figure 5.24 shows the variation of the stagnation point heat flux with flight altitude. The heat flux is contributed by the mixture heat conductivity, diffusion of the chemically reacting species, and the wall shear stress power. In accordance with boundary condition (5.76), it is compensated by the wall radiation. Curves A through C correspond to various degrees of wall catalycity. The full and hair-dotted curves depict calculations with the wall boundary conditions (5.79) and (5.80), respectively. The results obtained on the vibrational equilibrium assumption are plotted by the hair curves. Vibrational relaxation increases the gas translational temperature, which accounts for the higher heat flux values calculated including the vibrational non-equilibrium effects. The increase of the heat flux to the fully catalytic surface does not exceed 10 per cent. In the case of a non-catalytic wall, it can be as much as 25 per cent. The use of body surface material which is non-catalytic with respect to vibrational energy of the molecules (wall boundary condition (5.80)) reduces the heat flux. The reduction is about 10 per cent.

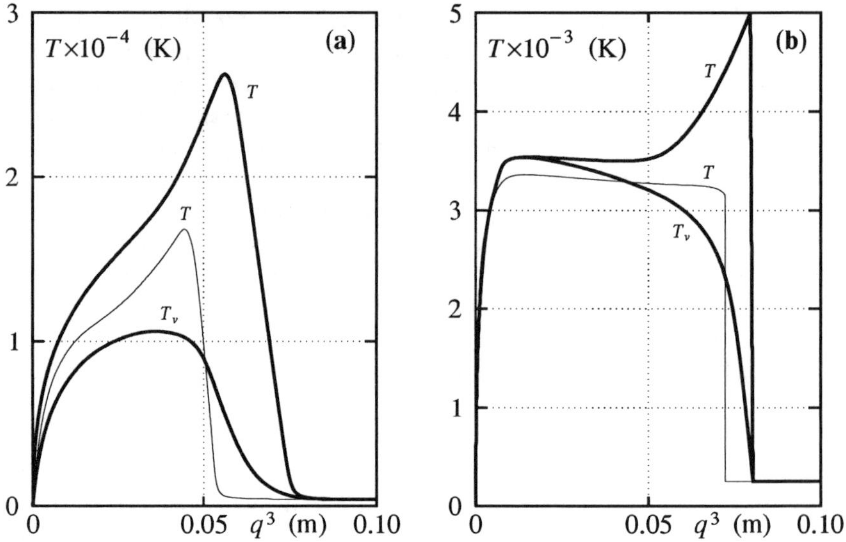

Figure 5.23 Translational and vibrational temperature profiles on the stagnation streamline at $H = 80$ km (a) and $H = 50$ km (b) for the flight trajectory of figure 5.20. $\kappa = 0.4$. Hair curves, vibrational equilibrium assumption ($T_v = T$).

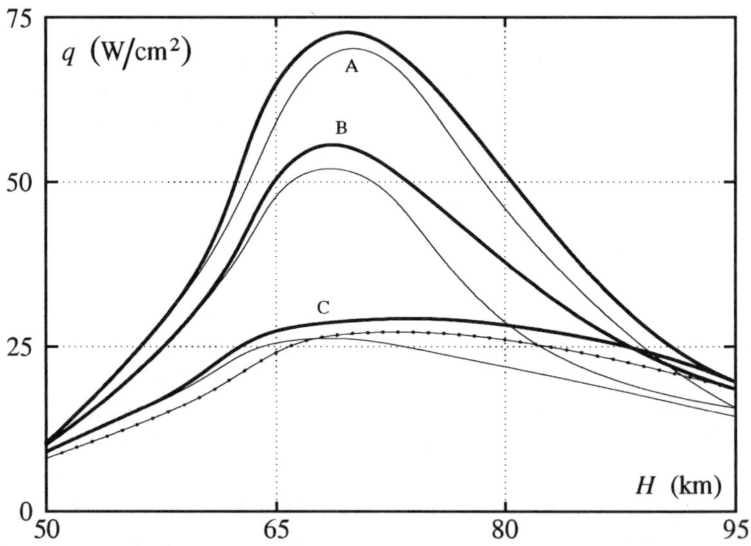

Figure 5.24 Variation of the stagnation point heat flux with flight altitude for the trajectory of figure 5.20. $\kappa = 0.4$; A, fully catalytic wall; B, wall catalycity (5.69); C, non-catalytic wall. Full curves, wall boundary condition (5.79); hair curves, vibrational equilibrium assumption ($T_v = T$); hair-dotted curve, wall boundary condition (5.80).

Consider now calculations of the three-dimensional non-equilibrium flows about the whole front of a blunt body. The problem is greatly simplified by using a thin viscous shock layer model. Below, its application to three-dimensional flow about a blunted body at angles of attack and side-slip is considered (Borodin *et al* 1990).

Assuming equilibrium excitation of the internal energy of all species, the governing equations for the curvilinear body-fitted coordinate system q^i $(i = 1, 2, 3)$ and non-dimensional variables are written as follows (see also subsection 4.4.1)

$$
\begin{aligned}
&\frac{\partial}{\partial q^i}\left(\rho u^i \sqrt{\frac{g}{g_{(ii)}}}\right) = 0 \qquad \frac{\partial p}{\partial q^3} = -\rho A^3_{\alpha\beta} u^\alpha u^\beta \\
&\rho\left(\frac{u^i}{\sqrt{g_{(ii)}}}\frac{\partial u^\alpha}{\partial q^i} + A^\alpha_{\beta\delta} u^\beta u^\delta\right) = -g^{\alpha\beta}\sqrt{g_{(\alpha\alpha)}}\frac{\partial p}{\partial q^\beta} \\
&\qquad + \frac{\partial}{\partial q^3}\left(\frac{\mu}{Re}\frac{\partial u^\alpha}{\partial q^3}\right) \\
&\rho c_p \frac{u^i}{\sqrt{g_{(ii)}}}\frac{\partial T}{\partial q^i} = 2\frac{u^\alpha}{\sqrt{g_{(\alpha\alpha)}}}\frac{\partial p}{\partial q^\alpha} + \frac{\partial}{\partial q^3}\left(\frac{\mu c_p}{Pr Re}\frac{\partial T}{\partial q^3}\right) \\
&\qquad - \frac{\partial T}{\partial q^3}\sum_{k=1}^{N} c_{pk} J_k - \sum_{k=1}^{N} h_k \dot{w}_k + \frac{2\mu}{Re}\psi_{\alpha\beta}\frac{\partial u^\alpha}{\partial q^3}\frac{\partial u^\beta}{\partial q^3} \\
&\rho\frac{u^i}{\sqrt{g_{(ii)}}}\frac{\partial c_j}{\partial q^i} + \frac{\partial J_j}{\partial q^3} = \dot{w}_j \qquad j = 1, \ldots, N - L \\
&\rho\frac{u^i}{\sqrt{g_{(ii)}}}\frac{\partial c^*_k}{\partial q^i} + \frac{\partial J^*_k}{\partial q^3} = 0 \qquad k = 1, \ldots, L - 1
\end{aligned}
\tag{5.81}
$$

Here $\psi_{\alpha\beta} = g_{\alpha\beta}/\sqrt{g_{(\alpha\alpha)}g_{(\beta\beta)}}$, $Re = \rho_\infty V_\infty R/\mu(T^\circ)$, the gas temperature is normalized by $T^\circ = 10^4$K, the specific enthalpy by $V^2_\infty/2$, the specific heat capacity by $V^2_\infty/2T^\circ$, and the other quantities are denoted and normalized in the same manner as in the above analysis of the flow in the vicinity of the stagnation line. The coefficients $A^3_{\alpha\beta}$, $A^\alpha_{\beta\gamma}$ are defined by the same formulae as in set (4.57). The diffusion fluxes are determined from the Stephan–Maxwell relationships (5.13).

The inflow boundary conditions are given by the modified Rankine–Hugoniot relationships taking into account a hypersonic flight speed and neglecting chemical reactions within the bow shock wave:

$$\rho\left(u^3 - \frac{u^\alpha}{\sqrt{g_{(\alpha\alpha)}}}\frac{\partial q_s^3}{\partial q^\alpha}\right) = u_\infty^3 \qquad p = (u_\infty^3)^2$$

$$u_\infty^3\,(u^\alpha - u_\infty^\alpha) = \frac{\mu}{Re}\frac{\partial u^\alpha}{\partial q^3}$$

$$u_\infty^3\left[\sum_{i=1}^{N} c_{i\infty}(h_i - h_{i\infty}) - \psi_{\alpha\beta}(u^\alpha - u_\infty^\alpha)(u^\beta - u_\infty^\beta) - (u_\infty^3)^2\right] = \frac{\mu c_p}{Pr Re}\frac{\partial T}{\partial q^3} \tag{5.82}$$

$$u_\infty^3(c_i - c_{i\infty}) + J_i = 0 \qquad i = 1, \ldots, N - L$$

$$u_\infty^3(c_j^* - c_{j\infty}^*) + J_j^* = 0 \qquad j = 1, \ldots, L - 1$$

When formulating the wall boundary conditions, all velocity components are taken to be zero on the body surface. The boundary conditions for the species continuity and energy balance equations are given by relationships (5.65) and (5.66). Besides, the diffusion fluxes of the chemical elements J_j^* ($j = 1, \ldots, L - 1$) are supposed to be equal to zero on the body surface, which means the absence of element production.

The numerical method used by Borodin *et al* (1990) is essentially the same as that described in subsection 4.4.2. The calculations were carried out for a non-equilibrium air flow about a triaxial ellipsoid with the axis ratio 10:7:3, the minimum radius at the critical point $R_1 = 0.5$ m and wall emissivity $\varepsilon = 0.85$. The flight conditions correspond to the glide trajectory shown in figure 5.20 with the free stream density distribution (5.78).

Calculations show a significant effect of the angles of incidence and side-slip on the heat transfer. Under the flight conditions considered, the maximum heat flux towards the body surface at $\alpha = \beta = 45°$ appears to be less by 20 per cent as compared with the flow at $\alpha = \beta = 0$. Figure 5.25 presents variation of the body surface temperature normalized by its stagnation point value with the circumferential coordinate. The figure demonstrates a relatively small wall catalycity effect on the wall temperature distribution. The same has been revealed for the heat flux distribution as well.

Keeping in mind the above property, Borodin *et al* (1990) suggested to evaluate the heat flux to the body surface in a three-dimensional non-equilibrium dissociating air flow in the following way

$$q\left(q^1, q^2, k_w\right) = q_0\left(k_w\right) q^0\left(q^1, q^2\right) \tag{5.83}$$

Here $q_0(k_w)$ is the stagnation point heat flux on the body with prescribed wall catalycity, $q^0(q^1, q^2)$ is the relative heat flux distribution over the body with a fully catalytic surface. The quantity $q_0(k_w)$ is obtained from the self-similar solution for the vicinity of the stagnation point. The relative heat flux distribution $q^0(q^1, q^2)$, at fairly high Reynolds numbers ($Re \gtrsim 10^3$), can be obtained from the boundary layer calculations or the approximate formulae of Bruikina *et al* (1982).

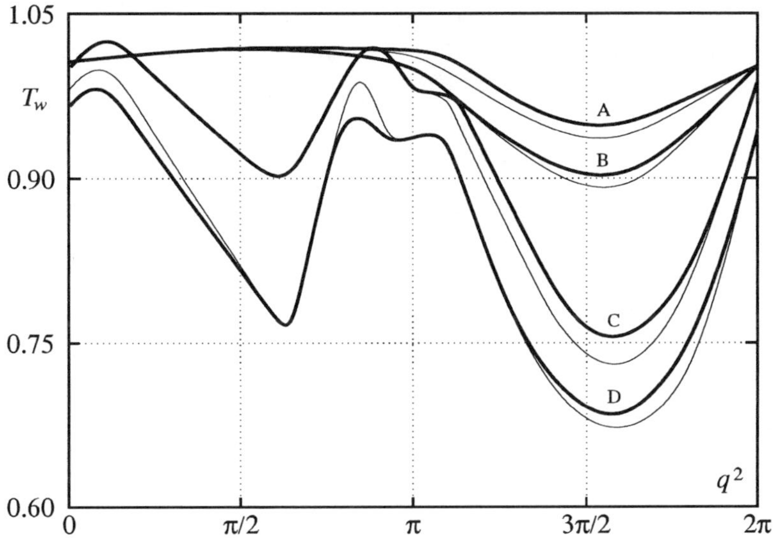

Figure 5.25 Wall temperature distributions over a triaxial ellipsoid for the flight trajectory of figure 5.20. $\alpha = \beta = 45°$; A, H = 90 km, q^1 = 0.1; B, H = 70 km, q^1 = 0.1; C, H = 90 km, q^1 = 0.4; D, H = 70 km, q^1 = 0.4. Full lines, fully catalytic wall; hair lines, non-catalytic wall.

The thin viscous shock layer model provides an accurate prediction for the non-separated flows about smooth bodies. However, its applicability is usually restricted to a rather small vicinity of the stagnation point. In a general case of a three-dimensional stationary blunt body flow, one has to solve the three-dimensional boundary value problem. Such calculations have been carried out, for instance, by Golovachov and Karyakin (1980) and Reddy and Fujiwara (1989).

Let us consider the problem following the first of the above papers in which three-dimensional non-equilibrium carbon dioxide flows about ellipsoids of revolution were considered taking into account the following chemical reactions

$$CO_2 + M \rightleftarrows CO + O + M \quad O_2 + M \rightleftarrows O + O + M \quad CO_2 + O \rightleftarrows CO + O_2$$

The rotational and vibrational states of the molecules are supposed to be in equilibrium with the translational temperature. The reaction rate constants are presented in table 5.1. The equilibrium constants and the species enthalpies are calculated using formulae (5.44) and (5.45); the viscosity and heat conductivity are calculated using formulae (5.16) and (5.17); the diffusion is described by means of the constant effective Schmidt number.

The shock layer flow is calculated using the reduced Navier–Stokes equations which involve only the largest terms responsible for the molecular mass, momentum, and energy transport. The numerical method is based on the implicit unidirectional finite-difference scheme described in subsection 4.2.2 with reference to shock layer flow of a gas with constant heat capacity.

The governing equations are written in tensor form. The mixture continuity and momentum balance equations are the same as those of (4.19). The energy balance and the species continuity equations read

$$\rho\frac{\partial e}{\partial t} + \rho(u^j - \omega^j)\nabla_j e + p\nabla_j u^j - g^{33}\left[\nabla_3\left(\lambda\nabla_3 T + \frac{\mu}{\mathrm{Sc}}\sum_{i=1}^{4} h_i\nabla_3 c_i\right) + \mu\nabla_3 u_j\nabla_3 u^j\right] = 0 \tag{5.84}$$

$$\rho\frac{\partial c_i}{\partial t} + \rho\left(u^j - \omega^j\right)\nabla_j c_i - g^{33}\nabla_3\left(\frac{\mu}{\mathrm{Sc}}\nabla_3 c_i\right) - \dot{w}_i = 0 \qquad (i = 1, 2) \tag{5.85}$$

where the notations are the same as those of subsection 4.2.2. The continuity equations are written for CO_2 and O_2 concentrations. The concentrations of other species are determined from known element composition of the mixture, which is assumed to be the same as in the free stream.

The stationary solution is obtained by the time-asymptotic technique. Each time step is fulfilled by means of iterations in which the one-dimensional sets of difference equations are solved along the lines drawn across the shock layer. With the modifications of subsection 5.2.1, the set of difference equations on each line is written as follows

$$\begin{aligned}&\mathbf{P}^{n(q)}_{l+1/2}\left(\Delta \boldsymbol{X}_{l+1} + \Delta \boldsymbol{X}_l\right) - \frac{\Delta t}{2}\mathbf{Q}^{n(q)}_{l+1/2}\left(\Delta \boldsymbol{X}_{l+1} + \Delta \boldsymbol{X}_l\right)\\ &+ \frac{\Delta t}{\Delta q^3}\mathbf{R}^{n(q)}_{l+1/2}\left(\Delta \boldsymbol{X}_{l+1} - \Delta \boldsymbol{X}_l\right)\\ &- g^{33}_{l+1/2}\frac{\Delta t}{(\Delta q^3)^2}\left[\mathbf{S}^{n(q)}_{l+1}\left(\Delta \boldsymbol{X}_{l+1} - \Delta \boldsymbol{X}_l\right)\right.\\ &\left.- \mathbf{S}^{n(q)}_{l}\left(\Delta \boldsymbol{X}_{l} - \Delta \boldsymbol{X}_{l-1}\right)\right] + \boldsymbol{H}^{n(q)}_{l+1/2} = 0\end{aligned} \tag{5.86}$$

where

$$
\boldsymbol{X} = \begin{Bmatrix} u_1 \\ u_2 \\ u^3 \\ T \\ \rho \\ c_1 \\ c_2 \end{Bmatrix} \qquad
\boldsymbol{H} = \begin{Bmatrix} H_1 \\ H_2 \\ H_3 \\ H_4 \\ H_5 \\ H_6 \\ H_7 \end{Bmatrix} \qquad
\mathbf{P} = \begin{Bmatrix}
\rho & 0 & 0 & 0 & 0 & 0 & 0 \\
0 & \rho & 0 & 0 & 0 & 0 & 0 \\
0 & 0 & \rho & 0 & 0 & 0 & 0 \\
0 & 0 & 0 & \rho e'_T & 0 & \rho e_1 & \rho e_2 \\
0 & 0 & 0 & 0 & 1 & 0 & 0 \\
0 & 0 & 0 & 0 & 0 & \rho & 0 \\
0 & 0 & 0 & 0 & 0 & 0 & \rho
\end{Bmatrix}
$$

$$
\mathbf{Q} = \begin{Bmatrix}
0 & 0 & 0 & 0 & 0 & 0 & 0 \\
0 & 0 & 0 & 0 & 0 & 0 & 0 \\
0 & 0 & 0 & 0 & 0 & 0 & 0 \\
0 & 0 & 0 & 0 & 0 & 0 & 0 \\
0 & 0 & 0 & 0 & 0 & 0 & 0 \\
0 & 0 & 0 & 0 & 0 & \frac{\partial \dot{\boldsymbol{W}}}{\partial \boldsymbol{C}} & \\
0 & 0 & 0 & 0 & 0 & &
\end{Bmatrix} \qquad
\mathbf{S} = \begin{Bmatrix}
\mu & 0 & 0 & 0 & 0 & 0 & 0 \\
0 & \mu & 0 & 0 & 0 & 0 & 0 \\
0 & 0 & 0 & 0 & 0 & 0 & 0 \\
0 & 0 & 0 & \lambda & 0 & \frac{\mu}{\mathrm{Sc}} h_1 & \frac{\mu}{\mathrm{Sc}} h_2 \\
0 & 0 & 0 & 0 & 0 & 0 & 0 \\
0 & 0 & 0 & 0 & 0 & \frac{\mu}{\mathrm{Sc}} & 0 \\
0 & 0 & 0 & 0 & 0 & 0 & \frac{\mu}{\mathrm{Sc}}
\end{Bmatrix}
$$

$$
\mathbf{R} = \begin{Bmatrix}
\rho V^3 & 0 & 0 & 0 & 0 & 0 & 0 \\
0 & \rho V^3 & 0 & 0 & 0 & 0 & 0 \\
0 & 0 & \rho V^3 & g^{33} p'_T & g^{33} p'_\rho & g^{33} p'_{c_1} & g^{33} p'_{c_2} \\
0 & 0 & p & \rho V^3 e'_T & 0 & \rho V^3 e_1 & \rho V^3 e_2 \\
0 & 0 & \rho & 0 & V^3 & 0 & 0 \\
0 & 0 & 0 & 0 & 0 & \rho V^3 & 0 \\
0 & 0 & 0 & 0 & 0 & 0 & \rho V^3
\end{Bmatrix}
$$

$$
e_1 = e_{\mathrm{CO}_2} - \frac{28}{44} e_{\mathrm{CO}} - \frac{16}{44} e_{\mathrm{O}} \qquad e_2 = e_{\mathrm{O}_2} - e_{\mathrm{O}} \qquad V^3 = u^3 - \omega^3
$$

$$
h_1 = h_{\mathrm{CO}_2} - \frac{28}{44} h_{\mathrm{CO}} - \frac{16}{44} h_{\mathrm{O}} \qquad h_2 = h_{\mathrm{O}_2} - h_{\mathrm{O}}
$$

$$
\dot{\boldsymbol{W}} = \begin{Bmatrix} \dot{w}_1 \\ \dot{w}_2 \end{Bmatrix} \qquad \boldsymbol{C} = \begin{Bmatrix} c_1 \\ c_2 \end{Bmatrix}
$$

$\boldsymbol{H}$ is the vector of residuals of the governing equations; e, h are the specific internal energy and specific enthalpy; subscripts 1,2 at the elements of $\dot{\boldsymbol{W}}$ and $\boldsymbol{C}$ correspond to CO_2 and O_2, respectively.

On each line q^3, set (5.86) falls into several independent subsets. Like a non-reacting flow calculation, the covariant velocity vector components u_1 and u_2 are obtained from two tridiagonal subsets with the same coefficients. Further, the species continuity equations are solved jointly using the block 'progonka' algorithm. The obtained elements of the increment vector $\Delta \boldsymbol{X}$ are excluded from the other three difference equations of (5.86), which are solved jointly by the block 'progonka' algorithm.

Some of the calculation results are presented in the following figures. They pertain to a flow about an ellipsoid of revolution with half-axes $a = 0.2$ m, $b = 0.3$ m at various angles of attack with respect to the smaller axis. All distances are normalized by a, the shear stress by $\rho_\infty V_\infty^2/2$; the heat flux and pressure are normalized by the conventional scales $\rho_\infty V_\infty^3$ and $\rho_\infty V_\infty^2$, respectively.

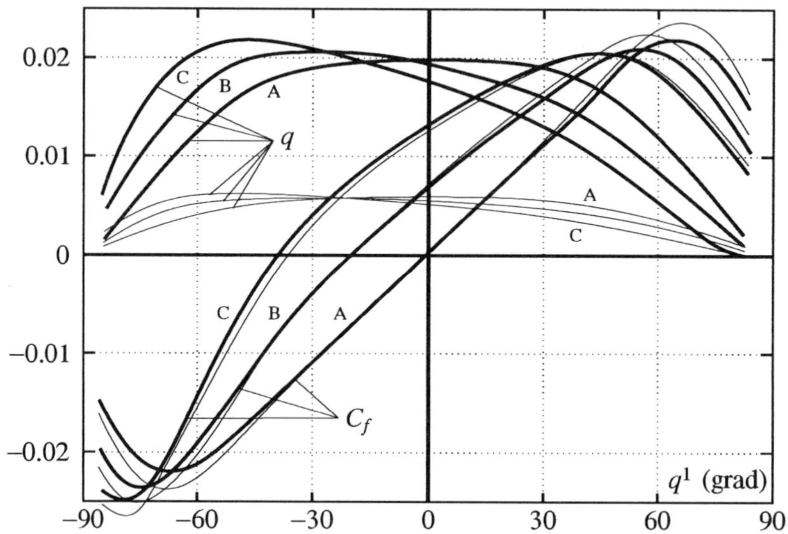

Figure 5.26 Heat flux and skin friction coefficient distributions in the symmetry plane of an ellipsoid of revolution in a non-equilibrium carbon dioxide flow. $V_\infty = 6$ km/s, $\rho_\infty = 2 \times 10^{-7}$ g/cm^3, $T_\infty = 140$K, $T_w = 2500$K; A, $\alpha = 0°$; B, $\alpha = 10°$; C, $\alpha = 20°$. Full lines, fully catalytic wall; hair lines, non-catalytic wall.

Figure 5.26 shows the skin friction coefficient and heat flux distributions in the symmetry plane $\varphi = 0, \pi$. It is seen that with increasing α, the maximum heat flux and zero skin friction points shift to the windward part of the body surface, the displacements being different and exceeding the angle of incidence. It is also worth to note an increase in the maximum heat flux values, which is accounted for by an increase of the body curvature at the maximum point. Reduction of the wall catalycity down to zero results in an over two-fold heat flux reduction on most of the body surface. Figure 5.27 depicts isolines of the skin friction coefficient (front view) and the heat flux and pressure isolines (side view). The arrows depict the limiting streamlines.

The solution of a three-dimensional boundary value problem requires a large amount of computations. For this reason, the above approach is usually applied to calculate flows about bodies of a relatively small length. When investigating stationary flows about the lateral part of elongated bodies, it is expedient to use the parabolized Navier–Stokes equations and the space-marching numerical methods.

This approach is illustrated below with reference to a non-equilibrium air flow about the windward side of a flat delta wing with cylindrical leading edges (Afonina *et al* 1984). Non-equilibrium air is considered as a five-component mixture of molecules and atoms (N_2, O_2, NO, N, O) with equilibrium excitation of the internal energy of all species. The mixture viscosity and heat conductivity are calculated by approximate formulae similar to (5.16) and (5.17), the diffusion mass fluxes are obtained from Stephan–Maxwell relationships (5.13). The shock layer flow

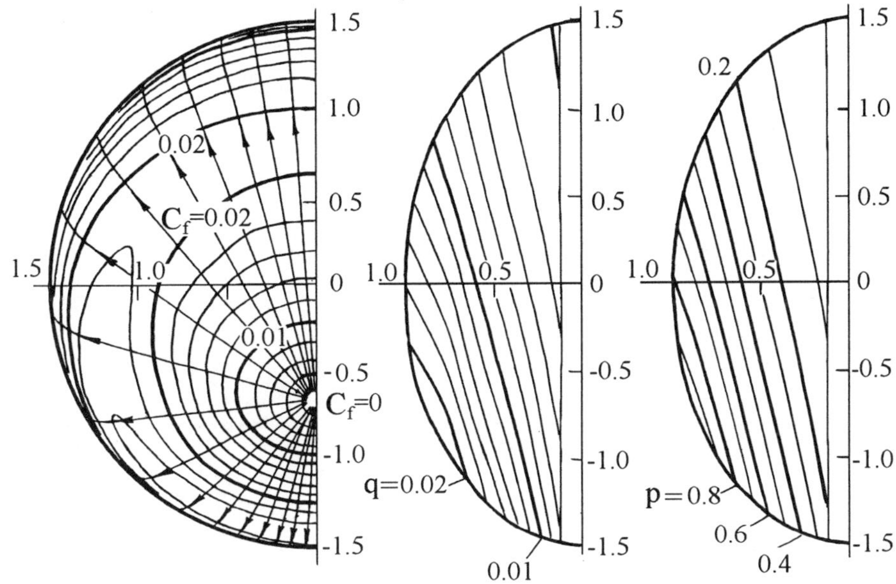

Figure 5.27 Isolines of the skin friction coefficient, heat flux and pressure on ellipsoid of revolution in a non-equilibrium carbon dioxide flow. $V_\infty = 6$ km/s, $\rho_\infty = 2\times10^{-7}$ g/cm^3, $T_\infty = 140$K, $T_w = 2500$K, $\alpha = 20°$, fully catalytic wall.

is calculated using the reduced Navier–Stokes equations similar to those of the preceding example. Formulation of the problem is analogous with that described in subsection 4.3.3, except that now equations (4.52) are complemented by three species continuity equations and the continuity equation for one of the chemical elements. The complementary equations include only those diffusion flux vector components which are normal to the body surface. The wall boundary conditions are formulated taking into account the catalytic atom recombination, velocity slip, and temperature jump. The gas temperature on the body surface is determined from the heat flux balance, assuming the wall emissivity $\varepsilon = 0.8$. The bow shock boundary conditions are given by modified Rankine–Hugoniot relationships. Having calculated the flow near the wing nose, the authors calculate the shock layer flow on the lateral part of the wing using the block marching method discussed in subsection 4.3.3. The calculations were carried out with a grid whose points were concentrated near the wing nose, in the boundary layer, and near the leading edges.

Figures 5.28 and 5.29 present the calculations of a flow about the windward side of the wing with a sweep angle $\beta = 75°$. The first figure shows the isotherms on the wing surface for the cases of a non-catalytic and fully catalytic wall. The temperature is plotted in Kelvins. It is seen that in both cases the maximum temperature is achieved in the vicinity of the stagnation point. Besides, there is an additional local temperature maximum on the wing edge, which is more

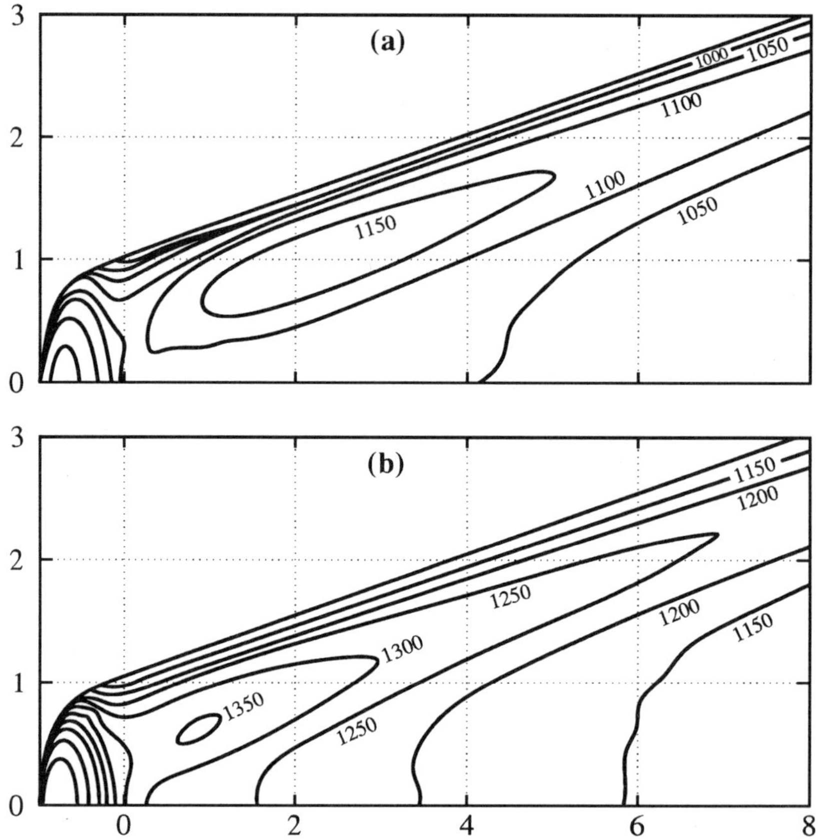

Figure 5.28 Isotherms on the windward side of a flat delta wing with blunt leading edges. Non-equilibrium air, $V_\infty = 7$ km/s, $p_\infty = 10^{-5}$ atm, $T_\infty = 210$K, $\alpha = 40°$; (a) non-catalytic wall; (b) fully catalytic wall.

distinct on the non-catalytic surface. The largest reduction of heat flux to the wing surface due to a decrease of its catalycity is achieved in the vicinity of the stagnation point. The wall catalycity effect on the heat transfer diminishes with a distance from the bluntness. Figure 5.29 depicts the pressure contours on the wing surface. It is seen that the maximum pressure is achieved at the stagnation point. The pressure is nearly constant on the flat part of the wing surface, except for the vicinity of the junction with spherical bluntness where there is a small local pressure minimum. The effect of the heterogeneous recombination rate on the wall pressure distribution was found to be small. The wall shear stress contribution to the drag and lift forces of the wing does not exceed a few per cent for the flight conditions considered.

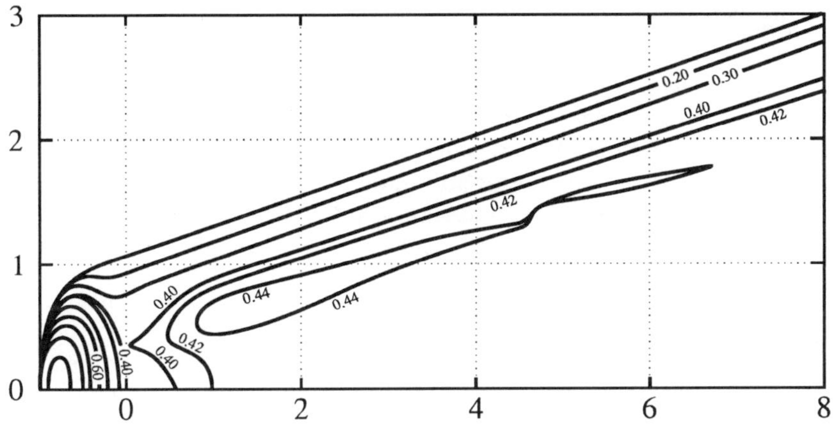

Figure 5.29 Pressure contours on the windward side of a flat delta wing with blunt leading edges. Non-equilibrium air. $V_\infty = 7$ km/s, $p_\infty = 10^{-5}$ atm, $T_\infty = 210$K, $\alpha = 40°$.

5.3 NON-EQUILIBRIUM WEAKLY IONIZED FLOW

The problem of non-equilibrium shock layer ionization is considered below for the flight conditions of the preceding section. Under these conditions, the charged particle concentration is small and the ionization effect on the shock layer flow field is negligible. Nevertheless, the charged particle concentration can be quite sufficient to impede propagation of the radio waves blocking the communication performance. For this reason, the study of non-equilibrium ionization in weakly ionized shock layer flows is a matter of interest.

The problem was investigated by many researchers. Most of them calculated air flows investigating, in particular, the kinetics of charged particle generation, the influence of vibrational relaxation, the difference between electron and heavy particle temperatures, and diffusion and impurities effects. The main results can be formulated as follows.

At the altitudes $H \lesssim 75$ km, the charged particles are generated in a shock layer predominantly due to the associative ionization reaction

$$\mathrm{N} + \mathrm{O} \rightleftarrows \mathrm{NO}^+ + e$$

which has a minimum activation energy. The calculations by Miner and Lewis (1975) witness for a predominant role of the above ionization process up to $H \approx$ 84 km. With increasing altitude, other ionization processes become important because of a lower dissociation degree and a higher gas translational temperature. This has been substantiated by the calculations of Kang *et al* (1973) and Zalogin (1976) and by comparison with flight measurements. Hayes and Rotman (1973) showed that vibrational relaxation of O_2 and N_2 molecules could significantly influence the electron concentration in a shock layer. Multicomponent diffusion and baro-diffusion effects manifest themselves to the greatest extent within a bow shock wave. They become significant at Reynolds numbers $Re_\infty \lesssim 10^3$ (Zalogin 1976). The results of Kang *et al* (1973) and Song *et al* (1985) show a rather small difference between the electron and heavy particle temperatures and its small effect on the electron concentration even at high altitudes. This agrees with the high efficiency of electron-molecule energy exchange and the fact that under the conditions considered, ionization occurs mainly through heavy particle collisions. Lewis and Song (1986) demonstrated significant effects of impurities and ablation of the heat protection material on the charged particle concentration.

5.3.1 Formulation of the problem

Non-equilibrium shock layer ionization is considered below with reference to axisymmetric air and $CO_2 + N_2$ flows about spherically blunted cones (Golovachov 1981a,b; Golovachov and Leontieva 1982). Bearing in mind a weak ionization effect on the shock layer flow field, the gas dynamic functions and neutral species concentrations are taken from the solutions which have been obtained ignoring the ionization processes; see subsection 5.2.4. In most of the calculations, the gas mixture is assumed to be quasi-neutral throughout the shock layer. The diffusion is supposed to be ambipolar and described by the Fick law with constant effective Schmidt number Sc_a. The electron temperature is taken to be equal to the translational temperature of the heavy particles.

With the above assumptions, the calculation of non-equilibrium shock layer ionization reduces to integration of the continuity equations for ions. In accordance with the shock layer model employed in the calculation of gas dynamic functions and neutral species concentrations, the continuity equations for ions include only those diffusion flux components which are normal to the body surface. For the body-fitted (s, n)-coordinate system, they are written as follows

$$\frac{\rho u}{H_s}\frac{\partial c_i}{\partial s} - \frac{\partial}{\partial n}\left(\frac{\mu}{\mathrm{Sc}_a}\frac{\partial c_i}{\partial n}\right) + \left[\rho v - \frac{\mu}{\mathrm{Sc}_a}\left(\frac{\kappa}{H_s} + \frac{\cos\theta}{H_\varphi}\right)\right]\frac{\partial c_i}{\partial n} = \dot{w}_i \qquad (5.87)$$

where c_i are the mass fractions of ions; H_s, H_φ are the Lamé coefficients, κ is the body generatrix curvature, and θ is the generatrix slope to the symmetry axis.

Equations (5.87) are of a parabolic type. The solution is sought for in the domain bounded by the body surface and the bow shock wave. The ion concentrations on the stagnation streamline are obtained from the ordinary differential equations derived from (5.87) taking into account the flow symmetry. The bow shock boundary conditions are given by the modified Rankine–Hugoniot relationships assuming ionization to be frozen within the shock wave. The body surface is supposed to be insulated, fully catalytic with respect to the ion recombination, and having a fixed temperature. The proper wall boundary conditions, ignoring possible violation of the mixture quasi-neutrality near the body surface, read

$$(c_i)_w = 0 \qquad (5.88)$$

Continuity equations (5.87) are integrated subject to above boundary conditions by the implicit space-marching finite-difference scheme of Crank–Nicolson type including linearization of the species production rates in accordance with subsection

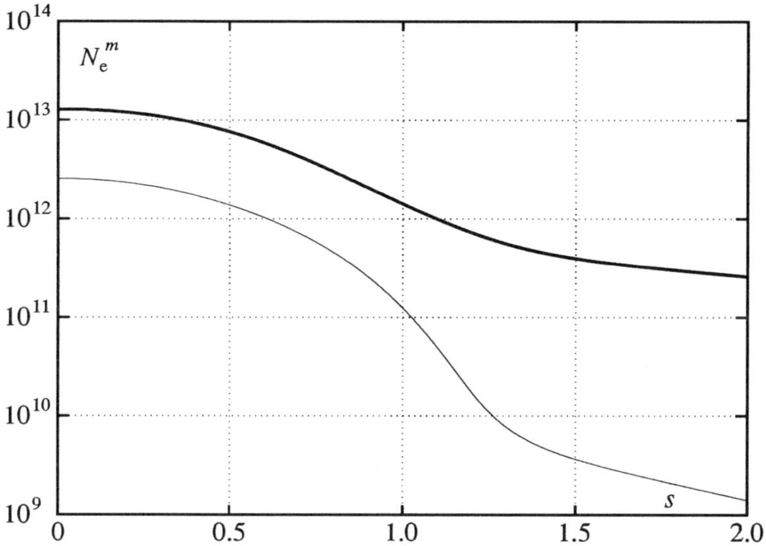

Figure 5.30 Variation of the electron number density along the shock layer on the spherically blunted cone. Air, $V_\infty = 5$ km/s, $H = 60$ km, $R = 1$ m, $\theta_c = 20°$, $T_w = 400$K. Hair line, local thermodynamic equilibrium assumption.

5.2.1. In each step in the s-coordinate, the difference equations are solved jointly by a block 'progonka' algorithm.

5.3.2 Air flows

When calculating air flows, the electrons are supposed to be produced in a shock layer due to associative ionization of nitrogen monoxide molecules. The ionization rate constants are given by the formulae (Dellinger 1971)

$$k_f = 6.4 \times 10^9 \sqrt{T} \exp\left(-\frac{32000}{T}\right) \frac{\mathrm{cm}^3}{\mathrm{mol} \cdot \mathrm{s}}$$
$$k_r = 7 \times 10^{18} T^{-0.9} \frac{\mathrm{cm}^3}{\mathrm{mol} \cdot \mathrm{s}} \tag{5.89}$$

The results discussed below correspond to a wall which is fully catalytic with respect to reactions between neutral species. The effective ambipolar Schmidt number was taken to be equal to 0.375.

Figure 5.30 shows the variation of the electron number density along the shock layer. Here N_e^m (cm^{-3}) is the maximum of the electron number density on a line $s = const$. The results are compared with those calculated on the local thermodynamic equilibrium assumption. One can see that the electron concentration substantially exceeds the equilibrium level. The difference increases with the distance from the stagnation point, which is accounted for by the freezing of the electron recombination in the expansion flow near the lateral part of the cone.

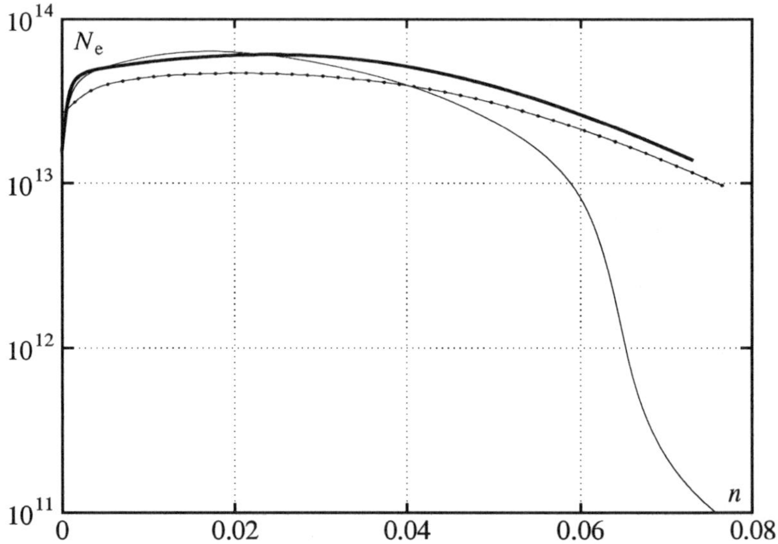

Figure 5.31 Electron number density profiles on the stagnation streamline. Non-equilibrium air, $V_\infty = 7.9$ km/s, $\rho_\infty = 2.2 \times 10^{-7}$ g/cm^3, $R = 3.05$ cm, $T_w = 1500$K. Full and hair-dotted lines, calculations of Golovachov and Leontieva (1982) with various reaction rate constants; hair line, self-similar solution of Dellinger (1971).

Figure 5.31 compares the electron number density on the stagnation streamline with the self-similar solution of Dellinger (1971) calculated by means of the bow shock capturing method. One can see a good agreement between the maximums of N_e. The discrepancy near the bow shock is caused by the difference in formulation of the problem and by the underestimation of the shock layer thickness in the self-similar solution.

Available data for the rate constants of chemical reactions differ from each other, which introduces some ambiguity in the shock layer calculations. Numerical investigation carried out by Golovachov and Leontieva (1982) with the use of various rate constants of associative ionization of nitrogen monoxide molecules have shown that under the flight conditions considered, the N_e^m value on the stagnation streamline varies nearly proportionally with the square root of the forward reaction rate constant and is nearly independent of the reverse rate constant. To estimate the effect of uncertainty of the data on chemical reaction rates for neutral species, the non-equilibrium ionization was calculated with the flow field data obtained using the rate constants of Evans *et al* (1974). The results are plotted in figure 5.31 by the hair-dotted curve. The discrepancy is about 20 per cent.

Figure 5.32 compares calculations of the electron number density with the shock tube measurements by Kaegi and Mc Menamin (1969). Shown in this figure are the N_e profiles on the line $s = const$ drawn from the sphere-cone junction point. The agreement between the viscous shock layer calculations and the experimental data seems to be satisfactory, especially if one keeps in mind that the experimental N_e

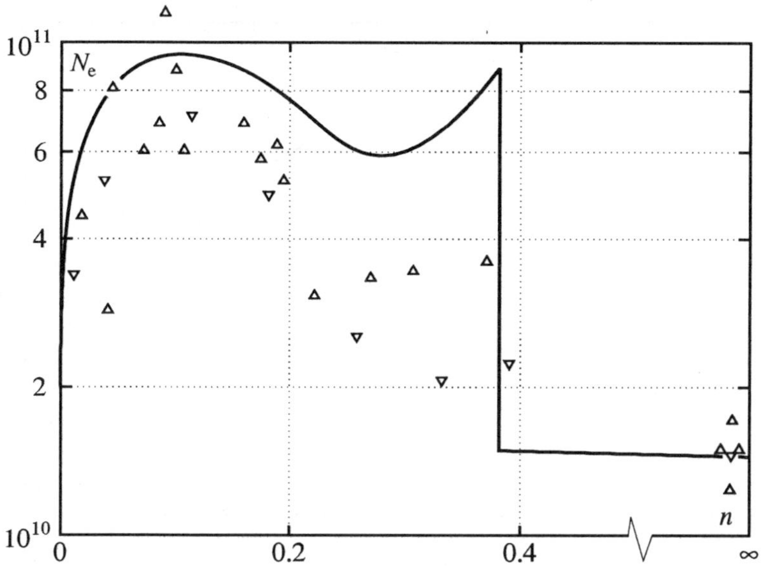

Figure 5.32 Comparison of the calculated and measured electron concentration profiles in flow about a spherically blunted cone. Non-equilibrium air, $V_\infty = 4.57$ km/s, $\rho_\infty = 1.8 \times 10^{-7}$ g/cm^3, $R = 7.6$ cm, $\theta_c = 10°$, $T_w = 300$K, $s = 1.396$.

values are somewhat underestimated because of the use of the collisionless plasma theory in the probe data processing (Dellinger 1971).

Figure 5.33 compares the calculation results with the flight measurement data presented by Hayes and Rotman (1973). The experiments were carried out with a spherically blunted cone with nose radius $R = 16.08$ cm and half-angle $\theta_c = 9°$. The cone was lifted up to the altitude $H = 322$ km by a rocket and then it fell down in a nearly vertical trajectory with an entry velocity $V_\infty = 5.34$ km/s.

The full line in figure 5.33 depicts the variation of the electron number density in the stagnation region calculated by Golovachov and Leontieva (1982). The measurement data are shown by the circles. At small altitudes, the discrepancy between the calculated and measured electron concentrations can be caused by the vehicle deceleration. At high altitudes, it results from neglecting some physical processes. The main of them are vibrational relaxation and ionization by the heavy particle collisions.

The vibrational relaxation effect on the concentration of the electrons produced by the associative ionization of nitrogen monoxide molecules is determined by two competing factors, namely, decrease of the dissociation degree and increase of the gas temperature. The first of them reduces the ionization rate because of a diminution of the atom concentration, while the second one increases it due to an increase of the ionization rate constant. To investigate vibrational non-equilibrium effects, Hayes and Rotman (1973) compared numerical solutions of the thin viscous shock layer equations with and without vibrational relaxation of

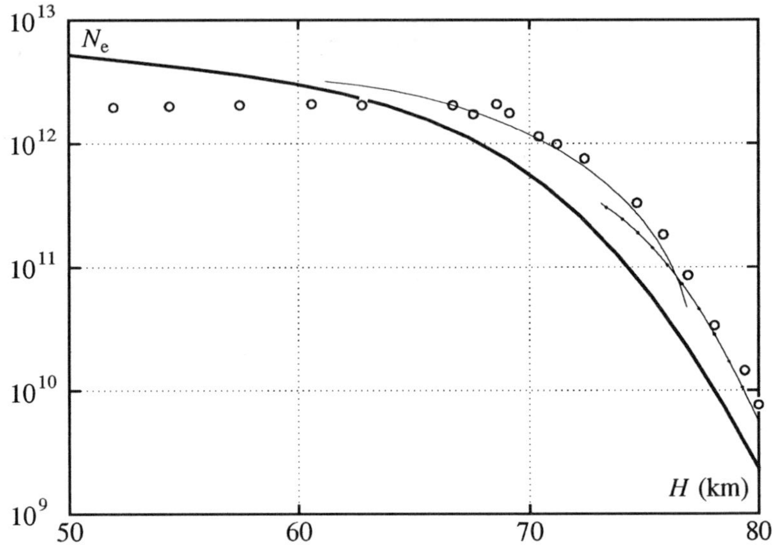

Figure 5.33 Comparison of calculated and measured electron number densities in the stagnation region of the shock layer. Full line, Golovachov and Leontieva (1982); hair line, Hayes and Rotman (1973); hair-dotted line, Zalogin (1976); circles, flight measurement data presented by Hayes and Rotman (1973).

O_2 and N_2 molecules. Their results, calculated with the vibrational equilibrium assumption, do not differ from those shown in figure 5.33 by the full line. The calculations taking into account the vibrational relaxation are plotted by the hair line. One can see that at the altitudes $H = 65 \div 75$ km, they provide higher values of the electron concentration, which are in excellent agreement with the measurements. However, at higher altitudes, the discrepancy between calculations and measurements rapidly increases. At $H \gtrsim 80$ km, it exceeds an order of magnitude, the calculations with vibrational relaxation providing a lower electron number density than those with the vibrational equilibrium assumption. Zalogin (1976) has shown that this discrepancy is caused by a change of the ionization kinetics with decreasing ambient density. At Reynolds numbers $Re_\infty \lesssim 3 \times 10^3$, the associative ionization is surpassed by the impact ionization:

$$N_2 + M \rightleftarrows N_2^+ + e + M \qquad O_2 + M \rightleftarrows O_2^+ + e + M$$

where M is N_2 or O_2 molecules. The hair-dotted curve in figure 5.33 depicts the calculations by Zalogin (1976), in which both associative and molecular collision ionizations were taken into account. One can see a good agreement of the calculations with the flight experiments at high altitudes.

Figure 5.34, borrowed from (Zalogin 1976), demonstrates a baro-diffusion effect on the electron number density on the stagnation streamline. The calculations taking into account the baro-diffusion result in a higher electron concentration

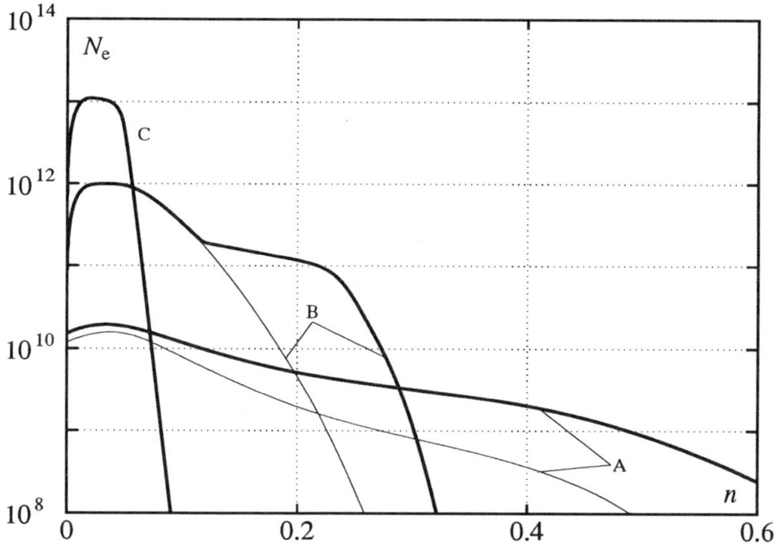

Figure 5.34 Profiles of electron number density on the stagnation streamline calculated with (full lines) and without (hair lines) baro-diffusion. Non-equilibrium air, $V_\infty = 7.5$ km/s; A, $Re_\infty = 150$; B, $Re_\infty = 480$; C, $Re_\infty = 1500$.

and a larger extent of the ionized flow region at small Reynolds numbers. As the Reynolds number increases, the baro-diffusion effect diminishes and manifests itself only in the vicinity of the bow shock wave. Together with an extension of the ionized flow region, the total number of electrons increases. This is accounted for by the species redistribution in the shock layer. Because of the baro-diffusion, the light species N and O are concentrated near the bow shock. This increases the electron production rate by associative ionization because of the increasing atom concentration in the high-temperature region of the shock layer.

5.3.3 $CO_2 + N_2$ flows

Formulation of the problem and the numerical method are analogous to those of the previous subsection. Keeping in mind a small content of nitrogen in the Martian and Venusian atmospheres, the gas dynamic functions and CO_2, CO, O_2, and C concentrations are borrowed from calculations of pure carbon dioxide flows discussed in subsection 5.2.4. Ion concentrations are obtained from the continuity equations (5.87). In the calculation of $CO_2 + N_2$ flows, they are complemented by the continuity equations for N_2, NO, and N concentrations. Some calculations took into account the shock layer contamination by sodium vapour produced by the destruction of the heat protection material. This being the case, the continuity equation for element Na was added to the set of governing equations.

Preliminary calculations of $CO_2 + N_2$ flows were carried out taking into account about 30 reactions between the charged particles including the ionization

Table 5.3

Reaction number	Forward rate constant ($cm^3/mol \cdot s$)	Reverse rate constant ($cm^3/mol \cdot s$)
1	$2.43 \times 10^{10}\sqrt{T}\exp(-32150/T)$	$4.0 \times 10^{20}T^{-1.2}$
2	$7.55 \times 10^{10}\sqrt{T}\exp(-33660/T)$	$1.08 \times 10^{21}T^{-1.2}$
3	$5.42 \times 10^{10}\sqrt{T}\exp(-80560/T)$	$7.2 \times 10^{20}T^{-1.2}$
4	$2.30 \times 10^{12}\sqrt{T}\exp(-55190/T)$	1.5×10^{13}
5	$1.09 \times 10^{12}\sqrt{T}\exp(-4630/T)$	$2.42 \times 10^{11}\sqrt{T}$
6	$1.07 \times 10^{12}\sqrt{T}\exp(-33745/T)$	$4.65 \times 10^{16}T^{-1/2}$
7	$8.00 \times 10^{24}T^{-2}\exp(-62400/T)$	$5.4 \times 10^{27}T^{-2}(cm^6/mol^2 \cdot s)$
8	$2.06 \times 10^{12}\sqrt{T}\exp(-79410/T)$	4.3×10^{14}
9	$2.23 \times 10^{11}\sqrt{T}\exp(-47710/T)$	4.6×10^{13}
10	$2.23 \times 10^{11}\sqrt{T}\exp(-100160/T)$	4.6×10^{13}

by electron and heavy particle impact, associative ionization, and charge exchange processes. The rates of impact ionization of C, O, and N atoms were evaluated using the theory of Biberman *et al* (1970). For reactions with sodium atoms and ions, the rate constants recommended by Jensen (1974) and Farragher *et al* (1969) were used. The rate constants of other processes were borrowed from the papers by Dunn (1971), Kang *et al* (1973) and Evans *et al* (1974). The calculations have shown that about 90 per cent of the electrons are produced in the following reactions

$$
\begin{array}{llll}
(1) & N + O \rightleftarrows NO^+ + e & (6) & CO + CO^+ \rightleftarrows C^+ + CO_2 \\
(2) & C + O \rightleftarrows CO^+ + e & (7) & Na + M \rightleftarrows Na^+ + e + M \\
(3) & O + O \rightleftarrows O_2^+ + e & (8) & O_2 + Na^+ \rightleftarrows O_2^+ + Na \\
(4) & CO + NO^+ \rightleftarrows CO^+ + NO & (9) & NO + Na^+ \rightleftarrows NO^+ + Na \\
(5) & CO + O^+ \rightleftarrows CO^+ + O & (10) & CO + Na^+ \rightleftarrows CO^+ + Na
\end{array}
$$

The rate constants of these reactions are presented in table 5.3.

At first, let us dwell on the calculation results which enable us to estimate a validity of the assumptions made in the formulation of the problem. Thus, comparison of the electron concentration profiles calculated with various ambipolar Schmidt numbers for the limits of fully catalytic and non-catalytic walls with regard reactions with the neutral species have shown a negligible dependence of the results on the above factors. So most calculations were carried out for the fully catalytic wall with $Sc_a = 0.25$.

The effects of thermal non-equilibrium between the electrons and heavy particles have been estimated with the use of numerical solutions obtained with the electron temperature taken to be equal to the vibrational temperature of CO_2 molecules. This is consistent with available data on high efficiency of the energy exchange between electrons and vibrational states of the molecules (Biberman *et al* 1970). The electron temperature $T_e = T_v$ was employed to evaluate the rates of reactions involving electrons. The difference in the electron density profiles calculated with $T_e = T_v$ and $T_e = T$ has been found to be negligible.

The applicability of approximate boundary condition (5.88) ignoring a violation of the mixture quasi-neutrality near the body surface has also been investigated. The above occurs within the sheath whose thickness is equal, to an order of magnitude, to the Debye length. Under the flight conditions considered, the sheath thickness can be either more or less than the mean free path of the particles. For a collisionless regime, the sheath effect can be taken into account with the boundary condition

$$\frac{\mu}{\mathrm{Sc}_a} \frac{\partial c_i}{\partial n} - 0.8\rho\sqrt{R^* T} c_i = 0 \tag{5.90}$$

which presents a continuity of the ion mass fluxes on the sheath edge obtained using the continuum and approximate kinetic considerations (Camac and Kemp 1963). The calculations showed that the use of wall boundary condition (5.90) instead of (5.88) noticeably changes the electron number density only at several grid points adjacent to the body surface, where the electron concentration is by far less than its maximum value. The applicability of boundary condition (5.88) to the collisional flow regime within the sheath has been shown by Knight (1971) and Nishida (1972).

Some of the calculation results are presented in figures 5.35 through 5.38. Figure 5.35 depicts the electron number density profiles on the stagnation streamline ahead of a spherically blunted cone in a pure carbon dioxide stream. The hair curve shows the N_e profile calculated assuming vibrational equilibrium for all molecules. One can see a significant effect of the carbon dioxide vibrational relaxation on the electron concentration in spite of a rather small length of the vibrational relaxation zone at the flight conditions considered; see figure 5.11.

Figure 5.36 demonstrates the ambient density and impurity effects on the N_e profile on the stagnation streamline. Evolution of the N_e profile with ambient density illustrates a transition from the frozen to the equilibrium flow regime. With increasing ρ_∞, the electron density maximum becomes more distinct and shifts towards the bow shock front. The full curves in figure 5.36 depict the results for a pure carbon dioxide free stream. The hair curve B1 corresponds to the free stream composition of 99% CO_2 + 1% N_2 by mass. The hair-dotted curve B2 shows the electron density profile calculated for the free stream composition of 99% CO_2 + 1%N_2 with the sodium impurity in the shock layer. According to the data presented by Evans *et al* (1974), the rate of Na production on the body surface was assumed to be 5×10^{-5} g/(cm$^2 \cdot$ s). Keeping in mind a small mass injection rate, the proper wall boundary condition was obtained by equating the mass diffusion flux of element Na to its production rate on the body surface. The calculations reveal a significant influence of the impurity on the shock layer ionization. Figure 5.37 presents the profiles of mass fractions of ions for the free stream composition of 99% CO_2 + 1% N_2.

Figure 5.38 presents the electron number density profiles at several distances from the stagnation streamline. The non-monotonic variation of the N_e maximum

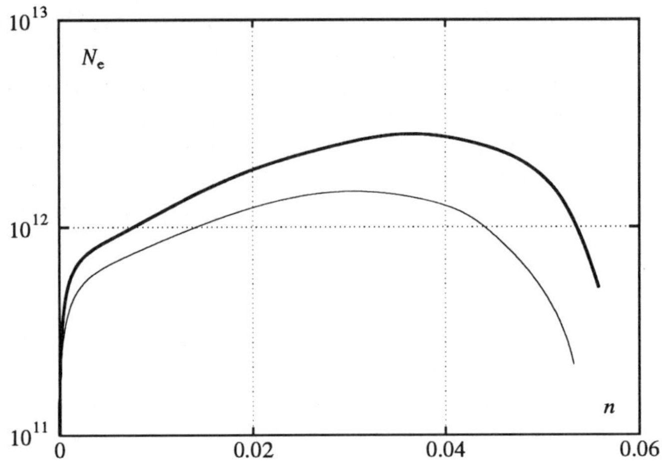

Figure 5.35 Profiles of electron number density on the stagnation streamline in non-equilibrium carbon dioxide flow about a spherically blunted cone. $V_\infty = 5.5$ km/s, $\rho_\infty = 2.1 \times 10^{-7}$ g/cm^3, $T_w = 1100$ K, $R = 0.2$ m, $\theta_c = 60°$. Hair curve, vibrational equilibrium assumption.

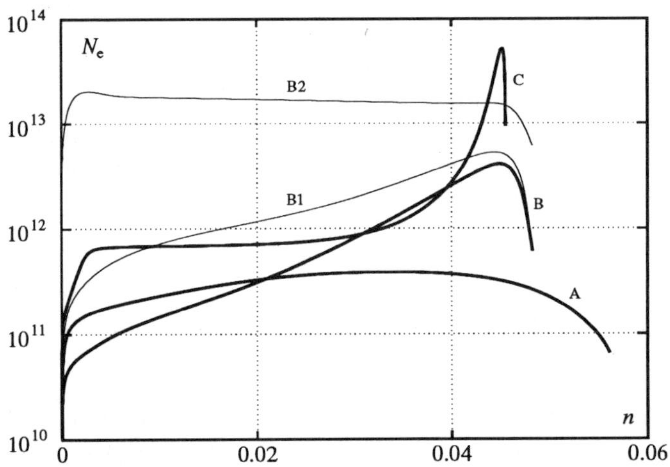

Figure 5.36 Profiles of electron number density on the stagnation streamline at various ambient densities and free stream compositions. $V_\infty = 5.5$ km/s, $T_w = 1100$ K, $R = 1$ m, $\theta_c = 60°$; A, $\rho_\infty = 3 \times 10^{-8}$ g/cm^3; B, $\rho_\infty = 3 \times 10^{-7}$ g/cm^3; C, $\rho_\infty = 3 \times 10^{-6}$ g/cm^3. Full lines, 100%CO_2; hair line B1, 99%CO_2 + 1%N_2; hair line B2, 99%CO_2 + 1%N_2 with Na impurity.

along the shock layer correlates with a non-monotonic variation of the bow shock slope to the cone axis.

In conclusion, it should be stressed that the above analysis of non-equilibrium

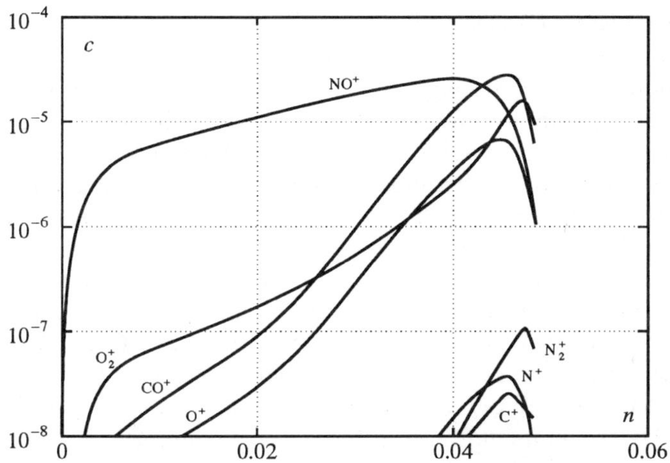

Figure 5.37 Profiles of mass fractions of ions on the stagnation streamline in a 99%CO_2 + 1%N_2 flow about a spherically blunted cone. $V_\infty = 5.5$ km/s, $\rho_\infty = 3 \times 10^{-7}$ g/cm^3, $T_w = 1100$ K, $R = 1$ m, $\theta_c = 60°$.

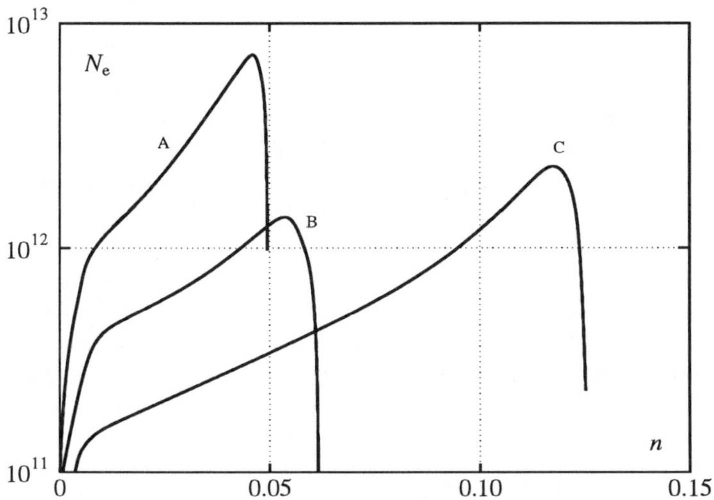

Figure 5.38 Profiles of electron number density on the lines s =const in the shock layer on a spherically blunted cone. 98%CO_2 + 2%N_2, $V_\infty = 5.5$ km/s, $\rho_\infty = 3 \times 10^{-7}$ g/cm^3, $T_w = 1100$ K, $R = 1$ m, $\theta_c = 60°$; A, $s = 0$; B, $s = 0.698$; C, $s = 1.92$.

ionization was carried out ignoring ionization effect on the shock layer flow field, which is admissible at a moderate hypersonic flight speed. In a general case, one has to solve the coupled gas dynamics-ionization problem. An example of such calculations can be found in the paper by Golovachov and Lunkin (1979).

5.4 RADIATING FLOW

Shock layer flow field and aerodynamic heating of vehicles entering the planetary atmospheres at high hypersonic speed can be strongly influenced by radiation energy transfer. The radiative heat flux to the vehicle depends on its dimensions and shape, the atmosphere chemical composition and the flight trajectory. For the Earth's atmosphere entry, radiative heating becomes significant at flight speed exceeding, approximately, 10 km/s. In this part of the entry trajectory, the shock layer gas is usually in local thermodynamic equilibrium. For the Venusian and Martian atmospheres, which consist mainly of carbon dioxide and nitrogen, the radiation energy transfer has to be taken into account at a lower flight speed because of the intensive molecular band system radiation of the shock layer. The chemical relaxation length of $CO_2 + N_2$ mixtures is several times larger than in air flows, and, therefore, non-equilibrium effects may be important, which is especially true for the Martian atmosphere entry. Shock layer radiation dominates the energy transfer in flows about vehicles entering the atmospheres of outer planets, for which the entry speed amounts to some tens of kilometres per second.

Most theoretical investigations of radiating shock layers were performed for flows in the vicinity of the stagnation point where the radiation effect is often a maximum while the flow analysis can be essentially simplified by using the thin viscous shock layer model and self-similar solutions.

Theoretical and experimental investigations carried out since the early 60's elucidated the main radiation processes in the gas mixtures corresponding to planetary atmospheres and revealed the importance of the shock layer radiation cooling, frequency dependence, and reabsorption of radiation. The effects of radiation-shock layer flow coupling and the ablation of heat protection material on convective and radiative heating were investigated as well.

The above results have been discussed in reviews by Goulard *et al* (1969), Anderson (1969), Karasev (1973), Olstad (1974), and the monographs by Agafonov *et al* (1972), Belotserkovskii *et al* (1974), and Pilugin and Tirskii (1989). It is worth noting that most of the calculations of two- and three-dimensional radiating shock layer flows have been performed with an inviscid gas model. Below, we shall focus our attention on numerical simulation of viscous radiating shock layer flows.

5.4.1 Calculation of the radiative heat flux and its divergence

It has been noted that the radiation influence on the flows under study can be taken into account by including the divergence of the radiative heat flux vector in the energy balance equation. In calculation of shock layer flows, the above quantity is often evaluated with a local one-dimensional tangent slab approximation in which the radiative heat flux vector and its divergence are given by formulae (5.28) and (5.29).

To take into account the radiation frequency dependence, the spectrum is divided into N intervals, each being characterized by a mean absorption coefficient independent of the radiation frequency. In this case, the formula for the divergence of the radiative heat flux vector takes the form

$$\operatorname{div}\boldsymbol{q}_r = 2\pi \sum_{i=1}^{N} \kappa_i \left[2S_i - \int_0^{\tau_{is}} S_i E_1 \left(|\tau_i - t_i| \right) dt_i \right] \tag{5.91}$$

where

$$S_i = \int_{(\Delta\nu_i)} S_\nu \, d\nu$$

Integrating by parts and taking into account the properties of integro-exponential functions (Shih I Pai 1966), one can rewrite (5.91) as follows

$$\begin{aligned} \operatorname{div}\boldsymbol{q}_r = 2\pi \sum_{i=1}^{N} \kappa_i \Bigg[& S_i(0)E_2(\tau_i) + S_i(\tau_{is})E_2(\tau_{is} - \tau_i) \\ & + \int_0^{\tau_i} \frac{dS_i}{dt_i} E_2(\tau_i - t_i) dt_i - \int_{\tau_i}^{\tau_{is}} \frac{dS_i}{dt_i} E_2(t_i - \tau_i) dt_i \Bigg] \end{aligned} \tag{5.92}$$

If we denote the grid point number by subscript j ($j = 0, 1, \ldots, M$) and assume a linear variation of S_i between the neighbouring grid points, we can obtain the following formula for the divergence of the radiative heat flux vector at the j-th grid point

$$\begin{aligned} (\operatorname{div}\boldsymbol{q}_r)_j = 2\pi \sum_{i=1}^{N} \kappa_{i,j} \Bigg\{ & S_{i,0} E_2(\tau_{i,j}) + S_{i,M} E_2(\tau_{i,M} - \tau_{i,j}) \\ & + \sum_{k=0}^{M-1} \frac{S_{i,k+1} - S_{i,k}}{\tau_{i,k+1} - \tau_{i,k}} \left[E_3 \left(|\tau_{i,j} - \tau_{i,k+1}| \right) - E_3 \left(|\tau_{i,j} - \tau_{i,k}| \right) \right] \Bigg\} \end{aligned} \tag{5.93}$$

The integro-exponential functions in (5.93) can be replaced by the ordinary exponential functions with the formulae

$$E_2(t) \approx \exp(-2t) \qquad E_3(t) \approx 0.5 \exp(-2t) \tag{5.94}$$

whose error does not exceed a few per cent (Shih I Pai 1968). A large amount of CPU time can be saved by evaluating the inner sum of (5.93) on a uniform grid in the optical coordinate followed by interpolation of the results into the points of the basic computational grid.

When the shock layer gas is in a local thermodynamic equilibrium, the source function S_ν turns into the Planck function B_ν. This being the case, the integrals $S_i = B_i$ can be evaluated using the Taylor series expansion of the equilibrium spectral intensity B_ν (Olstad 1971). Retaining two terms, one obtains

$$\int_\nu^\infty B_\nu \, d\nu \approx \frac{\sigma T^4}{\pi} f(z) \tag{5.95}$$

where σ is the Stephan–Boltzmann constant, $z = h\nu/kT$, and

$$f(z) = \frac{8}{51}\left[e^{-z}\left(z^3 + 3z^2 + 6z + 6\right) + \frac{1}{2}e^{-2z}\left(z^3 + \frac{3}{2}z^2 + \frac{3}{2}z + \frac{3}{4}\right)\right]$$

With (5.95), the quantities B_i are evaluated from the formula ($\nu_i > \nu_{i+1}$)

$$B_i = \int\limits_{\Delta\nu_i} B_\nu d\nu = \int\limits_{\nu_{i+1}}^{\infty} B_\nu d\nu - \int\limits_{\nu_i}^{\infty} B_\nu d\nu \approx \frac{\sigma T^4}{\pi}\left[f(z_{i+1}) - f(z_i)\right] \qquad (5.96)$$

For the shock layer flows under study, the accuracy of approximate formula (5.96) is within a few per cent.

The radiative heat flux at the j-th grid point is determined by the formula

$$\begin{aligned}(q_r)_j = \pi \sum_{i=1}^{N} \Big\{ & B_{i,M} \exp\left[-2\left(\tau_{i,M} - \tau_{i,j}\right)\right] - B_{i,0} \exp\left(-2\tau_{i,j}\right) \\ & - \frac{1}{2}\sum_{k=0}^{M-1}\left(B_{i,k+1} - B_{i,k}\right)\Big[\exp\left(-2\left(|\tau_{i,j} - \tau_{i,k+1}|\right)\right) \\ & + \exp\left(-2\left(|\tau_{i,j} - \tau_{i,k}|\right)\right)\Big]\Big\}\end{aligned} \qquad (5.97)$$

which gives the radiative heat flux on the body surface with $\tau_{i,j} = 0$.

5.4.2 Radiating flows of equilibrium gas mixtures

The set of equations governing an equilibrium flow involves the mass balance equations of the mixture and chemical elements, and the momentum and energy balance equations; see subsection 5.1.3. If the element concentrations are assumed to be constant, which is consistent with the binary diffusion model, the equilibrium composition of the gas mixture, its thermodynamic properties, the molecular transport coefficients, and the radiation properties become functions of only two thermodynamic variables, e.g. temperature and pressure. This being the case, the set of radiating viscous shock layer equations looks similar to that for a non-radiating gas with constant heat capacity, except for the energy balance equation, which contains a divergence of the radiative heat flux vector.

Equilibrium flows are characterized by fairly high Reynolds numbers, therefore, they are usually calculated in the domain bounded by the bow shock wave with the use of the conventional Rankine–Hugoniot relationships. The effects of free stream perturbation by the shock layer radiation have been investigated by Agafonov *et al* (1972) and Biberman *et al* (1972). According to their estimates, the above effects become significant in the Earth's atmosphere at an entry speed $V_\infty \gtrsim 18$ km/s.

Formulating the wall boundary conditions, one has to take into account, in a general case, the destruction of the heat protection material due to intensive aerodynamic heating, which essentially complicates the chemical composition of the shock layer gas and, therefore, calculation of the absorption coefficient. Fortunately, an exact calculation of the absorption coefficients including the foreign species effect is not mandatory, at least in the cases of the Earth and Venus entry. The matter is that the spectral dependence of absorption coefficients of typical destruction products is like that of the shock layer gas, which allows one to take into account the body destruction using available data on the absorption coefficients of high-temperature air and $CO_2 + N_2$ mixtures. Variation of the total radiative heat flux in a foreign gas layer is usually relatively small. For typical Earth and Venus entry conditions, it does not exceed 30 per cent, which is comparable with an error in the radiative heat flux prediction caused by the uncertainty of available data on the optical properties of high-temperature gases. The convective heat flux varies in the same way as in non-radiating flows with mass blowing.

Radiating viscous shock layer flows can be calculated with the same numerical methods which apply to the calculation of non-radiating flows. The divergence of the radiative heat flux vector is generally evaluated using the flow field data of the previous iteration or time step. Below, we discuss some calculations of axisymmetric blunt body flows in the Earth and Venusian atmospheres.

In the calculation of air flows (Golovachov 1972, Golovachov and Popov 1975a), the equilibrium composition of the shock layer gas and its thermodynamic functions were calculated using the analytical approximations by Mikhailov (1961) and Kraiko (1964). To save CPU time, the thermodynamic functions were determined by interpolation using the tables made beforehand with uniform spacing in temperature and pressure.

The viscosity and total heat conductivity were borrowed from the papers by Hansen (1959), Yos (1963), and Sokolova (1973). The divergence of the radiative heat flux vector was evaluated with formulae (5.93) through (5.96). The absorption coefficient for equilibrium air was calculated using the spectral model by Olstad (1971), which took into account the contributions of free-bound and free-free continua, atomic lines, and molecular band systems. The body surface was supposed to have a fixed temperature $T_w = 2500$K. The effects of its destruction were ignored.

The flows in the vicinity of the stagnation streamline were calculated within the framework of the thin viscous shock layer model using the numerical method of subsection 2.5.2. Shock layer radiation decreases the gas temperature and increases the density, which results in diminution of the bow shock distance from the body surface. Figure 5.39 shows the variation of the shock layer thickness on the stagnation streamline with the body nose radius. Here, T_s is the gas temperature just behind the bow shock related to the flight velocity by the implicit formula $V_\infty \approx \sqrt{2h_s(T_s, p_s)}$. The filled circle depicts the calculations by Wilson (1970). The hair line shows the calculations ignoring shock layer radiation. A slight decrease of the shock layer thickness at small R in this case is caused by a relatively low wall temperature; compare with figure 3.21.

Figures 5.40 and 5.41 demonstrate the dependence of the stagnation point convective and radiative heat fluxes on the body nose radius and shock layer pressure.

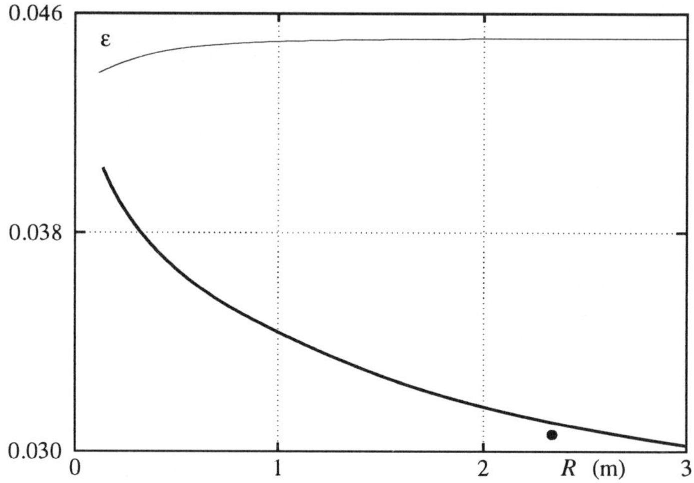

Figure 5.39 Variation of the radiating shock layer thickness on the stagnation streamline with the body nose radius. Equilibrium air, $T_s = 15000$K, $p = 1$ atm. Full line, Golovachov (1972); •, Wilson (1970); hair line, non-radiating shock layer.

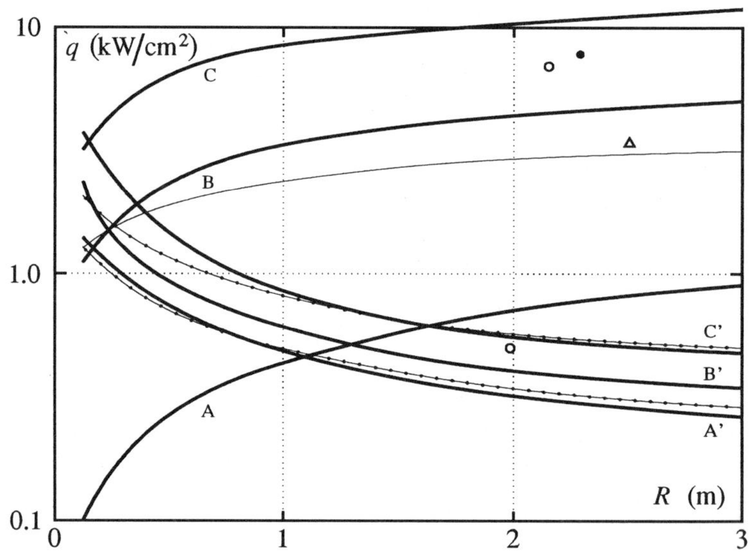

Figure 5.40 Variation of the stagnation point radiative ($A - C$) and convective ($A' - C'$) heat fluxes with the body nose radius. Equilibrium air, $p = 1$ atm; A, A' $T_s = 10000$K; B, B' $T_s = 12500$K; C, C' $T_s = 15000$K. Full lines, Golovachov (1972), hair line and bright circles, Page *et al* (1968); filled circle, Wilson (1970); triangle Chin (1969); hair-dotted curves, Fay and Riddell (1958).

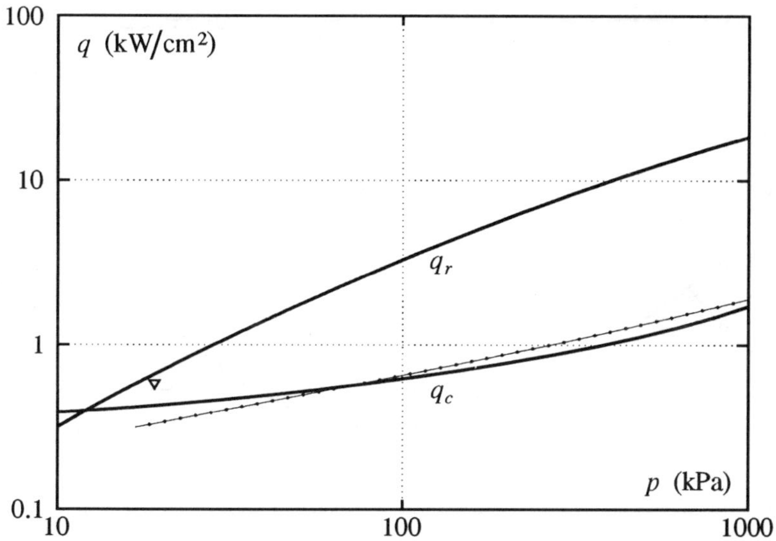

Figure 5.41 Variation of the stagnation point radiative and convective heat fluxes with shock layer pressure. Equilibrium air, T_s = 12500K, R = 1 m. Full lines, Golovachov (1972); ∇, Stulov and Shapiro (1970); hair-dotted curve, Fay and Riddell (1958).

The full curves A through C in figure 5.40 correspond to the radiative and convective heat flux values, respectively, calculated at various temperatures T_s. It is seen that with temperature T_s and pressure increasing, the radiative heat flux increases by far more rapidly than the convective one and becomes dominant at smaller body nose radii. The figures show also the radiative heat flux values calculated with other spectral absorption coefficient models of equilibrium air. It is worth noting that the discrepancy between the radiative heat flux values does not extend beyond the limit predetermined by the uncertainty of available data on the optical properties of high-temperature air.

The hair-dotted curves in figures 5.40 and 5.41 depict the convective heat flux values given by the Fay and Riddell (1958) formula which approximates the calculation results for non-radiating dissociating boundary layer flows. On the usual assumptions

$$Le = 1 \qquad \left(\frac{\rho_w \mu_w}{\rho_s \mu_s}\right)^{0.1} = 1 \qquad h_s \gg h_w \tag{5.98}$$

where Le is the Lewis number, the above formula takes the form

$$q_c = 0.763(Pr)^{-0.6}\sqrt{\rho_s \mu_s}h_s\sqrt{\frac{1}{R}\sqrt{\frac{2(p_s - p_\infty)}{\rho_s}}} \tag{5.99}$$

One can see in the figures that formula (5.99) provides the convective heat flux values which are close to the results of radiating viscous shock layer calculations.

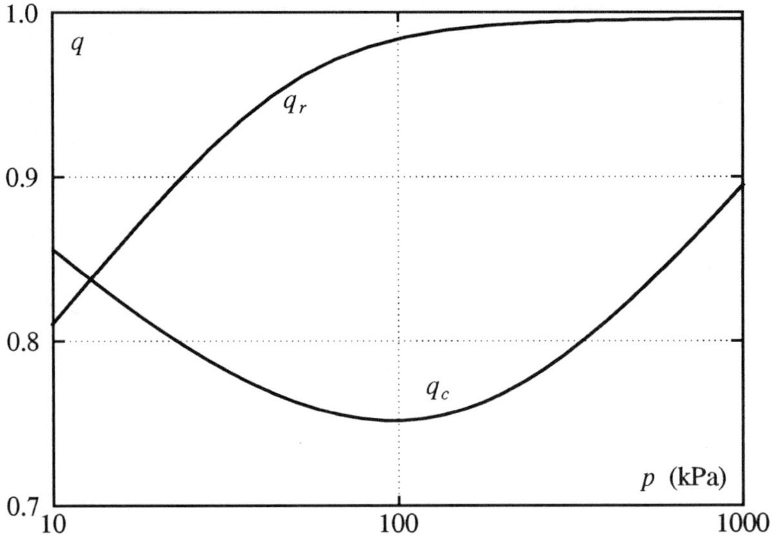

Figure 5.42 Viscosity and radiation effects on the stagnation point radiative and convective heat fluxes. Equilibrium air, $T_s = 12500$K, $R = 1$ m.

A probable reason is that the radiation effect on the convective heat flux is approximately equal to the error introduced by assumptions (5.98).

Figure 5.42 demonstrates the boundary layer effect on radiative heating and the radiation effect on the convective heat flux. Shown in this figure are the variation of the stagnation point radiative heat flux normalized by its value obtained from the inviscid flow calculations and variation of the convective heat flux normalized by its value obtained from the non-radiating flow calculations. It is seen that the radiative cooling can result in a significant reduction of convective heating. A non-monotonous dependence of this effect on the gas pressure is due to an increasing opacity of the shock layer. At relatively high pressure, the radiative heat flux is close to that provided by the inviscid flow calculations. As the pressure decreases, the inviscid gas model begins to overestimate the radiative heating.

Radiating gas flows about the whole front of blunt bodies were calculated using the reduced Navier–Stokes equations (set II of subsection 2.1.1). The solutions were obtained with the implicit unidirectional method of subsection 1.2.2. In most of the calculations, the computational grid contained 21 lines $s = const$ with 26 points on each line. The grid points were clustered to the body surface, mesh spacing in the s-coordinate increased with distance from the stagnation point.

The accuracy of the viscous shock layer model is demonstrated in figure 5.43 which compares solutions of the full and reduced Navier–Stokes equations for a non-radiating air flow. The heat flux and wall shear stress are normalized by $\rho_\infty V_\infty^3$ and $\rho_\infty V_\infty^2/2$, respectively.

Figure 5.44 shows the heat flux distributions over spherically blunted cones. It is seen that both convective and radiative heat fluxes rapidly decrease on the

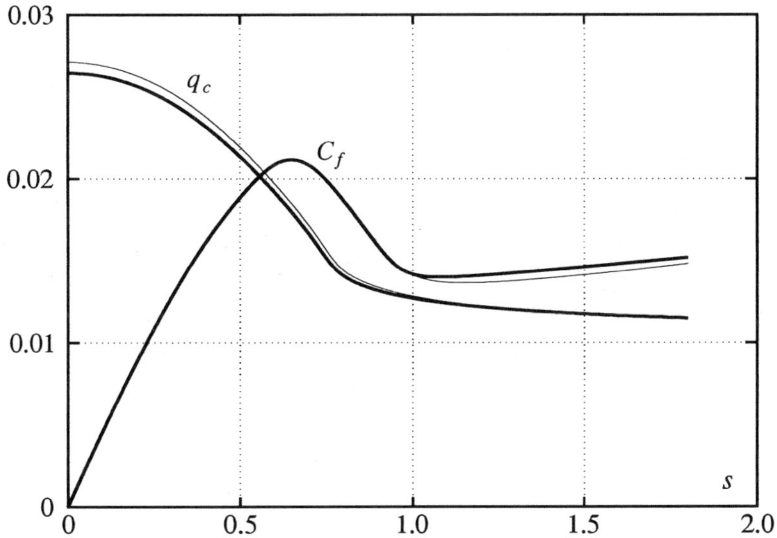

Figure 5.43 Variation of convective heat flux and skin friction coefficient over a spherically blunted cone. Equilibrium air, $V_\infty = 15$ km/s, $H = 61$ km, $R = 0.2$ m, $\theta_c = 45°$, $T_w = 2500$K. Full lines, reduced Navier–Stokes equations; hair lines, full Navier–Stokes equations.

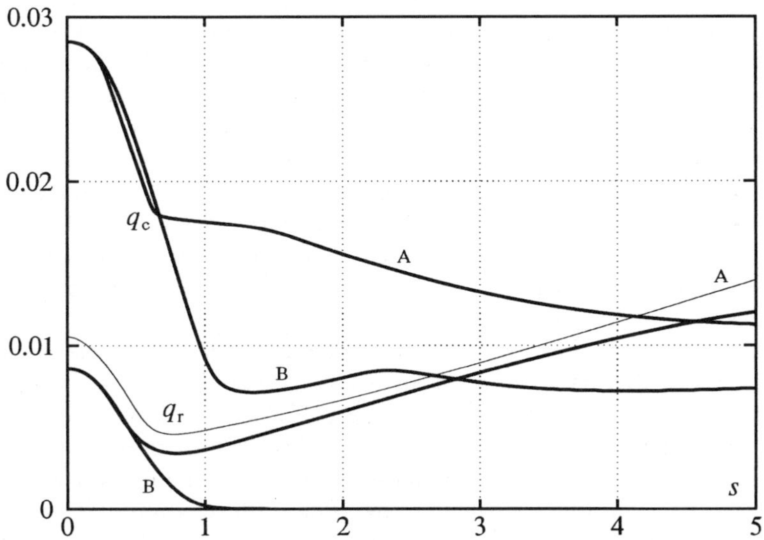

Figure 5.44 Convective and radiative heat flux distributions over spherically blunted cones. Equilibrium air, $V_\infty = 12.2$ km/s, $H = 61$ km, $R = 0.2$ m, $T_w = 2500$K; A, $\theta_c = 60°$; B, $\theta_c = 30°$. Hair line, inviscid flow calculations.

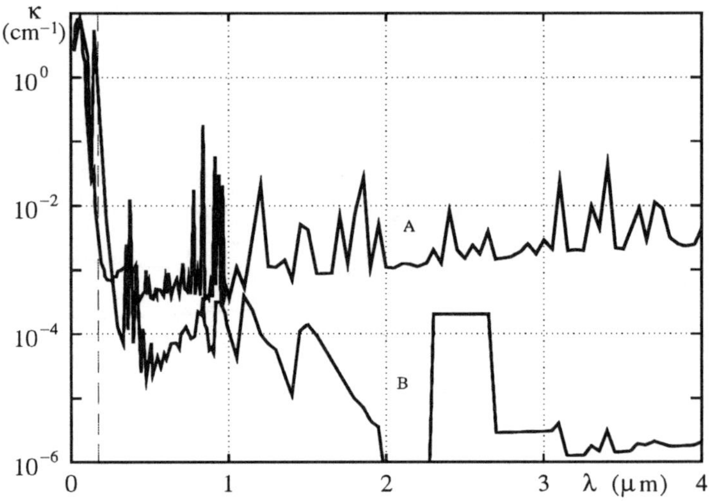

Figure 5.45 Wavelength dependence of the absorption coefficient of an equilibrium $90\%CO_2 + 10\%N_2$ mixture at $p = 1$ atm; A, $T = 10000$K; B, $T = 6000$K.

spherical part of the bodies. The convective flux continues to diminish on the lateral parts of both cones. The radiative heat flux to the lateral part of the cone with half-angle $\theta_c = 30°$ is by far less than its stagnation point value. Calculation of the radiating shock layer flow on the cone with half-angle $\theta_c = 60°$ reveals quite another behaviour of the radiative heat flux. An increase of the shock layer thickness, along with a small reduction of the gas temperature, result in a significant increase of the radiative heat flux to the lateral part of this cone. Far downstream, the radiative heat flux exceeds its stagnation point value and provides the main contribution to aerodynamic heating. The hair line in figure 5.44 depicts the inviscid flow calculations carried out by Callis (1971) with the same spectral model for the air absorption coefficient. The difference between these data and the results of viscous shock layer calculations demonstrates the heat-shielding effect of the boundary layer.

The entry speed in the Venusian atmosphere is usually within the range of $9 \div 12$ km/s. Below, some calculation results are presented, which pertain to blunt body flows of a gas mixture containing $90\%CO_2 + 10\%N_2$ by volume (Golovachov 1975, Golovachov and Popov 1975b). The calculations were carried out using the same viscous shock layer equations and the same numerical method as those employed in the air flow calculations. The thermodynamic functions of the equilibrium gas mixture and the species concentrations were calculated using the approximate formulae by Mikhailov (1961); the transport coefficients were borrowed from the paper by Zdunkevich and Leonas (1972).

The optical properties were determined using the data of Kamenschikov *et al* (1971) on the absorption coefficient of the equilibrium mixture under study in the wavelength range $0.02 \leqslant \lambda \leqslant 4$ μm at various pressures and temperatures.

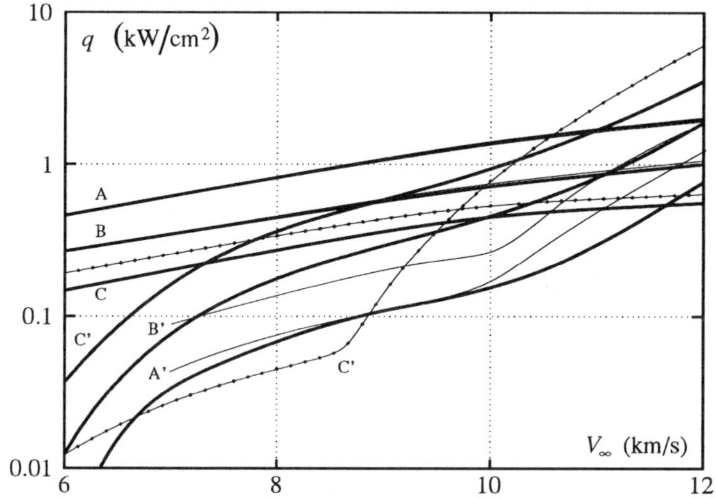

Figure 5.46 Dependence of the stagnation point convective ($A - C$) and radiative ($A' - C'$) heat fluxes on the flight velocity. Equilibrium $90\%CO_2 + 10\%N_2$ mixture, $p = 1$ atm; A, A' $R = 20.4$ cm; B, B' $R = 61$ cm; C, C' $R = 150$ cm. Full lines, Golovachov (1975); hair lines, Page and Woodward (1972); hair-dotted lines, air flow, Golovachov (1972).

The above data were calculated taking into account over forty elementary radiation processes in the continuous spectrum, spectral lines and molecular band systems. An example of these calculations is presented in figure 5.45. It demonstrates an abrupt reduction of the absorption coefficient at $\lambda \approx 0.175$ μm. Under the flight conditions considered, the radiation free path is by far less than the shock layer thickness at $\lambda \lesssim 0.175$ μm and significantly exceeds the shock layer thickness at $\lambda \gtrsim 0.175$ μm. Hence, the spectral dependence of the absorption coefficient was approximated by the step function with the Rosseland and Planck mean absorption coefficients in the wavelength intervals $0.02 \leqslant \lambda \leqslant 0.175$ μm and $0.175 \leqslant \lambda \leqslant 4$ μm, respectively. The integrals of the equilibrium radiation intensity over these spectral intervals were evaluated from formula (5.96) with $z_1 = 719650/T$, $z_2 = 82246/T$, $z_3 = 3598/T$.

Figure 5.46 shows the variation of the stagnation point convective and radiative heat fluxes with flight velocity. The results are compared with calculations by Page and Woodward (1972) carried out with a 9-step approximation of the spectral dependence of the absorption coefficient. Both calculations provide nearly the same values of the convective heat flux. Keeping in mind the use of different data on the optical properties, the agreement between the radiative heat flux values can be considered as satisfactory. The hair-dotted curves depict calculations of the air flow. One can see that the convective heating in $90\%CO_2 + 10\%N_2$ flows is somewhat lower, while the radiative heating at $V_\infty \lesssim 10$ km/s significantly exceeds the air flow values. The latter is caused by the intensive molecular band

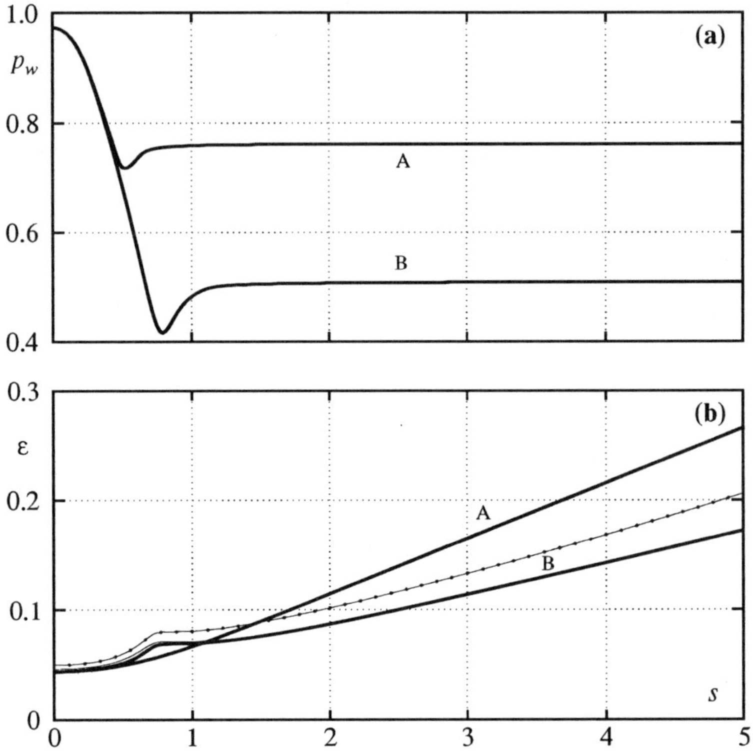

Figure 5.47 Wall pressure and shock layer thickness variations over spherically blunted cones. Equilibrium $90\%CO_2 + 10\%N_2$ mixture, $V_\infty = 9$ km/s, $\rho_\infty = 3.13 \times 10^{-6}$ g/cm^3, $R = 0.2$ m, $T_w = 2500$K; A, $\theta_c = 60°$; B, $\theta_c = 45°$. Hair line, radiation ignored; hair-dotted line, air flow.

system radiation of the $CO_2 + N_2$ mixture. With increasing flight velocity, the concentration of molecules in the shock layer decreases due to dissociation, and the radiation of the high-temperature $CO_2 + N_2$ mixture approaches the air radiation.

Figures 5.47 and 5.48 show the variation of the shock layer thickness, wall pressure, skin friction coefficient, and heat fluxes over two spherically blunted cones. The hair curves show the bow shock detachment and convective heat flux variations calculated while ignoring the radiation energy transfer. The hair-dotted curve in figure 5.47 presents the calculation of the air flow. One can see that the radiation effect on the shock layer thickness is small. Similar to the air flows, the shock layer radiation reduces the convective heat flux to the cone surface. In the vicinity of the stagnation point, the convective and radiative heat fluxes are nearly equal to each other. An increase of the radiative heat flux to the lateral part of the 60°-cone, which is caused by the same reasons as in the air flow, is by far more prominent.

The calculations carried out for $CO_2 + N_2$ mixtures with other free stream

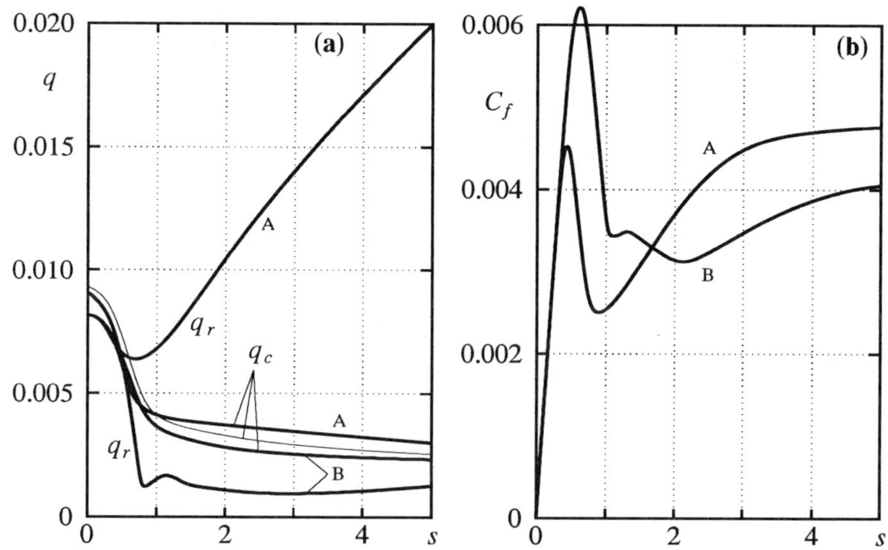

Figure 5.48 Heat flux and skin friction coefficient distributions over spherically blunted cones. Equilibrium 90%CO_2+10%N_2 mixture, $V_\infty = 9$ km/s, $\rho_\infty = 3.13\times10^{-6}$ g/cm^3, $R = 0.2$ m, $T_w = 2500$K; A, $\theta_c = 60°$; B, $\theta_c = 45°$. Hair line, radiation ignored.

chemical compositions have shown a significant influence of the nitrogen content on radiative heating and only a slight effect on convective heating, shear stress, wall pressure, and the bow shock shape.

5.4.3 Estimation of non-equilibrium radiative heating

As the flight velocity and ambient density decrease, the relaxation zone behind the bow shock extends and the shock layer flow becomes non-equilibrium. Radiative heating of the body surface in hypersonic non-equilibrium air and $CO_2 + N_2$ flows has been investigated by Zheleznyak *et al* (1970), Zalogin (1974), Grose and Nealy (1975), Golovachov (1978a, 1979b, 1981a). The radiative heating in the air flows has been found to be small as compared to the convective one, when the non-equilibrium effects are important. As concerns $CO_2 + N_2$ flows, they are characterized by a larger relaxation length, and their radiation at relatively low temperature $T \lesssim 8000$K is several times more intensive than the air radiation. Below, some calculation results of Golovachov (1978a, 1979b, 1981a) are considered, pertaining to the estimation of non-equilibrium shock layer radiation in $CO_2 + N_2$ mixtures at flight velocities $V_\infty = 4 \div 7$ km/s, which are typical for the Martian atmosphere entry.

According to theoretical and experimental data reviewed by Goulard *et al* (1969), non-equilibrium radiation of $CO_2 + N_2$ mixtures under the flight conditions considered is caused mainly by the electronic transitions in carbon monoxide and cyanogen molecules. Over 80 per cent of the total radiation is produced by the fourth positive CO, the violet and red CN band systems which correspond to quenching of the electronic states $CO(A^1\Pi)$, $CN(B^2\Sigma^+)$, and $CN(A^2\Pi_i)$.

To estimate non-equilibrium radiative heating, the simplest two-level model of non-equilibrium radiation (Shih I Pai 1968) is used. In this model, non-equilibrium radiation transport is described by the equation

$$\boldsymbol{\Omega} \cdot \nabla I_\nu = \kappa_\nu \left(S_\nu - I_\nu\right) \tag{5.100}$$

in which the source function S_ν is determined by formula (5.27) with the use of an effective electronic excitation temperature T^* instead of the translational temperature T, i.e.

$$S_\nu = \frac{2h\nu^3}{c^2}\left(e^{h\nu/kT^*} - 1\right)^{-1} \tag{5.101}$$

The temperature T^* is expressed by the formula

$$T^* = -\frac{E^*}{k\ln\left(g^\circ c^* / g^* c^\circ\right)} \tag{5.102}$$

where E^* is the energy of the excited electronic state, g is the degeneracy, k is the Boltzmann constant, c is the mass fraction, and the superscripts 0 and $*$ relate to the ground and excited states, respectively. The parameters of the electronic states have been borrowed from the monograph by Kamenschikov *et al* (1971). In calculating the effective electronic excitation temperature of cyanogen molecules, the state $B^2\Sigma^+$ is considered as the upper level.

The concentrations of the excited molecules are determined from the proper continuity equations. The requisite data on the flow field are taken from the numerical solutions ignoring the radiation transport. Specifying these data by taking into account the divergence of the radiative heat flux vector in the energy balance equation does not change noticeably the estimates of non-equilibrium radiative heating under the flight conditions considered. The above solutions are supposed to provide concentrations of carbon monoxide and cyanogen molecules in the ground electronic state.

The continuity equations for concentrations of the excited molecules for an axisymmetric blunt body flow look like (5.87). The effective Schmidt number is also assumed to be 0.5.

The mass production rate of the excited molecules is calculated taking account of binary particle collisions and radiative processes. Because of a lack of data on the kinetics of electronic excitation by heavy particle impact, the calculations provide only the estimates for non-equilibrium radiative heating. To obtain the upper estimation, the collisional quenching cross-section for the electronic states σ is assumed to be equal to the gas-kinetic one. With these collisional cross-sections of neutral particles, one may neglect the electron impact, whose quenching cross-section is about 10^{-17} cm^2 (Zeldovich and Rizer 1966, Mumma *et al* 1971). Thus, the mass production rate of the excited molecules is determined by the formula

$$\dot{w}^* = \rho\left[\frac{\rho k_c}{m}\left(Kc^\circ - c^*\right) + \alpha c^\circ - \beta c^*\right] \tag{5.103}$$

where ρ and m are the mixture density and molar mass; α is the photo-excitation rate constant; k_c and β are the collisional and radiative quenching rate constants; K is the equilibrium constant

$$K = g^*/g^\circ \exp\left(-\frac{E^*}{kT}\right)$$

With $\sigma = 10^{-15}$ cm^2, the collisional quenching rate constant is

$$k_c = N_A \overline{v} \sigma \approx 2.2 \times 10^{12} \sqrt{T} \frac{\mathrm{cm}^3}{\mathrm{mol} \cdot \mathrm{s}} \tag{5.104}$$

where N_A is the Avogadro number, $\overline{v}$ is the mean thermal velocity of the particles, σ is the effective quenching cross-section averaged over the Maxwellian distribution.

The radiative quenching rate constant is determined by the formula

$$\beta = 0.43 \times 10^8 (E^*)^2 \frac{g^0}{g^*} f_e \ 1/\mathrm{s} \tag{5.105}$$

where the excited state energy E^* is taken in electron-volts. The oscillator strength of the electronic transition f_e is assumed to be equal to 0.15 for the $CO(4^+)$ band system and to 0.027 for the CN violet band system (Menard *et al* 1968). This yields

$$\beta_{\mathrm{CO}} = 2.1 \times 10^8 \ 1/\mathrm{s} \qquad \beta_{\mathrm{CN}} = 1.2 \times 10^7 \ 1/\mathrm{s}$$

The photo-excitation rate constant is determined by the integral

$$\alpha = \int \frac{cU_\nu}{h\nu} \sigma_\nu d\nu \tag{5.106}$$

where c is the light speed, h is the Planck constant, σ_ν is the absorption cross-section, and U_ν is the spectral non-equilibrium radiation energy per unit volume:

$$U_\nu = \frac{1}{c} \int_{(4\pi)} I_\nu d\Omega$$

The integral (5.106) is evaluated over the spectral interval corresponding to the band system considered. The absorption cross-sections have been taken from (Kamenschikov *et al* 1971). The quantity cU_ν is obtained from the radiation transport equation (5.100) in which the source function is given by formula (5.101). Within the framework of the local-one-dimensional tangent slab approximation one obtains

$$cU_\nu = 2\pi \int_0^{\tau_{\nu s}} S_\nu(T^*) E_1 \left(|\tau_\nu - t_\nu|\right) dt_\nu \tag{5.107}$$

Having made the transformations analogous to those of subsection 5.4.1, it is easy to derive the following formula for the quantity cU_ν at the j-th point of the computational grid

$$(cU_\nu)_j = 2\pi\Bigg\{ 2S_\nu(\tau_{\nu j}) - S_{\nu,0}e^{-2\tau_{\nu,j}} - S_{\nu,M}e^{-2(\tau_{\nu,M}-\tau_{\nu,j})} - \frac{1}{2}\sum_{k=0}^{M-1}\frac{S_{\nu,k+1}-S_{\nu,k}}{\tau_{\nu,k+1}-\tau_{\nu,k}}\left[e^{-2(|\tau_{\nu,j}-\tau_{\nu,k+1}|)} - e^{-2(|\tau_{\nu,j}-\tau_{\nu,k}|)}\right]\Bigg\} \quad (5.108)$$

The radiative heat flux to the body surface is evaluated by the formula

$$q_r = 2\pi \int \int_0^{\tau_{\nu s}} S_\nu e^{-2t\nu} dt_\nu d\nu \quad (5.109)$$

The bow shock boundary conditions for the continuity equations are given by the modified Rankine–Hugoniot relationships. The concentration of excited molecules on the body surface is assumed to be equal to the equilibrium value at the wall temperature. The continuity equations on the stagnation streamline reduce to ordinary ones. To advance the solution along the shock layer, an implicit finite-difference scheme of Crank–Nicolson type is used. The reabsorption of radiation is taken into account by means of iterations.

Figure 5.49 shows the dependence of the stagnation point convective and radiative heat fluxes on the ambient density for the initial mixture composition $95\%CO_2 + 5\%N_2$ by volume. The heat fluxes are normalized by $\rho_\infty V_\infty^3$. The contributions of $CO(4^+)$, CN violet and red band systems are plotted by the curves A, B, C, respectively. The hair curves depict the convective and total radiative heat flux values calculated on the local equilibrium assumption. It is seen that the non-equilibrium radiative heat flux exceeds the equilibrium one approximately by two orders of magnitude. Note again that the calculation results strongly depend on the assumed electronic state quenching cross-sections in the molecule collisions and, therefore should be considered as estimates. The graphs of figure 5.49 indicate that, with the above assumptions on the electronic excitation kinetics, radiative heating in mixtures containing more than 95 per cent of carbon dioxide can be estimated taking into account only the $CO(4^+)$ band system radiation.

The next figures show calculations of non-equilibrium shock layer radiation in a pure carbon dioxide flow about a spherically blunted cone. Figure 5.50 presents the profiles of the gas translational temperature and effective electronic excitation temperature of CO molecules along the line $s = 0.5235$ drawn from the sphere-cone junction point. The hair and hair-dotted curves depict the translational temperature profile calculated on the local equilibrium assumption and the T^* profile obtained when assuming the shock layer to be transparent for the $CO(4^+)$ band system radiation. It is seen that reabsorption significantly increases the concentration of excited CO molecules near the body surface resulting in excess T^*

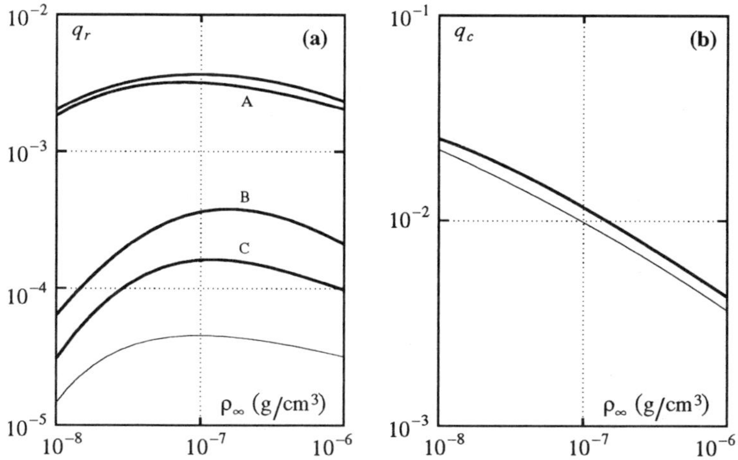

Figure 5.49 Variation of the stagnation point radiative (a) and convective (b) heat fluxes with ambient density. 95%CO_2 + 5%N_2, $V_\infty = 6$ km/s, $R = 1$ m, $T_w = 2900$K; A, CO(4^+); B, CN violet; C, CN red. Hair curves, local equilibrium assumption.

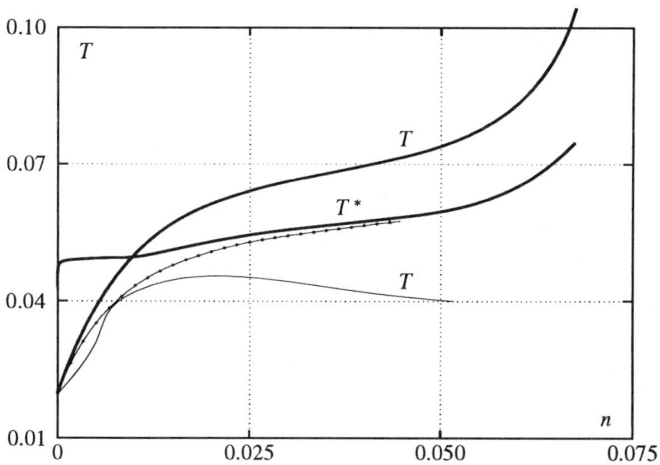

Figure 5.50 Temperature profiles along the line $s = 0.5235$ in an axisymmetric carbon dioxide flow past a spherically blunted cone. $V_\infty = 7$ km/s, $\rho_\infty = 2.07 \times 10^{-7}$ g/cm^3, $R = 0.2$ m, $\theta_c = 60°$, $T_w = 2500$K. Hair line, local equilibrium assumption; hair-dotted line, reabsorption ignored.

over the gas translational temperature. It is worth noting that the calculations of $CO_2 + N_2$ flows have shown a negligible role of reabsorption in the electronic excitation kinetics of CN molecules. The effective temperatures T^* of CN molecules calculated with and without reabsorption are found to be nearly the same.

It is also worthwhile to note that with the assumed collisional quenching cross-

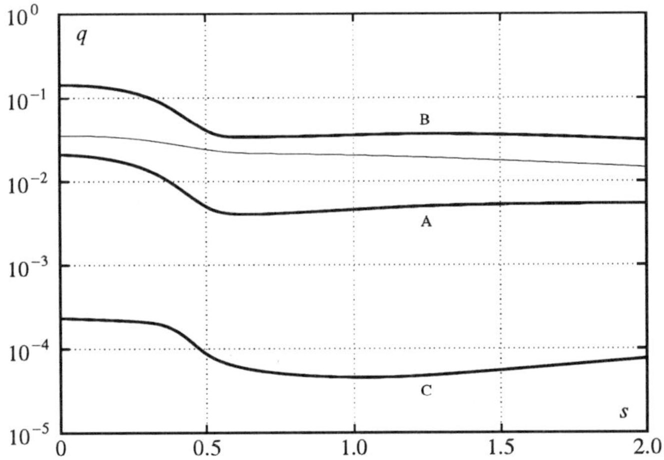

Figure 5.51 Heat flux distributions over a spherically blunted cone in a hypersonic non-equilibrium carbon dioxide flow. $V_\infty = 7$ km/s, $\rho_\infty = 2.07 \times 10^{-7}$ g/cm^3, $R = 0.2$ m, $\theta_c = 60°$, $T_w = 2500$K. Full lines, radiative heat flux; A, $q_r(T^*)$; B, $q_r(T)$; C, $q_r(T_{eq})$; hair line, convective heat flux.

section, only radiative processes violate the electronic equilibrium. All of the effective temperatures T^* calculated ignoring shock layer radiation do not differ from the translational one.

Figure 5.51 shows the heat flux distributions over a cone. The full lines depict the radiative heat flux values calculated taking account of the non-equilibrium excitation of the CO$A^1\Pi$ electronic state (A), assuming $T^* = T$ (B), and on the local thermodynamic equilibrium assumption (C). The hair line depicts the convective heat flux to the fully catalytic wall. The calculations do not demonstrate a drastic increase of the radiative heat flux to the lateral part of the cone, in contrast to the equilibrium radiating flows discussed in the previous subsection. The difference in the radiative heat flux distributions is due to a significant reabsorption of the CO(4$^+$) band system radiation.

The above results show the invalidity of the local thermodynamic equilibrium assumption for the evaluation of radiative heating of the body surface under the flight conditions considered. Assuming the electronic state excitation to be in equilibrium with the translational temperature ($T^* = T$), one overestimates the radiative heat flux. Account of the radiative processes in the electronic excitation kinetics yields more realistic radiative heat flux values, which have been found to be less than the convective ones.

The above examples pertain to the Earth, Venusian, and Martian entries. The viscous radiating shock layer approach applies also to the investigation of flows about vehicles entering the atmospheres of other planets. Relevant calculation results can be found, for instance, in the papers by Moss *et al* (1978), Tiwary and Szema (1978), Tiwary *et al* (1978), and Gupta *et al* (1985).

References

Adamson T C Jr and Messiter A F 1980 *Ann. Rev. Fluid Mech.* **12** 103–38

Afonina N E and Gromov V G 1978 *Izv. Akad. Nauk SSSR, Mekh. Zhidk. i Gaza* **4** 102–5

Afonina N E, Vlasov A Yu and Gromov V G 1983 *Chisl. Metod. Mekhan. Sploshn. Sred.* **3** 3–17

Afonina N E, Vlasov A Yu and Gromov V G 1984 *Izv. Akad. Nauk SSSR, Mekh. Zhidk. i Gaza* **5** 196–9

Agafonov V P, Vertushkin V K, Gladkov A A and Polyanskii O Yu 1972 *Neravnovesnuie Fiziko–Khimicheskie Protsessui v Aerodinamike* (Moscow: Mashinostroenie)

Agarwal R and Rakich J V 1978 *AIAA Paper* **78–0065**

Agarwal R and Rakich J V 1982 *AIAA J.* **20** 761–8

Ahouse D R and Bogdonoff S M 1969 *AIAA Paper* **69–0656**

Anderson D A, Tannehill J C and Pletcher R H 1984 *Computational Fluid Mechanics and Heat Transfer* (Washington: Hemisphere)

Anderson J D Jr 1969 *AIAA J.* **7** 1665–75

Anfimov N A 1962 *Izv. Akad. Nauk SSSR, Mekhanika i Mashinostroenie* **1** 25–31

Anfimov N A 1966 *Izv. Akad. Nauk SSSR, Mekh. Zhidk. i Gaza* **1** 22–31

Ankudinov A L 1970 *Izv. Akad. Nauk SSSR, Mekh. Zhidk. i Gaza* **3** 40–5

Ankudinov A L 1973 *Trudui TSAGI* **1448** 47–61

Ankudinov A L 1983 *Trudui TSAGI* **2203** 66–73

Ankudinov A L 1989 *Uch. Zap. TSAGI* **4** 37–45

Arkhipov V N and Polenov A N 1971 *Izv. Akad. Nauk SSSR, Mekh. Zhidk. i Gaza* **3** 74–81

Artamonov A K and Arkhipov V N 1979 *Izv. Akad. Nauk SSSR, Mekh. Zhidk. i Gaza* **6** 85–90

Avduevskii V S and Glebov G A 1970 *Inzh.–Fiz. Zhurnal* **18** 777–81

Avilova *et al* 1970 *Opticheskie Svoistva Goryachego Vozdukha* (Moscow: Nauka)

Babenko K I, Voskresenskii G P, Luibimov A N and Rusanov V V 1964 *Prostranstvennoye Obtekanie Gladkikh Tel Idealnuim Gazom* (Moscow: Nauka)

Baldwin B S and Lomax H 1978 *AIAA Paper* **78–0257**

Baldwin B S and Mac Cormack R W 1976 *AIAA Paper* **76–0350**

Barnett M and Davis R T 1986 *AIAA J.* **24** 1949–55

Batt R G and Legner H H 1983 *AIAA J.* **21** 7–22

Beam R and Warming R F 1978 *AIAA J.* **16** 393–402

Belotserkovskii O M *et al* 1967 *Obtekanie Zatuplennuikh Tel Sverkhzvukovuim Potokom Gaza. Teoreticheskoye i Eksperimentalnoie Issledovania* ed O M Belotserkovskii (Moscow: Computing Center Akad. Nauk SSSR)

Belotserkovskii O M *et al* 1974 *Chislennoye Issledovanie Sovremennuikh Zadach Gazovoi Dinamiki* (Moscow: Nauka)

Belotserkovskii O M, Buirkin A P, Mazurov A P and Tolstykh A I 1982 *Zhurnal Vychisl. Matem. i Matem. Fiz.* **22** 1480–90

Belousov V L and Golovachov Yu P 1992 *Gas Dynamics* (NY: Nova Science) 251–60

Belousov V L, Golovachov Yu P and Leontieva N V 1988 *Modelirovanie v Mekhanike* **2** 31–8

Belousov V L, Golovachov Yu P and Zemlyakov V V 1991 *Preprint of A F Ioffe Physical-Technical Inst. Akad. Nauk SSSR* **1560**

Belova O N, Kokoshinskaya N S, Luzyanina T B and Paskonov V M 1985 *Problemui Dinamiki Vyazkoi Zhidkosti* ed Yu A Berezin (Novosibirsk: Inst.Theor. and Appl. Mech.Sib. Div. Acad. Nauk SSSR) pp 24–7

Bertram M H and Everhart P E 1963 *NASA TR* R–153

Biberman L M, Bronin S Ya and Lagarkov A N 1972 *Izv. Akad. Nauk SSSR, Mekh. Zhidk. i Gaza* **5** 112–23

Biberman L M, Mnatsakanyan A Kh and Yakubov I T 1970 *Usp. Fiz. Nauk* **102** 431–62

Bird G A 1976 *Molecular Gas Dynamics* (Oxford: Clarendon Press)

Borodin A I and Peigin S V 1989 *Izv. Akad. Nauk SSSR, Mekh. Zhidk. i Gaza* **2** 150–8

Borodin A I, Kazakov V Yu and Peigin S V 1990 *Izv. Akad. Nauk SSSR, Mekh. Zhidk. i Gaza* **1** 143–50

Boylan D E 1971 *AIAA J.* **9** 1661–3

Brackbill J U and Saltzman J S 1982 *J. Comput. Phys.* **46** 342–68

Breyev I M and Golovachov Yu P 1970 *Zhurnal Prikl. Mekh. i Tekhn. Fiz.* **4** 151–3

Briley W R and Mc Donald H 1977 *J. Comput. Phys.* **24** 372–97

Bruikina I G 1978 *Zhurnal Vychisl. Matem. i Matem. Fiz.* **18** 154–66

Bruikina I G 1979 *Aerod. Giperzvuk. Techenii pri Nalichii Vduva* ed G A Tirskii (Moscow: Moscow University Press) pp 99–110

Bruikina I G and Gershbein E A 1979 *Izv. Akad. Nauk SSSR, Mekh. Zhidk. i Gaza* **2** 91–102

Bruikina I G and Rusakov V V 1988 *Izv. Akad. Nauk SSSR, Mekh. Zhidk. i Gaza* **2** 143–50

Bruikina I G and Rusakov V V 1989 *Zhurnal Prikl. Mekh. i Tekhn. Fiz.* **4** 16–22

Bruikina I G and Rusakov V V 1990 *Izv. Akad. Nauk SSSR, Mekh. Zhidk. i Gaza* **1** 117–22

Bruikina I G, Gershbein E A and Peigin S V 1980 *Izv. Akad. Nauk SSSR, Mekh. Zhidk. i Gaza* **5** 37–48

Bruikina I G, Gershbein E A and Peigin S V 1982 *Izv. Akad. Nauk SSSR, Mekh. Zhidk. i Gaza* **3** 49–58

Bruikina I G, Rusakov V V and Scherbak V G 1989 *Prikl. Matem. i Mekh.* **53** 956–62

Buirkin A P and Schennikov V V 1970 *Zhurnal Vychisl. Matem. i Matem. Fiz.* **10** 124–31

Buchmanova N C, Lapuigin V I and Lipnitskii Yu M 1973 *Izv. Akad. Nauk SSSR, Mekh. Zhidk. i Gaza* **6** 79–84

Bureyev A V and Zinchenko V I 1991 *Zhurnal Prikl. Mekh. i Tekhn. Fiz.* **6** 72–8

Bush W B 1964 *J. Fluid Mech.* **20** 353–67

Callis L B 1971 *AIAA J.* **9** 553–9

Camac M and Kemp N H 1963 *AIAA Paper* **63–0460**

Cebeci T and Smith A M O 1974 *Analysis of turbulent boundary layers* NY: Academic Press

Cebeci T, Chang K C and Whitelaw J H 1986 *AIAA J.* **24** 359–60

Chang J H and Sutton G W 1969 *AIAA J.* **7** 1110–4

Chaussee D S, Rizk Y M and Buning P C 1985 *Lect. Notes Phys.* **218** 148–53

Chen K K and Thyson N A 1971 *AIAA J.* **9** 821–7

Cheng H K 1961 *Proc. of the 1961 Heat Transfer and Fluid Mech. Inst.* (Stanford: Stanford University Press) pp 161–75

Cheng H K 1963 *IAS Paper* **63–0092**

Cheng P and Vincenti W G 1967 *J. Fluid Mech.* **27** 625–33

Cheraneva N A 1980 *Chislennuie Metodui v Aerodinamike* **4** (Moscow: Moscow University Press) pp 1–10

Chernii G G 1959 *Techenia Gaza c Bolshoi Sverkhzvukovoi Skorostyu* (Moscow: Fizmatgiz)

Chernii S G 1982 *Chisl. Metod. Mekhan. Sploshn. Sred.* **1** 132–46

Chernii S G 1983 *Preprint of Inst. Theor. and Appl. Mech. Sib. Div. Akad. Nauk SSSR* **7–83**

Chima R V and Johnson G M 1985 *AIAA J.* **23** 23–32

Chin J H 1969 *AIAA J.* **7** 1310–8

Chung P M 1961 *NASA TR* R–109

Chung P M, Holt J F and Liu S W 1968 *AIAA J.* **6** 2372–9

Chung P M, Talbot L and Touryan K J 1975 *Electric Probes in Stationary and Flowing Plasmas* (NY: Springer)

Coakley T J 1981 *AIAA Paper* **81–1257**

Daryin N A and Mazhukin V I 1988 *Zhurnal Vychisl. Matem. i Matem. Fiz.* **28** 454–60

Davis R L and Carter J E 1984 *AIAA Paper* **84–1613**

Davis R T 1970a *AIAA J.* **8** 843–51

Davis R T 1970b *AIAA Paper* **70–0805**

Davis R T and Flugge–Lotz I 1964 *J. Fluid Mech.* **20** 593–623

Degani D and Steger J L 1983 *AIAA J.* **21** 1604–6

Dellinger T C 1970 *AIAA Paper* **70–0806**

Dellinger T C 1971 *AIAA J.* **9** 262–9

Devoto R S 1966 *Phys. Fluids* **9** 1230–40

Devoto R S 1976 *Phys. Fluids* **19** 22–4

Dhavan S and Narasimha R 1958 *J. Fluid Mech.* **3** 418–36
Dunn M G 1971 *AIAA J.* **9** 2184–91
Dwyer H A and Onyejekwe O O 1985 *Lect. Notes Phys.* **218** 203–7
Dwyer H A, Kee R J and Sanders B R 1980 *AIAA J.* **18** 1205–12
Eaton R R and Kaestner P C 1973 *AIAA J.* **11** 1337–9
Eaton R R and Larson D E 1974 *AIAA Paper* **74–0599**
Edney B E 1968 *Aeronaut. Res. Inst. of Sweden. FFA Rept.* **115**
Eremeitsev I G and Pilugin N N 1981 *Izv. Akad. Nauk SSSR, Mekh. Zhidk. i Gaza* **4** 127–33
Eremeitsev I G and Pilugin N N 1986 *Izv. Akad. Nauk SSSR, Mekh. Zhidk. i Gaza* **2** 60–7
Eremeitsev I G, Pilugin N N and Yunitskii S A 1987 *Izv. Akad. Nauk SSSR, Mekh. Zhidk. i Gaza* **3** 154–9
Eremeitsev I G, Pilugin N N, Khlebnikov V S and Yunitskii S A 1988 *Issledovanie Aerodinamicheskikh Kharakteristik i Teploobmena Tel v Neravnomernuikh Sverkhzvukovuikh Potokakh Gaza* (Moscow: Moscow University Press)
Ermak Yu N 1983 *Zhurnal Prikl. Mekh. i Tekhn. Fiz.* **6** 94–101
Evans J S, Schexnayder C J and Grose W L 1974 *J. Spacecraft and Rockets* **11** 84–8
Farragher A L, Peden J A and Fite W L 1969 *J. Chem. Phys.* **50** 287–93
Favre A 1965 *Journal de Mecanique* **4** 361–90
Fay J A and Riddell F R 1958 *J. Aeronaut. Sci.* **25** 73–85
Feiereisen W J and Acharya M 1986 *AIAA J.* **24** 1642–9
Ferry A and Zakkay V 1962 *J. Aerospace Sci.* **29** 847–50
Gear W C 1971 *Numerical Initial Value Problems in Ordinary Differential Equations* (New Jersey: Prentice Hall)
Gershbein E A 1974 *Prikl. Matem. i Mekh.* **38** 1015–23
Gershbein E A 1978 *Nekotoruie Voprosui Mekhaniki Sploshn. Sred* (Moscow: Nauka) pp 144–56
Gershbein E A and Kolesnikov A F 1979 *Aerod. Giperzvuk. Techenii pri Nalichii Vduva* ed G A Tirskii (Moscow: Moscow University Press) pp 69–77
Gershbein E A and Peigin S V 1986 *Izv. Akad. Nauk SSSR, Mekh. Zhidk. i Gaza* **6** 27–38
Gershbein E A and Scherbak V G 1984 *Izv. Akad. Nauk SSSR, Mekh. Zhidk. i Gaza* **3** 150–4
Gershbein E A and Scherbak V G 1987 *Izv. Akad. Nauk SSSR, Mekh. Zhidk. i Gaza* **4** 134–42
Gershbein E A and Yunitskii S A 1979 *Prikl. Matem. i Mekh.* **43** 817–28
Gershbein E A and Yunitskii S A 1984 *Prikl. Matem. i Mekh.* **48** 768–75
Gershbein E A, Krupa V G and Schelin V S 1985a *Izv. Akad. Nauk SSSR Mekh. Zhidk. i Gaza* **6** 140–6
Gershbein E A, Schelin V S and Yunitskii S A 1981 *Giperzv. Prostr. Techenia pri Nalichii Fiz.-Khim. Prevraschenii* (Moscow: Moscow University Press) pp 72–92

Gershbein E A, Schelin V S and Yunitskii S A 1984a *Izv. Akad. Nauk SSSR, Mekh. Zhidk. i Gaza* **2** 104–8
Gershbein E A, Schelin V S and Yunitskii S A 1984b *Izv. Akad. Nauk SSSR, Mekh. Zhidk. i Gaza* **6** 127–35
Gershbein E A, Schelin V S and Yunitskii S A 1985b *Kosmich. Issled.* **23** 416–25
Gilinskii S M and Lebedev M G 1977 *Izv. Akad. Nauk SSSR, Mekh. Zhidk. i Gaza* **1** 117–24
Godunov S K, Zabrodin A V, Ivanov M Ya, Kraiko A N and Prokopov G P 1976 *Chislennoie Reshenie Mnogomernuikh Zadach Gazovoi Dinamiki* (Moscow: Nauka)
Golovachov Yu P 1972 *Izv. Akad. Nauk SSSR, Mekh. Zhidk. i Gaza* **6** 169–75
Golovachov Yu P 1975 *Teplofiz. Vysok. Temp.* **13** 1029–34
Golovachov Yu P 1978 *Teplofiz. Vysok. Temp.* **16** 1109–12
Golovachov Yu P 1979a *Izv. Akad. Nauk SSSR, Mekh. Zhidk. i Gaza* **5** 199–202
Golovachov Yu P 1979b *Izv. Akad. Nauk SSSR, Mekh. Zhidk. i Gaza* **6** 158–60
Golovachov Yu P 1979c *Trudui Tashkentskogo Politekhn. Inst.* **285** 45–8
Golovachov Yu P 1981a *Int. J. Heat Mass Tr.* **24** 649–57
Golovachov Yu P 1981b *Izv. Akad. Nauk SSSR, Mekh. Zhidk. i Gaza* **1** 185–8
Golovachov Yu P 1984 *Chisl. Metod. Mekhan. Sploshn. Sred.* **5** 63–8
Golovachov Yu P 1985 *Int. J. Heat and Mass Transfer* **28** 1165–71
Golovachov Yu P 1991 *Computers & Fluids* **19** 497–503
Golovachov Yu P and Fursenko A A 1981a *Pisma Zhurn. Tekhn. Fiz.* **7** 178–80
Golovachov Yu P and Fursenko A A 1981b *Zhurnal Vychisl. Matem. i Matem. Fiz.* **21** 1592–6
Golovachov Yu P and Kanailova A S 1987 *Inzh.-Fiz. Zhurnal* **53** 446–50
Golovachov Yu P and Karyakin V E 1980 *Chisl. Metod. Mekhan. Sploshn. Sred.* **6** 46–58
Golovachov Yu P and Koleshko S B 1988 *Chislennuie Metodui Reshenia Uravnenii Dinamiki Zhidkosti i Gaza* (Leningrad: NPO GIPKH)
Golovachov Yu P and Leontieva N V 1982 *Uch. Zap. TSAGI* **5** 40–8
Golovachov Yu P and Leontieva N V 1983 *Izv. Akad. Nauk SSSR, Mekh. Zhidk. i Gaza* **3** 176–9
Golovachov Yu P and Leontieva N V 1985 *Izv. Akad. Nauk SSSR, Mekh. Zhidk. i Gaza* **3** 143–8
Golovachov Yu P and Leontieva N V 1987 *Zhurnal Vychisl. Matem. i Matem. Fiz.* **27** 1739–44
Golovachov Yu P and Leontieva N V 1989a *Izv. Akad. Nauk SSSR, Mekh. Zhidk. i Gaza* **2** 186–9
Golovachov Yu P and Leontieva N V 1989b *Zhurnal Vychisl. Matem. i Matem. Fiz.* **28** 148–51
Golovachov Yu P and Leontieva N V 1989c *Zhurnal Vychisl. Matem. i Matem. Fiz.* **29** 791–3
Golovachov Yu P and Leontieva N V 1993 *Izv. Rus. Akad. Nauk, Mekh. Zhidk. i Gaza* **3** 169–72
Golovachov Yu P and Lunkin Yu P 1979 *Astronaut. Acta* **6** 545–55

Golovachov Yu P and Popov F D 1972a *Zhurnal Prikl. Mekh. i Tekhn. Fiz.* **5** 135–42

Golovachov Yu P and Popov F D 1972b *Zhurnal Vychisl. Matem. i Matem. Fiz.* **12** 1292–303

Golovachov Yu P and Popov F D 1975a *Inzh.–Fiz. Zhurnal* **29** 864–9

Golovachov Yu P and Popov F D 1975b *Izv. Akad. Nauk SSSR, Mekh. Zhidk. i Gaza* **2** 175–8

Golovachov Yu P and Timofeyev E V 1988a *Preprint of A F Ioffe Physical-Technical Inst. Akad. Nauk SSSR* **1254**

Golovachov Yu P and Timofeyev E V 1988b *Uch. Zap. TSAGI* **6** 106–9

Golovachov Yu P and Zemlyakov V V 1993a *Zhurnal Vychisl. Matem. i Matem. Fiz.* **33** 151–5

Golovachov Yu P and Zemlyakov V V 1993b *Zhurnal Vychisl. Matem. i Matem. Fiz.* **33** 1422–7

Golovachov Yu P and Zemlyakov V V 1994 *Int. J. Comput. Fluid Dyn.* 1422–7

Golovachov Yu P, Karyakin V E, Leontieva N V and Fursenko A A 1985 *Chisl. Metod. Mekhan. Sploshn. Sred.* **5** 29–37

Golovachov Yu P, Kuzmin A M and Popov F D 1973 *Zhurnal Vychisl. Matem. i Matem. Fiz.* **13** 1021–8

Golovachov Yu P, Zemlyakov V V and Matvienko E A 1994 *Izv. Rus. Akad. Nauk, Mekh. Zhidk. i Gaza* **6** 158–60

Goloviznin V P, Mishin G I, Serov Yu L and Yavor I P 1987 *Zhurnal Tekhn. Fiz.* **57** 1433–5

Gordeyev A N, Kolesnikov A F and Yakushin M I 1985 *Izv. Akad. Nauk SSSR, Mekh. Zhidk. i Gaza* **3** 166–72

Gorinov A S and Magomedov K M 1973 *Zhurnal Vychisl. Matem. i Matem. Fiz.* **13** 1264–72

Goulard R, Boughner R E, Burns R K and Nelson H F 1969 *Teplofiz. Vysok. Temp.* **7** 542–65

Griffith B J and Lewis C H 1964 *AIAA J.* **2** 438–44

Gromov V G and Kessel V L 1970 *Nauchn. Trudui Inst. Mekh. Moscow University* **1** 71–82

Grose W L and Nealy J E 1975 *AIAA J.* **13** 421–4

Gupta R N and Simmonds A L 1986 *AIAA Paper* **86–1349**

Gupta R N, Gnoffo P A and Mac Cormack R W 1985 *AIAA J.* **23** 723–32

Gurvich L V *et al* 1962 *Termodinamicheskie Svoistva Individualnuikh Veschestv* (Moscow: Akad. Nauk SSSR)

Gusev V N 1993 *Izv. Rus. Akad. Nauk, Mekh. Zhidk. i Gaza* **2** 142–52

Gusev V N and Nikolskii Yu V 1971 *Uch. Zap. TSAGI* **1** 122–5

Gusev V N *et al* 1970 *Trudui TSAGI* **1855** 3–47

Gusev V N, Kogan M N and Perepukhov V A 1970 *Uch. Zap. TSAGI* **1** 24–32

Hakkinen R J, Greber I, Trilling L and Abarbanel S S 1959 *NASA Memo.* 2–18–59 W

Hammerling P, Teare J D and Kivel B 1959 *Phys. Fluids* **2** 422–6

Hansen C F 1959 *NASA TR* **R–50**

Hawken D F 1987 *Progr. Aerospace Sci.* **24** 29–49

Hayes D T and Rotman W 1973 *AIAA J.* **11** 675–82

Hayes W D and Probstein R F 1959 *Hypersonic Flow Theory* (NY and London: Academic Press)

Helliwell W S, Dickinson R P and Lubard S C 1980 *AIAA Paper* **80–0064**

Hickman R S and Giedt W H 1963 *AIAA J.* **1** 665–72

Hirschfelder J O, Curtiss C F and Bird R B 1954 *Molecular Theory of Gases and Liquids* (NY: Wiley and Sons)

Hsieh T 1976 *AFDC–TR* **76–112**

Hussaini M Y and Zang T A 1987 *Ann. Rev. Fluid Mech.* **19** 339–67

Hussaini M Y, Kopriva D A, Salas M D and Zang T A 1985 *AIAA J.* **23** 64–70

Hussani M Y, Baldwin B S and Mac Cormack R W 1980 *AIAA J.* **18** 1014–6

Jain A C and Adimurthy V 1974 *AIAA J.* **12** 342–54

Jensen D E 1974 *J. Spacecraft and Rockets* **11** 736–7

Kaegi E N and Mc Menamin D L 1969 *AIAA Paper* **69–0081**

Kalinin A P, Leonas V B and Sermyagin A V 1971 *Teplofiz. Vysok. Temp.* **9** 1066–8

Kamenetskii V F and Turchak L I 1984 *Izv. Akad. Nauk SSSR, Mekh. Zhidk. i Gaza* **5** 141–7

Kamenschikov V A, Plastinin Yu A, Nikolayev V M and Novitskii L A 1971 *Radiatsionnuie Svoistva Vysokotemperaturnuikh Gazov* (Moscow: Mashinostroenie)

Kandil O A, Wong T C and Liu C H 1990 *AIAA Paper* **90–0598**

Kang S W and Dunn M G 1972 *AIAA J.* **10** 1361–3

Kang S W, Jones W L and Dunn M G 1973 *AIAA J.* **11** 141–9

Kao H C 1964 *AIAA J.* **2** 1892–906

Karasev A B 1973 *Teplo i Massoperenos* vol 10–1(Minsk: Inst. Teplo i Massoobmena Akad. Nauk Belorus. SSR) pp 244–61

Karatayev S G and Koterov V N 1990 *Zhurnal Vychisl. Matem. i Matem. Fiz.* **29** 586–600

Karyakin V E 1981 *Zhurnal Vychisl. Matem. i Matem. Fiz.* **21** 150–6

Karyakin V E and Popov F D 1977 *Zhurnal Vychisl. Matem. i Matem. Fiz.* **17** 1545–55

Katzen E D and Kaattari G E 1965 *AIAA J.* **3** 1230–7

Kazakov V Yu and Peigin S V 1988 *Teplofiz. Vysok. Temp.* **26** 901–8

Kim M D and Lewis C H 1983 *J. of Spacecraft and Rockets* **20** 101–7

Kiselev V Ya 1982 *Dokl.3 Vses. Shkolui po Metodam Aerofiz.Issled., Novosibirsk 1982* (Novosibirsk: Inst. Theor. and Appl. Mech. Sib. Div. Akad. Nauk SSSR) pp 111–3

Knight D D 1971 *AIAA J.* **9** 193–9

Kochin N E 1965 *Vektornoie Ischislenie i Nachala Tenzornogo Ischislenia* (Moscow: Nauka)

Kogan M N 1967 *Dinamika Razrezhennogo Gaza* (Moskow: Nauka)

Kokoshinskaya N S, Pavlov B M and Paskonov V M 1980 *Chislennoye Issledovanie Sverkhzvukovogo Obtekania Tel Vyazkim Gazom* (Moscow: Moscow University Press)

Kordulla W and Vinokur M 1983 *AIAA J.* **21** 917–8

Kovach E A and Tirskii G A 1970 *Dokl. Akad. Nauk SSSR* **190** 61–4

Kovenya V M and Chernii S G 1979 *Chisl. Metod. Mekhan. Sploshn. Sred.* **1** 71–87

Kovenya V M and Chernii S G 1983 *Zhurnal Vychisl. Matem. i Matem. Fiz.* **23** 1186–98

Kovenya V M and Yanenko N N 1981 *Metod Rrascheplenia v Zadachakh Gazovoi Dinamiki* (Novosibirsk: Nauka)

Kovenya V M, Tarnavskii G A and Chernii S G 1990 *Primenenie Metoda Rascheplenia v Zadachakh Aerodinamiki* Novosibirsk: Nauka

Kraiko A N 1964 *Inzh. Zhurnal* **4** 548–50

Krasilov N A, Levin V A and Yunitskii S A 1986a *Inzh.–Fiz. Zhurnal* **51** 940–6

Krasilov N A, Levin V A and Yunitskii S A 1986b *Izv. Akad. Nauk SSSR, Mekh. Zhidk. i Gaza* **1** 106–14

Krasilschikov A P and Nosov V V 1971 *Trudui TSAGI* **1341** 3–28

Kreis R I, Thames F C and Hassan H A 1986 *AIAA J.* **24** 404–10

Kulnev S S and Paskonov V M 1983 *Vychisl. Metod. i Program.* ed V M Paskonov and G S Roslyakov (Moscow: Moscow University Press) pp 22–38

Kumar A and Graves R A 1977 *AIAA J.* **15** 1061–2

Kumar A and Jain A C 1975 *Int. J. Heat Mass Tr.* **18** 1113–8

Kutler P, Pedelty J A and Pulliam T H 1980 *AIAA Paper* **80–0063**

Ladnova L A 1969 *Vestnik of Leningrad State University* **13** 105–12

Lapin Yu V 1982 *Turbulentnui Pogranichnui Sloi v Sverkhzvukovuikh Potokakh Gaza* (Moscow: Nauka)

Lapin Yu V and Strelets M Kh 1989 *Vnutrennie Techenia Gazovuikh Smesei* (Moskow: Nauka)

Launder B E and Spalding D B 1974 *Comput. Methods in Appl.Mech. and Eng.* **3** 269–89

Lebedev V I and Fomin V N 1969 *Zhurnal Vychisl. Matem. i Matem. Fiz.* **9** 655–63

Levinsky F S and Yoshihara H 1962 *Hypersonic Flow Research* ed F R Riddell (NY: Academic Press) pp 81–106

Lewis C H and Song D J 1986 *J. Spacecraft and Rockets* **23** 47–54

Lin T C and Rubin S G 1974 *AIAA J.* **12** 975–85

Lin T C and Rubin S G 1975 *AIAA Paper* **75–0853**

Lin T C and Rubin S G 1979 *AIAA Paper* **79–0205**

Lin T C and Rubin S G 1982 *AIAA J.* **20** 1500–7

Lin T C, Reeves B L and Siegelman D 1977 *AIAA J.* **15** 1130–7

Lipnitskii Yu M and Lyakhov V N 1976 *Uch. Zap. TSAGI* **4** 115–9

Li C P 1973 *AIAA Paper* **73–0202**

Li C P 1981 *AIAA Paper* **81–0191**

Loitsianskii L G 1987 *Mechanics of Liquids and Gases* (Moscow: Nauka)

Lomax H 1982 *AIAA J.* **20** 1033–43

Lubard S C and Helliwell W C 1974 *AIAA J.* **12** 965–74

Lubard S C and Rakich J V 1975 *AIAA Paper* **75–0149**

Lubimov A N and Rusanov V V 1970 *Techenia Gaza Okolo Tupuikh Tel* (Moscow: Nauka)

Lunev V V and Khramov N E 1970 *Izv. Akad. Nauk SSSR, Mekh. Zhidk. i Gaza* **3** 102–5
Mac Cormack R W 1969 *AIAA Paper* **69–0354**
Mac Cormack R W 1971 *Lect. Notes Phys.* **8** 151–63
Mac Cormack R W 1978 *SIAM – AMS Proc.* **11** 130–55
Mac Cormack R W 1982 *AIAA J.* **20** 1275–81
Mac Cormack R W 1985 *AIAA Paper* **85–0032**
Mac Cormack R W and Baldwin B S 1975 *AIAA Paper* **75–0001**
Magomedov K M 1970 *Izv. Akad. Nauk SSSR, Mekh. Zhidk. i Gaza* **2** 45–56
Markov A A 1974 *Izv. Akad. Nauk SSSR, Mekh. Zhidk. i Gaza* **1** 69–77
Markov A A 1980 *Izv. Akad. Nauk SSSR, Mekh. Zhidk. i Gaza* **5** 115–26
Markov A A 1984 *Izv. Akad. Nauk SSSR, Mekh. Zhidk. i Gaza* **3** 179–82
Markov A A 1987 *Izv. Akad. Nauk SSSR, Mekh. Zhidk. i Gaza* **5** 114–22
Markov A A 1988 *Izv. Akad. Nauk SSSR, Mekh. Zhidk. i Gaza* **1** 90–5
Markov A A 1989 *Preprint of Inst. Problem Mekh. Akad. Nauk SSSR* **428**
Marrone P V and Treanor C E 1963 *Phys. Fluids* **6** 1215–21
Martellucci A, Hahn J and Laggnelli A 1973 *SAMSO TR* **73–147**
Masek R V, Hender D and Forney J A 1973 *AIAA Paper* **73– 0737**
Mason E A and Saxena S C 1958 *Phys. Fluids* **1** 361–9
Mayne A W 1973 *AEDC–TR* **73–166**
Mc Donald H and Briley W R 1981 *Numerical and Physical Aspects of Aerodinamic Flows ed T Cebeci* (NY, Heidelberg, Berlin: Springer) pp 99–120
Mc Kenzie R L and Arnold J O 1967 *AIAA Paper* **67–0322**
Meleshko S V and Chernii S G 1985 *Preprint of Inst. Theor. and Appl. Mech. Sib. Div. Akad. Nauk SSSR* **32–85**
Menard W A, Thomas G M and Helliwell T M 1968 *AIAA J.* **6** 655–64
Mikhailov V V 1961 *Inzh. Sbornik* **31** 206–16
Miller D S and Wood R M 1985 *NASA TR* **2430**
Miner E W and Lewis C H 1975 *AIAA J.* **13** 80–8
Miner E W and Lewis C H 1976 *AIAA J.* **14** 64–9
Molodsov V K 1969 *Zhurnal Vychisl. Matem. i Matem. Fiz.* **9** 1211–7
Molodsov V K 1979 *Uch. Zap. TSAGI* **1** 122–6
Molodsov V K and Ryabov V V 1979 *Uch. Zap. TSAGI* **6** 30–6
Moon V J and Holt M 1989 *Lect. Notes Phys.* **323** 418–23
Moss J N and Bird G A 1984 *AIAA Paper* **84–0223**
Moss J N, Simmonds A L and Anderson E C 1978 *AIAA Paper* **78–1186**
Mozhayev Yu A 1967 *Trudui VVIA* **1198** 44–55
Mumma M J, Stone E J and Zipf E C 1971 *J. Chem. Phys.* **54** 2627–31
Murray A L and Lewis C H 1978 *AIAA Paper* **78–0844**
Murthy V S and Rose W C 1978 *AIAA J.* **16** 667–72
Newsome R W 1986 *AIAA J.* **24** 552–61
Newsome R W, Walters R W and Thomas J L 1987 *AIAA Paper* **87–1113–CP**
Neyland V Ya 1969 *Izv. Akad. Nauk SSSR, Mekh. Zhidk. i Gaza* **4** 53–7
Nishida M 1972 *Phys. Fluids* **15** 596–602
Olstad W B 1971 *AIAA J.* **9** 122–30

Olstad W B 1974 *Astronaut. and Aeronaut.* **12** 55–61
Page W A and Woodward H T 1972 *AIAA J.* **10** 1379–81
Page W A, Compton D L, Borucki W J, Ciffone D L and Cooper D M 1968 *AIAA Paper* **68–0784**
Pappas C C and Lee G 1970 *AIAA J.* **8** 954–6
Pavlov B M 1968 *Izv. Akad. Nauk SSSR, Mekh. Zhidk. i Gaza* **3** 128–33
Pavlov B M 1971 *Chisl. Metod. Mekhan. Sploshn. Sred.* **4** 61–6
Peigin S V 1985 *Izv. Akad. Nauk SSSR, Mekh. Zhidk. i Gaza* **6** 28–37
Peigin S V and Rusakov V V 1985 *Chisl. Metod. Mekhan. Sploshn. Sred.* **2** 118–33
Peigin S V and Tirskii G A 1988 *Itogi Nauki i Tekhniki. Mekhanika Zhidkosti i Gaza* **22** (Moscow: VINITI) pp 62–177
Petrova L I 1976 *Chislennuie Metodui v Aerodinamike* **1** (Moscow: Moscow University Press) pp 3–41
Petukhov I V 1964 *Chislennuie Metodui Reshenia Differentsialnuikh i Integralnuikh Uravnenii i Kvadraturnuie Formului* (Moscow: Nauka) pp 304–25
Pilugin N N and Talipov R F 1986 *Izv. Akad. Nauk SSSR, Mekh. Zhidk. i Gaza* **6** 120–5
Pilugin N N and Tirskii G A 1989 *Dinamika Ionizovannogo Izluchauschego Gaza* (Moscow: Moscow University Press)
Pinchukov V I 1979 *Issled. Pristen. Techenii Vyazkogo Gaza* (Novosibirsk: Inst. Theor. and Appl. Mech. Sib. Div. Akad. Nauk SSSR) pp 23–44
Pinchukov V I 1983 *Zhurnal Vychisl. Matem. i Matem. Fiz.* **23** 901–9
Provotorov V P and Ryabov V V 1981 *Uch. Zap. TSAGI* **5** 55–63
Pulliam T H and Steger J L 1978 *AIAA Paper* **78–0010**
Rai M M and Anderson D A 1982 *AIAA J.* **20** 496–502
Ramakrishnan S V and Rubin S G 1987 *AIAA J.* **25** 905–13
Reda D C and Murphy J D 1972 *AIAA Paper* **72–0715**
Reddy D R and Rubin S G 1984 *AIAA Paper* **84–1627**
Reddy K V and Fujiwara T 1989 *Numerical Methods in Fluid Dynamics Proc. of the Int. Symp. on Comput. Fluid Dynamics* vol II Nagoya, August 1989 ed M Yasuhara, H Daiguji and K Oshima (Japan Soc. of Comput. Fluid Dyn.) pp 573–9
Reyhner T A and Flugge–Lotz I 1968 *Int. J. Nonlinear Mech.* **3** 173–9
Rizzi A and Ericksson L E 1981 *Proc. of AIAA 5–th Comp. Fluid Dyn. Conf., June 1981* pp 43–68
Rubin S G 1981 *Numerical and Physical Aspects of Aerodynamic Flows* ed T Cebeci (NY, Heidelberg, Berlin: Springer) pp 171–85
Rubin S G and Lin A 1980 *Israel J. of Technology* **18** 21–31
Rudman S and Rubin S G 1968 *AIAA J.* **6** 1883–9
Russell D A 1968 *Phys. Fluids* **11** 1679–85
Saffman P G and Wilcox D C 1974 *AIAA J.* **12** 541–7
Saveliev A D and Tolstykh A I 1987 *Zhurnal Vychisl. Matem. i Matem. Fiz.* **27** 1709–24
Scala S M and Gordon P 1968 *AIAA J.* **6** 815–22
Scherbak V G 1987 *Izv. Akad. Nauk SSSR, Mekh. Zhidk. i Gaza* **5** 143–50

Scherbak V G 1989 *Zhurnal Prikl. Mekh. i Tekhn. Fiz.* **1** 49–56
Scherbak V G 1990 *Teplofiz. Vysok. Temp.* **28** 1164–70
Schiff L B and Steger J L 1980 *AIAA J.* **18** 1421–30
Schiff L B and Sturek W B 1980 *AIAA Paper* **80–0066**
Scott C D 1975 *AIAA J.* **13** 1271–8
Scott C D 1980 *AIAA Paper* **80–1477**
Scott C D 1981 *AIAA Paper* **81–1184**
Shang J S 1978 *AIAA J.* **16** 496–502
Shang J S 1985 *J. of Aircraft* **22** 353–70
Shapiro E G 1975 *Izv. Akad. Nauk SSSR, Mekh. Zhidk. i Gaza* **1** 82–5
Shevelev Yu D 1986 *Prostranstvennie zadachi vychislitelnoi aerohydrodinamiki* (Moscow:Nauka)
Shih I Pai 1966 *Radiation Gas Dynamics* (Wien, NY: Springer)
Sinchenko S P 1968 *Zhurnal Vychisl. Matem. i Matem. Fiz.* **8** 917–22
Smith R 1973 *Aeronaut. Quart.* **24** 241–51
Sokolova I A 1973 *Zhurnal Prikl. Mekh. i Tekhn. Fiz.* **2** 80–90
Sokolova I A and Tirskii G A 1988 *Zhurnal Prikl. Mekh. i Tekhn. Fiz.* **3** 14–28
Song D J, Swaminathan S and Lewis C H 1985 *J. Spacecraft and Rockets* **22** 614–9
Spaid F W and Zukoski E E 1968 *AIAA J.* **6** 205–11
Srivastava B N, Werle M J and Davis R T 1978 *AIAA J.* **16** 137–44
Srivastava B N, Werle M J and Davis R T 1979 *AIAA J.* **17** 107–10
Stewartson K and Williams P G 1969 *Proc. of the Royal Soc. London* **A 312** 181–206
Stone H L 1968 *SIAM J. Numer. Anal.* **5** 530–58
Stulov V P and Shapiro E G 1970 *Izv. Akad. Nauk SSSR, Mekh. Zhidk. i Gaza* **1** 154–60
Sturek W B and Schiff L B 1982 *AIAA J.* **20** 1724–31
Svehla R A 1961 *NASA TR* R–132
Tannehill J C, Holst T L and Rakich J V 1976 *AIAA J.* **14** 204–11
Tannehill J C, Venkatapathy E and Rakich J V 1981 *AIAA Paper* **81–0049**
Tarnavskii G A 1983 *Chislennuie Metodui Dinamiki Vyazkoi Zhidkosti* (Novosibirsk: Inst. Theor. and Appl.Mech. Sib. Div. Akad. Nauk SSSR) pp 282–7
Taylor R L and Bitterman S 1969 *Rev. Mod. Phys.* **41** 26–47
Thomas J L and Newsome R W 1989 *AIAA J.* **27** 1673–9
Thomas J L and Walters R W 1985 *AIAA J.* **23** 117–28
Tirskii G A and Scherbak V G 1990 *Izv. Akad. Nauk SSSR, Mekh. Zhidk. i Gaza* **1** 151–7
Tirskii G A, Bruikina I G, Rusakov V V and Scherbak V G 1989 *Inst. Mech. Moscow University TR* **3815**
Tiwary S N and Szema K Y 1978 *AIAA Paper* **78–0190**
Tiwary S N, Kumar A and Graves R A 1978 *AIAA Paper* **78–0909**
Tolstykh A I 1964 *Prikl. Matem. i Mekh.* **28** 553–6
Tolstykh A I 1966 *Zhurnal Vychisl. Matem. i Matem. Fiz.* **6** 113–20
Tolstykh A I 1978 *Zhurnal Vychisl. Matem. i Matem. Fiz.* **18** 139–53
Tolstykh A I 1990 *Kompaktnuie Raznostnuie Shemui i Ih Primenenie v Zadachakh*

Aerogidrodinamiki (Moscow: Nauka)
Tracy R R 1963 *California Inst. of Technology Techn. Memo.* **69**
Treanor C E and Marrone P V 1962 *Phys. Fluids* **5** 1022–6
Tugazakov R Ya 1980 *Uch. Zap. TSAGI* **4** 139–44
Turchak L I 1976 *Izv. Akad. Nauk SSSR, Mekh. Zhidk. i Gaza* **1** 166–70
Vasilievskii S A and Tirskii G A 1979 *Aerod. Giperzvuk. Techenii pri Nalichii Vduva* ed G A Tirskii (Moscow: Moscow University Press) pp 87–98
Vasilievskii S A and Tirskii G A 1987 *Issled. po Giperzvuk. Aerodinam. i Teploobmenu c Uchetom Neravn. Khimich. Reakz.* ed G A Tirskii (Moscow: Moscow University Press) pp 5–24
Vershinin I V, Tirskii G A and Utuzhnikov S V 1991 *Izv. Akad. Nauk SSSR, Mekh. Zhidk. i Gaza* **4** 40–4
Vidal R J and Wittliff C E 1963 *Rarified Gas Dynamics* vol II ed J A Laurmann (NY: Academic Press) pp 343–78
Viegas J R and Coakley T J 1977 *AIAA Paper* **77–0044**
Viegas J R and Horstman C C 1979 *AIAA J.* **17** 811–20
Vigneron Y C, Rakich J V and Tannehill J C 1978 *AIAA Paper* **78–1137**
Vinokur M 1983 *J. Comput. Phys.* **50** 215–34
Voronkin V G 1971 *Izv. Akad. Nauk SSSR, Mekh. Zhidk. i Gaza* **2** 144–7
Voronkin V G 1974 *Izv. Akad. Nauk SSSR, Mekh. Zhidk. i Gaza* **6** 99–105
Voronkin V G 1979 *Izv. Akad. Nauk SSSR, Mekh. Zhidk. i Gaza* **6** 15–20
Waskiewich J D, Murray A L and Lewis C H 1978 *AIAA J.* **16** 189–92
Wieting A R 1987 *NASA TM* **100484**
Wilke C R 1950a *Chem. Eng. Progr.* **46** 95–104
Wilke C R 1950b *J. Chem. Phys.* **18** 517–22
Wilson K H 1970 *AIAA Paper* **70–0203**
Wilson M R and Wittliff C E 1962 *ARS J.* **32** 275–6
Yanenko N N, Danayev N T and Liseikin V D 1977 *Chisl. Metod. Mekhan. Sploshn. Sred.* **4** 157–63
Yos J M 1963 *RAD–TM* **63–7**
Yun K S and Mason E A 1962 *Phys. Fluids* **5** 380–6
Zalogin G N 1974 *Izv. Akad. Nauk SSSR, Mekh. Zhidk. i Gaza* **6** 81–7
Zalogin G N 1976 *Izv. Akad. Nauk SSSR, Mekh. Zhidk. i Gaza* **3** 160–3
Zdunkevich M D and Leonas V B 1972 *Teplofiz. Vysok. Temp.* **10** 1110–2
Zeldovich Ya B and Raizer Yu P 1966 *Physics of Shock Waves and High Temperature Hydrodinamic Phenomena* (NY: Academic Press)
Zheleznyak M B, Mnatsakanyan A Kh and Yakubov I T 1970 *Izv. Akad. Nauk SSSR, Mekh. Zhidk. i Gaza* **4** 161–74

Index

Mechanics

FLUID MECHANICS AND ITS APPLICATIONS

Series Editor: R. Moreau

Aims and Scope of the Series

The purpose of this series is to focus on subjects in which fluid mechanics plays a fundamental role. As well as the more traditional applications of aeronautics, hydraulics, heat and mass transfer etc., books will be published dealing with topics which are currently in a state of rapid development, such as turbulence, suspensions and multiphase fluids, super and hypersonic flows and numerical modelling techniques. It is a widely held view that it is the interdisciplinary subjects that will receive intense scientific attention, bringing them to the forefront of technological advancement. Fluids have the ability to transport matter and its properties as well as transmit force, therefore fluid mechanics is a subject that is particularly open to cross fertilisation with other sciences and disciplines of engineering. The subject of fluid mechanics will be highly relevant in domains such as chemical, metallurgical, biological and ecological engineering. This series is particularly open to such new multidisciplinary domains.

1. M. Lesieur: *Turbulence in Fluids.* 2nd rev. ed., 1990 ISBN 0-7923-0645-7
2. O. Métais and M. Lesieur (eds.): *Turbulence and Coherent Structures.* 1991 ISBN 0-7923-0646-5
3. R. Moreau: *Magnetohydrodynamics.* 1990 ISBN 0-7923-0937-5
4. E. Coustols (ed.): *Turbulence Control by Passive Means.* 1990 ISBN 0-7923-1020-9
5. A.A. Borissov (ed.): *Dynamic Structure of Detonation in Gaseous and Dispersed Media.* 1991 ISBN 0-7923-1340-2
6. K.-S. Choi (ed.): *Recent Developments in Turbulence Management.* 1991 ISBN 0-7923-1477-8
7. E.P. Evans and B. Coulbeck (eds.): *Pipeline Systems.* 1992 ISBN 0-7923-1668-1
8. B. Nau (ed.): *Fluid Sealing.* 1992 ISBN 0-7923-1669-X
9. T.K.S. Murthy (ed.): *Computational Methods in Hypersonic Aerodynamics.* 1992 ISBN 0-7923-1673-8
10. R. King (ed.): *Fluid Mechanics of Mixing.* Modelling, Operations and Experimental Techniques. 1992 ISBN 0-7923-1720-3
11. Z. Han and X. Yin: *Shock Dynamics.* 1993 ISBN 0-7923-1746-7
12. L. Svarovsky and M.T. Thew (eds.): *Hydroclones.* Analysis and Applications. 1992 ISBN 0-7923-1876-5
13. A. Lichtarowicz (ed.): *Jet Cutting Technology.* 1992 ISBN 0-7923-1979-6
14. F.T.M. Nieuwstadt (ed.): *Flow Visualization and Image Analysis.* 1993 ISBN 0-7923-1994-X
15. A.J. Saul (ed.): *Floods and Flood Management.* 1992 ISBN 0-7923-2078-6
16. D.E. Ashpis, T.B. Gatski and R. Hirsh (eds.): *Instabilities and Turbulence in Engineering Flows.* 1993 ISBN 0-7923-2161-8
17. R.S. Azad: *The Atmospheric Boundary Layer for Engineers.* 1993 ISBN 0-7923-2187-1
18. F.T.M. Nieuwstadt (ed.): *Advances in Turbulence IV.* 1993 ISBN 0-7923-2282-7
19. K.K. Prasad (ed.): *Further Developments in Turbulence Management.* 1993 ISBN 0-7923-2291-6
20. Y.A. Tatarchenko: *Shaped Crystal Growth.* 1993 ISBN 0-7923-2419-6

Kluwer Academic Publishers – Dordrecht / Boston / London

Mechanics

FLUID MECHANICS AND ITS APPLICATIONS

Series Editor: R. Moreau

Aims and Scope of the Series

The purpose of this series is to focus on subjects in which fluid mechanics plays a fundamental role. As well as the more traditional applications of aeronautics, hydraulics, heat and mass transfer etc., books will be published dealing with topics which are currently in a state of rapid development, such as turbulence, suspensions and multiphase fluids, super and hypersonic flows and numerical modelling techniques. It is a widely held view that it is the interdisciplinary subjects that will receive intense scientific attention, bringing them to the forefront of technological advancement. Fluids have the ability to transport matter and its properties as well as transmit force, therefore fluid mechanics is a subject that is particularly open to cross fertilisation with other sciences and disciplines of engineering. The subject of fluid mechanics will be highly relevant in domains such as chemical, metallurgical, biological and ecological engineering. This series is particularly open to such new multidisciplinary domains.

1. M. Lesieur: *Turbulence in Fluids.* 2nd rev. ed., 1990 ISBN 0-7923-0645-7
2. O. Métais and M. Lesieur (eds.): *Turbulence and Coherent Structures.* 1991 ISBN 0-7923-0646-5
3. R. Moreau: *Magnetohydrodynamics.* 1990 ISBN 0-7923-0937-5
4. E. Coustols (ed.): *Turbulence Control by Passive Means.* 1990 ISBN 0-7923-1020-9
5. A.A. Borissov (ed.): *Dynamic Structure of Detonation in Gaseous and Dispersed Media.* 1991 ISBN 0-7923-1340-2
6. K.-S. Choi (ed.): *Recent Developments in Turbulence Management.* 1991 ISBN 0-7923-1477-8
7. E.P. Evans and B. Coulbeck (eds.): *Pipeline Systems.* 1992 ISBN 0-7923-1668-1
8. B. Nau (ed.): *Fluid Sealing.* 1992 ISBN 0-7923-1669-X
9. T.K.S. Murthy (ed.): *Computational Methods in Hypersonic Aerodynamics.* 1992 ISBN 0-7923-1673-8
10. R. King (ed.): *Fluid Mechanics of Mixing.* Modelling, Operations and Experimental Techniques. 1992 ISBN 0-7923-1720-3
11. Z. Han and X. Yin: *Shock Dynamics.* 1993 ISBN 0-7923-1746-7
12. L. Svarovsky and M.T. Thew (eds.): *Hydroclones.* Analysis and Applications. 1992 ISBN 0-7923-1876-5
13. A. Lichtarowicz (ed.): *Jet Cutting Technology.* 1992 ISBN 0-7923-1979-6
14. F.T.M. Nieuwstadt (ed.): *Flow Visualization and Image Analysis.* 1993 ISBN 0-7923-1994-X
15. A.J. Saul (ed.): *Floods and Flood Management.* 1992 ISBN 0-7923-2078-6
16. D.E. Ashpis, T.B. Gatski and R. Hirsh (eds.): *Instabilities and Turbulence in Engineering Flows.* 1993 ISBN 0-7923-2161-8
17. R.S. Azad: *The Atmospheric Boundary Layer for Engineers.* 1993 ISBN 0-7923-2187-1
18. F.T.M. Nieuwstadt (ed.): *Advances in Turbulence IV.* 1993 ISBN 0-7923-2282-7
19. K.K. Prasad (ed.): *Further Developments in Turbulence Management.* 1993 ISBN 0-7923-2291-6
20. Y.A. Tatarchenko: *Shaped Crystal Growth.* 1993 ISBN 0-7923-2419-6

Kluwer Academic Publishers – Dordrecht / Boston / London

Mechanics

FLUID MECHANICS AND ITS APPLICATIONS

Series Editor: R. Moreau

21. J.P. Bonnet and M.N. Glauser (eds.): *Eddy Structure Identification in Free Turbulent Shear Flows.* 1993 ISBN 0-7923-2449-8
22. R.S. Srivastava: *Interaction of Shock Waves.* 1994 ISBN 0-7923-2920-1
23. J.R. Blake, J.M. Boulton-Stone and N.H. Thomas (eds.): *Bubble Dynamics and Interface Phenomena.* 1994 ISBN 0-7923-3008-0
24. R. Benzi (ed.): *Advances in Turbulence V.* 1995 ISBN 0-7923-3032-3
25. B.I. Rabinovich, V.G. Lebedev and A.I. Mytarev: *Vortex Processes and Solid Body Dynamics.* The Dynamic Problems of Spacecrafts and Magnetic Levitation Systems. 1994 ISBN 0-7923-3092-7
26. P.R. Voke, L. Kleiser and J.-P. Chollet (eds.): *Direct and Large-Eddy Simulation I.* Selected papers from the First ERCOFTAC Workshop on Direct and Large-Eddy Simulation. 1994 ISBN 0-7923-3106-0
27. J.A. Sparenberg: *Hydrodynamic Propulsion and its Optimization.* Analytic Theory. 1995 ISBN 0-7923-3201-6
28. J.F. Dijksman and G.D.C. Kuiken (eds.): *IUTAM Symposium on Numerical Simulation of Non-Isothermal Flow of Viscoelastic Liquids.* Proceedings of an IUTAM Symposium held in Kerkrade, The Netherlands. 1995 ISBN 0-7923-3262-8
29. B.M. Boubnov and G.S. Golitsyn: *Convection in Rotating Fluids.* 1995 ISBN 0-7923-3371-3
30. S.I. Green (ed.): *Fluid Vortices.* 1995 ISBN 0-7923-3376-4
31. S. Morioka and L. van Wijngaarden (eds.): *IUTAM Symposium on Waves in Liquid/Gas and Liquid/Vapour Two-Phase Systems.* 1995 ISBN 0-7923-3424-8
32. A. Gyr and H.-W. Bewersdorff: *Drag Reduction of Turbulent Flows by Additives.* 1995 ISBN 0-7923-3485-X
33. Y.P. Golovachov: *Numerical Simulation of Viscous Shock Layer Flows.* 1995 ISBN 0-7923-3626-7

Kluwer Academic Publishers – Dordrecht / Boston / London